# CALMODULIN AND CELL FUNCTIONS

ANNALS OF THE NEW YORK ACADEMY OF SCIENCES
Volume 356

# CALMODULIN AND CELL FUNCTIONS

Edited by D. Martin Watterson and Frank F. Vincenzi

The New York Academy of Sciences
New York, New York
1980

**Library of Congress Cataloging in Publication Data**

Main entry under title:

Calmodulin and cell functions.

(Annals of the New York Academy of Sciences ; v. 356)
Includes bibliographical references.
1. Calmodulin—Physiological effect—Congresses.
2. Cell physiology—Congresses. I. Watterson, D. Martin. II. Vincenzi, Frank F. III. Series:

New York Academy of Sciences. Annals ; v. 356.
[DNLM: 1. Calcium-Binding proteins—Congresses.
2. Cells—Physiology—Congresses. W1 AN62YL v. 356 / QU 55 C146]
Q11.N5 vol. 356 [QP552.C28] 500s 80-29310
ISBN 0-89766-101-X [574.19'245]
ISBN 0-89766-102-8 (pbk.)

SP
*Printed in the United States of America*
**ISBN 0-89766-101-X (cloth)**
**ISBN 0-89766-102-8 (paper)**

ANNALS OF THE NEW YORK ACADEMY OF SCIENCES

VOLUME 356

December 15, 1980

# CALMODULIN AND CELL FUNCTIONS*

*Editors and Conference Chairmen*

D. MARTIN WATTERSON AND FRANK F. VINCENZI

CONTENTS



*This series of papers is the result of a conference held by The New York Academy of Sciences on May 21–23, 1980, which was entitled Calmodulin and Cell Functions.

**Part III. Protein Phosphorylation and Calmodulin**



**Part IV. Cyclic Nucleotides and Calmodulin**



**Part V. Membrane Transport and Calmodulin**



**Part VI. Cellular Receptors and Calmodulin**

**Part VII. Poster Session Papers**

**Financial assistance was received from:**

- AMERICAN CYANAMID COMPANY
- CYSTIC FIBROSIS FOUNDATION
- DOW CHEMICAL U.S.A.
- ENDO LABORATORIES, INC.
- FOGARTY INTERNATIONAL CENTER—NATIONAL INSTITUTES OF HEALTH
- HOFFMANN-LA ROCHE, INC.
- LILLY RESEARCH LABORATORIES
- McNEIL LABORATORIES
- MERCK SHARPE & DOHME RESEARCH LABORATORIES
- MUSCULAR DYSTROPHY ASSOCIATION, INC.
- NATIONAL INSTITUTE OF ARTHRITIS, METABOLISM, AND DIGESTIVE DISEASES
- NATIONAL SCIENCE FOUNDATION
- PENNWALT CORPORATION PHARMACEUTICAL DIVISION
- PFIZER, INC.
- REVLON HEALTH CARE GROUP
- RICHARDSON-MERRELL, INC.
- A. H. ROBINS CO.
- G. D. SEARLE & CO.
- SMITH KLINE & FRENCH LABORATORIES
- SYNTEX USA, INC.
- THE UPJOHN COMPANY

# INTRODUCTORY REMARKS

D. Martin Watterson

*The Rockefeller University*
*New York, New York 10021*

Frank F. Vincenzi

*Department of Pharmacology*
*University of Washington*
*Seattle, Washington 98195*

Calmodulin is the name proposed by Dr. Cheung[1] for a multifunctional calcium binding protein. The multiple names previously given to this protein reflect the multiple functions of calmodulin and the diverse backgrounds of the investigators. A brief summary of developments in this field will help introduce this meeting and its goals.

This year could be described as the tenth anniversary of the *definitive* discovery or description of calmodulin by Cheung[2] and Kakiuchi.[3] There are communications and preliminary reports in the literature prior to 1970 that describe the isolation of calcium binding protein fractions that contained calmodulin[4] and the activation of phosphodiesterase by snake venom.[5,6] As noted by Dr. Cheung[5,6] much of his data indicated that the activation of phosphodiesterase by snake venom "might be a consequence of proteolysis." However, the elegantly simple reconstitution experiments of Cheung[2] using brain phosphodiesterase and brain activating factor clearly demonstrated that this latter form of phosphodiesterase stimulation was due to an effector protein, which was subsequently shown to be calmodulin. A couple of years after the reconstitution experiments this activator protein was shown to be a calcium binding protein, and chemically homogeneous preparations were available by 1976. As the characteristics of the activator protein were elucidated, many investigators compared the phosphodiesterase activator to the proteins that they were studying. Eventually, these direct comparisons demonstrated that phosphodiesterase activator, calcium dependent regulator, erythrocyte ATPase activator, modulator protein, and brain troponin C-like protein were all calmodulin.

It is in this historical context that we can define one of the goals of this meeting: a summary of current research on calmodulin. Every laboratory that is studying calmodulin is not giving a presentation at this meeting. By select presentations we have tried to reflect the diverse approaches being used in the study of calmodulin. In this regard there are a number of specific questions that we want to continually address. For example, what are the strengths and limitations of the data and methods in this field? In molecular terms, what is calmodulin and how is calmodulin related to other calcium modulated proteins?

A second goal of this meeting is to define future research areas in studies of calmodulin and its possible roles in cell function. As Pavlov has noted, "Facts are the air of scientists. Without them you never can fly."[7] However, the testable hypothesis, based on facts, is the essence of science. Therefore, we encourage the formulation of testable hypotheses. We can ask at this time how calmodulin might be involved in regulating various cell processes? How does calmodulin and other calcium binding proteins influence or regulate each other? In this regard, we must keep in mind that the molecular basis of calcium regulation is not a simple equation consisting of calcium and calmodulin.

In conclusion, we would like to thank our colleagues for their participation in this meeting. The invited speakers and those who submitted posters did an excellent job on their presentations. We wisk to thank the governmental agencies, private foundations, and pharmaceutical and chemical corporations for their generous support. We also wish to acknowledge the competent staff of the New York Academy of Sciences for handling the details so important to this conference.

## REFERENCES

1. CHEUNG, W. Y. 1980. Science **207:**19–27.
2. CHEUNG, W. Y. 1970. Biochem. Biophys. Res. Commun. **38:**533–538.
3. KAKIUCHI, S., R. YAMAZAKI & H. NAKAJIMA. 1970. Proc. Jpn. Acad. **46:**587–592.
4. MOORE, B. W. & D. MCGREGOR. 1965. J. Biol. Chem. **240:**1647–1653.
5. CHEUNG, W. Y. 1967. Biochem. Biophys. Res. Commun. **29:**478–482.
6. CHEUNG, W. Y. 1969. Biochem. Biophys. Acta **191:**303–315.
7. PAVLOV, I. 1936. Bequest to Academic Youth of Soviet Russia (as cited in *The Great Quotations*, p. 339, by George Seldes.) Simon & Schuster. New York.

# STRUCTURE AND FUNCTION RELATIONSHIPS AMONG CALMODULINS AND TROPONIN C-LIKE PROTEINS FROM DIVERGENT EUKARYOTIC ORGANISMS*

G. A. Jamieson, Jr., D. D. Bronson, F. H. Schachat, and T. C. Vanaman†

*Departments of Microbiology and Immunology and Anatomy*
*Duke University Medical Center*
*Durham, North Carolina 27710*

## Introduction

The divalent cation, $Ca^{2+}$, has long been recognized[1-5] as a primary intracellular signal through which stimuli evoke cellular responses such as endo/exocytosis, motility, alterations in intermediary metabolism, and proliferation. However, the biochemical mechanisms through which stimulus-induced fluxes in free intracellular $Ca^{2+}$ concentration are coupled to cellular regulation have only recently been elucidated. This regulation is mediated through the action of a limited set of structurally related, $Ca^{2+}$-dependent regulatory proteins including, among others, calmodulins (CaM), muscle troponin Cs (TnC) and myosin light chains. Of these proteins, only CaM has a broad distribution within the cell and throughout different tissues and species. In addition, CaM alone is a multifunctional regulatory protein that can activate, in a $Ca^{2+}$-dependent manner, a number of the enzymes involved in the physiological responses mentioned above and discussed in detail in numerous other contributions to this Annal.

The studies discussed in this paper, together with those presented later in this volume by Van Eldik and Watterson[6] and Cormier and coworkers,[7] demonstrate that CaM has remained highly conserved in structural and functional properties throughout most of eukaryotic evolution, commensurate with its multiple regulatory capabilities. In addition, we present evidence that both CaM and a closely related, but distinct TnC-like protein (TnCLP) are present in the invertebrate *Caenorhabitis elegans* and that both may function in $Ca^{2+}$-regulation of actomyosin ATPase. This suggests that components for thin filament based regulation of actomyosin systems emerged early during eukaryotic evolution.

## Methods

Bovine brain and *Tetrahymena pyriformis* CaMs were purified by a combination of the standard biochemical procedures previously described by Watter-

*Supported by the National Institutes of Health (Grants NS 10123 (T.C.V.) and NS 15269 (F.H.S.)), and a Muscular Dystrophy Association grant (F.H.S.). G.A.J., Jr. is an N.R.S.A. predoctoral trainee supported by 5T32 CA 09111 from the National Institutes of Health.

†To whom all correspondence should be addressed.



0077-8923/80/0356-0001 $1.75/0 © 1980, NYAS

son *et al.*,[8] and $Ca^{2+}$-dependent affinity chromatography on a phenothiazine-Sepharose 4B conjugate as described by Jamieson and Vanaman.[9] Calmodulin from the coelenterate *Renilla reniformis* was supplied by Drs. Harold Jones and M. J. Cormier (Department of Biochemistry, University of Georgia) after purification by established procedures.[10] CaM and TnCLP were isolated from extracts of *C. elegans* prepared in urea and initially fractionated by the ethanol precipitation procedure described by Grand *et al.*[11] Ten grams of wild-type *C. elegans* (strain N 2), grown and stored as previously described,[12] were disrupted with a French press in 75 mM Tris-HCl, 15 mM 2-mercaptoethanol, 7.8 M urea (freshly deionized) pH 8.0 containing 1 mM phenylmethylsulfonyl fluoride (PMSF) to inhibit proteolysis. The soluble fraction of this extract, obtained by centrifugation ($8000 \times g$ for 30 min), was treated at successively higher ethanol concentrations to yield a final 80% (vol/vol) ethanol-precipitate fraction from which CaM and TnCLP were then prepared as described in RESULTS. Rabbit skeletal muscle troponin I (TnI)-Sepharose 4B was prepared and used as described by Head *et al.*[13] CaM-activatable bovine brain 3′:5′-cyclic nucleotide phosphodiesterase (PDE) was prepared and assayed as previously described.[8,14]

Detailed analytical procedures such as amino acid composition analysis, CNBr digestion and peptide mapping as well as procedures for polyacrylamide gel electrophoresis (PAGE) have been presented elsewhere.[8,15–17]

## RESULTS AND DISCUSSION

### *General Properties of Calmodulins*

TABLE 1 summarizes a number of the physical and functional properties that have been established for CaMs isolated from widely divergent eukaryotic organisms. Clearly, all CaMs are relatively small, very acidic proteins which gives them a characteristic, rapid mobility on PAGE run in alkaline buffer systems either in the presence or absence of denaturants. Detailed $Ca^{2+}$-binding studies have been performed on a limited number of mammalian CaMs, for which four high affinity sites have been determined.[8,18,19] As discussed in a later section, the amino acid sequences thought to form these sites were easily identified in the complete sequence of mammalian CaMs by comparison with the structure of other members of the $Ca^{2+}$-binding protein superfamily, as discussed by Kretsinger.[20,21] These domains are conserved in the invertebrate protein. It is therefore considered likely that all metazoan CaMs have four high affinity $Ca^{2+}$-binding sites. All CaMs isolated to date also have a number of demonstrable $Ca^{2+}$-dependent activities. $Ca^{2+}$-induced alterations in their spectral properties, electrophoretic mobilities, and other physical properties have also been observed, further demonstrating the fact that they are $Ca^{2+}$-binding proteins. As will be discussed in more detail below, these similarities in the physical properties of CaMs reflect the conservation of their amino acid compositions and primary structures. In addition to the conservation of those residues dictating the gross physical properties of the molecule, the CaMs that have been characterized in detail contain a single, fully trimethylated lysyl residue whose function is unknown.

The functional properties of eukaryotic CaMs also appear to be highly conserved. For example, all CaMs can activate bovine brain PDE, form complexes with rabbit fast-skeletal muscle TnI, and bind phenothiazines as noted in TABLE 1.* Recently, we have shown that bovine brain CaM $Ca^{2+}$-dependently activates the dynein ATPases prepared from the protozoan ciliate *T. pyriformis* equally as well as *T. pyriformis* CaM.[26,27] This conservation of CaMs functional properties again must reflect conservation in at least those structural features required for its activity. Support for this view has been gained from the detailed structural studies of vertebrate and invertebrate CaMs discussed below.

### *Structural Homology Among Metazoan Calmodulins and Their Relationship to Muscle Troponin Cs*

The results of detailed sequence analyses of four vertebrate CaMs and one invertebrate CaM coupled with detailed comparative studies of this protein from a number of other sources have demonstrated that its structure is extremely conserved in metazoans. Comparison of the primary structure of CaMs to those determined for other $Ca^{2+}$-binding proteins has also allowed the tentative identification of several important features of the linear sequence. The sequence shown in single-letter designation in FIGURE 1, line B is that of bovine brain CaM[15,28] arranged so as to depict the four internally homologous domains of the protein. These domains contain residues predicted, based on the criteria of Kretsinger,[20,21] to be involved in forming four helix-loop-helix $Ca^{2+}$-binding site structures. The loops consist of 12 amino acid residues, six of which have oxygen-containing side chains assumed to contribute oxygen ligands (noted by dots) to $Ca^{2+}$ at the vertices of an octahedron. These six bonding residues

TABLE 1
PROPERTIES OF CALMODULINS

| | Metazoan* | Protozoan† | Plant‡ |
|---|---|---|---|
| *Structural* | | | |
| Molecular Weight | 16,500–16,700 | ~15,000 | ~16,500 |
| Charge Properties | pI ~ 4 | acidic | acidic |
| U.V. Analysis | Phe-fine spectrum | Phe-fine spectrum | Phe-fine spectrum |
| $Ca^{2+}$-binding | 4 sites ($K_D \sim \mu M$) | + | + |
| *Functional* | | | |
| Activate 3′:5′-cyclic nucleotide phosphodiesterase | + | + | + |
| Formation of complexes with rabbit skeletal Troponin I | + | + | + |
| Phenothiazine binding | + | + | + |

*Watterson *et al.*,[15] Levin and Weiss,[44] Jones *et al.*[10]
†Jamieson *et al.*[17]
‡Anderson *et al.*,[45] and these proceedings.[6,7]

*For a discussion of these and other Ca·CaM regulated enzymatic activities, see other papers in this volume and several recent reviews.[22-25]

alternate in the sequence as shown, most often being interspersed with glycine or other small side chain amino acids. The glycyl residue found at the apex (between bonding positions 3 and 4) of each loop has been proposed to be especially important, allowing the loop to make a sharp bend at this position. Eight residues on either side of each loop are also predicted to form two helices, as found for the $Ca^{2+}$-binding sites in parvalbumin. Little of the linear sequence of CaM is found outside these predicted helix-loop-helix domains—only the 11 amino terminal residues and three 8-9 residues interdomain sequences assumed to be hinge regions.

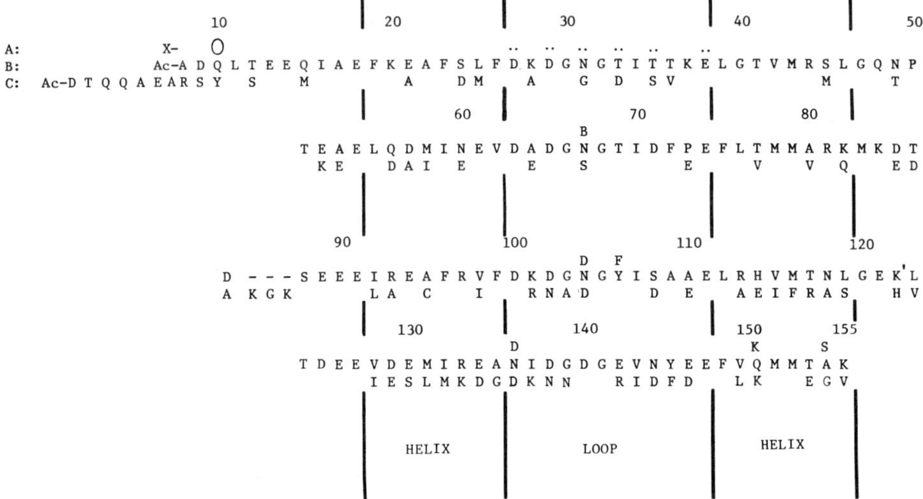


FIGURE 1. Comparison of calmodulin and muscle TnC amino acid sequences. The complete amino acid sequence of bovine brain CaM[15,28] is shown in single letter designation in line B. Differences *only* between this sequence and that of *Renilla reniformis* CaM[37] or rabbit skeletal muscle TnC[38] are shown in lines A and C, respectively. The numbers refer to positions in the rabbit TnC sequence. Regions of these sequences predicted to form the helix-loop-helix $Ca^{2+}$-binding site structures[20] are noted in the figure with predicted liganding residues in the loop denoted by a double dot (· ·) above the upper most set of lines. The single letter abbreviations used are: A = alanine; C = cysteine; D = aspartic acid; E = glutamic acid; F = phenylalanine; G = glycine; H = histidine; I = isoleucine; K = lysine; L = leucine; M = methionine; N = asparagine; P = proline; Q = glutamine; R = arginine; S = serine; T = threonine; V = valine; Y = tyrosine; K′ = trimethyllysine. Gaps required to keep sequences in register for maximum homology are indicated by dashes or an O.

Sequence studies have also been reported for vertebrate CaMs from two other sources, bovine uterine smooth muscle[29] and rat testis.[30] The sequence shown in FIGURE 1 is identical to that reported by Grand and Perry[29] for the bovine uterus protein except for the amidation states of the aspartyl residues at two positions, residues 24 and 97. Together, these bovine sequences differ at a number of positions from the sequence reported for the CaM from rat testis largely in amide assignments. As previously discussed, these assignments in the rat testis sequence must be considered tentative, owing to the preliminary nature of that study.[30]

Detailed comparative studies other than sequence analysis have been performed on several vertebrate CaMs. For example, Stevens and co-workers[31] found no differences between the bovine heart and brain proteins based on studies including tryptic peptide mapping. Similarly, the amino acid compositions, physiochemical properties, functional activities, and tryptic peptide maps of bovine, porcine, rabbit, rat, and chicken brain CaMs have been shown to be indistinguishable.[16,32] CaMs from various other tissues[11,33,34] as well as transformed chick embryo fibroblasts[35] and the murine cell line P388D$_1$[36] also appear to be indistinguishable from the brain proteins by similar criteria.

In view of the very highly conserved structures of vertebrate CaMs, we have recently[37] begun investigating the structure of this protein isolated from more divergent species. FIGURE 1 also presents the sequence of CaM from the coelenterate *Renilla reniformis* established by a direct sequence analysis of tryptic and thermolytic peptides and their placement by homology to the bovine brain protein. These sequences were identical to their counterparts in the bovine brain protein except for the 7 positions where other residues are indicated (FIGURE 1, line A). This limited number of differences between the vertebrate and invertebrate CaMs is striking, particularly in view of the fact that six of the seven positions differing in the two proteins contain pairs of residues that are functionally conserved and could have arisen by changing a single base in codons; except for the Ser/Ala interchange adjacent to the carboxyl terminus. Clearly, the highly conserved nature of these two CaM isolates is in accord with the lack of gross differences in their functional properties.

FIGURE 1 also presents a comparison of the sequences of CaMs and rabbit fast skeletal muscle TnC as determined by Collins *et al.*[38] Line C shows those residues in this TnC that differ from the bovine brain CaM sequence (line B) when the two proteins are aligned to obtain maximum homology. With this alignment, only a single three-residue gap must be introduced in the CaM sequence to maintain homology to the TnC. The level of structural similarity between the CaMs and TnC is clearly substantial. It is interesting to note, however, that the greatest variability between these structures occurs in the fourth $Ca^{2+}$-binding domain (residues 118–155). Also of interest is the fact that the sequence immediately adjacent to the fourth $Ca^{2+}$-binding domain contains the fully trimethylated lysyl residue found in all CaMs but not TnCs. It appears likely from these and other considerations[39] that this region of the CaM sequence provides it with the specific ability, not shared by TnC to any significant extent, to interact with and activate numerous enzymes.

## *Comparison of Metazoan and Protozoan Calmodulins*

In order to further examine the structural and phylogenetic relationships among eukaryotic CaMs, we have recently begun detailed studies of CaM and its biological role in the protozoan ciliate *T. pyriformis*. As shown by Jamieson *et al.*,[17] CaM can be purified from this organism by virtue of its $Ca^{2+}$-dependent binding to phenothiazines covalently linked to Sepharose 4B, a common property of CaMs. The amino acid composition of *T. pyriformis* CaM has many of the features of the vertebrate and invertebrate proteins discussed above (including the absence of Trp and Cys and the presence of trimethyllysine), but is clearly less rigidly conserved than the metazoan proteins as shown in TABLE 2. In addition, the protozoan CaM is slightly smaller (MW ~ 15,000) than the animal proteins and differs from them in at least 2 out of the 7 CNBr cleavage products detectable by gel electrophoretic analysis as shown in FIGURE 2. Despite these readily detectable differences in structure, the protozoan and metazoan CaMs appear to have several $Ca^{2+}$-dependent activities in common including: activation of bovine brain PDE (TABLE 3), formation of complexes with rabbit skeletal muscle TnI (TABLE 1), and binding to phenothiazines (TABLE 2).

The above results suggest that the regions of the CaM sequence responsible

TABLE 2

AMINO ACID COMPOSITIONS OF CALMODULINS AND TROPONIN CS

| Residue | CaM Bovine* | CaM *C. elegans*† | CaM *R. reniformis** | CaM *T. pyriformis*† | TnC Rabbit* | TnCLP *C. elegans*‡ |
|---|---|---|---|---|---|---|
| Lys | 7 | 7.1 | 8 | 7.72 | 9 | 11.3 |
| Tml | 1 | § | 1 | § | 0 | 0 |
| His | 1 | 1.2 | 1 | 2.3 | 1 | 1.3 |
| Arg | 6 | 5.8 | 6 | 5.4 | 7 | 6.2 |
| Asp | 23 | 26.8 | 23 | 23.5 | 22 | 20.1 |
| Thr | 12 | 9.8 | 12 | 10.6 | 6 | 5.9 |
| Ser | 4 | 5.3 | 5 | 4.7 | 7 | 6.2 |
| Glu | 27 | 25.9 | 25 | 26.5 | 31 | 32.7 |
| Pro | 2 | 2.6 | 2 | 2.3 | 1 | 1.2 |
| Gly | 11 | 11.8 | 11 | 12.1 | 13 | 13.5 |
| Ala | 11 | 9.8 | 10 | 10.7 | 13 | 11.9 |
| Val | 7 | 6.8 | 7 | 6.2 | 7 | 4.9 |
| Met | 9 | 10.5 | 9 | 8.1 | 10 | 4.7 |
| Ile | 8 | 6.4 | 8 | 9.2 | 9 | 10.7 |
| Leu | 9 | 8.3 | 9 | 13.6 | 9 | 12.5 |
| Tyr | 2 | 1.3 | 1 | 1.0 | 2 | 4.1 |
| Phe | 8 | 8.8 | 9 | 8.5 | 10 | 16.7 |
| 1/2 Cys | 0 | 0 | 0 | 0 | 1 | § |
| Trp | 0 | 0 | 0 | 0 | 0 | 0 |

*By sequence: bovine brain CaM,[15] *R. reniformis* (unpublished observation); Rabbit skeletal Tnc.[38]

†Based on 16,700 gm/mole.

‡Based on 18,000 gm/mole.

§Tml detected as a shoulder on lysine peak by double column methodology is reported together with lysine.

for $Ca^{2+}$-dependent interaction with other proteins, including muscle TnI are present even in the protozoan protein. This is even more intriguing in view of our recent demonstration[26,27] that CaM may play a role in regulating the ciliary dynein ATPase system in *T. pyriformis.* This microtubule-associated ATPase system appears to be responsible for the sliding of adjacent microtubules, the basis for flagellar and ciliary linked motility in numerous organisms. It is interesting to note that bovine brain CaM is equally effective in the activation of *T. pyriformis* 14S and 30S dynein ATPases while the skeletal muscle TnC is not. It appears likely that one of the earliest functions of CaM in eukaryotic organisms was the regulation of motility systems. However, the increasing complexities of

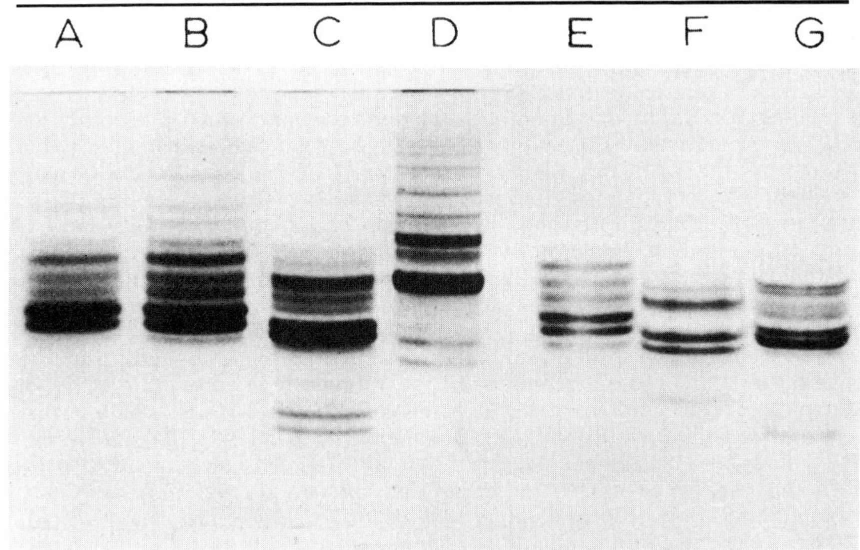


FIGURE 2. Cyanogen bromide peptide maps of metazoan and protozoan calmodulins, TnC, and TnC-like proteins. Freeze-dried cyanogen bromide digests were dissolved in 20 mM Tris (free base)-100 mM glycine-8 M urea, pH 8.3 and applied to a 15% acrylamide slab gel prepared and run in the same buffer. Electrophoresis was performed at 30 mA until bromphenol blue migrated to within 1 cm of the bottom of the gel. The Coomassie-blue stained bands shown in the figure were obtained with digests of: A—Bovine brain CaM, 75 $\mu$g; B—*C. elegans* CaM, 75 $\mu$g; C—Rabbit skeletal muscle TnC, 75 $\mu$g; D—*C. elegans* TnCLP, 75 $\mu$g; E—Bovine brain CaM, 50 $\mu$g; F—*T. pyriformis* CaM, 50 $\mu$g; G—Rabbit skeletal muscle TnC, 50 $\mu$g.

such systems in higher organisms, particularly the actomyosin-based motility apparatuses, demanded acquisition of additional $Ca^{2+}$-dependent regulatory elements including a dedicated $Ca^{2+}$-receptor protein such as troponin C. In agreement with this hypothesis, studies described in the following section have demonstrated that both CaM and a TnC-like protein are present in at least one

TABLE 3

ACTIVITY OF BOVINE BRAIN 3':5'-CYCLIC NUCLEOTIDE PHOSPHODIESTERASE IN THE PRESENCE OF:

| Additions to Assay Mixture* | Phosphodiesterase Activity† (Percent of Maximum Activity) |
|---|---|
| 1. None | 16 |
| 2. Bovine brain calmodulin (2.0 $\mu$g) | (100) |
| 3. *C. elegans* calmodulin (2.0 $\mu$g) | 88 |
| 4. *C. elegans* TnCLP (2.0 $\mu$g) | 28 |
| 5. *T. pyriformis* calmodulin (2.0 $\mu$g) | 101 |

*Assay mixture: 2 mM cAMP, 20 mM Tris-HCl (pH 8.0), 1 mM $Ca^{2+}$, 0.4 mM $Mn^{2+}$, 50 $\mu$g brain PDE, plus calmodulin as indicated.

†Incubation was for 30 min at 30°C and the 5'-AMP produced was determined.

invertebrate, where they may both be involved in the $Ca^{2+}$-dependent regulation of actomyosin contractile activity.

### *Characterization and Roles of Calmodulin and a Second $Ca^{2+}$-Binding Protein from the Nematode Caenorabditis Elegans*

In contrast to the almost identical primary structures of bovine brain and *R. reniformis* CaMs discussed earlier, Waisman *et al.*[40] have reported that CaM from the earthworm, *Lumbricus terrestris*, was quite dissimilar from the bovine protein in both its amino acid composition and tryptic peptide map. Recent studies in our laboratory of the $Ca^{2+}$-binding proteins isolatable from *C. elegans*, an invertebrate whose complexity is intermediate between *R. reniformis* and *L. terrestris*, provide a possible explanation for the differences in these results. We have purified and characterized two $Ca^{2+}$-binding proteins from *C. elegans*, one of which is CaM and the other a TnC-like protein (TnCLP) which may be involved in actomyosin thin filament regulation. As will be documented in more detail elsewhere (Schachat *et al.*, in preparation), the physicochemical properties of these two proteins are so similar that they copurify through most purification schemes and are poorly resolved, at best, in all PAGE systems. However, when the 80% (vol/vol) ethanol precipitate fraction, prepared from the soluble components of urea extracts of *C. elegans* (see METHODS), is subjected to ion-exchange chromatography on DEAE-cellulose in urea containing buffers under the conditions described by Grand *et al.*,[11] these two components can be separated by judicious pooling of column fractions. Both components were resolved from other contaminants by $Ca^{2+}$-dependent binding on TnI-Sepharose 4B and gel filtration on Sephadex G-100 to yield nearly homogeneous preparations (purity >90%). One of these proteins was clearly identifiable as CaM, based on its ability to activate bovine brain PDE (TABLE 3)—an activity in which the second protein was largely deficient. In addition to the fact that both were bound by TnI-Sepharose 4B in the presence of $Ca^{2+}$, they have also been shown to form $Ca^{2+}$-dependent complexes with rabbit skeletal muscle TnI and TnT by PAGE. These facts coupled with the physiochemical properties of these two proteins discussed below, led us to term the second protein, TnCLP or TnC-like protein.

The detailed structural properties of these two *C. elegans* proteins are clearly distinct. As shown in TABLE 2, the amino acid composition of *C. elegans* CaM determined is clearly similar to that of the *R. reniformis* protein, deduced from its sequence, while that of the TnCLP is distinct from the CaMs and bears a close resemblance to vertebrate muscle TnCs. Similar conclusions can be made based on the CNBr peptide maps of these two *C. elegans* proteins. As shown in FIGURE 2, the patterns obtained on 15% polyacrylamide alkaline-urea gels with CNBr digests of *C. elegans* (slot A) and bovine brain (slot B) CaMs are indistinguishable from each other and are different from rabbit skeletal muscle TnC (slot C). The CNBr fragments of *C. elegans* TnCLP (FIGURE 2, slot D) are clearly different from those of the CaMs, the overall pattern resembling those of vertebrate TnCs (slot C; Grand *et al.*[11]). The fact that this method can detect differences between the structure of CaMs is illustrated by comparing the patterns obtained for the bovine brain (FIGURE 2, slot E) and *T. pyriformis* (FIGURE 2, slot F) proteins.

The physiochemical properties of TnCLP and its ability to bind subunits of the muscle troponin complex suggested that the TnCLP might be involved in thin filament-based regulation of *C. elegans* actomyosin. In support of this hypothesis, we have now shown that TnCLP is associated with actin thin filaments in an

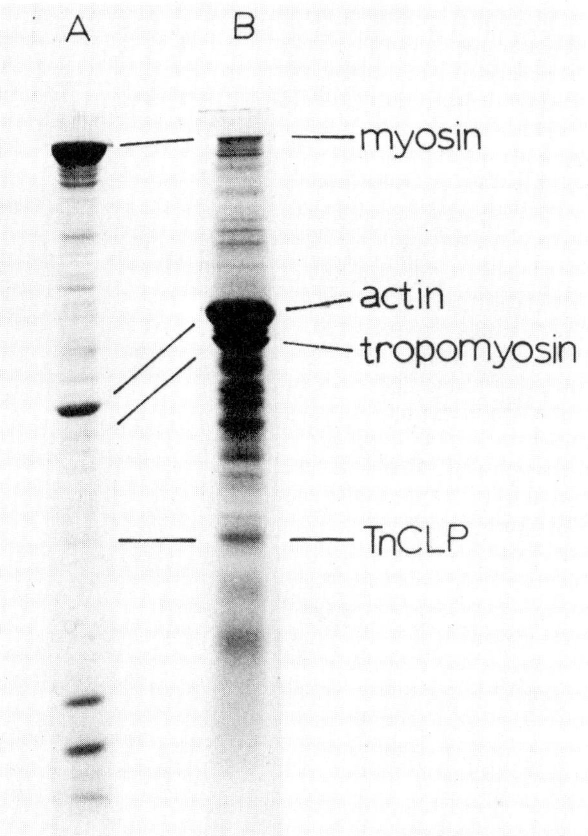


FIGURE 3. Localization of *C. elegans* TnCLP. SDS-PAGE analysis of preparations of *C. elegans* myofilaments and thin filaments were performed with discontinuous buffer systems as described by Laemmli.[46] *C. elegans* myofilament proteins were resolved on a 10% polyacrylamide gel (Gel A) while proteins of thin filament preparations were resolved on a 15% gel (Gel B). The positions of *C. elegans* myosin, actin, tropomyosin, and TnCLP, are indicated by lines connecting gel bands to labels on the right of the figure. Densitometric quantification of the ratio of tropomyosin ($M_r \sim 38{,}000$) to TnCLP ($M_r \sim 19{,}600$) in gel B was calculated by dividing the determined Coomassie-blue staining intensities of these two components by their apparent molecular weights.

appropriate stoichiometry. As shown in FIGURE 3, SDS-PAGE analysis of preparations[41] of *C. elegans* myofilaments (Track A) shows the presence of TnCLP, which copurifies with thin filaments (Track B). Based on quantitative scans of this and other gels, the molar ratio of TnCLP to tropomyosin is 0.92:1 in these preparations. Preliminary studies suggest that the TnCLP can be isolated as a complex with TnI and TnT-like components using standard procedures for isolating skeletal muscle troponin (Bronson and Schachat, unpublished observations).

The results presented above strongly suggest that $Ca^{2+}$-dependent regulation

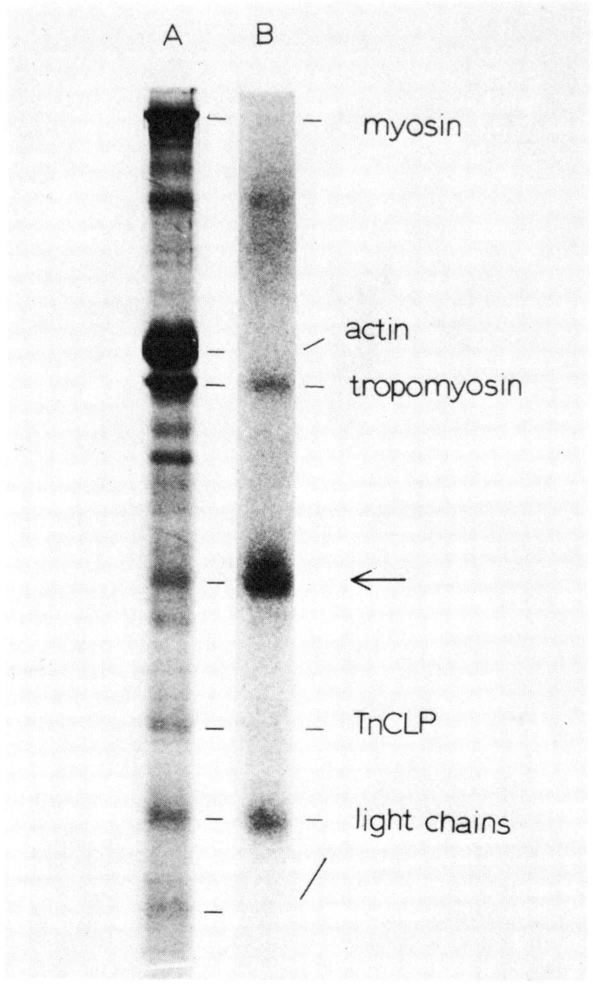


FIGURE 4. *C. elegans* myosin light chain phosphorylation. Ca·CaM dependent phosphorylation of components of *C. elegans* myofibril preparations was demonstrated by SDS-PAGE analysis of samples obtained after incubation in the presence of $\gamma-{}^{32}$P-ATP. Coomassie-blue staining components resolved from this preparation on a 15% acrylamide gel are shown in A, while B shows an autoradiograph of the same gel track. The sample was incubated at room temperature in 40 mM NaCl, 6.7 mM potassium phosphate, 1 mM EGTA, pH 6.0 for 5 minutes after addition of excess $CaCl_2$ (5 mM final concentration), bovine brain CaM (5 $\mu$g/100 $\mu$l) and $\gamma$-${}^{32}$P-ATP(5 mM; 9 × $10^4$ dpm/$\mu$mol.). Reaction was terminated by the addition of SDS and 2-mercaptoethanol to 1% (w/vol) each final concentration and then boiling for 2 minutes immediately prior to application to SDS-PAGE. The radioactive bands detected in the autoradiograph (B) were not present in samples incubated in the presence of 1 mM EGTA without the addition of excess $Ca^{2+}$, or if CaM was omitted in the presence or absence of added $Ca^{2+}$ (data not shown).

of *C. elegans* actomyosin occurs at least in part through actin thin-filament bound regulatory components. Evidence has been obtained suggesting that the now widely described CaM-activated myosin light-chain kinase system is also present in *C. elegans*. FIGURE 4 shows SDS-PAGE analyses of myofibrillar preparations from *C. elegans* incubated with $\gamma[^{32}P]$ ATP in the presence of added CaM. Track A shows the Coomasie-blue staining components present in these preparations identified by their characteristic molecular weights as shown in the figure. Track B shows an autoradiograph of the same gel. Two major phosphorylated components are observed, one of which comigrates with a *C. elegans* myosin light chain. The identity of the other major labeled component (denoted by the arrow) remains to be determined. The incorporation of $^{32}P$ into these proteins is dependent on the presence of $Ca^{2+}$ and CaM as would be expected for myosin light chain kinase.[42,43] We conclude that a CaM-dependent myosin light chain kinase may be involved in the regulation of actomyosin contractile activity in *C. elegans* in addition to a thin filament-based regulatory system.

## Concluding Remarks

From the studies presented here, it is clear that CaM is a protein of highly conserved structure that is present in all eukaryotes which have been examined. Although CaM has not been found in prokaryotes, structurally related proteins will most likely be found in these organisms.

Elucidation of the presence of at least two $Ca^{2+}$-binding proteins in *C. elegans* and their characterization is important for two major reasons. First, it demonstrates that extreme care must be exercised to insure that putative isolates of CaM from any source are free of other structurally related $Ca^{2+}$-binding proteins whose properties or activities might be incorrectly ascribed to the putative CaM. Second, demonstration that one of these proteins (TnCLP) may be involved in thin filament-based control of actomyosin ATPase in this organism is the first indication that troponin-like regulatory components may be present even in very primitive metazoans. This fact coupled with CaM's role in thick filament regulation through the myosin light chain kinase system and CaM's ability to function as a $Ca^{2+}$-dependent regulator of the ciliary motility apparatus of protozoans strongly suggests that the major role for the CaM/TnC pair has been regulation of motility throughout much of eukaryotic evolution. CaM's additional ability to regulate other enzymatic activities required for the initiation and termination of cellular responses further supports our view[28] that this ubiquitous $Ca^{2+}$-dependent regulatory protein may control the entire stimulus-response-relaxation cycle.

The rigidly conserved structure of CaM is no doubt due to the multiple specific interactions that it must make with both small molecules and proteins in order to properly regulate overall cellular activity. The enzyme systems that CaM regulates are so essential to cellular function that any significant alteration in CaM structure would probably be lethal to the organism. In fact, it would appear that tissue-specific forms of CaM cannot be tolerated, since even the closely related muscle TnCs (which might be considered more highly evolved, specialized forms of $Ca^{2+}$-dependent regulators) possess no significant CaM-like activity. Sets of CaM-regulated enzymes, whose presence in a given tissue could be genetically determined, would seem to provide the best mechanism for producing the required tissue and function-specific stimulus-response systems of complex eukaryotic organisms.

## References


1. Douglas, W. W. 1974. Biochem. Soc. Symp. **39:** 1–28.
2. Rubin, R. P. 1974. Calcium and the Secretory Process. Plenum Publishing Corp. New York, N. Y.
3. Berridge, M. J. 1975. Adv. Cyc. Nuc. Res. **6:** 1–98.
4. Rasmussen, H. & D. B. P. Goodman. 1977. Physiol. Rev. **57:** 421–509.
5. Harris, P. 1978. *In* Cell Cycle Regulation: Cell Biology Monograph Series. J. R. Jeter, I. L. Cameron, G. M. Padilla & A. M. Zimmerman, Eds.: 75–104. Academic Press, Inc. New York, N. Y.
6. Van Eldik, L. J. & D. M. Watterson. 1980. This volume.
7. Jarrett, H. W., H. Charbonneau, J. M. Anderson, R. O. McCann & M. J. Cormier. 1980. This volume.
8. Watterson, D. M., W. G. Harrelson, Jr., P. M. Keller, F. Sharief & T. C. Vanaman. 1976. J. Biol. Chem. **251:** 4501–4513.
9. Jamieson, G. A., Jr. & T. C. Vanaman. 1979. Biochem. Biophys. Res. Commun. **90:** 1048–1056.
10. Jones, H. P., J. C. Matthews & M. J. Cormier. 1979. Biochemistry **18:** 55–60.
11. Grand, R. J. A., S. V. Perry & R. A. Weeks. 1979. Biochem. J. **177:** 521–529.
12. Brenner, S. 1974. Genetics **77:** 71–94.
13. Head, J. F., R. A. Weeks & S. V. Perry. 1977. Biochem. J. **161:** 465–471.
14. Watterson, D. M. & T. C. Vanaman. 1976. Biochem. Biophys. Res. Commun. **73:** 40–46.
15. Watterson, D. M., F. S. Sharief & T. C. Vanaman. 1980. J. Biol. Chem. **255:** 962–975.
16. Watterson, D. M., P. A. Mendel & T. C. Vanaman. 1980. Biochemistry **19:** 2672–2676.
17. Jamieson, G. A., Jr., T. C. Vanaman & J. J. Blum. 1979. Proc. Natl. Acad. Sci. USA **76:** 6471–6475.
18. Dedman, J. R., J. D. Potter, R. L. Jackson, J. D. Johnson & A. R. Means. 1977. J. Biol. Chem. **252:** 8415–8422.
19. Wolff, D. J., P. G. Poirier, C. O. Brostrom & M. A. Brostrom. 1977. J. Biol. Chem. **252:** 4108–4117.
20. Kretsinger, R. H. 1980. This volume.
21. Kretsinger, R. H. 1976. Ann. Rev. Biochem. **45:** 239–266.
22. Wang, J. H. & D. M. Waisman. 1979. Curr. Top. Cell. Reg. **15:** 47–107.
23. Cheung, W.-Y. 1980. Science **207:** 19–27.
24. Means, A. R. & J. R. Dedman. 1980. Nature **285:** 73–77.
25. Klee, C. B., T. H. Crouch & P. G. Richman. 1980. Ann Rev. Biochem. **49:** 489–515.
26. Jamieson, G. A., Jr., A. Hayes, J. J. Blum & T. C. Vanaman. 1980. This volume.
27. Blum, J. J., A. Hayes, G. A. Jamieson, Jr. & T. C. Vanaman. 1980. J. Cell Biol. (In press.)
28. Vanaman, T. C., F. S. Sharief & D. M. Watterson. 1977. *In* Calcium Binding Proteins and Cellular Function. R. H. Wasserman, R. A. Corradino, E. Carafoli, R. H. Kretsinger, D. H. MacLennan & F. L. Seigel, Eds.: 107–116. American Elsevier, Inc. New York, N.Y.
29. Grand, R. J. A. & S. V. Perry. 1978. F.E.B.S. Lett. **92:** 137–142.
30. Dedman, J. R., R. L. Jackson, W. E. Schreiber & A. R. Means. 1978. J. Biol. Chem. **253:** 343–346.
31. Stevens, F. C., M. Walsh, H. C. Ho, T. S. Teo & J. H. Wang. 1976. J. Biol. Chem. **251:** 4495–4500.
32. Vanaman, T. C., F. Sharief, J. L. Awramik, P. A. Mendel & D. M. Watterson. 1976. *In* Contractile Systems in Non-Muscle Tissues. S. V. Perry, A. Margreth & R. S. Adelstein, Eds.: 165–176. Elsevier/North Holland. Amsterdam.
33. Van Eldik, L. J. & D. M. Watterson. 1979. J. Biol. Chem. **254:** 10250–10255.
34. Jarrett, H. W. & J. T. Penniston. 1978. J. Biol. Chem. **253:** 4676–4682.
35. Kuo, I. C. Y. & C. J. Coffee. 1976. J. Biol. Chem. **251:** 1603–1609.
36. Jamieson, G. A., Jr. & T. C. Vanaman. 1980. J. Immunol: **125:** 1171–1177.

37. VANAMAN, T. C. & F. S. SHARIEF. 1979. Fed. Proc. **38:** 788.
38. COLLINS, J. H., M. L. GREASER, J. D. POTTER & M. J. HORN. 1977. J. Biol. Chem. **252:** 6356–6362.
39. VANAMAN, T. C. 1980. *In* Calcium and Cell Function: Volume 1. Calmodulin. W.-Y. Cheung, Ed. Academic Press, Inc. New York, N.Y. (In press.)
40. WAISMAN, D. M., F. C. STEVENS & J. H. WANG. 1978. J. Biol. Chem. **253:** 1106–1113.
41. HARRIS, H. E., M.-Y. W. TSO & H. F. EPSTEIN. 1977. Biochemistry **16:** 859–865.
42. DABROWSKA, R., J. M. F. SHERRY, D. K. ARAMATORIO & D. J. HARTSHORNE. 1978. Biochemistry **17:** 253–258.
43. NAIRN, A. C. & S. V. PERRY. 1979. Biochem. J. **179:** 89–97.
44. LEVIN, R. M. & B. WEISS. 1978. Biochim. Biophys. Acta **540:** 197–204.
45. ANDERSON, J. M., H. CHARBONNEAU, H. P. JONES, R. O. MCCANN & M. J. CORMIER. 1980. Biochemistry **19:** 3113–3120.
46. LAEMMLI, U. K. 1970. Nature **227:** 680–685.

## DISCUSSION OF THE PAPER

DR. WATTERSON (*Rockefeller University, New York, NY*): Have you confirmed by sequence studies the presence of two histidinyl residues in *Tetrahymena* calmodulin?

DR. T. C. VANAMAN (*Duke University, Durham, NC*): The composition of *Tetrahymena* calmodulin was calculated using a molecular weight of 16,700. In fact, the molecular weight is around 15,000, so those values are artificially high. The initial data from tryptic peptide studies suggest there is only one histidine.

DR. WATTERSON: Are the amidation differences between the amino acid sequences of bovine brain calmodulin and *R. reniformis* calmodulin real, or due to differences in isolation methods?

DR. VANAMAN: The amide differences at sequence positions 24 and 97 may be due to deamidation.

DR. C. KLEE (*National Institutes of Health, Bethesda, MD*): Have you found calmodulin in yeast?

DR. VANAMAN: No.

# CRYSTALLOGRAPHIC STUDIES OF CALMODULIN AND HOMOLOGS

Robert H. Kretsinger

*Department of Biology*
*University of Virginia*
*Charlottesville, Virginia 22901*

## Introduction

Five years ago I published "Hypothesis: Calcium Modulated Proteins Contain EF-Hands."[1] The list of homologs now includes S-100, intestinal calcium-binding protein, parvalbumin, regulatory and essential light-chains of myosin, troponin C, and most important, calmodulin. A calcium-modulated protein is defined by its presence in the cell cytosol or on a membrane facing the cytosol and by its calcium affinity, $pK_d\,(Ca^{2+})$ 5 to 7 under cytosolic conditions. None of the calcium modulated proteins, characterized to date, are themselves enzymes. Yet by their interactions they confer calcium modulation on a wide variety of cytosolic enzymes and cell processes.

## EF-Hand Domain

In order to understand these interactions, we must know the structures of these calcium modulated proteins. To date, our only structural information about the homolog family of calcium modulated proteins comes from the crystal structure of parvalbumin.[2] It contains three isologous regions, two of which bind calcium. This structural similarity, considered together with the many amino acid sequences now available for parvalbumin and these other calcium modulated proteins, reveals the essential features of the basic evolutionary domain. This has been referred to as the "EF-hand" as illustrated in FIGURE 1. Alpha helix number 5 or "E" in parvalbumin is represented by the extended forefinger, helix "F", by the extended thumb. The loop containing the oxygen atoms that ligand the calcium ion is represented by the clenched middle finger.

I have recommended a standard-domain numbering scheme thirty-one residues long, numbers 0 through 30. Amino acids numbered 2, 5, 6, 9 and 22, 25, 26, 29 are on the inner aspects of $\alpha$-helices E and F, respectively. Their side chains form the hydrophobic core of the molecule.

The oxygen atoms of the calcium-liganding amino acids are found near the vertices of an octahedron. Residue 10 supplies the oxygen for the X vertex, 12 for the Y vertex, and so on, 14 Z, 18 $-$X, and 21 $-$Z (FIGURE 1). The oxygen atom at the $-$Y vertex, position 16, comes from the peptide oxygen; there is no constraint on the side group. In addition to the condition of hydrophobic side chains in positions 2,5,6,9,22,25,26,29 and Asp, Asn, Glu, Gln, Ser, or Thr in positions 10, 12, 14, 18 and 21, the conformation and packing of the loops impose two additional constraints. The bend at residue 15, the knuckle of the clenched middle finger, is usually so tight that only Gly can assume the required $\Psi$, $\Phi$ dihedral angles. Residue 17, usually Ilu, anchors the loop to the hydrophobic core of the molecule. We do not understand why position 1 is very often Glu. Of these sixteen



0077-8923/80/0356-0014 $1.75/0 © 1980, NYAS

characteristics positions (1,2,5,6,9,10,12,14,15,17,18,21,22,25,26,29) EF-hand domains score twelve or better.

Some EF-hand domains have lost the ability to bind calcium. They can be recognized because one or several of their oxygen ligands (positions 10, 12, 14, 18, 21) have been lost. We infer from the parvalbumin structure that these domains that no longer bind calcium do retain the same general EF-hand conformation. There are now examples of one, two, three, and four domain EF-hand proteins, with zero to four calcium binding domains:

| Protein | Domain 1 | Domain 2 | Domain 3 | Domain 4 |
|---|---|---|---|---|
| Calmodulin | Ca | Ca | Ca | Ca |
| Troponin C | Ca | Ca | Ca | Ca |
| TNC (cardiac) | 0 | Ca | Ca | Ca |
| Regulatory LC | Ca | 0 | 0 | 0 |
| Essential LC | 0 | 0 | 0 | 0 |
| Parvalbumin | X | 0 | Ca | Ca |
| Intestinal CaBP | X | X | 0 | Ca |
| S-100 | X | X | X | Ca |

For all these proteins, excepting S-100, the EF-hand domains comprise 70 to 90% of the amino acids. A bit is left over in loops connecting the domains and in a tail at the N-terminus. In S-100, the amino acid sequence of the N-terminal half of the molecule is not recognizable as an EF-hand. It may have evolved by splicing together two nonhomologous genes. One anticipates that evolutionarily successful recombinations and splicings would occur between domain coding regions; nontranscribed DNA would occur between domains.

In order for the hydrophobic surface of the domain to impart stability, it must press against another hydrophobic surface. In parvalbumin the palmar surface of the EF-hand fits the palmar surface of the CD-hand. Domains three (CD) and four (EF) are related by an approximate two-fold axis passing through the monomer. The four domain molecules, as calmodulin, appear to have evolved from a one domain precursor by gene duplication to form a CD–EF pair, as seen in parvalbumin. A second gene duplication produced the four domain molecules. The parvalbumin precursor deleted the first domain to become a three-domain molecule. Cladograms[3] based on available amino acid sequences support this interpretation.

## Calmodulin Model

In 1975, Kretsinger and Barry[4] published a predicted structure for troponin C and by inference for any four-domain calcium modulated protein. The four domains, recognizable in the amino acid sequences, each have the same conformation as does the EF-hand in parvalbumin. The domains are related in pairs, as are the CD- and the EF-hands of parvalbumin, and as indicated by their evolution—domains one and two pair, as do domains three and four. A pair of EF-hand domains has an exposed hydrophobic surface on the side opposite the two calcium-binding loops. We proposed that the hydrophobic surface of pair one–two fits against the corresponding surface of pair three–four. The local two-fold axis of one–two is coincident with, and antiparallel to, the local two-fold axis of the three–four pair.

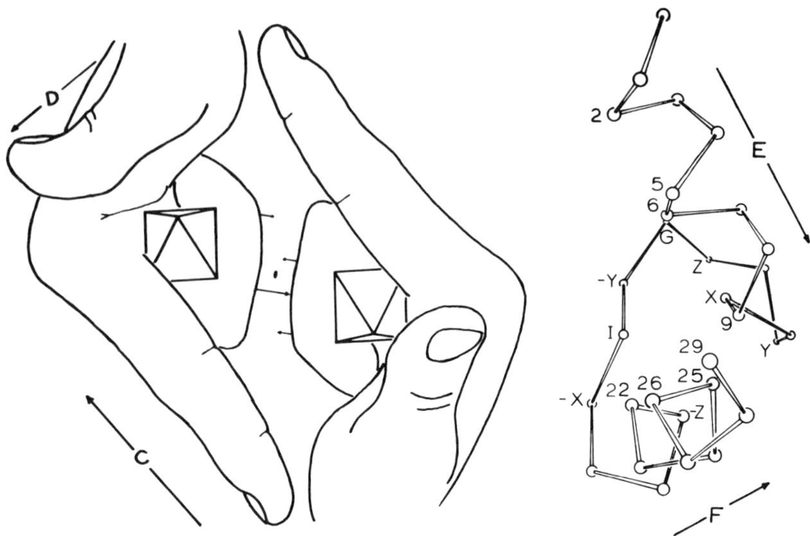


FIGURE 1. The CD- and EF-hand domains are similar and are related by an approximate two-fold axis passing through the monomeric protein. The octahedon represents the oxygen ligands of the calcium ion. The clenched middle finger is the loop about the calcium ion, residues 10 through 21. The sharp bend at position 15, which is usually Gly, is represented by the knuckle of the middle finger. The two $\alpha$-helices, residues 0 through 10 and residues 20 through 30, are represented by the extended forefinger and thumb. The palmar surfaces of the two forefingers and two thumbs face the hydrophobic core of the molecule. The (usually) Ilu at position 17 attaches the calcium-binding loop to the hydrophobic core. The two helices act as lever arms. As calcium is released their relative orientations change; conversely as the helices are modified or constrained the calcium affinity and selectivity of the loop varies.

The residues characteristic of the EF-hand are listed in terms of the domain numbering scheme: E, Glu; 'L', Phe, Leu, Ilu, Val, Met, Ala *or* Tyr; 'D', Asp, Asn, Glu, Gln, Ser *or* Thr; G, Gly; 'I', Ilu *or* Val.

| 0 | 1 | 2 | 3 | 4 | 5 | 6 | 7 | 8 | 9 | 10 | 11 | 12 | 13 | 14 | 15 | 16 | 17 | 18 | 19 | 20 | 21 | 22 | 23 | 24 | 25 | 26 | 27 | 28 | 29 | 30 |
|---|---|---|---|---|---|---|---|---|---|---|---|---|---|---|---|---|---|---|---|---|---|---|---|---|---|---|---|---|---|---|
| | E | 'F' | | | 'L' | 'L' | | | 'L' | 'D' | | 'D' | | 'D' | G | | 'I' | 'D' | | | E | 'L' | | | 'L' | 'L' | | | 'L' | |

We rotated the three–four model and translated it relative to the one–two pair to find an optimal fit (FIGURE 2). The first goal of our structural studies is to refute or confirm this model.

## CRYSTALLOGRAPHIC RESULTS

I summarize our preliminary crystallographic results[5] and those of Cook *et al.*[6] In addition to calmodulin, we are studying S-100 and crayfish sarcoplasmic calcium-binding protein.

The S-100 group of proteins is found in large quantities in brain and may be involved either in the differentiation of glial cells or in control of their permeability. Isobe and Okuyama[7] characterized the two predominant components and determined the amino acid sequence of S-100b.

Sarcoplasmic calcium binding protein[5] is found in the sarcoplasm of various invertebrates. Although the function of these proteins is not known, studies on crayfish sarcoplasmic calcium binding protein indicate a possible regulation of concentration of cellular calcium ions, and a possible functional similarity to the vertebrate muscle parvalbumins.[8] Its amino acid sequence is not known.

We have crystallized these three proteins by the hanging-drop distillation method. A 5–10 $\mu$l droplet of solution containing protein, counter ions, and

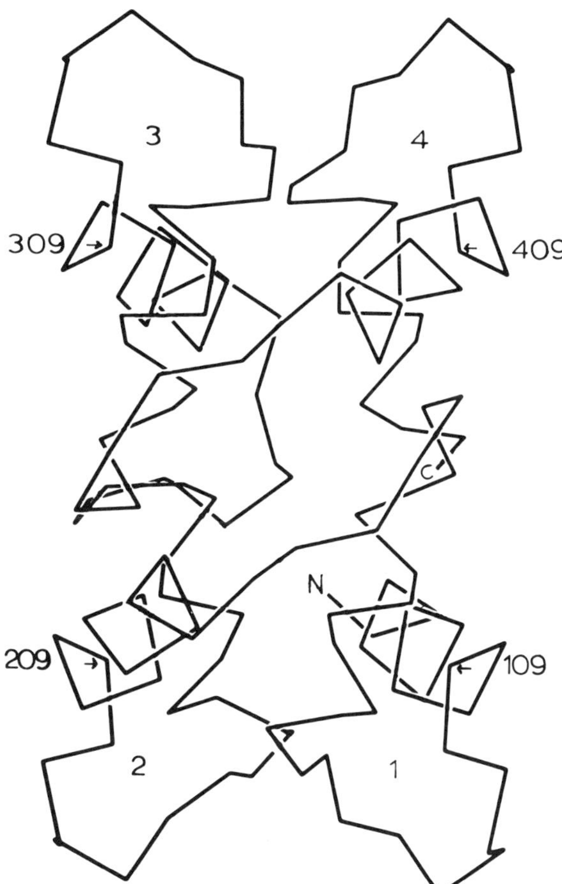


FIGURE 2. The predicted structure of calmodulin consists of two pairs of EF-hands. The local approximate two-fold axis relating domains one and two (bottom of the figure) is assumed to be collinear with, and antiparallel to, the local two-fold relating domains three and four (top of the figure). "N" and "C" denote the N- and C-termini of the model. The numbers—1,2,3 & 4—represent the approximate positions of the four calcium ions. Residue nine in each domain is indicated.

precipitating agent is placed on a siliconized cover slip. It is then inverted over a reservoir containing a ml of precipitating agent at higher concentration. We use this method both to survey crystallization conditions and to grow larger crystals for X-ray diffraction studies. Good crystals of S-100 and of the crayfish sarcoplasm calcium binding protein grow readily. We have worked several years

TABLE 1

SUMMARY OF CRYSTAL AND PROTEIN CHARACTERISTICS

| | Calmodulin[6] | Calmodulin[5] | S-100b[5] | Crayfish Sarcoplasmic[5] Calcium-Binding Protein |
|---|---|---|---|---|
| Range of Crystallization Conditions | | | | |
| Droplet | | | | |
| pH | 6.0 | 5.1–5.3 | 4.7–4.9 | 6.5–7.0 |
| Buffer | 20 mM CAC | 10 mM CAC or PIPES | 10 mM PIPES | 10 mM PIPES |
| [Ca] added | 3 mM | 5–20 mM | 0.1–0.3 mM | none added |
| [Mg] added | 0 | 0–10 mM | 1–3 mM | none added |
| Ppt. agent | 22% MPD | 5–8% PEG | 10% PEG | 20–25% MPD |
| Reservoir | | | | |
| Ppt. agent | 55% MPD | 12–25% PEG | 12–20% PEG | 50–55% MPD |
| Buffer | 50 mM CAC | 20 mM CAC | 20 mM PIPES | 20 mM PIPES |
| $\beta$-ME | 0 | 0.5 mM | 0.5 mM | 0.5 mM |
| Crystallographic Data | | | | |
| Space Group | Pl | $P2_1$ | $P4_1$ | $P2_12_12_1$ |
| a A | 29.8 | 61.8 | 56.0 | 58.9 |
| b A | 53.7 | 56.7 | 56.0 | 68.5 |
| c A | 24.8 | 40.0 | 112.8 | 116.1 |
| $\alpha$ | 93.5° | | | |
| $\beta$ | 97.0° | 92.7° | | |
| $\delta$ | 89.1° | | | |
| Content per Asymmetric Unit: | | | | |
| | 1 monomer* | 2 monomers* | 2 dimers* | 1 dimer* |
| | 52% protein | 58.7% protein | 58.3% protein | 46% protein |
| | 2.35 Å$^3$/dalton | 2.09 Å$^3$/dalton | 2.11 Å$^3$/dalton | 2.7 Å$^3$/dalton |
| Protein Characteristics | | | | |
| Molecular Weight | | 16,680/monomer | 21,014/dimer | 44,000/dimer |
| $Ca^{2+}$/monomer | | 4 | 1 | 3 |
| EF-Hand Homolog Domains/monomer† | | 4 | 1 | ? |
| pI | | 4.1 | 4.1 | 4.7 |

*Presumed.

†As deduced from amino acid sequence homology.

PEG = polyethylene glycol MW = 6000; CAC = cacodylic acid $(CH_3)_2As(O)OH$; PIPES = 1,4-Piperazinediethanesulfonic acid; MPD = 2-methyl-2,4-pentanediol; $\beta$-ME = $\beta$-Mercaptoethanol.

to get medium-quality crystals of calmodulin. Cook *et al.*[6] have recently told us of their calmodulin crystals. In TABLE 1 we summarize their and our results.

We recently measured a 4.5 Å resolution sphere of calmodulin X-ray diffraction data and a 4.0 Å sphere of S-100 data. We are preparing heavy-atom derivatives of both.

In a rotation, or Patterson superposition function calculation, one can find the relative orientation of two similar molecules within a crystallographic asymmetric unit or the orientation of two similar molecules within a crystal relative to a known model.[9] One compares the diffraction pattern with itself following a rotation of one set of data relative to another. A large value in such a calculation indicates not only a correct alignment of the two sets of data but also a reasonable similarity of the two structures. Our preliminary rotation function calculations have yielded such a large value. We hope this will be the first step in determining the high-resolution structure of calmodulin and ultimately details of its interactions with its many target molecules.

## References


1. Kretsinger, R. H. 1975. Hypothesis: calcium modulated proteins contain EF-hands. *In* Calcium Transport in Contraction and Secretions. E. Carafoli, F. Clementi, W. Drabikowski & A. Margreth, Eds.: 469–478. Elsevier North-Holland Publishing Co. Amsterdam.
2. Moews, P. C. & R. H. Kretsinger. 1975. Refinement of the structure of carp muscle calcium-binding parvalbumin by model building and difference fourier analysis. J. Mol. Biol. **91:** 201–228.
3. Goodman, M., J.-F. Pechère, J. Haiech, & J. G. Demaille. Evolutionary diversification of structure and function in the family of intracellular calcium-binding proteins. J. Mol. Evol. **13:** 331–352.
4. Kretsinger, R. H. & C. D. Barry. 1975. The predicted structure of the calcium-binding component of troponin. Biochim. Biophys. Acta **405:** 40–52.
5. Kretsinger, R. H., S. E. Rudnick, L. A. Sneden & V. B. Schatz. 1980. Calmodulin, S-100, and crayfish sarcoplasmic calcium binding protein crystals suitable for X-ray diffraction studies. J. Biol. Chem. (In press.)
6. Cook, W. J., J. R. Dedman, A. R. Means & C. E. Bugg. 1980. Crystallization and preliminary X-ray investigation of calmodulin. J. Biol. Chem. (In press.)
7. Isobe, T. & Okuyama, T. 1978. The amino-acid sequence of S-100 protein (PAP I-b protein) and its relation to the calcium-binding proteins. Eur. J. Biochem. **89:** 379–38.
8. Wunk, W., J. A. Cox, L. G. Kohler & E. A. Stein. 1979. Calcium and magnesium binding properties of a high affinity calcium-binding protein from crayfish sarcoplasm. J. Biol. Chem. **254:** 5284–5289.
9. Rossman, M. G., Ed. 1971. The Molecular Replacement Method. International Science Review Series. Gordon & Breach. New York, N.Y.

# SPIN LABELED CALMODULIN: A NEW PROBE FOR STUDYING $Ca^{2+}$ AND MACROMOLECULAR INTERACTIONS*

Paula B. Hewgley and David Puett†

*Department of Biochemistry*
*Vanderbilt University*
*Nashville, Tennessee 37232*

## INTRODUCTION

As discussed elsewhere in this volume calmodulin (CaM) is a highly conserved protein that, when $Ca^{2+}$ activated, interacts with several enzymes and modulates a variety of cellular processes. In addition to these accompanying chapters, two recent reviews on CaM have appeared.[1,2]

A number of chemical and physicochemical studies have been reported that delineate to some extent the nature of the conformational changes that CaM undergoes in response to $Ca^{2+}$ binding.[3-6] By and large, these studies are restricted to purified, simple systems, e.g., CaM with and without bound $Ca^{2+}$. In an attempt to introduce a new probe for biophysical studies and to develop a modified protein that can be used in complex systems involving enzymes, cytoskeletal elements, and membranes, we have prepared a spin labeled (SL) derivative of CaM that exhibits biological function in response to $Ca^{2+}$. Electron paramagnetic resonance (EPR) of SL-CaM thus adds a new dimension for investigating conformational changes in response to $Ca^{2+}$ and permits the use of preparations where the calmodulin-binding component(s) is not necessarily in a purified state.

## MATERIALS AND METHODS

### *Supplies*

The spin label reagent 3-(2-iodoacetamido)-2,2,5,5-tetramethyl-1-pyrrolidinyl oxyl, was purchased from Syva Corp. (Palo Alto, Calif.). Crotalus Atrox venom and 3′,5′-cyclic guanosine monophosphate (cGMP) were from Sigma Chemical Co. (St. Louis, MO), and [$^3$H]cGMP was from New England Nuclear Corp. (Boston, MA). All other chemicals and reagents were obtained from standard suppliers.

### *CaM Preparation*

CaM was purified from bovine brain using the method of Watterson *et al.*[7] with modifications (T. C. Vanaman, personal communication). All operations

*Supported by the National Institutes of Health (Grants BRSG RR05424 and AM15838).

†Address correspondence to: Dr. J. David Puett, Department of Biochemistry, 555 Light Hall, Sta. 17, Vanderbilt University, Nashville, TN 37232.



0077-8923/80/0356-0020 $1.75/0 © 1980, NYAS

were conducted at 0–4°C. Six brains were trimmed of dura mater and brain stem. The tissue (1.5 kg) was then homogenized in a Waring blender using 2 liters of the following buffer: 0.1 M sodium acetate containing 1 mM dithioerythrytol (DTE); 1 mM ethylene glycol bis($\beta$-aminoethyl ether)-N,N,N′,N′-tetraacetic acid (EGTA); 10 mM benzamidine; and 0.125 mg/ml soybean trypsin inhibitor, pH 7.2. The homogenate was centrifuged for 30 min at 16,000 × g, and the supernatant was combined with that obtained by rehomogenization and centrifugation of the pellet. The combined supernatant (1.9 liters) was made 50% in ammonium sulfate and the pH adjusted to 7.4. The suspension was stirred for 16 hours and then centrifuged for 1 hour at 10,000 × g. The supernatant was titrated to pH 4.2 with 1 N $H_2SO_4$ containing 50% ammonium sulfate and stirred for 5 hours. Following centrifugation (1 hour at 10,000 × g) the pellet was suspended in 50 ml water and the pH adjusted to 7.0. Following dialysis against water, the sample was lyophilized, then dissolved and dialyzed against the starting buffer (10 mM Tris-HCl, 1 mM DTE, 1 mM EGTA, pH 7.5, containing 0.2 M NaCl) for the first chromatographic step.

The dialyzed CaM-fraction was centrifuged and the supernatant was applied to a diethylaminoethyl (DEAE)-Sephadex A-50 column, equilibrated and developed with the above buffer and then the same buffer but with 0.3 M NaCl. A linear salt gradient was then established from 0.3 M NaCl to 0.7 M NaCl. Peak tubes from the ultraviolet (UV)-absorbing fractions were assayed using an activator-depleted smooth muscle phosphodiesterase (PDE) preparation.[8] The most active CaM fraction was pooled, dialyzed against water, and lyophilized to give 90 mg of material. This fraction was then chromatographed on Sephadex G-100, and the CaM fractions were pooled, dialyzed against water, and lyophilized to give 50 mg of purified CaM.

### *Preparation of SL-CaM*

The protein was dissolved (7 mg/ml) in 0.1 M sodium succinate buffer, pH 5.8, containing 0.1 mM $CaCl_2$. The SL reagent, 3-(2-iodoacetamido)-2,2,5,5-tetramethyl-1-pyrrolidinyl oxyl (FIGURE 1), was first taken up in ethanol (100 mg/ml)

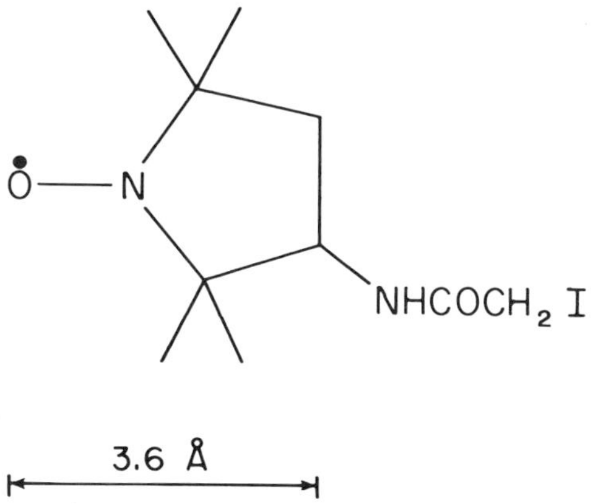

Ȯ—N
NHCOCH$_2$ I
3.6 Å

FIGURE 1. The SL reagent used to modify CaM, 3-(2-iodoacetamido)-2,2,5,5-tetramethyl-1-pyrrolidinyl oxyl.

and then 0.1 ml of this solution was added to 1 ml of the $Ca^{2+}$-containing succinate buffer. This was slowly added to 3 ml of the CaM solution, which was then flushed with nitrogen and incubated for 24 hours at 37°C. Following exhaustive dialysis against 0.1 M ammonium bicarbonate to remove free spin label, SL-CaM was obtained by lyophilization.

### *PDE Preparation and Assay Conditions*

A bovine brain PDE fraction was prepared using the method of Watterson *et al.*[7] except that DTE and EGTA were used instead of mercaptoethanol and ethylenediaminetetraacetic acid, respectively, and a smooth muscle (chicken gizzard) PDE preparation[8] was kindly provided by Dr. J. N. Wells. PDE activity was determined using cGMP and the tritiated substrate method as described by Wells *et al.*[8]

### *CD and EPR Spectroscopy*

The circular dichroic (CD) spectra were obtained using a Cary 60 spectropolarimeter equipped with a CD attachment. A single cell of 2 mm pathlength was used for all measurements, and all spectra were recorded in duplicate or triplicate. The EPR spectra were determined with a Varian E-112 spectrometer equipped with a $TE_{102}$ microwave cavity (Varian E-231). The instrument was operated at a microwave power of 5 mwatts and X-band frequency (9.5 $GH_z$) was used. Routinely, scans were made at either 40 or 100 gauss. All solutions used for spectral measurements contained 50 mM Tris-HCl, pH 7.5, and effective $Ca^{2+}$ concentrations were obtained by adjusting the EGTA:$Ca^{2+}$ ratios[9] at constant pH under conditions where ionic strength changes were small. All spectral measurements were made at 25°C.

### *Other Methods*

Isoelectric focusing gel electrophoresis was based on a modification[10] of the technique described by O'Farrell.[11] Protein hydrolysis, performic acid oxidation, and amino acid analysis were performed using standard techniques.

## RESULTS AND DISCUSSION

The ion exchange and gel exclusion chromatograms obtained during the purification of CaM are shown in FIGURE 2. The major CaM-containing fraction is indicated in each chromatrogram but other fractions with less CaM activity were also noted.

CaM was modified with [$^{14}$C]-iodoacetamide using the same conditions under which spin labeling was accomplished. The $^{14}$C-labeled protein was acid hydrolyzed and the hydrolysate was chromatographed on the acidic column of a

Beckman 121 amino acid analyzer. Column fractions were collected, and separate aliquots were counted and analyzed for amino acids by an *o*-phthalaldehyde assay.[12] A single peak of radioactivity eluted just after proline indicating the presence of either S-carboxymethylhomocysteine or 1-carboxymethylhistidine; these derivatives of methionine and histidine exhibit almost identical $R_f$s under the chromatographic conditions used.

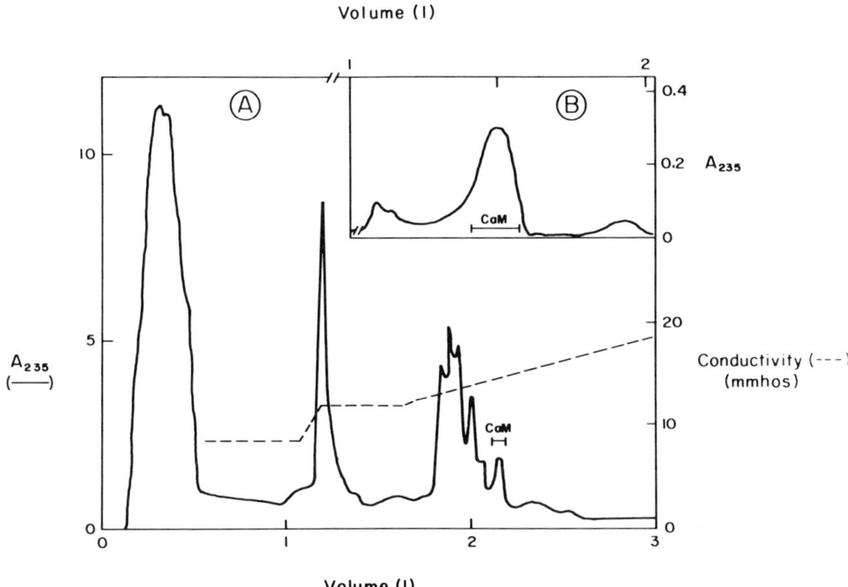


FIGURE 2. A. Ion exchange chromatograph of the bovine brain CaM-containing precipitate from the 50% ammonium sulfate fraction at pH 4.2. The material was applied to a 4.8 × 33 cm column of DEAE-Sephadex A-50 equilibrated and developed at about 15 ml/hr for 500 ml with the 0.2 M NaCl-containing buffer described in the text. The column was then developed for 500 ml with the same buffer but with 0.3 M NaCl. A linear gradient was then established from 0.3 M NaCl (1 liter) to 0.7 M NaCl (1 liter). The absorbance at 235 nm and the conductivity of the effluent (fraction size = 10 ml) are shown on the left and right ordinates, respectively. The major CaM-containing fractions were pooled as indicated, dialyzed against water, and lyophilized.

B. The CaM-fraction was then chromatographed on a 5 × 142 cm column of Sephadex G-100 equilibrated and developed with 5 mM Tris-HCl, 1 mM DTE, 1 mM EGTA, pH 7.4. The CaM fractions were pooled as indicated, dialyzed against water, and lyophilized. Purity was established by a variety of chemical and physiochemical methods.

In order to distinguish between methionyl and histidyl modification, or to ascertain if both were spin labeled, SL-CaM was performic acid oxidized, acid hydrolyzed, and the amino acid composition determined on a Beckman 121 2-column analyzer. The results indicated that about 2 moles of methionine were modified per mole of CaM (TABLE 1), and there was no evidence of other

TABLE 1

AMINO ACID ANALYSIS OF SL-CAM AND PERFORMIC ACID OXIDIZED SL-CAM*

| Amino Acid | Experimental Mole %† | Expected Mole %‡ |
|---|---|---|
| (SL-CaM) | | |
| methionine | 4.5 | 6.1 |
| methionine derivatives | 1.8 | 0 |
| (performic acid oxidized SL-CaM) | | |
| methionine | —§ | 0 |
| methionine sulfone | 3.9 | 6.1 |

*The results, expressed as mole % of total residues, are shown for methionine and derivatives only since the content of the other amino acids were the same for CaM and SL-CaM; also, these were in excellent agreement with the values expected from the amino acid sequence. In particular, analysis of several hydrolyzates showed that the histidyl content of SL-CaM was within ± 5% of that expected in CaM.

†The results for SL-CaM are from FIGURE 3, and the data on performic acid oxidized SL-CaM were obtained using a standard hydrolysate and the long column of a Beckman 121 amino acid analyzer. In the latter case, the results were normalized to aspartic acid to obtain the overall mole % of methionine. (For example, the ratio of methionine sulfone to aspartic acid was 0.25 in performic acid oxidized SL-CaM. The ratio of methionine to aspartic acid was 0.28 in SL-CaM while the expected ratio from the sequence is 0.39; a ratio of 0.35 was found experimentally in CaM.) Since CaM contains 9 methionyl residues, these results indicate that an average of 2–3 methionines are modified.

‡From the amino acid sequence of bovine brain CaM (*cf.* Vanaman, this volume).

§Trace amounts were detected.

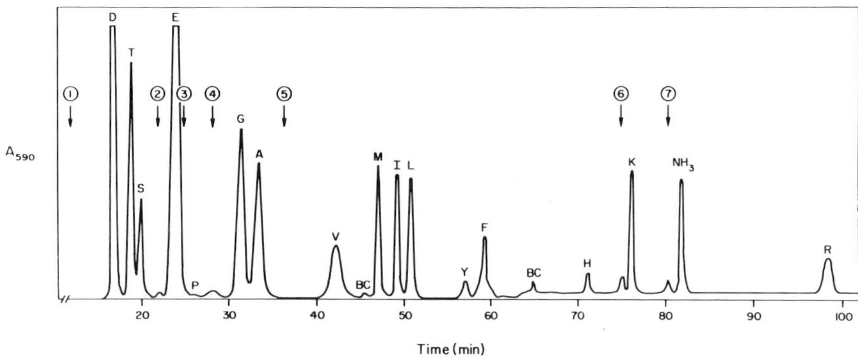


FIGURE 3. Chromatograph of acid-hydrolyzed (6 N HCl, 24 hrs, 110°C) SL-CaM on a single-column Durrum amino acid analyzer where the absorbance at 590 nm (ninhydrin reaction) is plotted as a function of retention time. The amino acids are identified by the single letter abbreviation and BC denotes buffer change. Elution positions of standards are indicated by the arrows and the appropriate identification is given below. These data were obtained in two separate experiments by modifying N-acetylhistidine and N-acetylmethionine under the same spin labeling conditions used for CaM. The reaction products were acid hydrolyzed and chromatographed; comparison with published standards permitted final assignments. It can be seen that there is no evidence for histidine modification since no components elute at positions corresponding to dicarboxymethylhistidine or either 1-or 3-carboxymethylhistidine. All nonstandard amino acids can be accounted for as methionine derivatives, e.g., homoserine, S-carboxymethylhomocysteine, and homoserine lactone. Also, the component eluting between lysine and ammonia has not been identified but it was observed in the hydrolyzed SL-N-acetylmethionine standard and not in the hydrolyzed SL-N-acetylhistidine standard. The methionine derivatives were assumed to have the same color constant as methionine. 1) dicarboxymethylhistidine, 2) homoserine, 3) 1-carboxymethylhistidine, 4) S-carboxymethylhomocysteine, 5) 3-carboxylmethylhistidine, 6) homoserine lactone, and 7) unidentified methionine derivative.

modified residues. This was also demonstrated using a single column Durrum analyzer (AAA Laboratory, WA) to analyze acid hydrolyzed SL-CaM (FIGURE 3). These data show that the sum of the hydrolysis products of S-carboxymethylmethionine, homoserine, homoserine lactone, and S-carboxymethylhomocysteine is just under 2 moles/mole SL-CaM and that 6.5 moles of unreacted methionines/mole SL-CaM can be identified (TABLE 1). Thus, under the conditions used, spin labeling of CaM seems highly specific for methionyl residues and the amino acid analyses indicate that an average of two groups are modified.

Isoelectric focusing gels were run on CaM, SL-CaM, and a mixture of the two.

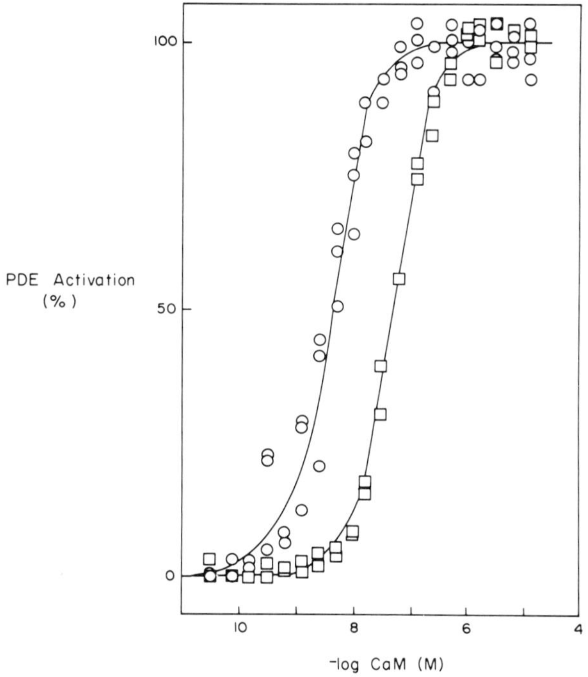


FIGURE 4. Stimulation of a bovine brain PDE fraction by CaM (○) and SL-CaM (□). The results of several assays are shown and the assay conditions are given in Reference 8. The data are plotted as % stimulation above basal and there was no significant difference in the apparent $V_{max}$ of CaM and SL-CaM (e.g., the fold-stimulations averaged 4.2 and 4.3, respectively).

There was no detectable unmodified protein following spin labeling; the apparent isoelectric pHs of CaM and SL-CaM were 4.2 and 4.4, respectively.

The apparent $K_m$ describing the interaction of $Ca^{2+}$-saturated SL-CaM with activatable brain PDE is higher than that of CaM, although the $V_{max}$ is not changed by spin labeling (FIGURE 4). Similar results were obtained using an activatable smooth muscle PDE preparation (data not shown). Thus, the iodoacetamide-based spin label hinders but does not abolish the interaction of the CaM-derivative with PDE.

The near UV CD spectra of CaM in the presence and absence of $Ca^{2+}$ is shown in FIGURE 5A. The relatively sharp CD bands between 255–270 nm arise mainly from phenylalanyl residues, and the band(s) above 280 nm can be assigned to tyrosine. There is a slight enhancement in the magnitude of the various negative bands concomitant with $Ca^{2+}$ binding. The changes begin at about 0.1 μM $Ca^{2+}$ and occur over a very narrow concentration of $Ca^{2+}$ (FIGURE 5B). These results are consistent with a $Ca^{2+}$-mediated change in the microenvironment of at least some of the aromatic residues in CaM. A similar conclusion was reached by others;[3-6] however, the CD changes above 270 nm found in this study are quite different from those in another report.[4] A comparable investigation was attempted with SL-CaM but excessive turbidity at concentrations required for these measurements (*ca.* 0.18 mM) precluded the acquisition of meaningful near UV CD spectra.

In contrast, far UV CD spectra could be obtained for both CaM and SL-CaM since much lower protein concentrations were used in this spectral region, e.g.,

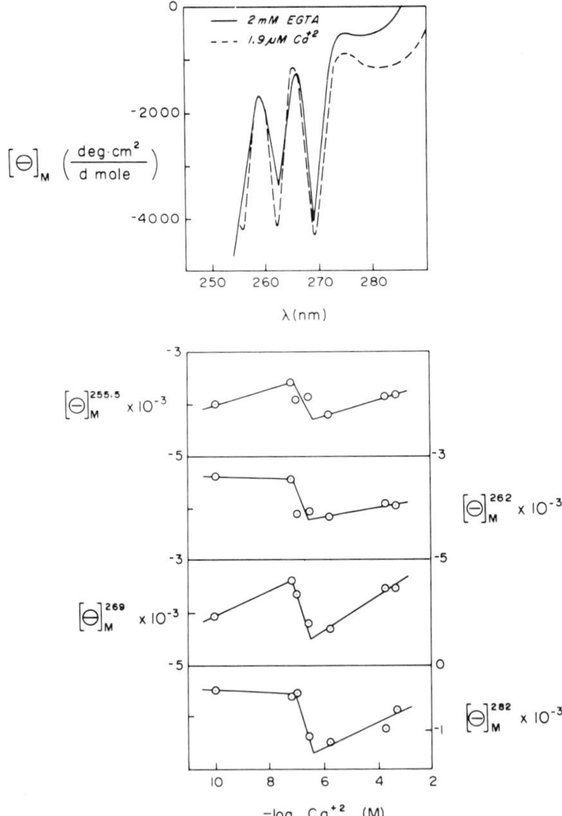


FIGURE 5. *Upper panel.* Near UV CD spectra of CaM in the presence of 2 mM EGTA and of 1.9 μM $Ca^{2+}$ at a protein concentration of 3 mg/ml. *Lower panels.* The molar ellipticity at various wavelengths is plotted at different $Ca^{2+}$ concentrations.

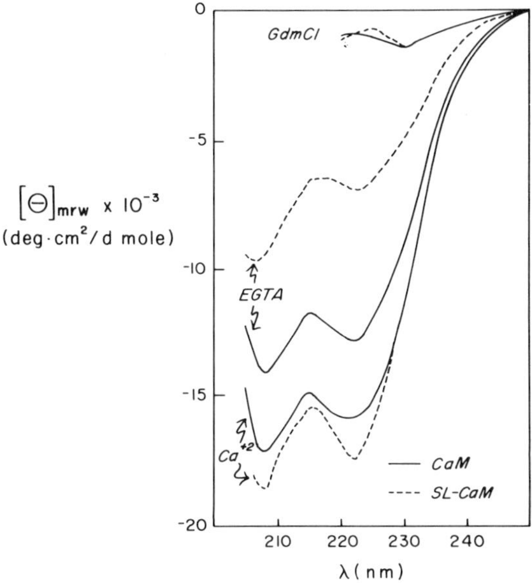


FIGURE 6. Far UV CD spectra of CaM and SL-CaM in the presence of 7 M GdmCl (containing 20 mM EGTA), EGTA (2 mM), and $Ca^{2+}$ (1.2 mM for CaM and 0.2 mM for SL-CaM).

concentrations of about 12 μM were employed. As shown in FIGURE 6, CaM and SL-CaM exhibit similar CD spectra in the presence of EGTA plus the denaturant 7 M GdmCl. These spectra are typical of unfolded proteins or proteins devoid of secondary structure.[13] Under nondenaturing conditions the apoproteins exhibit spectra with the characteristic double minima of α-helical regions arising from peptide $n-\pi^*$ and $\pi-\pi^*$ transitions. Interestingly, CaM appears to contain more helicity than SL-CaM in the presence of EGTA. In contrast, CaM and SL-CaM have similar CD spectra in the presence of $Ca^{2+}$, although the magnitude of the ellipticity of the latter is somewhat greater than that of the native protein. These results argue for great similarity in the secondary structure of native and spin labeled CaM following $Ca^{2+}$ binding.

Typical X-band EPR spectra are shown in FIGURE 7 for SL-CaM in the presence and absence of $Ca^{2+}$. Concomitant with $Ca^{2+}$ binding, a narrowing of line widths and a slight increase in the ratio of the high field peak to the center field peak are observed. However, the most dramatic change occurring with $Ca^{2+}$ binding is a significant decrease in the total spectral intensity which is indicative of spin-spin interactions.[14] Control spectra of free spin label in the presence and absence of $Ca^{2+}$ showed no differences in spectral characteristics. Moreover, measurements at various concentrations of SL-CaM ± $Ca^{2+}$ showed linearity between spectral intensity and protein concentration; hence, the effects appear to be intramolecular.

We also obtained EPR spectra of SL-CaM in the presence and absence of 1 mM $Mn^{2+}$, a paramagnetic ion which might be a potent relaxing species. It was found that $Mn^{2+}$ had no significant effect on the peak height ratio of either the free spin label or SL-CaM. A small decrease was observed in spectral intensities

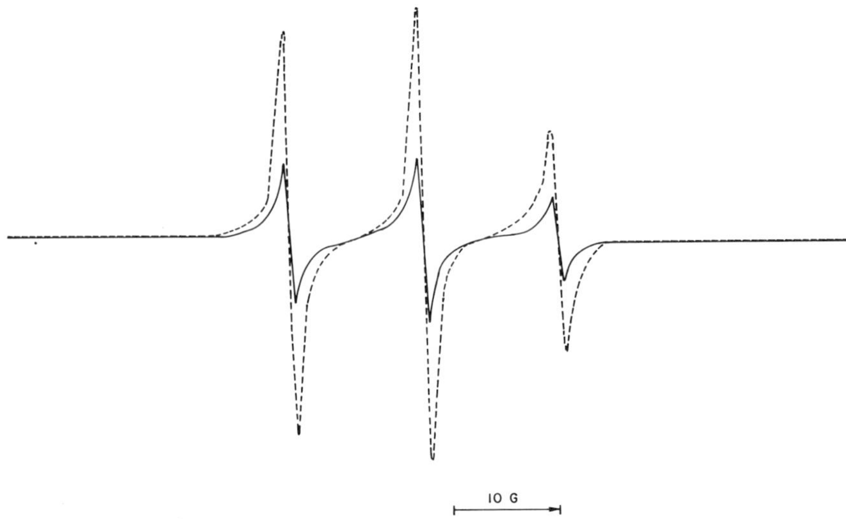


FIGURE 7. X-band EPR spectra of SL-CaM in the presence of EGTA (10 mM) and of $Ca^{2+}$ (1 mM).

but the free spin label and SL-CaM did not differ appreciably in the magnitude of this change.

These EPR results suggest that the effects of $Ca^{2+}$ are not due to an enhancement of spin-lattice relaxation resulting from $Ca^{2+}$ binding. Rather, $Ca^{2+}$ binding appears to induce localized changes in the relative proximity of two or perhaps three spin labeled methionyl residues.

Changes in several spectral parameters (e.g., $[\theta]$ at 222 nm for CaM and SL-CaM; the height of the high field EPR peak of SL-CaM) were measured at various $Ca^{2+}$ concentrations as were changes in the activation of a smooth muscle PDE preparation for both CaM and SL-CaM. To facilitate comparison the results have been normalized to the fractional change $f$ in either the spectral parameter or the enzymic stimulation such that $f = 0$ at $Ca^{2+}$ concentrations lower than 0.1 $\mu$M and $f = 1.0$ at 1 $\mu$M $Ca^{2+}$. The following equation relates $f$ to the experimental observable y (i.e., spectral intensity or degree of PDE stimulation) at any $Ca^{2+}$ concentration,

$$f = (y - y_0)/(y' - y_0), \qquad (1)$$

where $y_0$ denotes the magnitude of the response at $Ca^{2+}$ concentrations between $pCa^{2+} = 10$ and $pCa^{2+} = 7$ and $y'$ is the corresponding value at $pCa^{2+} = 6$.

The experimental points plotted in this manner are shown in FIGURE 8. Several items of interest emerge from this plot. First, it is clear that CaM and SL-CaM respond to $Ca^{2+}$ in a similar if not identical fashion. Second, conformational changes as monitored by CD or EPR are closely coupled with each other and with the stimulation of PDE activity. Third, at high $Ca^{2+}$ concentrations the stimulation of PDE activity by both CaM and SL-CaM diminishes slightly and a second transition is noted with the EPR signal of SL-CaM. Interestingly, no changes are observed in $[\theta]$ (222 nm) at millimolar $Ca^{2+}$ concentrations for either CaM or SL-CaM.

The inset to FIGURE 8 shows the data in the major transition region (i.e., between a $pCa^{2+}$ of 6 and 7) plotted according to the theory of linked functions.[15] This model considers a macromolecule that can exist in two states; one of which binds $n$ moles of ligand per mole of macromolecule. Letting the conformation of CaM (or SL-CaM) in the absence of $Ca^{2+}$ denote one such state and the conformation at micromolar $Ca^2$ denote the other state with $n$ $Ca^{2+}$ ions bound, the following relationship is established.

$$n = \delta \log K_{app}/\delta \log(Ca^{2+}), \quad (2)$$

where $K_{app}$ is the apparent equilibrium constant for formation of the $CaM \cdot Ca_n^{2+}$ complex. Thus $n$ can be obtained from a Hill-type plot that uses physicochemical or enzymic properties rather than direct binding parameters. The data in FIGURE 8 are consistent with an $n$ of 3 or 4 and clearly not 1 or 2. One caveat is that Equation 2 properly refers to the free $Ca^{2+}$ concentration whereas the plot in FIGURE 8 is based on the free $Ca^{2+}$ plus protein-bound $Ca^{2+}$. Under the experimental conditions used, this will be a small correction and its inclusion will tend to increase somewhat the experimental value of $n$.

EPR spectra were also obtained of $Ca^{2+}$-saturated SL-CaM in the presence of a smooth muscle PDE preparation,[8] neurofilaments (10 nm filaments) which contain an activatable PDE[16] and were kindly provided by Drs. M. S. Runge and R. C. Williams, Jr., and the drug trifluoperazine which binds to CaM.[17] These compounds also alter the EPR spectrum of SL-CaM, but the changes are small compared to those induced by $Ca^{2+}$ (data not shown).

The particular residues modified by the spin label are not known, but recent

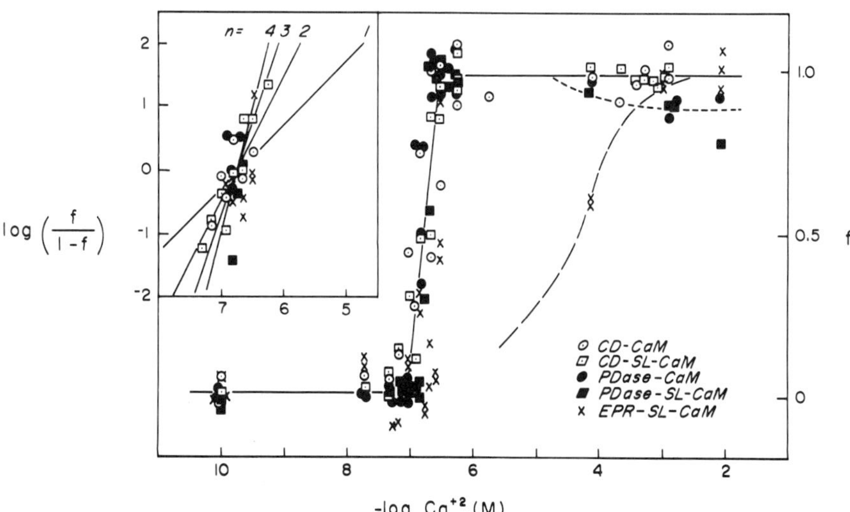


FIGURE 8. The fractional change from $pCA^{2+} = 10$ ($f = 0$) to $pCa^{2+} = 3$ (with $f = 1.0$ being defined at $pCa^{2+} = 6$) is shown for CaM and SL-CaM. Various parameters including CD, EPR, and PDE stimulation were used to obtain the data. The inset shows a linked function, Hill-type plot of these data in the major transition region, and the solid lines indicate the expected relationships for 1, 2, 3, or 4 $Ca^{2+}$ ions bound/protein molecule. All spectral data were collected in 50 mM Tris-HCl, pH 7.5, containing EGTA and $CaCl_2$.

studies by Walsh and Stevens[18] are perhaps relevant to our work. They modified methionyl residues with N-chlorosuccinimide and found that 2 residues were oxidized; tentatively, these were identified as any two of residues 71, 72, 76, or perhaps 109. It is not unlikely that the iodoacetamide-based spin label used herein would react with the same methionines that are accessible to N-chlorosuccinimide.

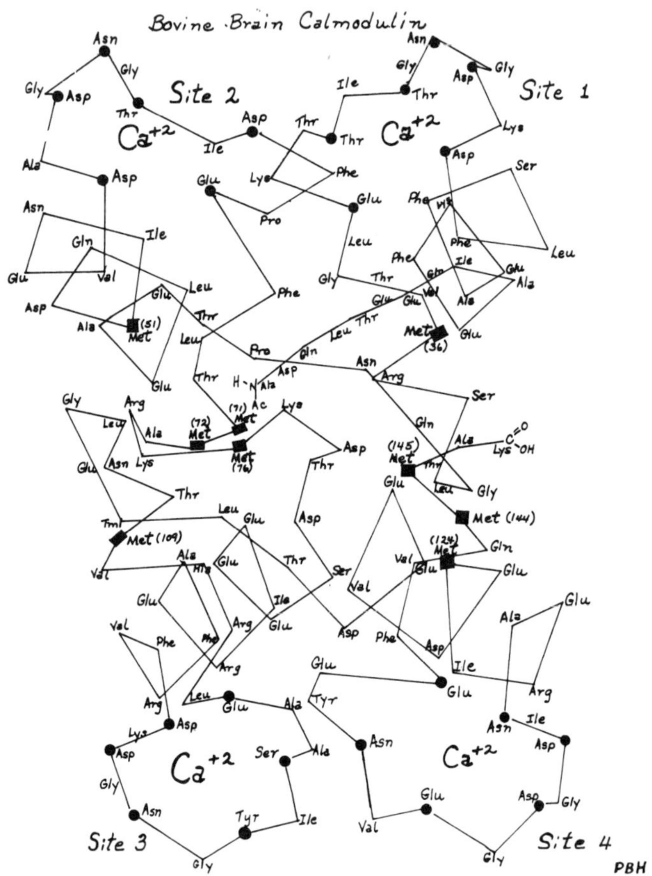


FIGURE 9. Tentative sketch of CaM from the model predicted by Kretsinger and Barry for troponin C (*cf.* text for details).

We have added the sequence of bovine brain CaM[19] to the model proposed by Kretsinger and Barry[20] for troponin C, which was based on the crystal structure of parvalbumin,[21] and obtain the structure shown in FIGURE 9. This is of course highly tentative and serves simply as a working model. It is of interest that within the limitations of this structure two clusters of methionyl residues can be identified. One such cluster involves residues 71, 72, and 76 and is flanked by methionines 51 and 109; the other cluster contains residues 124, 144, and 145 and is in proximity to methionine 36. It is likely that the spin labeled methionyl residues occur in one of these clusters. It is also noteworthy that in this model the methionine clusters are not particularly close to either of the four $Ca^{2+}$ binding sites; this agrees with our EPR results on SL-CaM.

## Conclusions

We have demonstrated that CaM can be spin labeled at methionyl residues and the derivative is both responsive to $Ca^{2+}$ and capable of stimulating the activity of PDE. CD and EPR spectral changes occur when $Ca^{2+}$ binds to SL-CaM. The latter appear to be mediated via $Ca^{2+}$-induced conformational changes and do not seem to result from direct ion-free radical interactions. SL-CaM should be suited for additional studies involving CaM interactions with membranes and various enzymes.

## Acknowledgments

We are pleased to acknowledge the support of necessary instrumentation by Dr. Howard E. Smith under the auspices of the Organic Chemistry Core Facility of Center grant HD05797. P.B.H. is a Cellular and Molecular Biology Predoctoral Trainee (GM07319) and D.P. is a Research Career Development Awardee (AM00055). It is a pleasure to thank Drs. Albert Beth, Larry Dalton, Ray Perkins, Thomas Vanaman, and Jack Wells for many helpful discussions. We also thank Dr. Harry Green of Smith Kline & French for providing the phenothiazine sample.

## References

1. Cheung, W. Y. 1980. Science **207:** 19–27.
2. Means, A. R. & J. R. Dedman. 1980. Nature **285:** 73–77.
3. Klee, C. B. 1977. Biochemistry **16:** 1017–1024.
4. Wolff, D. J., P. G. Poirier, C. O. Brostrom & M. A. Brostrom. 1977. J. Biol. Chem. **252:** 4108–4117.
5. Dedman, J. R., J. D. Potter, R. L. Jackson, J. D. Johnson & A. R. Means. 1977. J. Biol. Chem. **252:** 8415–8422.
6. Seamon, K. B. 1980. Biochemistry **19:** 207–215.
7. Watterson, D. M., W. G. Harrelson, Jr., P. M. Keller, F. Sharief & T. C. Vanaman. 1976. J. Biol. Chem. **251:** 4501–4513.
8. Wells, J. N., C. E. Baird, Y. J. Wu & J. G. Hardman. 1975. Biochim. Biophys. Acta **384:** 430–442.
9. Holloway, J. H. & C. N. Reilley. 1960. Anal. Chem. **32:** 249–256; Owen, J. D. 1976. Biochim. Biophys. Acta **451:** 321–325.
10. Berkowitz, S. A., J. Kalagire, H. K. Bender & R. C. Williams, Jr. 1977. Biochemistry **16:** 5610–5617.
11. O'Farrell, P. H. 1975. J. Biol. Chem. **250:** 4007–4021.
12. Butcher, E. C. & O. H. Lowry. 1976. Anal. Biochem. **76:** 502–523.
13. Holladay, L. A. & D. Puett. 1976. Biopolymers **15:** 43–59.
14. Likhtenshtein, G. I. 1976. *In* Spin Labeling Methods in Molecular Biology. pp. 37–65. John Wiley & Sons, Inc. New York, N.Y.

15. WYMAN, J., JR. 1964. Adv. Prot. Chem. **19:** 223–286; *see also* TANFORD, C. 1970. Adv. Prot. Chem. **24:** 1–95.
16. RUNGE, M. S., P. B. HEWGLEY, D. PUETT & R. C. WILLIAMS, JR. 1979. Proc. Natl. Acad. Sci. USA **76:** 2561–2565.
17. LEVIN, R. M. & B. WEISS. 1977. Molecular Pharmacology **13:** 690–697.
18. WALSH, M., F. C. STEVENS, K. OIKAWA & C. M. KAY. 1978. Biochemistry **17:** 3924–3930.
19. WATTERSON, D. M., F. SHARIEF & T. C. VANAMAN. 1980. J. Biol. Chem. **255:** 962–975.
20. KRETSINGER, R. H. & C. D. BARRY. 1975. Biochim. Biophys. Acta **405:** 40–52.
21. MOEWS, P. C. & R. H. KRETSINGER. 1975. J. Mol. Biol. **91:** 201–228.

# CALMODULIN-ACTIVATED PHOSPHODIESTERASE FROM VERTEBRATE SMOOTH MUSCLE*

James F. Head, Robert J. Birnbaum, and Benjamin Kaminer

*Department of Physiology*
*Boston University School of Medicine*
*Boston, Massachusetts 02118*

The potential roles of calmodulin in vertebrate smooth muscle, as in other tissues, are evidently diverse. Our previous studies have shown that calmodulin constitutes about 0.4% of the total protein of chicken gizzard,[1] this represents a calcium-sensitive "switching" mechanism of considerable capacity. One of the potential functions of calmodulin in smooth muscle is to regulate cyclic nucleotide levels by altering phosphodiesterase activity. Cyclic nucleotides may influence smooth muscle contraction[2] and the mechanisms controlling their levels may therefore be of considerable importance in the regulation of smooth muscle activity.

The cyclic nucleotide phosphodiesterases of vertebrate smooth muscle may be separated from one another to some extent, and from calmodulin by DEAE-cellulose chromatography of tissue extracts.[3,4] In the presence of EDTA (20 mM Tris HCl, 1 mM EDTA, 15 mM 2-mercaptoethanol pH 7.8) the enzymes separate into two principal peaks of enzymic activity when eluted with a linear 0–0.5 M NaCl gradient in the same buffer. The peaks of enzyme activity elute at approximately 0.15 and 0.25 M NaCl. Acrylamide gel electrophoresis of eluate fractions show calmodulin to elute at about 0.3 M NaCl. Only the first of the two peaks of enzymic activity is activated by addition of calmodulin in the presence of calcium.

Nondenaturing alkaline acrylamide gels of samples from peak I enzyme show two bands when stained for phosphodiesterase activity. If the stain incubation is performed in the presence of calmodulin then the stain intensity of only one of the two bands is increased (FIGURE 1). This suggests that only this form of the enzyme is calmodulin activated. Gels of a comparable peak I enzyme obtained from a similiar preparation of bovine brain also showed two bands of activity only one of which is increased in intensity by incubation with calmodulin (FIGURE 1). The band corresponding to the calmodulin-activated phosphodiesterase has similar electrophoretic mobility in both the chicken gizzard and the bovine brain peak I enzyme fractions, however, the second band present in the peak I fractions from these tissues has considerably different mobility.

If the peak I enzyme from the gizzard preparation is passed over an affinity column of calmodulin-Sepharose in the presence of calcium then a considerable proportion of the enzymic activity remains bound to the column. This enzyme may be eluted subsequently by application of an EGTA-containing buffer. The enzyme obtained in this way is activated by calmodulin in the presence of calcium and shows only a single band on acrylamide gels stained for phosphodiesterase activity (FIGURE 2). Thus at least three forms of phosphodiesterase have been identified in the gizzard, only one of which is activated by calmodulin.

SDS gels of the affinity column purified enzyme show principally an 80,000

*Supported by the National Institutes of Health (Grant AM 18207).



0077-8923/80/0356-0033 $1.75/0 © 1980, NYAS

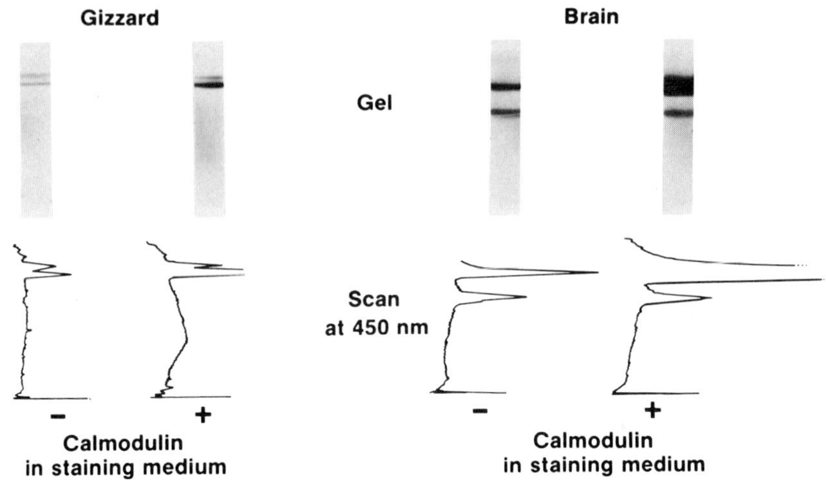


FIGURE 1. Acrylamide gels of gizzard and brain peak I phosphodiesterase stained for phosphodiesterase activity. Electrophoresis performed in 8% polyacrylamide gels containing 40% glycerol and using a continuous buffer of 25 mM Tris, 80 mM glycine pH 8.6. Gels incubated for 12 hrs in 80 mM Tris-maleate, 2.5 mM $MgSO_4$, 3 mM $Pb(NO_3)_2$, 2 mM c-AMP, pH 7.0 with 1 unit/ml of alkaline phosphatase[5] with (+) or without (−) 0.1 mg/ml calmodulin. Washed in distilled water for 2 hrs and stained with 5% ammonium sulfide for 2 min.

dalton protein and several other components. The 80,000 dalton protein apparently does not correspond to the phosphodiesterase since the same protein may also be obtained, without any phosphodiesterase activity, by similar chromatography using a bovine serum albumin-Sepharose affinity column. The 80,000 dalton protein therefore does not bind specifically to calmodulin.

Further details of the structure and properties of smooth muscle calmodulin-activated phosphodiesterase and its relationship with the brain enzyme remain to be determined.

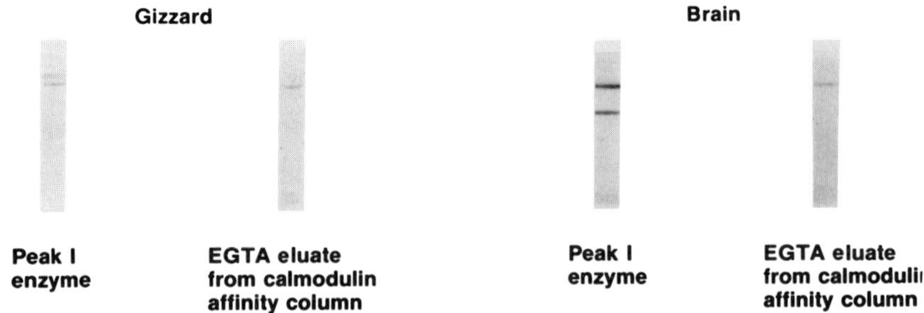


FIGURE 2. Alkaline gels of peak I phosphodiesterase before and after purification on calmodulin affinity column. Gels stained for phosphodiesterase activity as in FIGURE 1. Affinity column sample and wash buffer—200 mM NaCl, 20 mM Tris HCl, 1 mM $CaCl_2$, 15 mM 2-mercaptoethanol, pH 7.8. Elution buffer—same but 5 mM EGTA in place of $CaCl_2$.

## References


1. Head, J. F., S. Mader & B. Kaminer. 1977. Troponin-C-like modulator protein from vertebrate smooth muscle. *In* Calcium-binding Proteins and Calcium Function. R. H. Wasserman, R. A. Carradino, E. Carafoli, R. H. Kretsinger, D. H. MacLennan & F. L. Siegel, Eds.: 275–277. North-Holland. New York, N.Y.
2. Diamond, J., 1978. Role of cyclic nucleotides in control of smooth muscle contraction. *In* Advances in Cyclic Nucleotide Research. W. J. George & L. J. Ignarro, Eds. Vol. **9:** 327–340. Raven Press. New York, N.Y.
3. Wells, J. N., C. E. Baird, Y. J. Wu & J. G. Hardman. 1975. Cyclic nucleotide phosphodiesterase activities of pig coronary arteries. Biochim. Biophys. Acta **384:** 430–442.
4. Murtaugh, T. J. & R. C. Bhalla.1979. Multiple forms of cyclic nucleotide phosphodiesterase from bovine carotid artery smooth muscle. Arch. Biochem. Biophys. **196**(2): 465–474.
5. Goren, E. N., A. H. Hirsch & O. Rosen. 1974. Activity stain for the detection of cyclic nucleotide phosphodiesterase in polyacrylamide gels. Methods Enzymol. **38:** 259–261.

# COMPARATIVE BIOCHEMISTRY OF CALMODULINS AND CALMODULIN-LIKE PROTEINS*

Linda J. Van Eldik, Gianni Piperno, and D. Martin Watterson

*Department of Cell Biology*
*The Rockefeller University*
*New York, New York 10021*

Calmodulin is the name proposed[1] for a brain phosphodiesterase activator protein which has been shown to be indistinguishable from the unique chemical structure previously termed modulator protein or brain troponin C-like protein.[2] Bovine brain calmodulin is a multifunctional, calcium-modulated protein that has a calculated molecular mass of 16,680 g/mol and is structurally homologous to troponin C.[3] As summarized by Dr. Vanaman in these proceedings,[4] all vertebrate, invertebrate, and plant proteins that have been called calmodulins will activate bovine brain phosphodiesterase. However, a definition of the term calmodulin has not been formulated and calmodulin is a member of a family of proteins with extremely similar physical, chemical and, sometimes, functional properties.

In many biological systems where calcium regulation is being studied, large amounts of material are not always available, thus precluding detailed structure–function analysis. As a result, various limited criteria have been used to demonstrate the presence of calmodulin in supramolecular structures and to implicate calmodulin in the regulation of cell processes. In this report we will summarize studies on the comparative biochemistry of calmodulins from plant and vertebrate species, and of a calmodulin-like protein from *Chlamydomonas* flagella. This and other reports[2,3,5-8] more fully define what is and is not a unique characteristic of calmodulins, and indicate the inherent limitations of many procedures used in studies on the role of calmodulin in cell function.

## Materials and Methods

Culture of cells, labeling with [$^{35}$S]sulfuric acid, preparation of flagella, and two-dimensional electrophoresis of polypeptides were performed as previously described.[8] Calmodulins and flagellar calmodulin-like protein were isolated using previously described procedures.[6-8] Parvalbumin and S100b were generous gifts of Dr. R. Kretsinger (Charlottesville, Va.).

Antibody against calmodulin was prepared and iodination of calmodulin was done as described elsewhere in these proceedings.[9] Competition radioimmunoassays were performed in RIA buffer (0.1 M NaCl, 0.001 M EDTA, 0.02 M Tris-HCl, pH 7.6) containing 0.2% (w/vol) bovine serum albumin. Reaction mixtures containing a limiting dilution of anti-calmodulin sera (that amount determined by direct radioimmunoassay to give 50% precipitation of the $^{125}$I counts) and varying concentrations of competing antigens were incubated at 4°C overnight. $^{125}$I-labeled bovine brain calmodulin (1 ng; approx. 50,000 cpm) was

*Supported by the Andrew Mellon Foundation, the Cystic Fibrosis Foundation, and the National Institutes of Health (grants GM26383, GM17132, and CA06381).



0077-8923/0356-0036 $1.75/0 © 1980, NYAS

then added and the solution incubated at 4°C overnight. Goat anti-rabbit serum (Cappel) was added and the mixture incubated at 4°C for an additional 8–24 hrs. The mixture was washed with RIA buffer and centrifuged at 900 × g for 10 min, the pellet washed again with RIA buffer, and radioactivity of the resultant pellets determined in a Packard gamma counter.

Direct radioimmunoassays were done as above, except that no competing antigens were added and the initial incubation mixture consisted of normal rabbit serum, $^{125}$I-labeled calmodulin, and various dilutions of anti-calmodulin sera.

Troponin was isolated from rabbit skeletal muscle by the procedure of Ebashi *et al.*,[10] and troponin C was subsequently purified by the method of Greaser and Gergely.[11] Trypsin digestions were performed in the presence of [ethylenebis(oxyethylene-nitrilo)] tetraacetic acid (EGTA) as previously described.[6] Column peptide maps of trypsin digests were done as described elsewhere[8] on a Whatman Partisil M9 ODS-2 column (9.4 × 250 mm) using a modified Hewlett-Packard 1084B liquid chromatography system. Amino acid analyses were done and protein concentrations were determined as previously described.[6,7] Polyacrylamide gel electrophoresis in the presence and absence of sodium dodecyl sulfate was performed exactly as described.[7,8] Phosphodiesterase was prepared and activator assays were done as previously described.[12]

## Results and Discussion

### *Spinach Leaf Calmodulin*

We have recently reported[7] on the isolation and limited characterization of spinach leaf calmodulin. As evidenced by its isolation using calcium-dependent, affinity-based adsorption chromatography on phenothiazine-Sepharose conjugates, spinach calmodulin will bind phenothiazines. Identical protocols have been used for the isolation of calmodulin from several vertebrate tissues.

It has been reported[7,8,13–15] that the mobility of calmodulins during polyacrylamide gel electrophoresis in the presence of sodium dodecyl sulfate is altered by the presence of calcium or chelator in the sample. When calcium is added to the sample, buffer, and gel, calmodulins will have an anomalously high electrophoretic mobility. Therefore, estimated molecular weights under these conditions are artifactually low. When EGTA is added to the sample, buffer, and gel, calmodulin will migrate considerably slower relative to molecular weight standards; under these conditions, molecular weight estimates are anomalously high. This altered mobility relative to molecular weight standards has been termed a calcium-dependent electrophoretic mobility shift. Burgess *et al.*[13] have reported that other proteins in the same family of calcium-modulated proteins, such as troponin C, parvalbumin, and S100b, will not undergo this large mobility shift. We have routinely observed a slight mobility shift in troponin C. However, the shift in troponin C mobility is minor when compared to that observed with calmodulins.

Spinach calmodulin will undergo this large mobility shift similar to, but less pronounced than, that observed with vertebrate calmodulins. While spinach leaf calmodulin will comigrate with vertebrate calmodulins during polyacrylamide gel electrophoresis under a variety of conditions, it will migrate faster than vertebrate calmodulins if electrophoresis and sample preparation are done in the presence of EGTA. The difference in electrophoretic mobility represents an apparent molecular weight difference of 1,000. It is not known from these data if this difference in electrophoretic mobility is due to spinach calmodulin having a shorter amino acid sequence, different calcium binding properties, or an amino acid sequence change that results in altered sodium dodecyl sulfate binding. Preliminary amino acid sequence analysis indicates that spinach calmodulin is not significantly shorter than vertebrate calmodulin; therefore, the apparent molecular weight difference of 1,000 cannot be explained solely by spinach calmodulin having a shorter amino acid sequence.

When examined for immunoreactivity with an antiserum prepared against vertebrate calmodulin, spinach leaf calmodulin was found to quantitatively compete with $^{125}$I-labeled bovine brain calmodulin. As noted elsewhere in these proceedings,[9] even though the antiserum was produced in response to injection of performic acid oxidized calmodulin, the antiserum reacted with unoxidized calmodulin. The iodinated antigen was unoxidized calmodulin and the competition standard was unoxidized bovine brain calmodulin.

Purified spinach leaf calmodulin was also examined for the presence of $N^{\epsilon}$-trimethyllysine since the majority of calmodulins isolated have been found to contain this unusual amino acid. Employing exactly the same procedures used for the analysis of vertebrate calmodulins, spinach leaf calmodulin was found to contain 1 mole of $N^{\epsilon}$-trimethyllysine per mole of protein.

Spinach leaf calmodulin was also examined for its ability to quantitatively stimulate bovine brain phosphodiesterase in the presence of excess calcium. An activation curve for purified bovine brain calmodulin was done in parallel. As previously reported,[7] the spinach leaf calmodulin quantitatively activated bovine brain phosphodiesterase. Under the conditions used, the activation curve for spinach calmodulin was indistinguishable from that of bovine brain calmodulin.

### *Translation Calmodulin*

A poly(A)$^+$ mRNA fraction isolated from spinach leaf tissue was translated in a cell-free wheat germ system and translation calmodulin isolated from the reaction products.[7] Translation calmodulin was isolated using the phenothiazine-Sepharose chromatography protocol used in the isolation of spinach leaf and vertebrate calmodulins. Translation calmodulin comigrated during gel electrophoresis with spinach leaf calmodulin and underwent a calcium-dependent mobility shift indistinguishable from that of spinach leaf calmodulin. Affinity-based precipitations of translation mixtures using antiserum to vertebrate calmodulin also demonstrated a major polypeptide that comigrated with calmodulin. Thus, translation calmoduli was indistinguishable from tissue-isolated calmodulin using the following criteria: 1) ability to be isolated by a calmodulin purification protocol including calcium-dependent interaction with phenothiazine-Sepharose conjugates, 2) mobility during polyacrylamide gel electrophoresis under a variety of conditions including the calcium-dependent mobility shift, and 3) ability to react with antiserum prepared against vertebrate calmodulin.

*In vitro* translations are done using isotopically labeled amino acids and all isolations and characterizations are done on the basis of radioactivity. While it is possible to generate a large amount of label in a polypeptide, the total mass of polypeptide is very low. This precluded the determination of phosphodiesterase activator activity or of competition radioimmunoassay activity. When translations were done in the presence of [$^3$H]lysine, it was possible to determine the presence of isotopic lysine and its methyl derivatives. Using the identical conditions that had shown the presence of trimethyllysine in spinach and vertebrate calmodulins by ninhydrin detection, it was shown that trimethyllysine was not present isotopically.[7] Thus, although translation calmodulin lacked trimethyllysine it would still bind to phenothiazine-Sepharose and undergo the calcium-dependent mobility shift characteristic of calmodulins.

### *Chlamydomonas Flagellar Protein*

*Chlamydomonas reinhardtii* is a unicellular, biflagellated algae which swims in different modes. It has been shown[16,17] that flagellar movement depends on calcium concentration. The effects of calcium have been observed in isolated flagellar apparatus and in whole cells subjected to photostimulation. However, the molecular targets of calcium in *Chlamydomonas* flagella have not been described. During our attempts to isolate calmodulin from *Chlamydomonas* flagella, we isolated[8] a protein with many of the characteristics of a calmodulin. However, this flagellar protein did not stimulate phosphodiesterase, lacked trimethyllysine, and did not undergo the calcium-dependent mobility shift in gel electrophoresis. The properties of this flagellar protein are discussed below.

The $^{35}S$-labeled *Chlamydomonas* flagellar protein was isolated using a calmodulin isolation protocol that employed a calcium-dependent, affinity-based adsorption chromatography step using phenothiazine-Sepharose conjugates.[8] While this flagellar protein bound to phenothiazine-Sepharose in the presence of calcium and was step-eluted with EGTA, it did not comigrate with plant or vertebrate calmodulins during polyacrylamide gel electrophoresis. The flagellar protein had a mobility in several gel systems between that of troponin C and vertebrate calmodulin. Further, the flagellar protein did not undergo the large calcium-dependent mobility shift seen with vertebrate and spinach calmodulin. Thus, the order of polyacrylamide gel electrophoretic mobility of these proteins in the presence of sodium dodecyl sulfate and EGTA is: spinach calmodulin $>$ vertebrate calmodulin $>$ flagellar protein $>$ troponin C.

Calmodulin is troponin C-like in structure.[3] The flagellar protein might be another member of this class of troponin C-like proteins or might be an altered form of calmodulin. Because comparative tryptic peptide maps have been shown to readily contradistinguish vertebrate calmodulins from troponin C, comparative tryptic peptide mapping of calmodulin, the flagellar protein, and troponin C was done. The $^{35}S$-labeled *Chlamydomonas* protein and unlabeled chicken gizzard calmodulin were mixed, digested with trypsin, and analyzed by column peptide mapping. Six $^{35}S$-labeled peaks were readily detected and three isotopic peaks exactly comigrated with calmodulin tryptic peptides. Amino acid analysis of an acid hydrolysate of the flagellar protein demonstrated that the majority of the $^{35}S$-label was in methionine. The approximate number of methionine-containing tryptic peptides expected for a calmodulin digest were found in the peptide analysis of the flagellar protein. Further, methionine was detected in acid hydrolysates of the three calmodulin peptide pools that comigrated with the labeled flagellar peptides. Therefore, $^{35}S$-labeled methionine-containing tryptic peptides had retention times on reverse-phase columns similar to, but distinct from, calmodulin. When a trypsin digest of troponin C was analyzed in an identical manner, a profile with very little similarity to the flagellar protein was obtained. To reiterate, the $^{35}S$-labeled flagellar protein resembled calmodulin in its interaction with phenothiazine-Sepharose conjugates and in comparative peptide maps; based on its small calcium-dependent mobility shift in polyacrylamide gel electrophoresis, the flagellar protein resembled troponin C.

Because isolated flagella are the starting material in the purifications described above, severe restrictions are placed on the total amount of protein that can be purified from a single flagella preparation. Therefore, flagella preparations were accumulated over several weeks, stored at $-20°C$, then subjected to the purification protocol used for the high specific activity/low protein preparations described above. The purified protein was: 1) analyzed by polyacrylamide gel electrophoresis, 2) tested for phosphodiesterase activation activity, 3) tested for calmodulin immunoreactivity, and 4) analyzed for the presence of $N^{\epsilon}$-trimethyllysine.

The larger scale preparation yielded the same protein and allowed a further demonstration of the similarities and dissimilarities between calmodulin and the flagellar protein. First, upon electrophoretic analysis all Coomassie-blue staining material exactly comigrated with the $^{35}S$-labeled protein. Second, the flagellar protein did not stimulate bovine brain phosphodiesterase under the conditions used. Bovine brain calmodulin was tested in parallel as a positive control. When the flagellar protein was mixed with calmodulin and then tested for activator activity, the calmodulin was still able to activate phosphodiesterase. Third, the flagellar protein quantitatively competed with bovine brain calmodulin in radioimmunoassay. While all vertebrate and plant calmodulins tested showed competition curves similar to from that of bovine brain calmodulin, other structurally and functionally related proteins such as parvalbumin and S100b did not compete. Skeletal muscle troponin C competed only at relatively high protein concentrations. This level of troponin C competition may be due to: 1) its high degree of structural similarity to calmodulin,[3] or 2) a minor calmodulin contamination in the troponin C preparation analogous to that found in parvalbumin preparations.[18] Finally, when an acid-hydrolyzed aliquot of the flagellar protein was analyzed for the presence of $N^{\epsilon}$-trimethyllysine, none was detected. Vertebrate and plant calmodulins isolated and analyzed under identical conditions have been shown to contain trimethyllysine.

We conclude from these studies that *Chlamydomonas* flagella contain a protein that is structurally, functionally, and immunologically similar to, but not identical to, calmodulin. *Chlamydomonas* flagella may also contain calmodulin. However, under the conditions used in this study a calmodulin-like protein, but not calmodulin, was isolated from flagella. It should be noted that when *Chlamydomonas* cell bodies were subjected to an identical purification protocol, a polypeptide with electrophoretic properties similar to calmodulin was isolated as well as a polypeptide resembling the flagellar protein. This mixture of polypeptides stimulated bovine brain phosphodiesterase.

It is possible that flagellar motility may be regulated by a protein such as the one described in this report. The calmodulin-like protein described here is certainly a major flagellar polypeptide because: 1) it is one of the major $^{35}S$-labeled components resolved by two-dimensional gel electrophoresis, 2) based on isotopic recovery the flagellar protein constitutes 0.6–1.0% of the extracted protein, and 3) the flagellar preparations used in these studies are not contaminated by cellular debris. However, it should be emphasized that the localization and function of this protein in *Chlamydomonas* flagella are not presently known. Careful, quantitative suborganellar localization studies will be required in order to establish the localization of this protein in the flagella and any proteins that it might regulate.

The studies summarized in this report demonstrate that caution is necessary: 1) in identifying a protein as a calmodulin, 2) in using phenothiazines or antisera directed against vertebrate calmodulins as specific probes for calmodulin, and 3) in the interpretation of experiments on biological systems where calmodulin is substituted for the homologous calmodulin-like protein.

## Summary

As summarized in Table 1, a number of properties of vertebrate and plant calmodulins and of a flagellar calmodulin-like protein have been directly compared in this report. These data demonstrate that the presence of trimethyllysine is not required for phenothiazine binding, calcium-dependent mobility shift in gel electrophoresis, or cross-reactivity with anti-calmodulin antisera. These data also demonstrate that it is possible for proteins other than calmodulin to bind to phenothiazine-Sepharose conjugates and to react with antisera prepared against vertebrate calmodulin. The *Chlamydomonas* flagellar protein described in this study binds phenothiazine, shows calmodulin immunoreactivity, and has some chemical similarity to calmodulin; however, the flagellar protein does not activate bovine brain phosphodiesterase. Clearly, more extensive information on

Table 1

Comparison of Calmodulins and Flagellar Protein

| Characteristic | Vertebrate Calmodulin | Spinach Leaf Calmodulin | Translation Calmodulin | Flagellar Protein |
|---|---|---|---|---|
| Binds phenothiazines | + | + | + | + |
| PDE activator activity | + | + | N.D. | − |
| CaM immunoreactivity | + | + | + | + |
| Mobility shift | + | + | + | − |
| Trimethyllysine | + | + | − | − |

N.D. = not determined.
PDE = phosphodiesterase.
CaM = calmodulin.

the structural and functional characteristics of the flagellar protein, calmodulin from nonvertebrate species, and calmodulin from other subcellular structures is required before it can be unequivocally stated that the flagellar protein is not a calmodulin. Because of the multiple activities attributed to calmodulin, it becomes increasingly important that the characteristics which define calmodulin and contradistinguish it from other structurally and functionally related proteins be elucidated. This information will, in turn, allow for the unambiguous interpretation of biological studies.

## References

1. Cheung, W. Y. 1980. Science **207:** 19–27.
2. Watterson, D. M., W. G. Harrelson, P. M. Keller, F. Sharief & T.C. Vanaman. 1976. J. Biol. Chem. **251:** 4501–4513.
3. Watterson, D. M., F. S. Sharief & T. C. Vanaman. 1980. J. Biol. Chem. **255:** 962–975.
4. Vanaman, T. C. 1980. Ann. N.Y. Acad. Sci. This volume.

5. WATTERSON, D. M., P. A. MENDEL & T. C. VANAMAN. 1980. Biochemistry **19:** 2672–2676.
6. VAN ELDIK, L. J. & D. M. WATTERSON. 1979. J. Biol. Chem. **254:** 10250–10255.
7. VAN ELDIK, L. J., A. R. GROSSMAN, D. B. IVERSON & D. M. WATTERSON. 1980. Proc. Natl. Acad. Sci. USA **77:** 1912–1916.
8. VAN ELDIK, L. J., G. PIPERNO & D. M. WATTERSON. 1980. Proc. Natl. Acad. Sci. USA. (In press.)
9. VAN ELDIK, L. J. & D. M. WATTERSON. 1980. Ann. N.Y. Acad. Sci. This volume.
10. EBASHI, S., T. WAKABAYASHI & F. EBASHI. 1971. J. Biochem. **69:** 441–445.
11. GREASER, M. L. & J. GERGELY. 1971. J. Biol. Chem. **246:** 4226–4233.
12. WATTERSON, D. M., D. B. IVERSON & L. J. VAN ELDIK. 1980. J. Biochem. Biophys. Methods **2:** 139–146.
13. BURGESS, W.H., D. JEMIOLO & R. H. KRETSINGER. 1980. Biochim. Biophys. Acta **623:** 257–270.
14. GRAB, D. J., K. BERZINS, R. S. COHEN & P. SIEKEVITZ. 1979. J. Biol. Chem. **254:** 8690–8696.
15. KLEE, C. B., T. H. CROUCH & M. H. KRINKS. 1979. Proc. Natl. Acad. Sci. USA **76:** 6270–6273.
16. HYAMS, J. S. & G. G. BORISY. 1978. J. Cell Sci. **33:** 235–253.
17. SCHMIDT, J. A. & R. ECKERT. 1976. Nature **262:** 713–715.
18. LeDONNE, N. C. & C. J. COFFEE. 1979. J. Biol. Chem. **254:** 4317–4320.

## DISCUSSION OF THE PAPER

QUESTION: Did you check whether the flagellar protein would activate the erythrocyte ATPase?

DR. L. J. VAN ELDIK: Not yet.

DR. BARRY: Have you isolated calmodulin from the basal apparatus or the cell body?

DR. VAN ELDIK: Those studies are in progress.

# CONCERTED ROLE OF CALMODULIN AND CALCINEURIN IN CALCIUM REGULATION

Claude B. Klee and Jacques Haiech

*Laboratory of Biochemistry*
*National Cancer Institute*
*National Institutes of Health*
*Bethesda, Maryland 20205*

## Introduction

Calmodulin was discovered ten years ago as a heat- and acid-stable activator of cyclic nucleotide phosphodiesterase[1] specific for the $Ca^{2+}$-dependent form of the enzyme[2] and has recently been shown to play a unique role among the intracellular $Ca^{2+}$ receptor proteins as defined by Kretsinger.[3] These proteins, which include the troponins, the parvalbumins, S-100 protein, the $Ca^{2+}$ intestinal binding proteins and probably several others which remain to be identified, have as their unique function, in the cell, to translate the $Ca^{2+}$ signal into cellular responses. Their structure has been predicted by Kretsinger on the basis of the known structure of parvalbumin.[4] All of them bind $Ca^{2+}$ specifically at physiological concentration ($10^{-5}$–$10^{-7}$ M) in the presence of the normally large intracellular concentration of $Mg^{2+}$ ($10^{-3}$ M) and $K^{+}$ ($10^{-1}$ M).[3] Prior to the discovery of calmodulin, most of these proteins were known to be tissue- and species-specific, performing specific functions.[5] In contrast, calmodulin, which is almost perfectly conserved throughout evolution, is a ubiquitous and multifunctional protein that has recently been recognized as the regulator of a large number of $Ca^{2+}$-dependent intracellular processes (reviewed in References 6–10). Calmodulin can therefore serve as a receptor of $Ca^{2+}$, acting as a "second messenger" as proposed by Rasmussen.[11] It was suggested that calmodulin serves an analogous role towards $Ca^{2+}$ as does protein kinase towards cAMP.[7]

## Calmodulin: $Ca^{2+}$-Binding Properties

The unique role of calmodulin resides in its ability to bind $Ca^{2+}$ and subsequently to interact with a large number of enzymes and proteins whose properties are modified by this interaction. There is general agreement that calmodulin contains four $Ca^{2+}$-binding sites that correspond to the four $Ca^{2+}$-binding domains identified on the basis of the primary structure of the protein.[12,13] Most studies report the existence of two different classes of sites or of negative cooperativity between the sites.[14-20] Recently we described positive cooperative behavior that can be observed between the first two sites that are occupied by $Ca^{2+}$.[20] The variability of the $Ca^{2+}$-binding properties of calmodulin (the data presently available are summarized in Table 1) is, at least in part, due to the pronounced effect of $Mg^{2+}$ and other cations on the affinity of the protein for $Ca^{2+}$ (J. Haiech, C. B. Klee and J. G. Demaille, manuscript in preparation). Although the four sites have similar affinities for $Ca^{2+}$ under physiological conditions they are not equivalent, a finding that is consistent with the different amino acid compositions of the four $Ca^{2+}$-binding domains.[12]

It was recognized very early that calmodulin undergoes large conformational



0077-8923/0356-0043 $1.75/0 © 1980, NYAS

TABLE 1

$Ca^{2+}$-BINDING PROPERTIES OF CALMODULIN

| Experimental Conditions* | Number of Specific $Ca^{2+}$-binding Sites | Dissociation Constants† |
|---|---|---|
| 25 mM Tris-HCl, 25 mM Imidazole 3 mM $Mg^{2+}$; 25°C (14) | 3 | $10^{-6}$ M (1), 1.2 × $10^{-5}$ M (2) |
| 25 mM Tris-HCl, pH 8.0; 4°C (15) | 4 | 3.5 × $10^{-6}$ M (3), 1.8 × $10^{-5}$ M (1) |
| 100 mM Imidazole, pH 7.0; 24°C (16) | 4 | 1.1 × $10^{-6}$ M (2), 8.6 × $10^{-4}$ M (2) |
| 10 mM Tris, pH 7.4; 25°C (17) | 4 | 2 × $10^{-7}$ M (3), $10^{-6}$ M (1) |
| 10 mM Tris, pH 7.4, 1 mM $Mg^{2+}$; 25°C (17) | 3 | 3 × $10^{-6}$ M (3) |
| 10 mM HEPES, pH 6.5, 100 mM KCl 0.1 mM EGTA; 4°C (18) | 4 | 2.4 × $10^{-6}$ M (4) |
| 10 mM Imidazole, 100 mM NaCl pH 7.0; 4°C (19) | 4 | 7.6 × $10^{-6}$ M (4) |
| 10 mM HEPES, pH 7.5, 10 mM KCl, 25°C (20) | 4 | 3.3 × $10^{-6}$ M (1), $10^{-6}$ M (1) 8.6 × $10^{-6}$ M (1), 2 × $10^{-5}$ M (1) |
| 10 mM HEPES, pH 7.5, 100 mM KCl 3 mM $Mg^{2+}$; 25°C (20) | 4 | 5 × $10^{-6}$ M (1), 4 × $10^{-6}$ M (1) 2.5 × $10^{-5}$ M (1), 4 × $10^{-5}$ M (1) |

*The reference number is indicated in parentheses.

†The numbers in parenthesis indicate the number of sites with the corresponding dissociation constant.

changes upon interaction with $Ca^{2+}$.[17,18,20-24] These structural transitions occur in distinct steps according to the degree of $Ca^{2+}$ occupancy.[23,25,26] The positive cooperativity, which accompanies the binding of the first $Ca^{2+}$, can also be taken as evidence of a conformational change. This transition has not been distinguished from the major structural change that follows the binding of two moles of $Ca^{2+}$. This major change has been detected by fluorescence, UV absorption, circular dichroic and proton nuclear magnetic resonance studies.[17,18-26] Seamon was able to detect yet another conformational change affecting the environment of Tyr 138 upon occupancy of the last two $Ca^{2+}$-binding sites.[26] Taken together these observations indicate that $Ca^{2+}$ binds to calmodulin in a sequential manner and that the protein exists in different conformations according to the extent of $Ca^{2+}$ occupancy as shown in FIGURE 1. Since the $Ca^{2+}$-binding domains are far from identical in primary structure, it is likely that binding of $Ca^{2+}$ is also ordered. In any case, the existence of different calmodulin$\cdot Ca^{2+}_{1-4}$ conformers, which could be recognized by different target proteins, confers upon calmodulin the ability to translate quantitative differences in $Ca^{2+}$ levels into qualitatively different cellular responses.[9]

## INTERACTION OF CALMODULIN WITH ITS TARGET PROTEINS

The activation of the enzymes under the control of calmodulin occurs in the two steps described below:

$$CaM + nCa^{2+} \rightleftharpoons CaM\cdot Ca^{2+}_n \rightleftharpoons {}^*CaM\cdot Ca^{2+}_n \quad (1)$$

$$^{*}\mathrm{CaM}\cdot\mathrm{Ca}_n^{2+} + \mathrm{Enz} \rightleftharpoons (^{*}\mathrm{CaM}\cdot\mathrm{Ca}_n^{2+})\cdot\mathrm{Enz} \rightleftharpoons (^{*}\mathrm{CaM}\cdot\mathrm{Ca}_n^{2+})\cdot\mathrm{Enz}^{*} \qquad (2)$$

$$n = 1\text{–}4$$

The asterisk indicates the active conformation of the protein. This model is based on the fact that, in general, no interaction is detected between calmodulin and its target proteins in the absence of $Ca^{2+}$ (in the presence of EGTA).[6] However there are at least two exceptions to this rule. Phosphorylase kinase exists as a tight calmodulin·enzyme complex even when isolated in the presence of EGTA.[27] In the absence of added calmodulin $Ca^{2+}$ only is required for the full activation of the enzyme. Troponin I, as shown in FIGURE 2, can also interact with calmodulin in the absence of $Ca^{2+}$. A crosslinked dimer of calmodulin and troponin I ($M_r = 39{,}000$) is formed during incubation in the presence of dimethylsuberimidate by the method of Davies and Stark.[28] A much stronger interaction is observed in the presence of $Ca^{2+}$, where the totality of troponin I is converted to the heterodimeric species (FIGURE 2). Under similar conditions (in the presence of EGTA) no complex formation was observed between calmodulin and cAMP-dependent phosphodiesterase[29] C. B. Klee, unpublished observation) or smooth muscle myosin light-chain kinase.[30] It is not yet known if the activation of different calmodulin-dependent enzymes actually requires different levels of $Ca^{2+}$ occupancy. Phosphodiesterase,[20,31] adenylate cyclase,[31] and myosin light-chain kinase[32] have all been shown to be activated by the calmodulin·$Ca_{3-4}^{2+}$ complex. This requirement for fully liganded calmodulin necessitates large concentrations of free $Ca^{2+}$ under the conditions used for the *in vitro* experiments ($10^{-5}$ M). *In vivo*, the concentration of calmodulin-binding proteins and of calmodulin is much larger than those used *in vitro* and consequently the concentration of $Ca^{2+}$ needed for activation should be lowered according to the dissociation constant of the enzyme·calmodulin complexes (usually $10^{-9}$ M).[6-9] The interaction of phosphodiesterase with the calmodulin·$Ca_{3-4}^{2+}$ complex results in a very strong positive cooperativity in the $Ca^{2+}$ dependence of activation, allowing the enzyme to respond to a very narrow $Ca^{2+}$ threshold.[20]

Some enzymes do not require the same $Ca^{2+}$ concentration for interaction and for activation. It is therefore possible that, as with the troponins, a lower $Ca^{2+}$

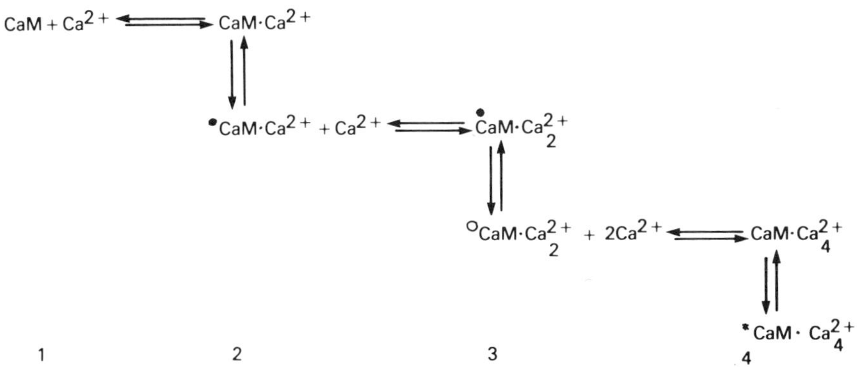


FIGURE 1. $Ca^{2+}$-dependent conformational transitions of calmodulin. The numbers 1–4 at the bottom of the figure indicate four postulated conformers CaM, •CaM·$Ca^{2+}$, °CaM·$Ca_2^{2+}$ and *CaM·$Ca_4^{2+}$.

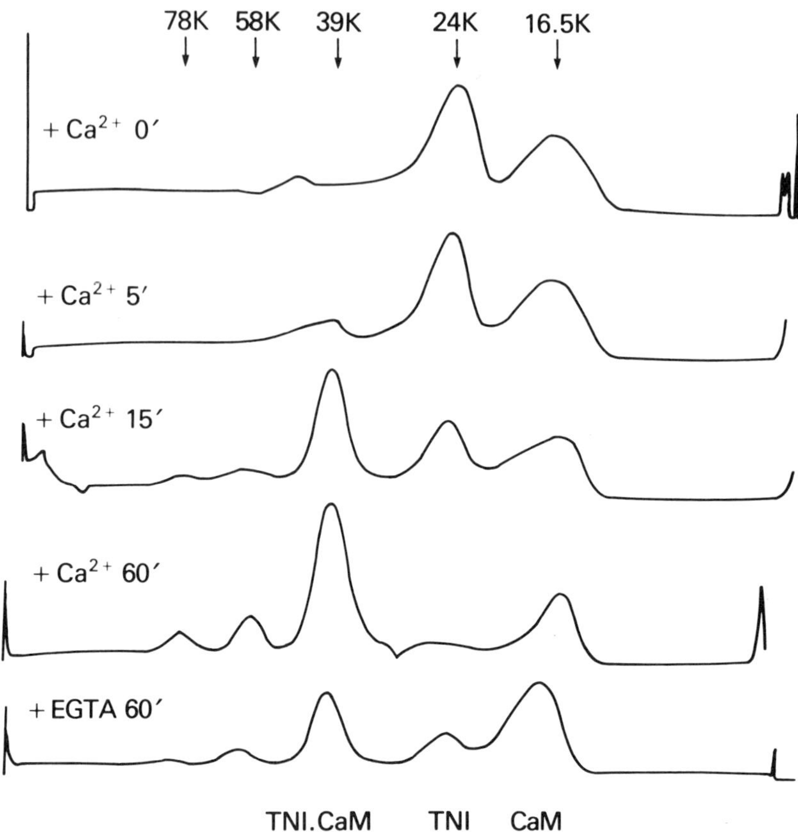


FIGURE 2. Cross-linking of calmodulin·troponin I complex. A mixture of calmodulin (3 μg) and troponin I (3 μg) in 0.2 M triethanolamine-HCl buffer, pH 8.5, containing 1 mM $Mg^{2+}$ and 2 mM dithiothreitol was incubated at 25°C for the times indicated on the figure in the presence of 0.5 mM $Ca^{2+}$ or 1 mM EGTA and 1 mg/ml dimethylsuberimidate. The products of the reaction were resolved by SDS gel electrophoresis as described by Davies and Stark[28] and the $M_r$ of the polypeptides were determined by comparison with those of the crosslinked products of β-lactoglobulin and lactic dehydrogenase. When radiolabeled calmodulin was used the 39,000 dalton polypeptide was radioactive indicating that it contained calmodulin (P. G. Richman and C. B. Klee, manuscript in preparation). Under the same conditions the $M_r$ of calmodulin alone was unchanged.

occupancy is sufficient to insure interaction of calmodulin with its targets than that required for their activation. Thus, at the intermediate $Ca^{2+}$ levels observed in the resting cell ($10^{-7}$–$10^{-8}$ M) calmodulin would not be free but partially complexed with some of the calmodulin-binding proteins. These complexes however would be inactive. A temporary increase in $Ca^{2+}$ concentration will saturate the remaining free $Ca^{2+}$-binding sites in a fast, diffusion-controlled interaction and lead to an immediate triggering of enzyme activity. The existence of preformed calmodulin·enzyme complexes would thereby eliminate the delay in activation due to the slow diffusion rate of macromolecules. Dissociation of

$Ca^{2+}$ from such an activated complex could be fast as opposed to the slow dissociation of $Ca^{2+}$ in the bound form (complexed with calmodulin) according to equation (2) (this slow dissociation can be predicted on the basis of the very small dissociation constant of the calmodulin·enzyme complex, $10^{-9}$ M). As shown in FIGURE 3, in the case of phosphodiesterase, when the formation of AMP is followed as a function of time, an immediate (less than 15 sec) decrease of the reaction rate is observed when $Ca^{2+}$ is removed by addition of EGTA. As discussed above this experiment is more easily explained on the basis of direct interaction of $Ca^{2+}$ with an inactive calmodulin·enzyme intermediate than by the two-step mechanism proposed earlier. It is also evident that changes in $Ca^{2+}$ concentrations can result in immediate activation and deactivation of the cellular processes.

## Modulation of Calmodulin Action: A Possible Role for Calcineurin

In contrast to calmodulin and the troponins C, which have a stimulatory function, other cytoplasmic $Ca^{2+}$-binding proteins such as parvalbumins may act

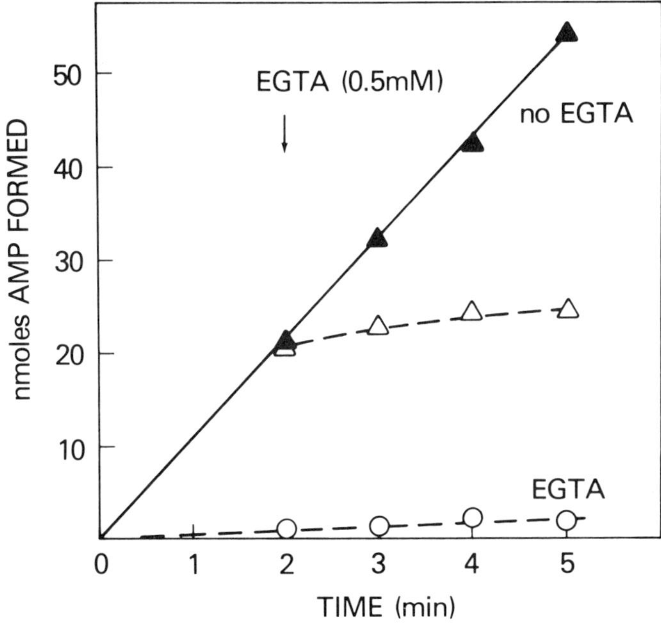


FIGURE 3. Fast inhibition of calmodulin-dependent stimulation of cAMP phosphodiesterase upon removal of $Ca^{2+}$. The formation of AMP was measured under standard conditions[23] in the presence of $10^{-6}$ M calmodulin and $10^{-3}$ M cAMP, at 30°C. The reaction was started by addition of 0.1 μg of enzyme per 100 μl of incubation mixture and at the indicated times aliquots were removed for AMP determinations. o, EGTA (0.5 mM); ▲, $Ca^{2+}$ (0.1 mM); $\Delta Ca^{2+}$ (0.1 mM) was present but EGTA was added, as shown by the arrow, prior to measure AMP formation (the observed inhibition could be fully reversed by addition 1 mM $Ca^{2+}$).

as inhibitors of the $Ca^{2+}$ signal.[33] Bovine brain contains a specific $Ca^{2+}$-binding protein, calcineurin[34-36] in addition to low levels of parvalbumin.[37] This protein, described as an inhibitor of calmodulin-dependent stimulation of cAMP phosphodiesterase by Wang and Desai,[34] was independently purified in our laboratory as a major calmodulin-binding protein in brain extracts.[35] For these reasons it was originally called inhibitor protein of phosphodiesterase or modulator binding protein.[34,35] More recently, we presented evidence that this protein is itself a $Ca^{2+}$-binding protein[38] and since the protein appears to be predominantly found in the nervous system[39] it was called calcineurin.

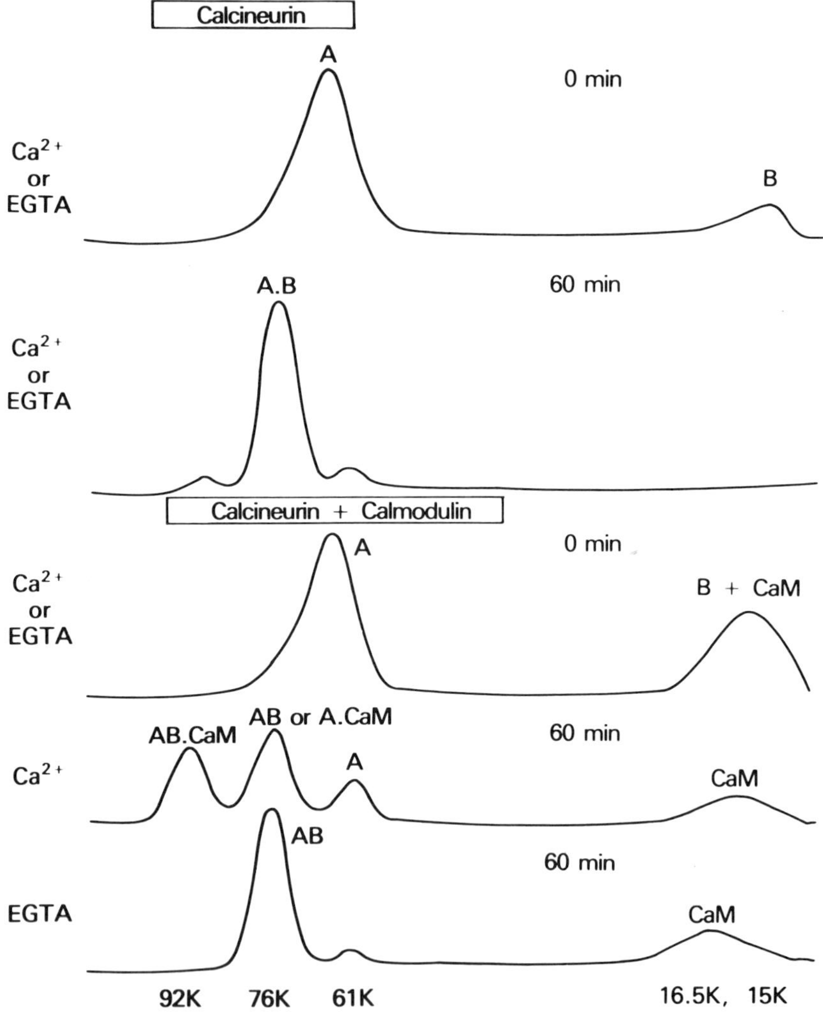


FIGURE 4. Cross-linking of calmodulin·calcineurin complex. The experimental conditions were as described in legend to FIGURE 2.

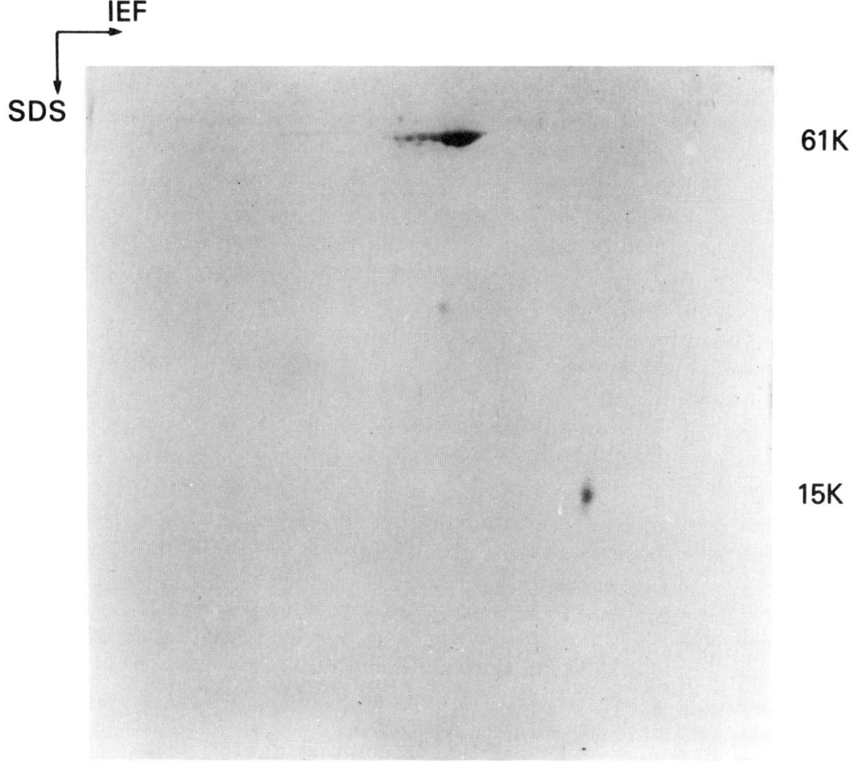


FIGURE 5. Two-dimensional gel electrophoresis of bovine brain calcineurin. Electrophoresis was carried as described by Cabral and Schatz.[48]

Calcineurin is composed of two subunits, which in the native state are tightly bound to each other in the presence or absence of $Ca^{2+}$ or $Mg^{2+}$.[35,38] Crosslinking experiments with dimethylsuberimidate according to Davies and Stark[28] indicate that calcineurin exists as heterodimer of 76,000 daltons (FIGURE 4). As shown in FIGURE 5, the two subunits can be resolved under denaturing conditions. The large polypeptide with a molecular weight of 61,000 (calcineurin A) has an isoelectric point of 5.5 and the small, 15,000 $M_r$, subunit (calcineurin B) has an isoelectric point of 4.8. The two-subunit complex binds 3–4 moles of $Ca^{2+}$ per mole with a very high affinity ($K_d < 10^{-6}$ M) measured in the presence of 0.1 M $K^+$ and 1 mM $Mg^{2+}$. $Ca^{2+}$ is bound by the small subunit, calcineurin B,[38] since in the presence of sodium dodecyl sulfate, calcineurin B exhibits different electrophoretic mobilities in the presence and absence of $Ca^{2+}$. Such behavior is characteristic of most cytoplasmic $Ca^{2+}$ receptor proteins analyzed by this method.[38,40-42] Calcineurin B has also been separated from calcineurin A by DEAE cellulose chromatography in the presence 6 M urea and 5 mM dithiothreitol. The amino acid composition of the isolated subunit is shown in TABLE 2 together with those of bovine brain calmodulin, bovine cardiac and rabbit skeletal muscles troponins C, and frog muscle parvalbumin. These proteins of

TABLE 2

AMINO ACID COMPOSITION OF INTRACELLULAR $Ca^{2+}$-BINDING PROTEINS*

| | Calmodulin | Troponin C | | Calcineurin B† | Parvalbumin |
|---|---|---|---|---|---|
| | | residues/mol | | | |
| Lys | 7 | 13 | 9 | 10 | 11 |
| His | 1 | 0 | 1 | 2 | 0 |
| Arg | 6 | 4 | 7 | 5 | 3 |
| Asp | 23 | 27 | 22 | 21 | 13 |
| Thr | 12 | 7 | 6 | 4 | 2 |
| Ser | 4 | 5 | 7 | 9 | 10 |
| Glu | 27 | 29 | 31 | 18 | 12 |
| Pro | 2 | 2 | 1 | 3 | 0 |
| Gly | 11 | 12 | 13 | 12 | 9 |
| Ala | 11 | 7 | 13 | 6 | 15 |
| Val | 7 | 8 | 7 | 9 | 7 |
| Met | 9 | 11 | 9 | 4 | 0 |
| Ile | 8 | 9 | 10 | 7 | 7 |
| Leu | 9 | 13 | 9 | 11 | 9 |
| Tyr | 2 | 3 | 2 | 3 | 1 |
| Phe | 8 | 9 | 10 | 8 | 9 |
| Cys | 0 | 2 | 1 | ND‡ | 0 |
| Try§ | 0 | 0 | 0 | + | 0 |
| Tml¶ | 1 | 0 | 0 | 0 | 0 |

*Amino acid compositions of bovine brain calmodulin[13] troponin C from bovine cardiac muscle[45] first column, rabbit skeletal muscle[46] second column, and frog skeletal muscle parvalbumin[47] are derived from their primary structure.

†Calcineurin B from bovine brain.

‡ND, not determined.

§The presence of tryptophan was deduced from the UV absorption spectrum.

¶Tml, trimethyllysine.

similar but not identical molecular weights each contain large numbers of acidic amino acid residues and exhibit a low tyrosine to phenylalanine ratio. Calcineurin B lacks the trimethyllysyl residue characteristic of calmodulin and has a high serine to threonine ratio of 9 to 4 as opposed to 4 to 12 for calmodulin. It also contains a tryptophanyl residue which is rarely observed in this class of $Ca^{2+}$-binding proteins. Calcineurin B fails to activate cAMP phosphodiesterase. Like calmodulin, calcineurin undergoes a large conformational transition upon binding of $Ca^{2+}$ as detected by a blue shift of its aromatic residues including the tryptophanyl residue. On the other hand, calcineurin B shows some similarities to the parvalbumins: those include the high serine to threonine ratio and the low methionine content. These common characteristics of parvalbumins and calcineurin B may be reflections of the shared high affinity of the two proteins for $Ca^{2+}$ ($K_d < 10^{-6}$ M) as opposed to the low affinity of calmodulin ($5 \times 10^{-5}$ M). Calcineurin A is believed to be the calmodulin-binding subunit. As shown in FIGURE 4, when calcineurin is crosslinked in the presence of calmodulin, a new, larger protein complex is detected with a molecular weight of 92,000. On the basis of its molecular weight it has been identified as a one-to-one complex of calcineurin and calmodulin. The formation of this complex requires the presence of $Ca^{2+}$ since it was not detected when the crosslinking reaction was performed in the presence of EGTA (FIGURE 4, bottom). The absence of a detectable crosslinked complex with a molecular weight corresponding to that of a complex of calcineurin B and calmodulin (32,000) suggests that the calmodulin interaction site of calcineurin is located on calcineurin A. Direct demonstration of those complexes has recently been accomplished with the aid of radiolabeled calmodulin (P. G. Richman and C. B. Klee, manuscript in preparation).

Calcineurin, by its $Ca^{2+}$-dependent interaction with calmodulin, prevents activation of calmodulin-dependent processes and can therefore exert an inhibitory action directly at the level of calmodulin-enzyme interaction. On the other hand, the high affinity of calcineurin for $Ca^{2+}$ (the affinity of calcineurin for $Ca^{2+}$ is about 100-fold greater than that of calmodulin under physiological conditions) confers upon it the ability to also act as an inhibitor of the $Ca^{2+}$ signal by lowering intracellular $Ca^{2+}$ concentrations. A similar role has been proposed for parvalbumins.[43] If the association of $Ca^{2+}$ with calmodulin and calcineurin is diffusion-controlled then shortly after a stimulus $Ca^{2+}$ will be predominantly complexed with the protein present in the largest concentration, calmodulin. In brain extracts the calmodulin concentration is about $10^{-5}$ M[13] and that of calcineurin is near $10^{-6}$ M.[44] As a steady state is approached $Ca^{2+}$ will redistribute according to its relative affinities for the two proteins. The cation will therefore be complexed primarily with calcineurin at long times (100 msec) after the $Ca^{2+}$ signal. If the temporarily increased $Ca^{2+}$ concentration is not greater than that of calcineurin, the interaction of $Ca^{2+}$ with calcineurin will result in a lowering of free $Ca^{2+}$ and

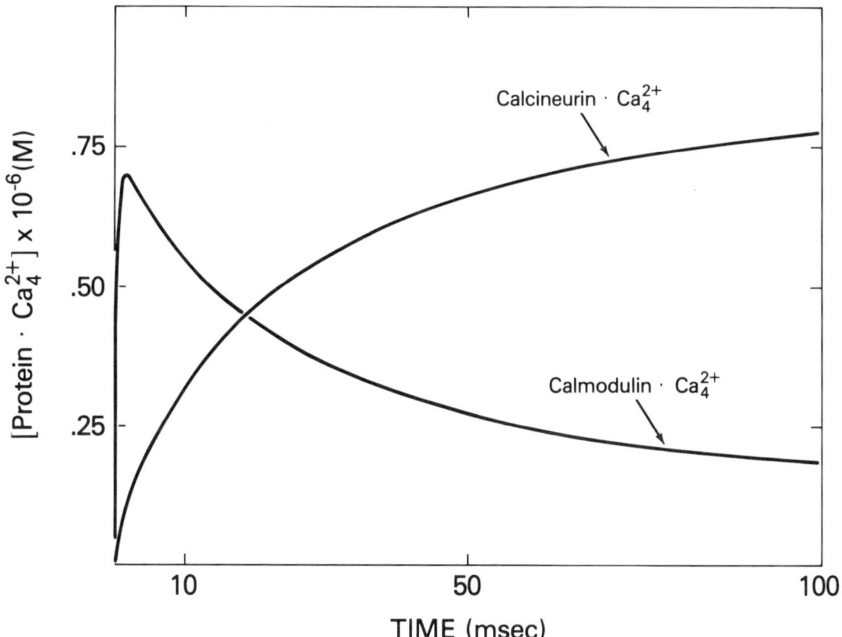


FIGURE 6. Computer-simulated distribution of $Ca^{2+}$ on calmodulin and calcineurin after increased $Ca^{2+}$ concentration to $10^{-6}$ M. The model is based on the following assumptions: association of $Ca^{2+}$ with calmodulin and calcineurin is diffusion controlled ($K_{on} = 5 \times 10^7$ $M^{-1}sec^{-1}$) the concentration of calmodulin sites is $4 \times 10^{-5}$ M with a $K_{off}$ for $Ca^{2+}$ of $5 \times 10^2$ $sec^{-1}$, and that of calcineurin sites is $4 \times 10^{-6}$ M with a $K_{off}$ for $Ca^{2+}$ of 1 $sec^{-1}$.

subsequent inactivation of the many calmodulin-dependent enzymes within an extremely short time period. A computer-simulated model of the time-dependent distribution of $Ca^{2+}$ between the two proteins is shown in FIGURE 6. This model is certainly oversimplified since it does not take into account the interactions of calmodulin with its target proteins and with calcineurin. An increased affinity of calmodulin for $Ca^{2+}$ is brought about by its tight association with calmodulin-dependent enzymes. Interaction with calmodulin is needed for calcineurin to preserve the difference in $Ca^{2+}$ binding properties of the two proteins and to enable calcineurin to act as an inhibitor of the $Ca^{2+}$ signal. As opposed to the mechanism proposed for parvalbumins[43] in which the inhibitory activity is limited to the action of parvalbumins as a $Ca^{2+}$ sink (the rate of inhibition is much slower and controlled by the off rate of $Mg^{2+}$ from the protein) this kinetic, diffusion-controlled regulation of a cellular stimulus allows an extremely fast on-off switch for biological processes such as those observed in the nervous system.

The model proposed here for the time course of calmodulin, calcineurin, and $Ca^{2+}$ interactions may have quite general applicability. The initial association of any ligand with cellular components will be governed by statistical considerations rather than affinity. Subsequent redistribution, as equilibrium is approached, will reflect the relative affinities of the sites available. It is generally observed that membranes exhibit a large number of nonspecific binding sites for any particular ligand along with a much smaller number of high-affinity receptor sites. Clearly, at very early times almost all the bound ligand will be associated with the nonspecific sites that may thus function to funnel ligands to their specific sites.

## REFERENCES

1. CHEUNG, W. Y. 1970. Biochem. Biophys. Res. Commun. **38:** 533–538.
2. KAKIUCHI, S., R. YAMAZAKI & H. NAKAJIMA. 1970. Proc. Jpn. Acad. **46:** 587–592.
3. KRETSINGER, R. H. 1980. C.R.C. Crit. Rev. Biochem. **8:** 119–174.
4. KRETSINGER, R. H. & C. D. Barry. 1975. Biochim. Biophys. Acta **405:** 40–52.
5. GOODMAN, M., J. F. PECHERE, J. HAIECH & J. G. DEMAILLE. 1979. J. Mol. Evol. **13:** 331–352.
6. WOLFF, D. J. & C. O. BROSTROM. 1979. Adv. Cyclic Nucleotide Res. **11:** 28–88.
7. WANG, J. H. & D. M. WAISMAN. 1979. Curr. Top. Cell Regul. **15:** 47–107.
8. CHEUNG, W. Y. 1980. Science **207:** 19–27.
9. KLEE, C. B., T. H. CROUCH & P. G. RICHMAN. 1980. Ann. Rev. Biochem. **49:** 489–515.
10. MEANS, A. R. & J. R. DEDMAN. 1980. Nature **285:** 73–77.
11. RASMUSSEN, H. 1970. Science **170:** 404–412.
12. VANAMAN, T. C., F. SHARIEF & D. M. WATTERSON. 1977. *In* Calcium Binding Proteins and Calcium Function. R. Wasserman, A. Cardino, E. Carafoli, R. H. Kretsinger, D. W. MacLennan & F. H. Siegel, Eds.: 63–72. American Elsevier Inc., New York, N.Y.
13. WATTERSON, D. M., F. SHARIEF & T. C. VANAMAN. 1980. J. Biol. Chem. **255:** 962–975.
14. TEO, T. S. & J. H. WANG. 1973. J. Biol. Chem. **248:** 5950–5955.
15. LIN, Y. M., Y. P. LIU & W. Y. CHEUNG. 1974. J. Biol. Chem. **249:** 4943–4954.
16. WATTERSON, D. M., W. G. HARRELSON, P. M. KELLER, F. SHARIEF & T. C. VANAMAN. 1976. J. Biol. Chem. **251:** 4501–4513.
17. WOLFF, D. J., P. G. POIRIER, C. O. BROSTROM & M. A. Brostrom. 1977. J. Biol. Chem. **252:** 4108–4117.
18. DEDMAN, J. R., J. D. POTTER, R. L. JACKSON J. D. JOHNSON & A. R. MEANS. 1977. J. Biol. Chem. **252:** 8415–8422.

19. JARRETT, H. W. & J. KYTE. 1979. J. Biol. Chem. **254:** 8237–8244.
20. CROUCH, T. H. & C. B. KLEE. 1980. Biochemistry **19:** 3692–3698.
21. WANG, J. H., T. S. TEO, H. C. HO & F. C. STEVENS. 1975. Adv. Cyclic Nucleotide Res. **5:** 179–194.
22. LIU, Y. P. & W. Y. CHEUNG. 1976. J. Biol. Chem. **251:** 4193–4198.
23. KLEE, C. B. 1977. Biochemistry **16:** 1017–1024.
24. YASAWA, M., H. KUWAYAMA & K. YAGI. 1978. J. Biochem. (Tokyo) **87:** 1313–1320.
25. MCCUBBIN, W. D., M. T. HINCKE & C. M. KAY. 1979. Can. J. Biochem. **57:** 15–20.
26. SEAMON, K. B. 1980. Biochemistry **19:** 207–215.
27. COHEN, P., A. BURCHELL, J. G. FOULKES, P. T. W. COHEN, T. C. VANAMAN & A. NAIRN. 1978. F.E.B.S. Lett. **92:** 287–293.
28. DAVIES, G. E. & G. R. STARK. 1970. Proc. Natl. Acad. Sci. USA **66:** 651–656.
29. LAPORTE, D. C., W. A. TOSCANO, Jr. & D. R. STORM. 1979. Biochemistry **18:** 2820–2825.
30. ADELSTEIN, R. S. & C. B. KLEE. 1980. *In* Calcium Binding Proteins as Cellular Regulators. W. Y. Cheung, Ed. (In press.)
31. WOLFF, D. J., M. A. BROSTROM & C. O. BROSTROM. 1977. *In* Calcium Binding Proteins and Calcium Function. R. Wasserman, A. Cardino, E. Carafoli, R. H. Kretsinger, D. H. MacLennan & F. H. Siegel, Eds.: 97–106. American Elsevier, Inc. New York, N.Y.
32. STULL, J. T., D. R. MANNING, C. W. HIGH & D. K. BLUMENTHAL. 1980. Fed. Am. Soc. Exp. Biol. **39:** 1552–1557.
33. DEMAILLE, J. G., J. HAIECH & M. GOODMAN. 1980. *In* Biological Fluids. (In press.)
34. WANG, J. H. & R. DESAI. 1977. J. Biol. Chem. **252:** 4175–4184.
35. KLEE, C. B. & M. H. KRINKS. 1978. Biochemistry **17:** 120–126.
36. WALLACE, R. W. & W. Y. CHEUNG. 1979. J. Biol. Chem. **254:** 377–382.
37. BARON, G., J. G. DEMAILLE & E. DUTRUGE. 1975. F.E.B.S. Lett. **56:** 156–160.
38. KLEE, C. B., T. H. CROUCH & M. H. KRINKS. 1979. Proc. Natl. Acad. Sci. USA **76:** 270–273.
39. WALLACE, R. W., E. A. TALLANT & W. Y. CHEUNG. 1980. Biochemistry **19:** 1831–1837.
40. GRAB, D. J., K. BERZINS, R. S. COHEN & P. SIEKEVITZ. 1979. J. Biol. Chem. **254:** 8690–8696.
41. BURGESS, W. H., D. K. JEMIOLO & R. H. KRETSINGER. 1980. Biochim. Biophys. Acta **623:** 257–270.
42. AUTRIC, F., C. FERRAZ, M. C. KILHOFFER, J. C. CAVADORE & J. G. DEMAILLE. 1980. Biochim. Biophys. Acta. (In press.)
43. PECHERE, J.-F. 1977. *In* Calcium Binding Proteins and Calcium Function. R. H. Wasserman, A. Cardino, E. Carafoli, R. H. Kretsinger, D. H. MacLennan & F. H. Siegel, Eds.: 213–221. American Elsevier, Inc. New York, N.Y.
44, SHARMA, R. K., R. DESAI, D. M. WAISMAN & J. H. WANG. 1979. J. Biol. Chem. **254:** 4276–4282.
45. VAN EERD, J.-P & K. TAKAHASHI. 1976. Biochemistry **15:** 1171–1180.
46. COLLINS, J. H., J. D. POTTER, M. J. HORN, G. WILSHIRE & N. JACKMAN. 1973. F.E.B.S. Lett. **36:** 268–272.
47. CAPONY, J.-P, J. G. DEMAILLE, C. PINA & J.-F PECHERE. 1975. Eur. J. Biochem. **56:** 215–227.
48. CABRAL, F. & G. SCHATZ. 1978. Methods Enzymol. **56:** 602–613.


## DISCUSSION OF THE PAPER

DR. D. R. STORM (*University of Washington, Seattle, WA*): We do not find positive cooperativity in our calcium binding data.

DR. E. CARAFOLI (*Swiss Federal Institute of Technology—ETH, Zurich*): Your dissociation constants are one to two orders of magnitude away from the recognized calcium activity of the cytosol.

DR. C. KLEE: Yes, but regulation by calcium depends on the concentration of free calmodulin. As I proposed here, you could activate at a much lower concentration of calcium if calmodulin was in excess. I have no information about the concentration of calmodulin in the cell and therefore we don't know the calcium concentration needed for activation of the calmodulin-enzyme complex *in vivo*.

DR. J. D. POTTER (*University of Cincinnati, Cleveland, OH*): If there are four distinct calcium binding sites A, B, C, and D, are you saying that it is possible to get A and B occupied without having occupancy of C and D?

DR. KLEE: Yes.

DR. K. SEAMON (*National Institutes of Health, Bethesda, MD*): Based on Dr. Klee's binding constants, it would be possible to occupy approximately 70% of the first two sites with only 20% of the third site occupied. Thus, it would be possible to get occupancy of two "high affinity" sites with a low occupancy of the "low affinity" sites. Have you looked at the effect of pH on calcium binding?

DR. KLEE: In some preliminary experiments, Dr. Haiech has observed that calcium binding was decreased below pH 6.

DR. R. H. KRETSINGER: You showed the results of a calculation dealing with kinetics of calcium binding to calmodulin and to calcineurin. In order to calculate such curves one has to break the equilibrium constant into a $K_{on}$ and a $K_{off}$. What were the assumtions made?

DR. KLEE: It was assumed that the $K_{on}$ was diffusion controlled for the two proteins. This would be a value of $5 \times 10^7$ $M^{-1}$ $sec^{-1}$ for calcium binding to the four sites.

# THE PRESENCE AND FUNCTIONS OF CALMODULIN IN THE POSTSYNAPTIC DENSITY

Dennis J. Grab, Richard K. Carlin, and Philip Siekevitz

*Department of Cell Biology*
*Rockfeller University*
*New York, New York 10021*

## INTRODUCTION

The first observations of a distinctive structure lying along the cytoplasmic side of the postsynaptic membrane were made in the 1950s by electron microscopy by Palade,[1] de Robertis,[2,3] and Palay.[4] This thickened region, first thought to be a membrane specialization or differentiation, is now thought to be a separate subcellular organelle,[5,6] and is currently termed the postsynaptic density (PSD) (FIGURE 1). Various attempts have been made to isolate this structure, using different detergents to separate it from its membrane attachment. Cotman *et al.*[7] used N-lauroyl sarcosinate, Matus and Walters[8] used deoxycholate, while Cohen *et al.*[9] used Triton X-100. Matus and Taff-Jones[10] have recently concluded that the Triton X-100-derived PSD is more like its *in situ* counterpart than either the N-lauroyl sarcosinate-derived or the deoxycholate-derived preparation. A review of the isolation procedures and properties of these three preparations will soon appear in press.[11]

Using Triton X-100, we have isolated from canine cerebral cortex a PSD preparation that contains some 10 major and at least 20 minor proteins.[9] No membranes are visible by electron microscopy (cf. FIGURE 2) while the phospholipid content is 1% or less.[9] The structure seems to be composed of protein, for no nucleic acid was found.[9] Based on enzymatic marker studies and on radioactive mixing experiments, it appears that there may be from 0.1% to 6% contamination of the PSD preparation by proteins from mitochondria and from synaptic vesicle, plasma, and myelin membranes.[9] Of the distinctive and known PSD protein bands detected on denaturing gels, only myelin[9] and intermediate filament protein[12] are contaminants. Of the enzymatic activities tested, there are no $Mg^{2+}$-, or $Ca^{2+}$-ATPases[9] nor adenylate kinase[13] activities. There are in the PSD preparation, however, a cAMP-dependent protein kinase activity and the two proteins which are chiefly phosphorylated by this kinase.[14] Actin has been identified to be one of the major proteins in the Triton X-100-derived PSD preparation, by amino acid composition, and immunology,[15] and by gel mobility;[10] Kelly and Cotman[16] have identified the actin in their N-lauroyl sarcosinate-derived preparation to be the $\beta$ and $\gamma$ forms. The $\alpha$ and $\beta$ subunits of tubulin were tentatively identified by gel mobility to be present in the Triton X-100-derived preparation,[10,15] and by tryptic digest maps in the N-lauroyl sarcosinate-derived preparation.[16] The major protein in both these preparations, of $M_r$ 50,000–51,000, which was first thought to be tubulin[17,18] and later to be neurofilament protein;[15,19] is now recognized as one of unknown identity.[12,16]

## PRESENCE OF CALMODULIN IN THE PSD

We would like to now review all that we know at present concerning calmodulin in the PSD. In our first attempt to characterize the proteins of the



0077-8923/80/0356-0055 $1.75/0 © 1980, NYAS

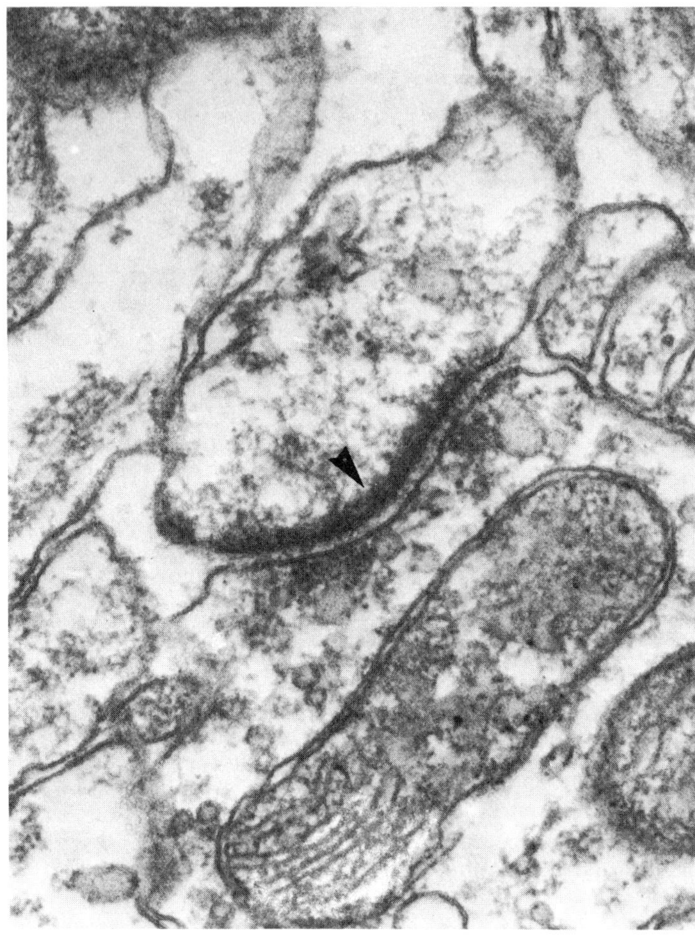

FIGURE 1. A section of canine cerebral cortex cut normal to the plane of the synaptic junctions. Arrow indicates PSD. × 75,000. Photograph courtesy of Dr. Rochelle Cohen, Dept. Anatomy, Northeastern University Medical School.

PSD, we had noticed that a protein band on an SDS gel, with a $M_r$ of 18,000, was removed by EGTA, migrated with a similar mobility to possibly one of the troponin components of a muscle actomyosin preparation, and seemingly bound radioactive calcium; we called it then a troponin C-like protein.[15] It has now been identified in published work[13,20] to be calmodulin, and the evidence for this can be briefly summarized. The 18,000 $M_r$ protein was purified from an EGTA extract of PSD by DEAE-Sephadex A-25 column chromatography[13] and was compared to calmodulin purified from whole porcine brain (gift of D. M. Watterson), or canine brain.[13] The PSD protein has been identified as calmodulin by the following criteria: 1) the 18,000 $M_r$ protein from PSD is similar to both purified canine and porcine brain calmodulin in stimulating a partially purified

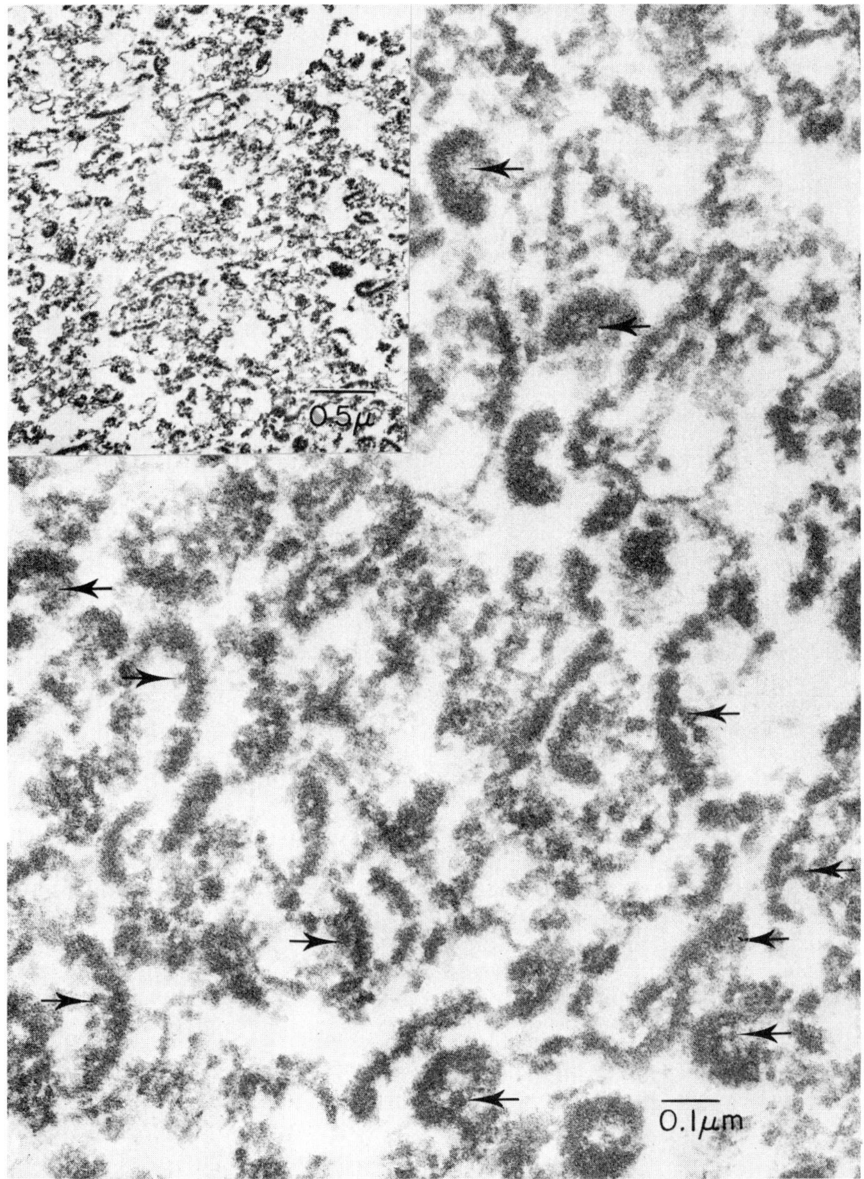


FIGURE 2. PSD preparation from cerebral cortex isolated by Triton X-100 method. Arrows indicate subsynaptic bodies or subsynaptic material. Figure reproduced from Cohen *et al.*,[9] with permission of the *Journal of Cell Biology*. × 100,000; insert × 20,000.

cyclic nucleotide phosphodiesterase in a calcium-dependent manner; 2) all three proteins comigrate as a single band on SDS polyacrylamide gels; 3) the amino acid composition of the 18,000 $M_r$ protein is similar to that of whole brain calmodulin, and contains ϵ-N-trimethyllysine; 4) the protein from the PSD and canine brain and porcine brain calmodulin all exhibit an enhanced migration rate on polyacrylamide gels in the presence of calcium, but not of magnesium, indicating strong $Ca^{2+}$ binding. In addition, purified calmodulin from porcine brain, canine brain, and PSDs could all be reconstituted in the presence of calcium, but not of magnesium, into a PSD preparation rendered deficient in calmodulin by EGTA treatment. This biochemical localization of calmodulin in the PSD structure has been confirmed by Wood *et al.*,[22] and by Lin *et al.*[23] using an immunohistochemical *in situ* method.

## Functions of Calmodulin in the PSD

What is the function, or functions, of the calmodulin in the PSD? We would like to summarize the evidence that it can activate a cyclic nucleotide phosphodiesterase which is also found in the PSD; that it can activate a protein kinase which is present there; that it binds to certain PSD proteins; and that it may be involved in excitatory responses at certain types of brain synapses.

### *Cyclic Nucleotide Phosphodiesterase Activation*

Our PSD preparation contains a cyclic nucleotide phosphodiesterase (PDE) activity,[20,24] confirming earlier work on PSDs isolated by the N-lauroyl sarcosinate method,[7] and confirming the much earlier histochemical localization.[25] The activity in the PSD preparation is about three-fold more active on cAMP as substrate than on cGMP, and the specific activity is some 5–6-fold less than in the parent synaptosome fraction.[24] However, there are several reports in the literature (cf. Reference 26) describing the existence of multiple forms of the PDE in brain, dependent on substrate specificity and cation activation, so one cannot make a valid comparison between the specific activities in the synaptosome and PSD until the exact nature of the brain PDE activities are known. Since our PSD preparation contains both a cyclic nucleotide PDE activity and calmodulin it was assumed that this preparation contained a calmodulin-activatable activity. Table 1 indicates that this apparently is not the case. Native PSDs showed no increase in the activity when calmodulin was added, nor did they show a decrease when the calmodulin was removed by EGTA treatment (calmodulin-deficient PSD).

Because of the above results, the nature of the cofactor requirement of the enzyme was examined. It was found that $Mn^{2+}$ was the cation most effective. When care was taken to rid the preparations of contaminating cations by EDTA treatment, it was found that, at as low a concentration of $Mn^{2+}$ as 0.5 $\mu$M, the activation was maximal at almost twenty-fold, with $Mg^{2+}$ giving a two-fold increase and $Ca^{2+}$ showing no effect at this concentration. At 10 $\mu$M concentration $Mg^{2+}$ addition resulted in a five-fold activation, while $Ca^{2+}$ addition gave a two-fold activation. It appeared that the enzyme in the PSD was activatable by $Mn^{2+}$ or by $Mg^{2+}$ but not by $Ca^{2+}$; earlier assays on the brain enzyme[27] had indicated a need for $Mg^{2+}$ or $Mn^{2+}$ in the assay mixture.

To get a further insight into the PSD enzyme, attempts were made to solubilize and purify it. It was found that vigorous sonication of a dilute PSD

TABLE 1

CYCLIC AMP PHOSPHODIESTERASE ACTIVITY OF NATIVE AND CALMODULIN-DEFICIENT PSD PREPARATIONS FROM CANINE CEREBRAL CORTEX IN THE PRESENCE OF EXOGENOUS CALMODULIN

| Sample | Additions | Cyclic AMP Phosphodiesterase Activity* | |
|---|---|---|---|
| | | Experiment A | Experiment B |
| Native PSD | None | 9.8 | 13.3 |
| Native PSD | Canine brain calmodulin | 8.7 | 14.3 |
| Native PSD | Porcine brain calmodulin | — | 14.7 |
| Calmodulin-deficient PSD | None | 10.9 | 13.8 |
| Calmodulin-deficient PSD | Canine brain calmodulin | 10.3 | 14.1 |
| Calmodulin-deficient PSD | Porcine brain calmodulin | — | 14.1 |

*nmole inorganic phosphate produced/minute/mg PSD protein at 30°C.

suspension solubilized about 60% of the activity, while similar sonication of a concentrated PSD suspension solubilized about 20% of the activity. However, as TABLE 2 shows, the solubilization of the enzyme did not affect the magnitude of the activation by the cations. A comparison (TABLE 2) of the activities in the sonicated supernatant with the activities in the pellet of the untreated sample indicates that in both cases $Mn^{2+}$ was the preferable activating cation and that

TABLE 2

SOLUBILIZATION BY SONICATION OF THE CYCLIC AMP PHOSPHODIESTERASE ACTIVITY OF NATIVE PSD

| Sample | Additions* | Cyclic AMP Phosphodiesterase Activity† | | | |
|---|---|---|---|---|---|
| | | Pellet | | Supernatant | |
| | | Expt. A | Expt. B | Expt. A | Expt. B |
| Native PSD | None | 152 | 93 | 72 | 64 |
| | $Mn^{2+}$ | 316 | 343‡ | 55 | 34 |
| | $Mg^{2+}$ | 196 | 260 | 39 | 62 |
| | $Ca^{2+}$ | 80 | 70 | 0 | 9 |
| | $Ca^{2+}$ + calmodulin | N.D. | 80 | 0 | 19 |
| | $Ca^{2+}$ + $Mg^{2+}$ + calmodulin | N.D. | 243 | N.D. | 19 |
| | $Ca^{2+}$ + $Mn^{2+}$ + calmodulin | N.D. | 330 | N.D. | 48 |
| Sonicated PSD | None | 37 | 0 | 39 | 0 |
| | $Mn^{2+}$ | 76 | 13 | 212 | 227 |
| | $Mg^{2+}$ | 14 | 0 | 170 | 174 |
| | $Ca^{2+}$ | 62 | 0 | 44 | 44 |
| | $Ca^{2+}$ + calmodulin | 0 | 0 | 69 | 37 |
| | $Ca^{2+}$ + $Mg^{2+}$ + calmodulin | N.D. | 0 | N.D. | 153 |
| | $Ca^{2+}$ + $Mn^{2+}$ + calmodulin | N.D. | 9 | N.D. | 256 |

*Each cation was at 1 mM. Where added, 2 μg chick brain calmodulin was used.
†pmole inorganic phosphate produced/minute at 30°C.
‡13.7 nmoles inorganic phosphate produced/minute/mg PSD protein at 30°C.

calmodulin addition was also ineffective. Thus the soluble enzyme behaved, as in the case of the PSD enzyme, like a $Mn^{2+}/Mg^{2+}$-activatable one, and not like a $Ca^{2+}$/calmodulin-activatable one.

Since multiple forms of the enzyme activity have been described in the literature (cf. Reference 26), an attempt was made to try to separate by gel filtration the possible different forms in the PSD. Sepharose-6B was first used, but the length of time necessary for the runs resulted in a great loss of enzyme activity, even in the cold. However, the much shorter runs affordable with S-300 Sephacryl columns resulted in the data shown in FIGURE 3. Two cyclic nucleotide PDE peaks emerged, one in the void volume (Fraction I) and one at approximately 215,000 $M_r$ (Fraction II), both more active against cyclic AMP than against

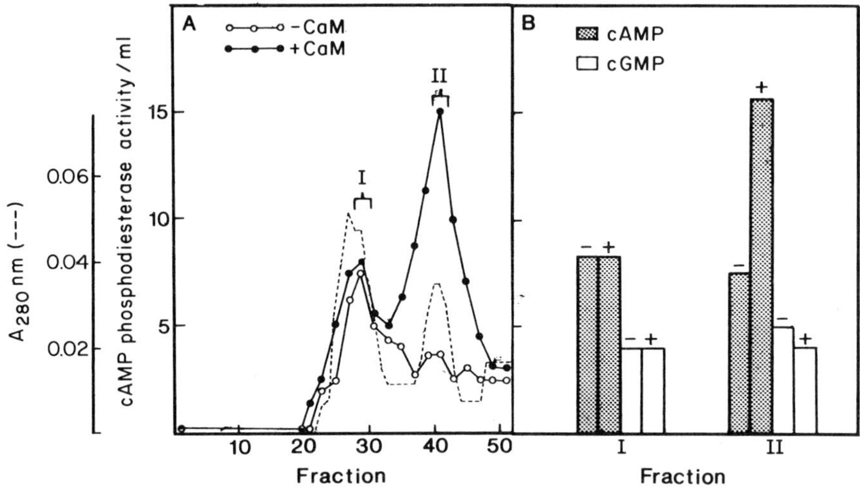


FIGURE 3. S-300 Sephacryl column chromatography of the sonicated supernatant fraction from PSDs isolated from cerebral cortex. A) Elution profile of cyclic nucleotide phosphodiesterase activity in the absence or presence of 1 mM $Ca^{2+}$ + 1.5 $\mu$g bovine calmodulin. B) Column Fractions I and II noted in (A) were pooled as indicated by the brackets in FIGURE 1A, and were incubated in either 2 mM cAMP or 2 mM cyclic GMP both in the absence or presence of $Ca^{2+}$ and bovine calmodulin as described in (A). Activity is expressed as nmole inorganic phosphate produced /30 min at 30°C.

cyclic GMP (FIGURE 3B), when the cyclic nucleotide concentrations were 2 mM. The addition of calmodulin plus $Ca^{2+}$ to Peak I resulted in no change, but did increase in Peak II the cAMP activity, and not the cGMP activity, some 2–3-fold. However, when the Peak II enzyme was tested against a concentration range of cAMP (20–1,000 $\mu$M) and cGMP (2–100 $\mu$M), it was found that the enzyme was active on both these nucleotides, with a $V_{max}$ (cGMP) = 2.1 units and a $V_{max}$ (cAMP) = 7.7, and with a $K_m$ (cGMP) = 19 $\mu$M and a $K_m$ (cAMP) = 152 $\mu$M. These values agree with those in the literature for the whole brain enzymatic activity (cf. Reference 28). The addition of $Ca^{2+}$ plus calmodulin increased the $V_{max}$ to 4.1 and 15.1 for cGMP and cAMP, and decreased the $K_m$ to 9 $\mu$M and 122 $\mu$M for cGMP and cAMP. A crude whole brain calmodulin-dependent cyclic nucleotide

PDE activity eluted on the Sephacryl column in the same position as Peak II. Thus it appears that Fraction II from the PSD contains a calmodulin-activatable cyclic nucleotide PDE activity, and that this activity resides in a complex similar to that which can be obtained from the entire cerebral cortex. This enzyme is probably intrinsic to the PSD since it was found there by histochemical methods,[25,29] it is present in PSDs obtained by the ionic detergent sarkosyl,[7] and it is not removed by EGTA treatment and requires vigorous sonication to remove a part of the activity.

Why is this calmodulin-activatable PDE activity obscured in the PSD and in the sonicated solubilized fraction? We really do not know but there are several possibilities. One is that the heat-labile calmodulin inhibitor protein[30,31] is found in the PSD, detected by immunocytochemical methods.[22] However, addition of excess calmodulin should have overcome this inhibition, but it did not; and furthermore, addition of calmodulin did increase a protein kinase activity in the PSD, as will be mentioned below. Another possibility is that the PSDs contain the myelin basic protein as a contaminant[9] and this protein has been implicated in the inhibition of whole brain PDE, and has no effect on calmodulin activation of myosin light-chain kinase.[32] However, antibody to this protein (obtained from S. Chou, F. Chou, and R. F. Kibler, Emory University) had a variable effect on the ability of calmodulin to increase the PDE activity in the intact PSD fraction, in the sonicated extract, or in Fraction I; in some cases a stimulation was observed while in other cases no stimulation was observed. In addition, additive-type experiments indicated that there was no free calmodulin inhibitor in Fraction I. Our present position is that there is more than one form of the PDE enzyme in the PSD, and that one of these forms is activatable by the calmodulin that is also present there.

### *Protein Kinase Activation*

Further examination of the possible function of calmodulin in the PSD was prompted by the demonstration by several groups of the existence of calmodulin-dependent protein kinases. Calmodulin has been shown to be a subunit of the $Ca^{2+}$-dependent myosin light-chain kinase,[33-40] as well as activating muscle[41] and platelet[42] phosphorylase kinase, glycogen synthase kinase,[43] and the phosphorylation of histone.[44] It has been implicated in the phosphorylation of brain cytosol proteins[45] and of synaptic vesicle proteins.[46] More to the point, Schulman and Greengard[47,48] recently showed that calmodulin was involved in stimulation of a $Ca^{2+}$-dependent phosphorylation of two proteins of 51,000 $M_r$ and 62,000 $M_r$ in lysed synaptosomes and a crude synaptic membrane fraction. We now have evidence that it is the PSD contained within the synaptic membrane fraction that is responsible for this phosphorylation.[49]

When a PSD preparation is incubated with labeled ATP many of the proteins are phosphorylated. The reaction reaches a maximum by 3 min at 30°C, then slowly decreases, whether the ATP concentration is 5 $\mu$M or 1 mM. It should be mentioned that there is no ATPase activity in our preparation. Upon addition of $Ca^{2+}$ plus calmodulin, there is an increase of 50–100% in the total phosphorylation of the PSD proteins. When the phosphorylated proteins are electrophoresed on an SDS gel system, only a few of these PSD proteins have their phosphorylation increased by the $Ca^{2+}$ plus calmodulin addition. As can be seen (FIGURE 4), the addition of $Ca^{2+}$ by itself does not do much, but the addition of calmodulin gives an increased phosphorylation in predominantly three PSD bands, of 62,000,

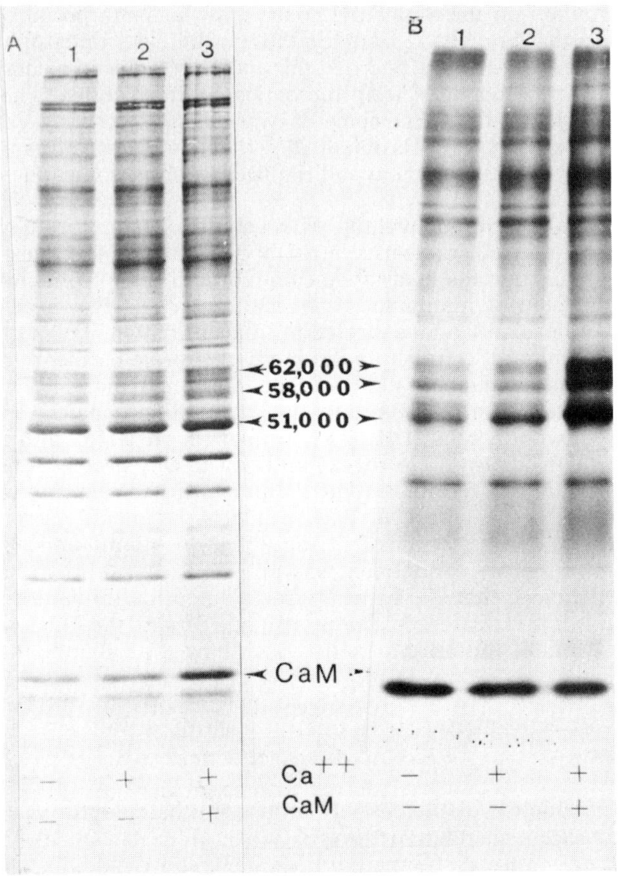


FIGURE 4. Increase in phosphorylation by $Ca^{2+}$ plus calmodulin of proteins from PSDs isolated from cerebral cortex. The three gel patterns on the left (A) are the Coomassie-blue stained patterns of a PSD preparation, while the autoradiograms (B) of the same are shown on the right. The standard reaction mixture contained in a final volume of 0.1 ml: 50 mM PIPES buffer, pH 7.0; 10 mM $MgCl_2$; 1 mM DTT: 0.1 mM EGTA; plus 120 μg PSD protein. When added, $Ca^{2+}$ was at 0.5 mM and calmodulin at 4 μg. After a one-minute preincubation at 30°C, 5 μM $\gamma$-$^{32}$P-ATP(1.6 Ci/mMol) was added and the mixture further incubated for 3 min at 30°C.

58,000, and 51,000 $M_r$. When the PSD preparation is freeze-thawed and sonicated, even the addition of $Ca^{2+}$ alone gives an increase, sometimes as much as when the $Ca^{2+}$ plus calmodulin are added, indicating, under these conditions, the increased availability of enzyme and protein substrates to the endogenous calmodulin. By repeating the experiments many times, it was estimated that the increase (by 70–90%) of phosphorylations of the 62,000 and 51,000 $M_r$ bands were the only ones uniformly repeatable and of significance. The $M_r$ of these proteins are the same as those found by Schulman and Greengard[47,48] in their synaptic membrane fractions. Yamauchi and Fujisawa[45] found brain cytosol proteins of

$M_r$ 45,000 and 60,000 to be so increased by $Ca^{2+}$ plus calmodulin, while De Lorenzo *et al.*[46] found synaptic vesicle proteins of $M_r$ 51,000–54,000 and 62,000–63,000 to be so increased.

When the $Ca^{2+}$ concentration in the reaction mixture is reduced by the addition of 1 mM EGTA, or when 100 $\mu$M chlorpromazine is included, there is no increase in the phosphorylation of these protein bands due to $Ca^{2+}$ plus calmodulin. $Mg^{2+}$ is necessary for the increased phosphorylation, as is $Ca^{2+}$, as shown by the EGTA experiment. That calmodulin is necessary is indicated by the suppression by chlorpromazine, an antipsychotic agent known to bind to calmodulin in the presence of $Ca^{2+}$,[50] and by the necessity for calmodulin when calmodulin-deficient PSDs are used in the reaction mixture (FIGURE 5). The increased phosphorylation in this calmodulin-deficient PSD system reached the same phosphorylation plateau when calmodulin was added, as in the case of native PSDs containing endogenous calmodulin. Incidentally, these phosphorylation experiments indicate that the calmodulin, which was previously shown to be able

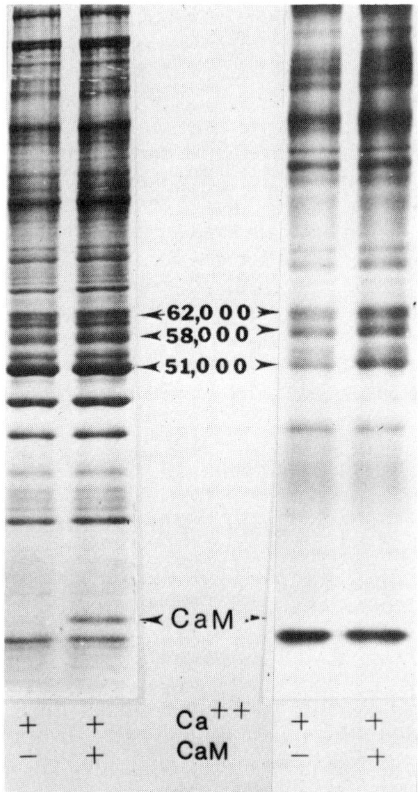


FIGURE 5. Effect of calmodulin on phosphorylation of calmodulin-deficient PSDs isolated from cerebral cortex. PSDs made deficient in calmodulin[13] were incubated in the standard reaction mixture, as given in the legend to FIGURE 4, without or with 4 $\mu$g calmodulin. The two gels on the left are Coomassie-blue stained, and the two on the right are the autoradiographs.

to be reconstituted into the PSD,[13] was probably reconstituted into its native position in the PSD. Experiments comparing the $Ca^{2+}$ plus calmodulin increased phosphorylation of the 62,000 and 51,000 $M_r$ proteins in a PSD preparation versus a synaptic membrane preparation would indicate that the kinase and/or substrates are much more concentrated in the PSD. Thus all the evidence indicates that there exists a protein kinase activatable by $Ca^{2+}$ plus calmodulin in the PSD, that it is intrinsic to the PSD, and that the substrates for this kinase are in the PSD.

What are the identities of the PSD proteins whose phosphorylation is increased by the addition of $Ca^{2+}$ plus calmodulin? One possibility is that the 51,000 $M_r$ protein corresponds to an intermediate brain filament of the same $M_r$, and that the 58,000 $M_r$ protein corresponds to the alpha subunit of tubulin. The intermediate filament protein is found in PSD preparations but is believed to be a contaminant,[12,16] while tubulin seems to be an intrinsic component of PSDs.[10,15,16,18] However, as seen in FIGURE 6A, the 51,000 $M_r$ component of the brain intermediate preparation moves slightly ahead of the major 51,000 $M_r$ phosphorylated band; on some gels, it was found that the 51,000 $M_r$ intermediate filament band moved slightly behind the major phosphorylated band (not shown). In FIGURE 6B it can be seen that the $\alpha$- and $\beta$-subunits of tubulin move between the 62,000 and 58,000 $M_r$ and between the 58,000 and 51,000 $M_r$ phosphorylated bands, respectively. Thus it appears that none of the phosphorylated proteins are identifiable as intermediate filament or tubulin proteins. Moreover, we do have evidence that the phosphorylation in the 62,000 $M_r$ region may be due to the PDE that is found in the PSD. In FIGURE 7, both purified calcineurin A (the heat-labile calmodulin binding protein) and purified PDE (both obtained through the courtesy of C. Klee, National Institutes of Health) migrate to this region (slots 5 and 6). However, the calmodulin-increased phosphorylation of PSD proteins in this region appears to be due to the PDE, rather than due to the calcineurin, for the increased phosphorylation occurs at the faster migrating band in this region (slot 2) which is probably the PDE (slots 6 and 7) and not the calcineurin (slots 4 and 5). We will discuss below our ideas as to the possible function of the 51,000 $M_r$ protein, but it is at least clear that the calmodulin in the PSD interacts with it.

### *Interaction of Calmodulin with PSD Proteins*

The binding interaction of calmodulin with PSD proteins is difficult to do by conventional means, since some of the proteins, including the major 51,000 $M_r$ one, are highly insoluble. Thus we have devised another method[51,52] for the specific binding of calmodulin to some of these proteins of the PSD. Calmodulin binding was determined by running a SDS-PAGE gel of the PSD proteins; washing out the SDS;[51,52] incubating the gel for 12 hours at room temperature with calmodulin which was radioiodinated by the lactoperoxidase procedure[53] and which allowed for a still active calmodulin as indicated by its ability to activate a PDE preparation; washing the unbound iodinated calmodulin from the gel; and finally performing autoradiography after drying the gel. Details of this method are given in a communication in this *Annal.*[52] This method gives us a qualitative measure of the specific binding, when controls are also run, but it is difficult to quantitate the results, since radioiodinated calmodulin only binds to the surface of the gel.

FIGURE 8 gives the results. Slot 1 shows the Coomassie-blue stained pattern, while slot 2 shows the autoradiograph of the sample after incubating as above.

The major radioactive regions are the 51,000 $M_r$ band and the 60,000 $M_r$ region (this region contains more than one band), with less radioactivity being found at 68,000, 140,000, 195,000, and 230,000 $M_r$. When 2 $\mu$M EGTA (Slot 3) or 1 mM chlorpormazine (slot 4) were coincubated with the radioactive calmodulin, no binding was observed. One mM EDTA or 0.1 mM chlorpormazine were also

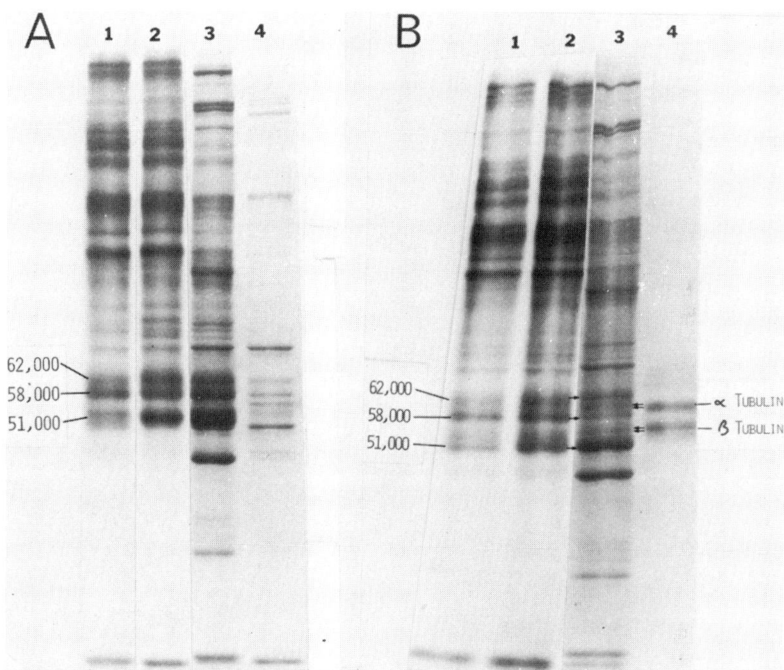


FIGURE 6. Mobilities on gels of proteins from PSDs, isolated from cerebral cortex, that are phosphorylated by a calmodulin-dependent protein kinase: comparison with tubulin and intermediate filament proteins. A) Comparison with intermediate filaments, using 5–15% polyacrylamide gradient gels: slot 1 = autoradiograph of PSD proteins after incubation of PSDs with standard phosphorylation medium (cf. FIGURE 4); Slot 2 = same as 1 but 0.5 mM $Ca^{2+}$ plus 4 $\mu$g calmodulin were added; slot 3 = Coomassie-blue stained gel of PSD proteins; Slot 4 = Coomassie-blue stained gel of purified brain intermediate filaments. B) Comparison with tubulin, using 5–10% polyacrylamide gradient gels: slot 1 = autoradiograph of PSD proteins after incubation of PSDs with standard phosphorylation medium; Slot 2 = same as 1 but 0.5 mM $Ca^{2+}$ plus 4 $\mu$g calmodulin were added; Slot 3 = Coomassie-blue stained gel of PSD proteins; Slot 4 = Coomassie-blue stained gel of purified brain tubulin.

effective in inhibiting the binding. Furthermore, preincubation for 12 hours at room temperature with 1 mg nonradioactive calmodulin, but not preincubation with 5 mg bovine serum albumin, reduced the binding to zero (data not shown). If calmodulin was iodinated by the chloramine-T method, a method which destroyed its activating activity,[54] and then incubated with the gel, no binding was observed (slot 5). Since calmodulin is an acidic protein, it is instructive (slot 2) that

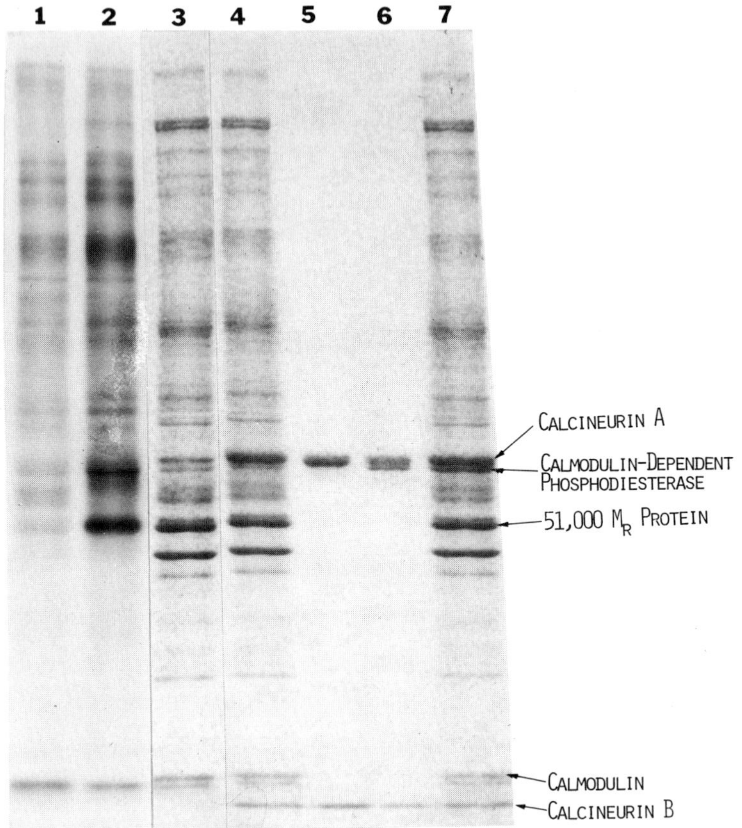


FIGURE 7. Evidence for phosphorylation of cyclic nucleotide phosphodiesterase in cerebral cortex PSDs by a calmodulin-dependent protein kinase. Phosphorylations were performed as given in legend to FIGURE 4. Slot 1 = autoradiograph of PSD protein phosphorylation using standard reaction mixture containing $\gamma$-$^{32}$P-ATP; slot 2 = same as (1) with addition of 0.5 mM $Ca^{2+}$ plus 4 $\mu$g calmodulin; slot 3 = Coomassie-blue stained gel of PSD proteins; slot 4 = same as slot 3 but with addition of the calmodulin-binding protein, calcineurin, containing calcineurin A and B (obtained from C. Klee of the National Institutes of Health); slot 5 = purified brain calcineurin; slot 6 = purified brain cyclic nucleotide phosphodiesterase, also containing calcineurin A and B (obtained from C. Klee of the National Institutes of Health); slot 7 = same as slot 3, but with addition of purified cyclic nucleotide phosphodiesterase.

it did not bind to the myelin basic protein, 17,000 $M_r$, a containment of our PSD preparation. It did bind to a preparation of histones run on a gel, but the inactive chloramine-T iodinated sample did also. All of the above results indicate that the washing out of the SDS from the gel allows for certain proteins to specifically bind calmodulin.

It is gratifying to observe that the PSD protein, 51,000 $M_r$, which is the major substrate for the calmodulin-activatable protein kinase (cf. above), is also the protein that binds calmodulin most effectively. We still have no evidence as to its functions in the PSD, but we have some ideas that will be presented below. The other protein region where bound calmodulin is found, at 60,000–62,000 $M_r$, contains both calcineurin, the heat-labile binding protein,[30,31] and the cyclic nucleotide PDE, both found in the PSD (cf. above). The major band of a partially purified cyclic nucleotide PDE (obtained from D. M. Watterson of the Rockefeller University and from C. Klee of the National Institutes of Health), as well as the purified inhibitor (obtained from R. W. Wallace of the University of Tennessee and from C. Klee of the National Institutes of Health) both migrate to the

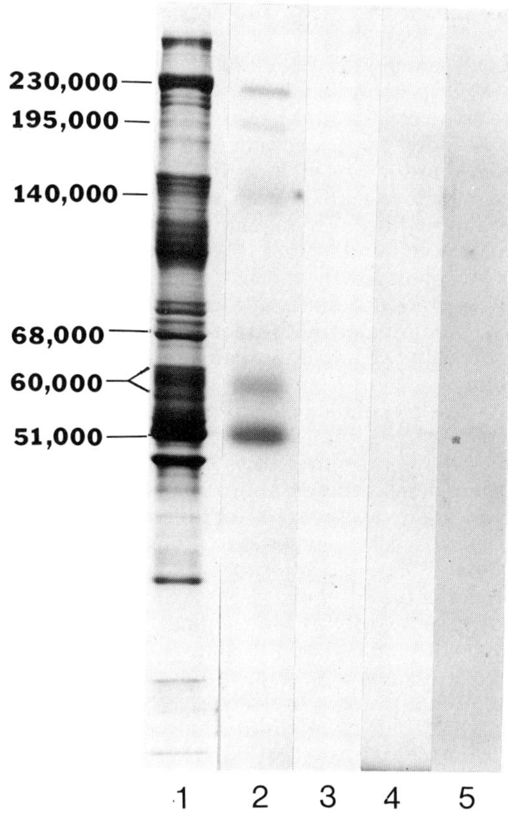


FIGURE 8. Specific binding on SDS-PAGE gel of iodinated calmodulin to proteins of PSDs isolated from cerebral cortex. The experiments were performed as described in the text. Approximately 175 μg PSD proteins were run on the gels and the gels were then incubated for 12 h at room temperature with 50 μg (if six slots were used) radioiodinated[54] calmodulin (sp. act., 150,000 cpm/μg) in a total volume of 50 ml. The gels were washed[52] and dried and autoradiography was performed. Slot 1 = Coomassie-blue stained gel; Slot 2 = radioautograph after calmodulin binding; Slot 3 = same as (2) but in the coincubating presence of 2 μM EGTA; Slot 4 = same as (2) but in the coincubating presence of 1 mM chlorpromazine; Slot 5 = same as (2) but calmodulin radioiodinated by the chloramine-T method[55] was used instead.

same 60,000–62,000 $M_r$ position of the SDS gels. In addition, the radioiodinated calmodulin was found to bind the heat-labile inhibitor, or calcineurin, previously run on an SDS gel (not shown). Thus we tentatively conclude that the protein that binds calmodulin in this region is calcineurin, the heat-labile inhibitor, though it is still possible that the cyclic nucleotide PDE also binds. It is gratifying, as a possible control for this type of binding experiment, that a protein known to interact with calmodulin is being bound to the radioactive calmodulin. Also, it is instructive that the calmodulin does not seem to bind to the PSD tubulins in this gel system, nor to the contaminating intermediate filaments.

### *Calmodulin in PDSs from Various Brain Areas*

All of the above results were obtained from PSDs prepared from cerebral cortex. Since the PSD preparation is most likely a mixture of densities from different types of synapses, it would be instructive to try to isolate PSDs from a known synapse type. However, because of the nature of the neuronal network in the brain and because of the low yield of PSDs, this was not possible. The next best procedure was to isolate PSDs from several brain areas, cerebral cortex, cerebellum, midbrain, and brain stem, and to observe any similarities or differences among them. It was found[12,55] that PSDs from cerebral cortex and midbrain were very similar morphologically and biochemically, that the PSD preparation from cerebellum was quite different, and that the PSD preparations from brain stem were grossly contaminated with neurofilaments. We would like to focus here on the differences between cerebral and cerebellar PSDs, particularly with regard to calmodulin.

FIGURE 9 shows the morphological difference between these two preparations. The thin-sectioned preparations (FIGURES 9A and 9C) indicate two gross points of difference: the cerebrum PSD is approximately 58 nm thick with the cerebellar PSDs being much thinner (approximately 33 nm). FIGURES 9B and 9D, showing replica views, point out two other differences: the presence of a large central hole in cerebral PSDs as opposed to a number of much smaller perforations in the cerebellar PSDs, and the presence of 20–30 nm particulates in the cerebral PSDs and not in the cerebellar PSDs.

FIGURE 10 shows some of the biochemical differences between the two preparations. FIGURE 10A, slots 1 and 2, show that the cerebellar PSDs almost totally lack the 51,000 $M_r$ major protein of the cerebral PSDs, and have about one-half of the calmodulin of cerebral PSDs. FIGURE 10A, slots 3 and 4, shows that the addition of $Ca^{2+}$ plus calmodulin stimulated the phosphorylation of the 51,000 $M_r$ and 62,000 $M_r$ of the cerebral PSD; in the case of cerebellar PSDs, the 62,000 $M_r$ was about equally phosphorylated, but the major increase in phosphorylation occurred at 58,000 $M_r$, with a cerebellar protein at 48,000 $M_r$ being uniquely phosphorylated. The calmodulin binding studies point up the same differences between these two preparations. A comparison of slots 1 and 2 of FIGURE 10B indicates that outside of the much reduced binding of calmodulin to the 51,000 region of cerebellar PSDs, reflecting the much reduced amount of this protein in this preparation as compared to cerebral PSDs, the other bindings of calmodulin were about the same in cerebral and cerebellar PSDs. In addition to the above differences between these two preparations, the cAMP-dependent phosphorylation of Proteins IA and IB is more intense in cerebral PSDs than in cerebellar PSDs.[12]

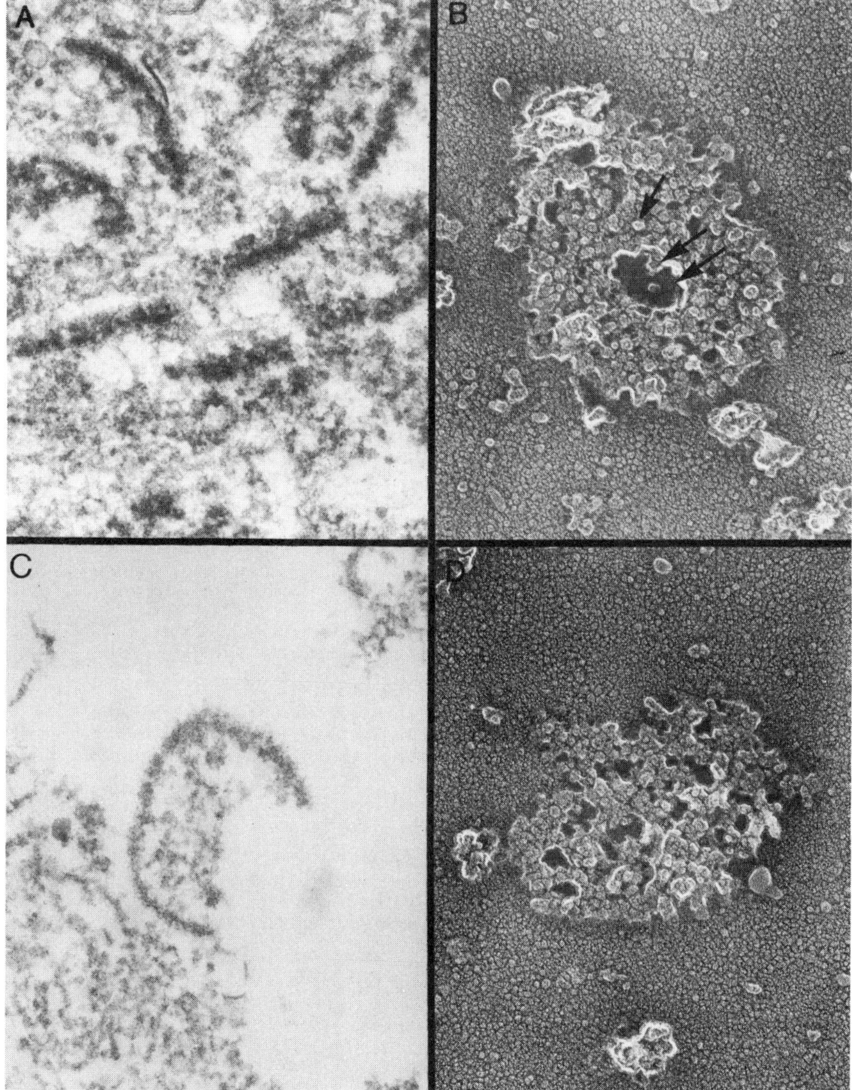


FIGURE 9. Electron microscope images of PSDs isolated from cerebral cortex and from cerebellum. A) Thin section, showing cross-sectional view of PSD preparation from cerebral cortex. × 85,000. B) Replica preparation, with rotary shadowing, of PSDs from cerebral cortex. × 100,000 C) Thin section, showing cross-sectional view of PSD preparation from cerebellum. × 85,000. D) Replica preparation, with rotary shadowing, of PSDs from cerebellum. × 100,000. Double arrows point to large central hole in (B), while single arrows point to 20–30 nm particulate bodies.

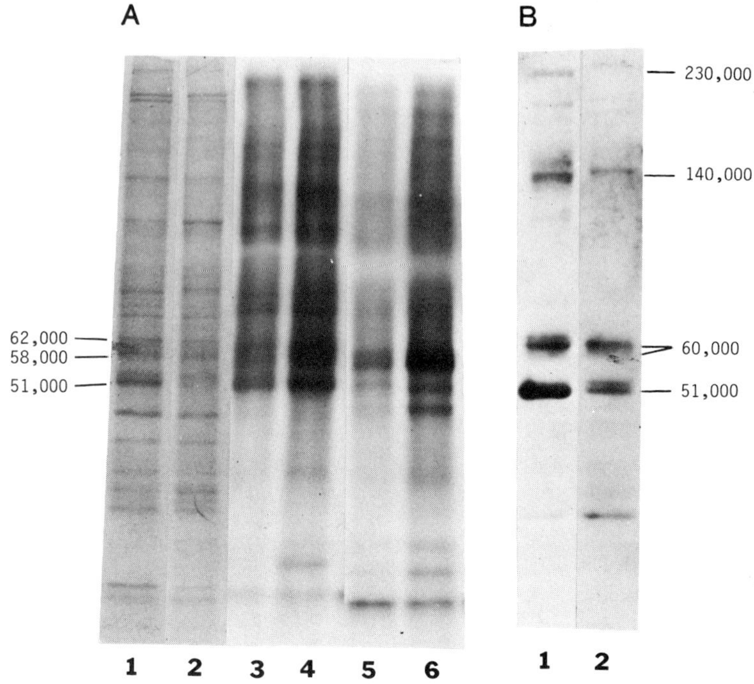


FIGURE 10. Comparison of gel profiles, $Ca^{2+}$ plus calmodulin-activated phosphorylation, and calmodulin binding between cerebral and cerebellar PSD preparations. A) 50 $\mu$g protein were run on the gels. Slot 1 = Coomassie-blue stain of cerebral cortex PSDs; Slot 2 = Coomassie-blue stain of cerebellar PSDs; Slot 3 = Autoradiograph of gel of preparation shown in slot 1, after incubation in basic phosphorylation medium (cf. legend to FIGURE 4); Slot 4 = same as slot 3, but incubation was done in presence of 0.5 mM $Ca^{2+}$ and 4 $\mu$g calmodulin; Slot 5 = autoradiograph of gel of preparation shown in slot 2 after incubation in basic phosphorylation medium; Slot 6 = same as 5, but incubation was done in presence of 0.5 mM $Ca^{2+}$ plus 4 $\mu$g calmodulin. B) 175 $\mu$g protein were run on the gels. Slot 1 = autoradiograph of gel of preparation shown in slot 1, after binding of radioiodinated calmodulin, performed as described in text and in legend to FIGURE 8; Slot 2 = autoradiograph of gel of preparation shown in slot 2, after binding of radioiodinated calmodulin.

## *Function of Calmodulin in Synaptic Excitation*

Can we say anything about these apparently two types of synapses and the role of calmodulin thereby? The appearance of the PSDs from cerebral cortex, particularly their thickness, is very reminiscent of the density seen *in situ* in the Type I synapse of gray[56] or the asymmetric synapse of Colonnier,[57] whereas the PSDs from cerebellum, particularly as to their thinness, are reminiscent of the density in the Type II[56] or symmetric synapses.[57] At present we have little idea as to why we have an enrichment of these presumably Type I PSDs from cerebral cortex and an enrichment of these presumably Type II PSDs from cerebellum, since both types of synapses are found in both these brain areas. The general classification of synapses as Type I and Type II has led Eccles[58] to propose that

Type I synapses mediate excitation responses and Type II synapses mediate inhibitory responses. There has been some confirmation of this hypothesis, but it is by no means proven. Assuming the theory to be correct, we can visualize that calmodulin is intimately involved in the excitation process at the synapse. It is found in large amounts (5–10% total PSD protein) in PSDs presumably occurring at this type of synapse. It binds to the major protein, 51,000 $M_r$, found in these PSDs, and in addition to binding, activates the phosphorylation of this protein by a protein kinase found in these PSDs. It can also activate a cyclic AMP-phosphodiesterase in these presumably excitatory PSDs, and it is noteworthy that these PSDs contain a larger amount of the cAMP-dependent protein kinase and/or the proteins IA and IB substrates for the kinase than do cerebellar PSDs presumably from inhibitory synapses. While calmodulin is found in cerebellar PSDs, the larger occurrence of it, together with the unique occurrence of the 51,000 $M_r$ protein, in cerebral cortex PSD, would suggest that the complex of these two proteins has a significant role in the excitation response.

## References

1. Palade, G. E. 1959. Anat. Rec. **118:** 335.
2. de Robertis, E. 1955. Acta Neurol. Lat. Amer. **1:** 3.
3. de Robertis, E. 1956. J. Biophys. Biochem. Cytol. **2:** 503.
4. Palay, S. L. 1956. J. Biophys. Biochem. Cytol. **2** (Suppl.): 193.
5. Akert, A., H. Moor, K. Pfenninger & C. Sandri. 1969. Prog. Brain Res. **31:** 223.
6. Boom F. E. 1970. *In* Neurosciences, Second Study Program. F. O. Schmitt, Ed.: 729. Rockefeller University Press. New York, N.Y.
7. Cotman, C. W., G. Banker, L. Churchill & D. Taylor. 1974. J. Cell Biol. **63:** 441.
8. Matus, A. I. & B. B. Walters. 1975. J. Neurocytol. **4:** 369.
9. Cohen, R. S., F. Blomberg, K. Berzins & P. Siekevitz. 1977. J. Cell Biol. **74:** 181.
10. Matus, A. I. & D. H. Taff-Jones. 1978. Proc. R. Soc. Lond. Ser. B. **203:** 135.
11. Siekevitz P. *In* Research Methods in Neurochemistry Vol. 3. N. Marks and R. Rodnight, Eds. Plenum Press, New York, N.Y. (In press.)
12. Carlin R., D. J. Grab, R. S. Cohen & P. Siekevitz. 1980. J. Cell Biol. **86:** 831.
13. Grab D. J., K. Berzins, R. S. Cohen & P. Siekevitz. 1979. J. Biol. Chem. **254:** 8690.
14. Ueda, T., P. Greengard, K. Berzins, R. S. Cohen, F. Blomberg, D. J. Grab & P. Siekevitz. 1979. J. Cell Biol. **83:** 308.
15. Blomberg, F., R. S. Cohen & P. Siekevitz. 1977. J. Cell Biol. **74:** 204.
16. Kelly, P. T. & C. W. Cotman. 1978. J. Cell Biol. **79:** 173.
17. Walters, B. B. & A. I. Matus. 1975. Nature **257:** 496.
18. Feit, H., P. Kelly & C. W. Cotman. 1977. Proc. Natl. Acad. Sci. USA **74:** 1047.
19. Yen, S.-H., P. Kelly, R. Liem, C. W. Cotman & M. L. Shelanski. 1977. Brain Res. **132:** 172.
20. Grab, D. J., K. Berzins, R. S. Cohen & P. Siekevitz. 1978. J. Cell Biol. **79:** 96a.
21. Watterson, D. M., W. G. Harrelson, Jr. P. M . Keller, F. Sharief & T. C. Vanaman. 1976. J. Biol. Chem. **251:**4501.
22. Wood, J. G., R. W. Wallace, J. N. Whitaker & W. Y. Cheung. 1980. J. Cell Biol. **84: 66.**
23. Lin, C.-H., K. R. Dedman, B. R. Brinkley & A. R. Means. 1980. J. Cell Biol. **85:** 473.
24. Grab, D. J., R. K. Carlin & P. Siekevitz. (To be submitted.)
25. Florendo, N. T., R. J. Barrnett, & P. Greengard. 1971. Nature **173:** 745.
26. Daly, J. W. 1977. Int. Rev. Neurobiol. **20:** 105.
27. Cheung, W. Y. 1967. Biochemistry **6:** 1079.
28. Filburn, C. R., F. Colpo & B. Sacktor. 1978. J. Neurochem. **30:** 337.
29. Ariano, M. A. & M. M. Appleman. 1979. Brain Res. **177:** 301.
30. Wang, J. H. & R. Desai. 1977. J. Biol. Chem. **252:** 4175.

31. KLEE, C. B. & M. H. KRINKE. 1978. Biochem. **17:** 120.
32. GRAND, R. J. A. & S. V. PERRY. 1979. Biochem. J. **183:** 285.
33. DRABOWSKA, R., O. D. AROMATORIO, J. M. F. SHERRY & R. J. HARTSHORNE. 1977. Biochem. Biophys. Res. Commun. **78:** 1263.
34. DRABOWSKA, R., J. M. F. SHERRY, R. K. AROMATORIO & R. J. HARTSHORNE. 1978. Biochem. **17:** 253.
35. YAGI, K., M. YAZAWA, S. KAKIUCHI. M. OSHIMU & K. UENISHI. 1978. J. Biol. Chem. **253:** 1338.
36. YAZAWA, M., H. KUWAYAMA & K. YAGI. 1978. J. Biochem. (Japan) **84:** 1253.
37. WALSH, M. P., B. VALLET, F. AUTRIC & J. G. DEMAILLE. 1979. J. Biol. Chem. **254:** 12136.
38. HATHAWAY, D. R. & R. S. ADELSTEIN. 1979. Proc. Natl. Acad. Sci. USA **76:** 1653.
39. NAIRN, A. C. & S. V. PERRY. 1979. Biochem. J. **179:** 89.
40. WALSH, M. P., B. VALLET, J.-C. CAVADORE & J. G. DEMAILLE. 1980. J. Biol. Chem. **255:** 335.
41. COHEN, P., A. BURCHELL, J. G. FOULKES, T. W. COHEN, T. C. VANAMAN & A. C. NAIRN. 1978. F.E.B.S. Lett. **92:** 287.
42. GERGELY, P. A. G. CASTLE & N. CRAWFORD. 1980. Biochim. Biophys. Acta **612:** 50.
43. SRIVASTAVA, A. K., D. M. WAISMAN, C. O. BROSTROM & T. R. SODERLING. 1979. J. Biol. Chem. **254:** 583.
44. WAISMAN, D. M., T. J. SINGH & J. H. WANG. 1978. J. Biol. Chem. **253:** 3387.
45. YAMAUCHI, T. & T. FUJISAWA. 1979. Biochem. Biophys. Res. Commun. **90:** 1172.
46. DELORENZO, R. J., D. D. FREEDMAN, W. B. YOHE & S. C. MAURER. 1979. Proc. Natl. Acad. Sci. USA **76:** 1838.
47. SCHULMAN, H. & P. GREENGARD. 1978. Nature **271:** 478.
48. SCHULMAN, H. & P. GREENGARD. 1978. Proc. Natl. Acad. Sci. USA **75:** 5432.
49. GRAB, D. J. & P. SIEKEVITZ. 1979. J. Cell Biol. **83:** 131a; and to be submitted.
50. LEVIN, R. M. & B. WEISS. 1976. Mol. Pharmacol. **12:** 581.
51. CARLIN, R. K., D. J. GRAB & P. SIEKEVITZ. 1980. Fed. Proceed. **39:** 1658.
52. CARLIN, R. K., D. J. GRAB & P. SIEKEVITZ. 1980. This Annal; and to be submitted.
53. RICHMAN, P. & C. B. KLEE. 1978. J. Biol. Chem. **253:** 6323.
54. THIRY, P., A. VANDERMEERS, M.-C. VANDERMEERS-PIRET, J. RATHE & J. CHRISTOPHE. 1980. Eur. J. Biochem. **103:** 409.
55. CARLIN, R. K., D. J. GRAB, P. SIEKEVITZ & R. S. COHEN. 1979. J. Cell Biol. **83:** 140a.
56. GRAY, E. G. 1959. J. Anat. **93:** 420.
57. COLONNIER, M. 1968. Brain Res. **9:** 268.
58. ECCLES, J. C. 1964. The Physiology of Synapses. Springer-Verlag. New York, N.Y.

# THE BINDING OF RADIO-IODINATED CALMODULIN TO PROTEINS ON DENATURING GELS

Richard K. Carlin, Dennis J. Grab, and Philip Siekevitz

*Department of Cell Biology*
*The Rockefeller University*
*New York, New York 10021*

## Calmodulin Iodination

Calmodulin was iodinated by the method of Richman and Klee[1] with the following slight modifications. Iodination was performed (25°C) in a reaction volume of 1 ml that contained 0.05 M sodium phosphate (pH 7.0), 1 mg of calmodulin, 12 μg lactoperoxidase, and 0.4 mM $CaCl_2$. Two mCi $Na^{125}I$ were added to the reaction mixture and the reaction was initiated with 5 μl $H_2O_2$ (diluted 1:500). After 7 min, 5 μl of $H_2O_2$ (1:500) was again added to the reaction mixture and the reaction was allowed to proceed for an additional 7 min. The iodinated calmodulin was isolated as described by Richman and Klee.[1] The specific activity of the calmodulin was approximately 1,000,000 cpm/μg (as determined by liquid scintillation counting). Calmodulin treated with chloramine-T has been shown to be inactivated,[2] and calmodulin iodinated by this method is inactive in binding. We have not tried but believe calmodulin iodinated with the Bolton-Hunter reagent will also work, since it has been shown by others that calmodulin so treated can still activate cyclic nucleotide phosphodiesterase.

## Calmodulin Binding

$^{125}I$-calmodulin binding was developed from previous ligand binding techniques.[3,5-7] Proteins in 2% SDS were electrophoresed in linear 5–15% polyacrylamide slab gels (2mm × 2 cm × 18 cm) containing 0.1% SDS. After electrophoresis the gels were fixed with 25% isopropanol-10% acetic acid for 12 hours with at least four changes by shaking the gels in plastic trays. All washing and binding was done at 25°C. The gels were then washed for 5–10 min in distilled water and then washed with buffer A (50 mM Tris, pH 7.6; 0.2 M NaCl; 1 mM $CaCl_2$) for 12 hours with at least four changes of buffer A. The gels were further washed for 2 hours with buffer A containing 1 mg/ml bovine serum albumin (BSA). For incubation with iodinated calmodulin, 10 ml of buffer A (no BSA) plus 10 μg iodinated calmodulin (the amounts were proportionally larger for large gels) were inserted into Sears boilable cooking pouches and were sealed with a Sears Seal-N-Save. The pouches were shaken for 12 hours on a rotary shaker. The gels, after removal from the pouch, were again placed in plastic trays and were washed for 12 hours with buffer A with at least 4 changes. The gels were then stained for two hours in 0.25% Coomassie blue, 50% methanol, and 7% acetic acid and destained in 25% methanol and 7% acetic acid. The gels were dried and exposed at 25°C for 48 hours on Dupont Cronex 2DC X-ray film or at −90°C for 5 hours using a Dupont Cronex Lighting-Plus YH enhancing screen. It was found that the addition of BSA increased specific binding 10–100 times, and under these



0077-8923/80/0356-0073 $1.75/0 © 1980, NYAS

conditions 100 ng of a calmodulin-binding protein could be detected. This technique was also found to work with two-dimensional SDS-PAGE. The results showing the specificity of the binding on the postsynaptic density proteins, with the requisite controls, are given in the accompanying paper in this Annal. Further characterization of this binding will be presented subsequently.[4]

## References

1. Richman, P. G. & C. B. Klee. 1978. J. Biol. Chem. **253:** 6323.
2. Thiry, P., A. Vandermeers, M. Vandermeers-Piret, J. Rathe & J. Christophe. 1980. Eur. J. Biochem. **103:** 409.
3. Adair, W. S., D. Jurivich & U. W. Goodenough. 1978. J. Cell Biol. **79:** 281.
4. Carlin, R. K., D. J. Grab & P. Siekevitz. J. Cell. Biol. (submitted).
5. Siekevitz, P. 1973. J. Supramol. Struct. **1:** 471.
6. Burridge, K. 1976. Proc. Natl. Acad. Sci. USA **73:** 4457.
7. Bigelis, R. & K. Burridge. 1978. Biochem. Biophys. Res. Commun. **82:** 322.

# IMMUNOCYTOCHEMICAL LOCALIZATION OF CALMODULIN IN REGIONS OF RODENT BRAIN*

John G. Wood,† R. W. Wallace,‡§ J. N. Whitaker,¶** and W. Y. Cheung‡§

*Departments of †Anatomy, ‡Biochemistry, and ¶Neurology*
*University of Tennessee Center for the Health Sciences*
*Memphis, Tennessee 38163*

*§The Department of Biochemistry*
*St. Jude Children's Research Hospital*
*Memphis, Tennessee 38101*

***Research and Neurology Services*
*Veterans Administration Hospital*
*Memphis, Tennessee 38104*

## Introduction

Clues regarding the possible function of calmodulin and a heat-labile calmodulin-binding protein, CaM-$BP_{80}$, in nervous tissue may be inferred from the cellular and subcellular distribution of the proteins. We have examined this distribution at the morphological level using antisera to calmodulin and CaM-$BP_{80}$ together with peroxidase labeling techniques to visualize the sites of specific immunoglobulin binding in tissue slices. Emphasis in this report will be on recent immunocytochemical studies of the distribution of calmodulin in different brain regions.

### *Fixation and Immunocytochemical Procedures*

Anesthetized rodents are perfused through the heart for 10 min with a fixative containing freshly prepared 4.0% paraformaldehyde and 0.1% glutaraldehyde in 0.12 M Millonigs phosphate buffer.[1] The brain and eye are removed and stored overnight at 4°C in phosphate buffered paraformaldehyde only. The tissue is sectioned using an Oxford Vibratome (40–50 $\mu$m slices) or an International Equipment Company Cryostat (16 $\mu$m slices). The slices are treated with rabbit antisera to calmodulin or CaM-$BP_{80}$ for 30–60 min, and, after extensive washing, to horseradish peroxidase labeled Fab fragments of goat anti-rabbit IgG.[2] The peroxidase is visualized with 3,3′-diaminobenzidine (DAB)[3] and the tissue is postfixed with osmium tetroxide prior to processing for light or electron microscopy. The slices are viewed directly for light microscopy and superficial ultrathin sections *en face* to the original cut surface are viewed for electron microscopy. The immunological specificity of the cytochemical results is ascertained by exposing tissue slices in the first step to either nonimmune serum or immune serum, which is first neutralized with an excess of the antigen.

*Supported in part by NS-12590 (J. G. W.), NS-08059 (W. Y. C.), The Sloan Foundation (J. G. W.) and Veterans' Administration (Research Projects Grant 9351-03 (J. N. W.)). R.W.W. is the recipient of a U.S. Public Health Service Service Fellowship AM 05689.



0077-8923/80/0356-0075 $1.75/0 © 1980, NYAS

### *Caudate-Putamen Localization*

We have reported the immunocytochemical localization of calmodulin and CaM-$BP_{80}$ in the rodent caudate-putamen[4] and the results are only briefly summarized here. One of the striking features is that the localization patterns for these two proteins are essentially identical indicating that they probably share the same cellular compartment. At the light microscopic level, calmodulin and CaM-$BP_{80}$ are found within the cytoplasm and processes of large cells. At the electron microscopic level, immunocytochemical labeling is found only within neurons and is apparently restricted to postsynaptic sites within neuronal somata and dendrites. Within the dendrites the label is associated predominantly with the postsynaptic density (PSD) and microtubules. The lack of label associated with presynaptic structures after either calmodulin or CaM-$BP_{80}$ antibody cytochemistry is surprising in view of the finding that calmodulin governs the release of neurotransmitter from synaptosomes.[5] Calmodulin and CaM-$BP_{80}$ may be present in presynaptic terminals either at levels below the limit of sensitivity of our methods or in forms that are not accessible to the antibody reagents. In order to study this problem further we have performed a regional distribution immunocytochemical study of calmodulin and CaM-$BP_{80}$ with particular emphasis on tissue areas that contain identified presynaptic regions. For simplicity, only the results of calmodulin localization will be illustrated although, in all cases, the CaM-$BP_{80}$ localization pattern was very similar.

### *Deep Cerebellar Nuclei*

The deep nuclei of the cerebellum provide a population of easily identified large neurons that receive an input from the Purkinje neurons of the cerebellar cortex. The presynaptic terminals of the Purkinje axons are easily identified when labeled in immunocytochemical experiments. In the deep nuclei of the rat (primarily nucleus interpositus) the characteristic large neurons of this structure contain low but detectable immunocytochemical label for calmodulin in their cytoplasm and a number of cell processes coursing through the neuropil in this region contain label as well (FIGURE 1). In general the label within the processes is more intense than the cell body label. Many of the labeled processes are fairly large and are most likely to be dendrites of the deep cerebellar neurons. Although we cannot rule out the possibility that some of these labeled processes are Purkinje cell axons approaching their target neurons, we see no evidence for a punctate distribution of reaction product surrounding the deep neurons indicating that the Purkinje cell axons do not contain detectable quantities of calmodulin in the presynaptic terminal region.

### *Spinal Cord*

The ventral horn of the spinal cord provides another region of the central nervous system where identified neurons that receive large presynaptic boutons are present. The motor neuron cell bodies contain immunocytochemical label for calmodulin (FIGURE 2) as do a number of large processes passing between cells.

Again no indication of a punctate deposition of reaction product surrounding either the cell bodies or the large processes is observed.

## *Retina*

The layered structure of the retina allows a comparison of the distribution of calmodulin in several different but functionally interrelated neurons. Although a

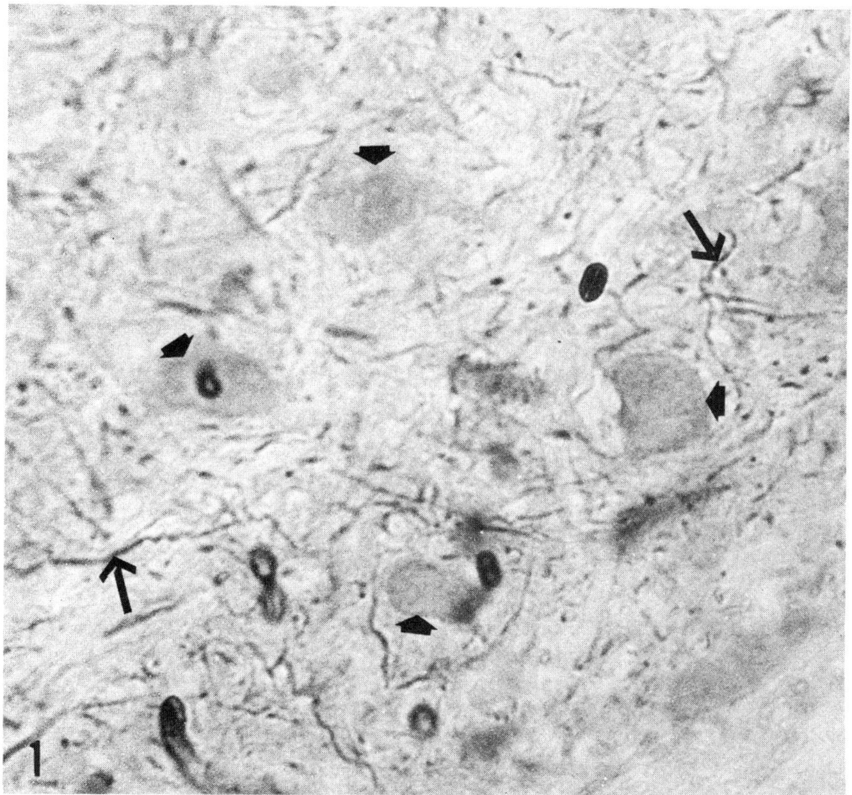


FIGURE 1. Light micrograph of calmodulin localization in a deep cerebellar nucleus of the rat. Immunocytochemical label is seen in the cytoplasm of large neurons (short arrows) and in processes (long arrows) coursing through the neuropil.

low level of immunoperoxidase label is apparent in the different retinal layers the most striking label is a diffuse product within the cytoplasm of retinal ganglion cells and a punctate distribution of product in the inner plexiform layer (FIGURE 3), which largely consists of the processes of retinal ganglion and amacrine cells in synaptic relation. This staining pattern is not altered by

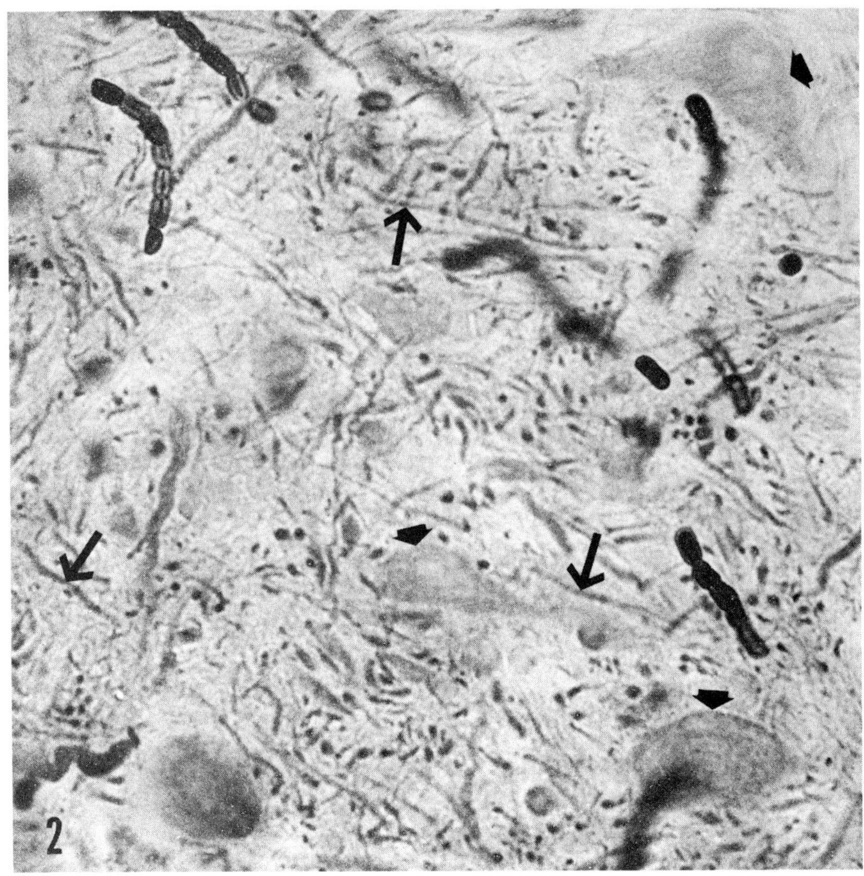


FIGURE 2. Light micrograph of calmodulin localization in the ventral horn of upper cervical spinal cord of the rat. Label is seen in the cytoplasm of motor neuron cell bodies (short arrows) and in longitudinal and cross section views (long arrows) of processes in the neuropil.

treatment of the retinal slices with 0.1–0.5% Triton-X 100 prior to performing the immunocytochemical procedures (results not shown).

## DISCUSSION

Results presented in this communication and in our previous publications support the hypothesis that a pool of calmodulin (and CaM-$BP_{80}$) is localized at postsynaptic sites in a variety of neuronal types. Electron microscopic examination of the rodent caudate-putamen demonstrates that the postsynaptic density and dendritic microtubules are the primary organelles containing immunocytochemical reaction product for calmodulin and CaM-$BP_{80}$. The association of calmodulin with the PSD is in agreement with the biochemical results of Grab *et al.*[6] who showed the presence of calmodulin in isolated PSDs. The association of calmodulin with microtubules implies a role for the protein in microtubular function as is suggested by the work of Marcum *et al.*[7] and Welsh *et al.*[8] The virtually identical localization of calmodulin and CaM-$BP_{80}$ in the basal ganglia and other brain regions indicates that these two proteins might interact *in vivo*

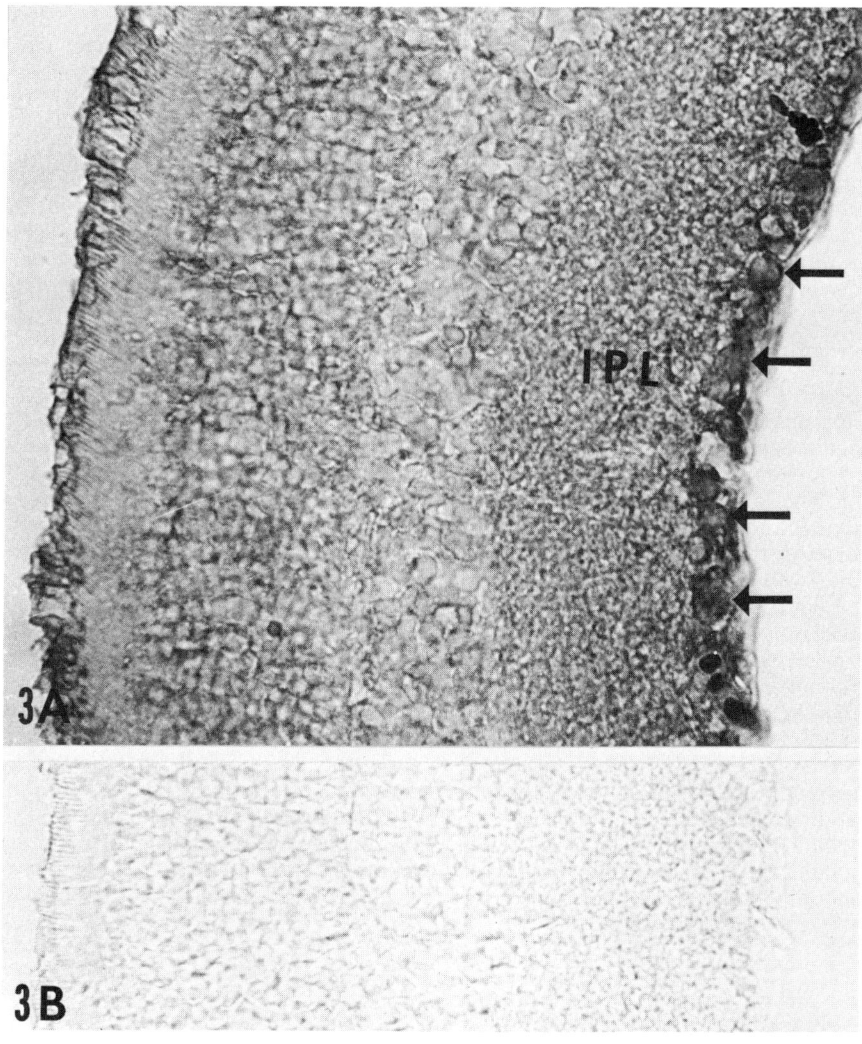


FIGURE 3. A. Light micrograph of calmodulin localization in the retina of Long-Evans rats. Label is most pronounced in the cytoplasm of retinal ganglion cells (arrows) and as punctate profiles in the inner plexiform layer (IPL). B. Light micrograph of the immunocytochemical control for the calmodulin localization shown in FIGURE 3A. The tissue is refractile but no specific reaction product for peroxidase is observed.

and this interaction could be very important in the calcium-dependent regulation of various enzyme activities involved in synaptic transmission.

In order to provide more evidence regarding a possible presynaptic localization of calmodulin we have studied several brain regions that contain known populations of presynaptic terminals that may be easily identified at the light microscopic level if they contain immunocytochemical reaction product. We found that neither the Purkinje cell presynaptic terminals on deep cerebellar neurons nor any of the massive presynaptic input to motor neurons of the spinal cord ventral horn showed reaction product for peroxidase. In both the deep cerebellar nucleus and the ventral spinal cord, a number of immunocytochemically labeled processes, which strongly resemble dendrites, may be observed and this result further supports the hypothesis that a pool of calmodulin exists at postsynaptic sites in different brain regions.

The major sites of immunocytochemical labeling of calmodulin in the retina are within retinal ganglion cells and the inner plexiform layers. The retina is a brain region that is characterized by a capacity of the constituent "neurons" to enter into reciprocal synaptic relationships where each cell may be considered both pre- and postsynaptic. The retinal ganglion cell is interesting in this regard because it is one of the few retinal neurons that exists in essentially only postsynaptic relationship with other neurons. Much of the input to the ganglion cell takes place in the inner plexiform layer from synaptic relationships with the amacrine cells. Although it is not possible from light microscopy to rule out the possibility that some of the punctate label in the inner plexiform layer represents immunocytochemically stained presynaptic terminals, the retinal results are again consistent with the hypothesis that one or more pools of calmodulin exists at postsynaptic sites in neurons.

The immunocytochemical results obtained to date cannot be used to suggest that calmodulin or CaM-$BP_{80}$ only exist at postsynaptic sites in brain. The tissue used in these experiments has been fixed and it is quite possible that fixation alters the antigenic sites of calmodulin so that some pools are no longer recognizable by antibody. This would be a particular problem if presynaptic pools of calmodulin, for example, are present in lower amounts than the postsynaptic pools so that a detectable level of the postsynaptic pool would be retained under conditions that inactivated other pools. It is also quite possible that the postsynaptic pool of calmodulin exists in a macromolecular environment that favors interaction with antibody and other potential pools exist in environments that hinder this interaction. Clearly the question of whether or not calmodulin is present in pools other than the postsynaptic area will require additional biochemical, pharmacological, and morphological data. One way in which the immunocytochemical data may be used to approach this problem would be to perform the immunoctyochemistry after vigorous extractions or modifications of the tissue or on tissue that has been subjected to extremes of physiological stimulation.

## References


1. Millonig, G. 1961. Advantages of a phosphate buffer for $OsO_4$ solution in fixation. J. Appl. Phys. **32:** 1637 (abstract).
2. Mendell, J. R., J. N. Whitaker & W. K. Engel. 1973. The skeletal muscle binding sites of antistriated muscle antibody in myasthenia gravis. An electron microscopic immu-

nohistochemical study using peroxidase conjugated antibody fragments. J. Immunol. **111:** 847–856.

3. GRAHAM, R. C. & M. J. KARNOVSKY. 1966. The early stages of absorption of injected horseradish peroxidase in the proximal tubules of mouse kidney. Ultrastructural cytochemistry by a new technique. J. Histochem. Cytochem. **14:** 291–320.
4. WOOD, J. G., R. W. WALLACE, J. N. WHITAKER & W. Y. CHEUNG. 1980. Immunocytochemical localization of calmodulin and a heat-labile calmodulin-binding protein (CaM-$BP_{80}$) in basal ganglia of mouse brain. J. Cell Biol. **84:** 66–76.
5. DE LORENZO, R. J., S. D. FREEDMAN, W. B. YOBE & S. C. MAURER. 1979. Stimulation of $Ca^{+2}$ dependent neurotransmitter release and presynaptic nerve-terminal protein phosphorylation by calmodulin and a calmodulin-like protein from synaptic vesicles. Proc. Natl. Acad. Sci. USA **96:** 1836–1842.
6. GRAB, D. J., K. BERZINS, R. S. COHEN & P. SIEKEVITZ. 1979. Presence of calmodulin in postsynaptic densities isolated from canine cerebral cortex. J. Biol. Chem. **254:** 8690–8696.
7. MARCUM, J. M., J. R. DEDMAN, B. R. BRINKLEY & A. R. MEANS. 1978. Control of microtubule assembly-disassembly by calcium-dependent regulator protein. Proc. Natl. Acad. Sci. USA **75:** 3771–3775.
8. WELSH, M. J., J. R. DEDMAN, B. R. BRINKLEY & A. R. MEANS. 1978. Calcium dependent regulator protein: localization in mitotic apparatus of eukaryotic cells. Proc. Natl. Acad. Sci. USA **75:** 1867–1871.

## DISCUSSION OF THE PAPER

QUESTION: Have you looked at neuromuscular junctions?

DR. J. G. WOOD: No.

DR. M. T. PIASCIK (*University of Cincinnati, Cincinnati, OH*): Do you have any statement as to the quantitative differences between the caudate and the cortex?

DR. WOOD: No. The differences, to me, are not striking. There is an interesting thing about the distribution that makes interpretation difficult. I could have shown you cells in the caudate which morphologically appear to be the same type of neuron and are adjacent to one another. One of these will be fairly intensely labeled and the adjacent cell will be very lightly labeled. That's characteristic of most of the regions we have looked at. This makes it even more difficult to interpret differences between different regions.

DR. SCHMIDT: Do you have any specific localization in photoreceptor outer segments?

DR. WOOD: We've looked. We can't find it.

DR. R. H. KRETSINGER: What is the control that shows that another acidic, calcium-modulated protein wouldn't also give a positive response in these localizations?

DR. WOOD: If that acidic protein had sequence homology with calmodulin so that the antisera would recognize both, it would give the same response.

DR. W. Y. CHEUNG (*St. Jude Children's Research Hospital, Memphis, TN*): These antisera do not recognize troponin C.

DR. L. RUBEN (*Yale University, New Haven, CT*): It appeared from your micrographs that the mitochondrial outer membrane was staining with your calmodulin antisera. Would you care to comment on that?

DR. WOOD: We have looked at a number of proteins by this peroxidase antibody methodology and I have yet to see localization studies of a soluble protein where you do not have outer mitochondrial membrane labeling. The most likely explanation for this is these are probably not mitochondrial proteins, but that the effect of fixation is to precipitate soluble proteins and they are going to precipitate on membranes that are in the vicinity. That precipitation, then, would give an apparent localization.

DR. F. F. VINCENZI (*University of Washington, Seattle, WA*): To what extent does the apparent distribution of calmodulin occur as a consequence of nonphysiological calcium distribution during the fixation?

DR. WOOD: I really can't answer that question. We have done fixations without calcium and there is no difference. We have done one experiment where EGTA was present and we looked only at the light microscopic level. It appeared the same as we see without EGTA. In every case the tissue was fixed. It would be ideal to use these methods on unfixed tissue.

DR. MURTAGH: In light of your comments about the precipitation of soluble proteins during fixation process, how much credence can we give specific membrane or microtubule localizations.

DR. WOOD: The problem is most serious when you have intense label in the cytosol. In the localizations I showed you, the only label was essentially on the PSD (postsynaptic density) and a little bit of microtubule localization. Also, ours are not the only data indicating localization in the PSD.

# POSSIBLE FUNCTIONS OF CALMODULIN IN PROTOZOA*

Birgit H. Satir, Robert S. Garofalo, Diana M. Gilligan, and Nita J. Maihle

*Department of Anatomy*
*Albert Einstein College of Medicine*
*Bronx, New York 10461*

## INTRODUCTION

From the beginnings of modern cell biology, calcium has been thought to be an extremely important ion affecting a large variety of biological processes.[9,20] Developments in many fields within the last decades have aided in bringing into focus questions as to the exact mechanism by which a cell can control and regulate either the influx or the local effective cytoplasmic concentration of calcium ion. The latest development that has spurred on a multitude of important investigations is the finding that calcium ion may not act primarily in its ionic form, but may act instead through binding to a series of highly specific proteins[3,11,16,17] including troponin C and recently, and perhaps more commonly, calmodulin.

One of the very interesting aspects of the calcium-binding protein, calmodulin, is that it is seemingly involved in many different physiological reactions within the same cell (see other authors in this volume). It is not clear how physiological and biochemical specificity is imposed on the molecule; one possibility is that the apparent affinity for each of its four calcium-binding sites can vary and perhaps different reactions require different numbers of bound $Ca^{2+}$. Alternatively, the local distribution of calmodulin in the presence of unstimulated cell levels of $Ca^{2+}$ may be controlled in ways not yet elucidated (perhaps in connection with the cytoskeleton?). A final possibility is that local $Ca^{2+}$ is rigidly specified and that calmodulin acts in an on-off manner dependent on a narrow range of $Ca^{2+}$ change. The diffusion of free $Ca^{2+}$ away from any point of entry in the cytoplasm is probably highly limited.[21,23]

In the absence of direct sequence information for the molecule in the systems we describe, we use the following criteria to define the presence of calmodulin in the ciliates: (1) it is a protein with an apparent molecular weight of 17,000 daltons; (2) it undergoes a conformational change when calcium is bound and exhibits a $Ca^{2+}$-dependent mobility shift in the presence of SDS. This conformational change is required for calmodulin to regulate many enzyme systems; (3) it stimulates brain cAMP phosphodiesterase activity; and (4) it has affinity for anti-calmodulin, antibody prepared against rat testes calmodulin.

In this paper we deal specifically with the current status of the presence, localization, and possible function of calmodulin in two protozoa, *Paramecium tetraurelia* and *Tetrahymena thermophila*. These cells have been characterized as models for several independent biological processes such as secretion and

*Supported by U.S. Public Health Service (GMS Grant 24724, 27298, 27859, 5T32GM7288, and 5T32CM07128).



0077-8923/80/0356-0083 $01.75/0 © 1980, NYAS

ciliary motility, processes which we and others have shown are regulated by external calcium ion influx[18,28] and therefore perhaps by calmodulin.

The great advantage of protozoan cell systems in looking for a possible connection between calmodulin and calcium in these processes is that well characterized mutants are available, which can be used to attempt to dissect both functional and biochemical reactions involved in some of the control steps.

## Calcium and the Secretory Process

In 1968, W. Douglass[6] coined the following term for the secretory process: "stimulus-secretion-coupling" in analogy to excitation-contraction-coupling, whereby a stimulus causes membrane depolarization followed subsequently by an influx of $Ca^{2+}$ raising the cytoplasmic concentration from $10^{-7}$ M to $10^{-6}$ M. A well known and characterized secretory system following this pathway is the synapse. The rise in cytoplasmic $Ca^{2+}$ at the presynaptic membrane causes the synaptic vesicle membrane to fuse with the axon membrane leading to release of transmitter.[5]

It is important to remember that an influx of $Ca^{2+}$ following stimulation can lead to several different responses in a single cell. This is well illustrated in *Paramecium* where $Ca^{2+}$ influx produces four distinct physiological responses in the cell, separated in time. These are: (1) ciliary stoppage and reversal, (2) cell contraction, (3) trichocyst release, and (4) deciliation (FIGURE 1). Ciliary stoppage, which occurs at $10^{-6}$ M $Ca^{2+}$,[18] is a very important indicator of internal calcium ion concentration. It is the first effect seen upon stimulation and we use it routinely as a biological monitor for cytoplasmic $Ca^{2+}$ concentration. In order to produce such spatially and temporally different responses, normal influx has to occur locally and site-specifically.

The question that arises is how can such specificity be insured? Satir[22,26,27] suggested that perhaps this could be brought about via membrane mosaicism in the form of local intramembrane particle arrays revealed with the freeze fracture technique. Specific arrays are seen in this case within the plasma membrane in *Paramecium* and *Tetrahymena* in connection with secretion (fusion rosettes) and within the ciliary membrane (ciliary necklaces and patches).[10,22,24,25,29]

Satir and Oberg[28] working with a temperature-sensitive *Paramecium* secretion mutant lacking fusion rosettes,[1] called nd9, found that secretion could be induced at the nonpermissive temperature by adding a divalent cation ionophore, A23187, in the presence of external calcium ion. Upon stimulation with picric acid (routine stimulus for secretory assay) in the absence of ionophore no release was induced and release could also be blocked in the presence of ionophore with high external concentration of magnesium ion. FIGURE 1 shows that all four calcium responses may be elicited within the span of 1 minute in nd9 at the nonpermissive temperature by micropipetting of ionophore near the cell in the presence of 5 mM $Ca^{2+}$. The nd9 mutant cells grown at the nonpermissive temperature have no fusion rosettes but otherwise appear normal. Ionophore-induced release in these cells is not dependent on reassembled rosettes, since 10 seconds after ionophore-induced release we do not find assembled rosettes in freeze fracture replicas of the cell membrane.[8] It seems likely that all of the ionophore-induced responses of this cell bypass the normal membrane stimulus mechanism and that the main feature of such a mechanism is to insure site-specific increase in cytoplasmic free $Ca^{2+}$.

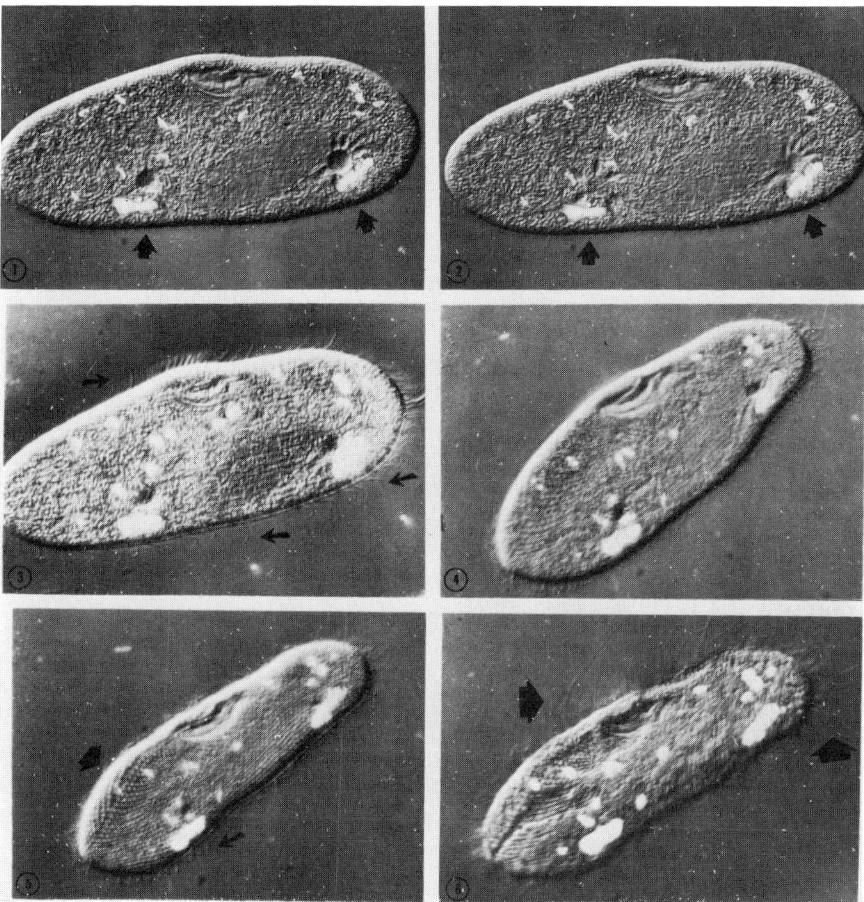

FIGURE 1. A living nd9 cell grown at the nonpermissive temperature (27°C), trapped in a drop of buffer containing 5 nM $CaCl_2$, and immobilized by withdrawal of solution with a micromanipulation needle. 10 $\mu$M A23187 and 5 mM $CaCl_2$ are added by a second needle. Light micrographs were taken on the Zeiss Axiomat. Frames 1 and 2 show the cell is alive—cilia are blurred since they are beating, and the contractile vacuoles are active (arrow). Frame 3 illustrates ionophore addition that first leads to stopping of ciliary beating (arrows). Frames 4 and 5 show the second response—contraction of the entire cell; Frame 6 shows release of trichocysts (arrows) and some deciliation taking place. All events occur within one minute. ×500.

## CALMODULIN IN PARAMECIUM AND TETRAHYMENA

We now wish to explore the question of whether $Ca^{2+}$ acts via a calmodulin-mediated pathway to produce the responses described in FIGURE 1. The first evidence we obtained of the presence of calmodulin in *Paramecium* came indirectly through a series of experiments in which we fractionated cytoplasm of

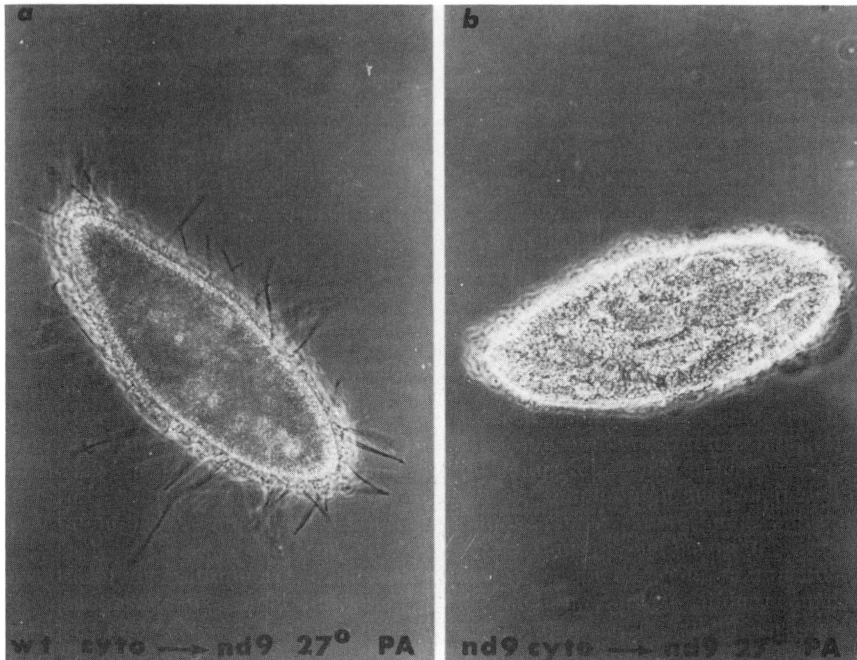


FIGURE 2. FIGURE 2a shows an nd9 mutant cell grown at nonpermissive temperature (27°C), microinjected with wild type cytoplasm which results in recovery of secretion and release of trichocysts when stimulated with picric acid. The trichocysts are visible as 20–40 μm long needlelike structures surrounding the cell. FIGURE 2b. An nd9 mutant grown at the nonpermissive temperature does not secrete in response to picric acid after microinjection with mutant (27°C) cytoplasm; no trichocysts are evident. ×680.

wild type (wt) and mutant *Paramecium.* Upon microinjection into nd9 cells grown at the restricted temperature both wild type cytoplasm and certain fractions obtained from wild type including a postmicrosomal high molecular weight fraction[7,27] restored secretory capacity in the mutant, while comparable mutant extracts had no effect (FIGURE 2). In collaboration with J. Dedman and A. R. Means, we found that calmodulin is one of the components in the competent extract; however, since calmodulin is also present in the mutant grown at both restrictive and nonrestrictive temperatures, we conclude that it is not calmodulin *per se* but factors indirectly associated with calmodulin that are active in restoring secretory capacity.

We measure the concentration of calmodulin in *Tetrahymena* and *Paramecium* cytoplasm to be about 5 ng/μg protein. Both *Paramecium* and *Tetrahymena* calmodulin are capable of stimulating bovine brain cAMP phosphodiesterase; the stimulation is inhibited by 18 μM trifluoperazine. *Paramecium* and *Tetrahymena* calmodulin are heat-stable proteins of molecular weight *ca.* 17 K daltons whose apparent mobility in 15% SDS PAGE shifts in EGTA versus calcium chloride.

The presence of calmodulin in whole cell extracts does not implicate it in any of the $Ca^{2+}$-regulated processes mentioned earlier. In order to explore this we attempted to localize calmodulin using an indirect immunofluorescent technique.[14,15] An antibody (goat) to rat testes calmodulin was given to us by Dedman and Means; its characteristics and purification have been published previously.[4] The antibody to calmodulin localizes in both protozoa (1) along the entire length of the cilia (FIGURE 3) and (2) in a punctate pattern on the cell surface, corresponding to the ciliary basal bodies. Fluorescence is also localized in *Paramecium* in 10–15 μm diameter vesicles, which may be calcium-containing crystals or stem from some other currently unspecified structures. The presence

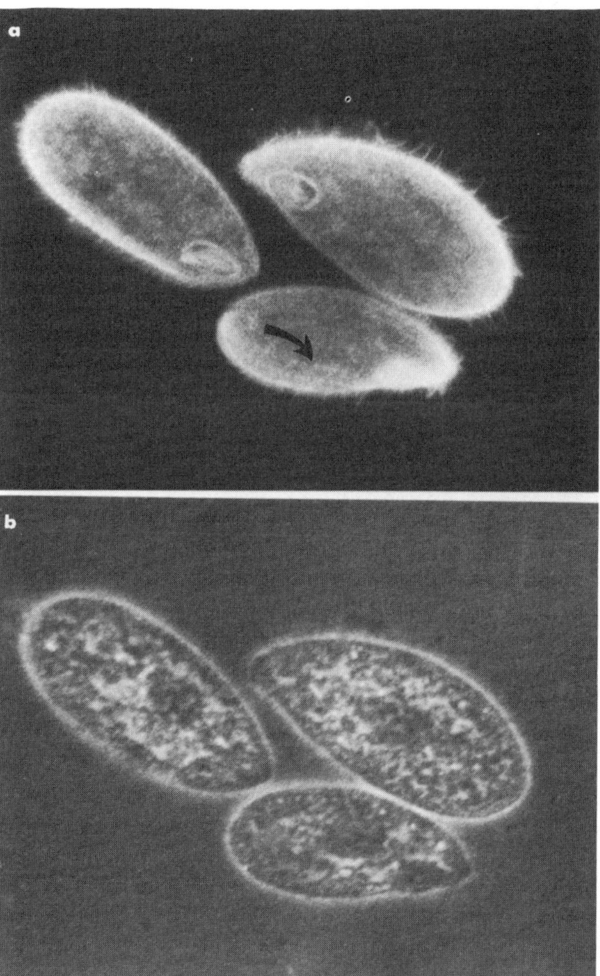


FIGURE 3. FIGURE 3a shows immunofluorescent localization of calmodulin in a group of *Tetrahymena* cells and FIGURE 3b the corresponding light micrograph. Fluorescence is seen along the length of cilia, in a rim along the oral groove, and just discernible as a punctate pattern along the 1° kineties (arrow). ×360.

of calmodulin in cilia came as no surprise to us since Dedman and Means earlier (unpublished) had assayed our *Tetrahymena* and *Paramecium* cell fractions, including isolated cilia, by radioimmunoassay and found significant concentrations of calmodulin.

The functional role of calmodulin in cilia has been pursued by Reed and P. Satir.[19] Presumably, one possible function is the formation of a $Ca^{2+}$-calmodulin complex that is part of an axonemal switch that produces ciliary stoppage in many ciliated cells, and ciliary reversal in protozoa. Also, the presence of calmodulin at the base of the cilium, near the ciliary necklace, both in normal and deciliated cells suggests a role for this molecule in the deciliation process.

We are unable to localize calmodulin at the secretory site by current immunofluorescent techniques. The apparent lack of fluorescence associated with secretory sites may be due to several factors: 1) the amount of calmodulin at one site may be below the sensitivity of our present methods; 2) preparation for immunofluorescence could destroy sites perhaps by inducing secretion; or 3) no calmodulin may be associated with secretory sites. We believe this last possibility to be unlikely in view of experiments now to be discussed.

## Effect of Trifluoperazine on Paramecium

Trifluoperazine or stelazine is a phenothiazine derivative and is used as an antipsychotic. It has been shown selectively to inhibit calmodulin activation of phosphodiesterase in a calcium-dependent fashion.[13] As mentioned above, 18 $\mu$M trifluoperazine inhibits the *Paramecium* extract activation of mammalian phosphodiesterase.

This result prompted us to determine if exposure to trifluoperazine would affect any of the $Ca^{2+}$ regulated processes of *Paramecium*. Figure 4 shows the effect of continuous exposure to 14 $\mu$M trifluoperazine on secretion by wt *Paramecium*. The drug is added at time zero to cells established in a salt-buffer medium.[2] At various times after drug addition, trichocyst release is assayed by challenging an aliquot of cells with a saturated solution of picric acid. The percentage of cells releasing 50 or more trichocysts is calculated. Figure 4 shows a pronounced inhibition of trichocyst secretion after 2 min exposure to the drug. The number of nonsecretors (cells releasing less than ten trichocysts) rises in a complementary manner. Controls without drug addition show a constant percentage of release for many hours (4–6).

## Discussion

We are currently examining the effect of trifluoperazine on ciliary motility, secretion, contraction, and deciliation. In the presence of trifluoperazine, the cells qualitatively swim more slowly and appear somewhat contracted. After prolonged exposure to the drug, some deciliation is observed. The effect of trifluoperazine on secretion in *Paramecium* is consistent with an involvement of calmodulin in stimulus-secretion coupling in these cells. Recently, Larsen and Vincenzi[12] reported that calmodulin stimulated the calcium ATPase-dependent transport of $^{45}Ca^{2+}$ into inside-out isolated red blood cell ghosts. Does calmodulin stimulate $Ca^{2+}$ pumps in *Paramecium* cell membrane? If so, the addition of trifluoperazine should lead to a turning-off, or inhibition of such calcium pumps and an increase in the cytoplasmic calcium ion concentration. In this case,

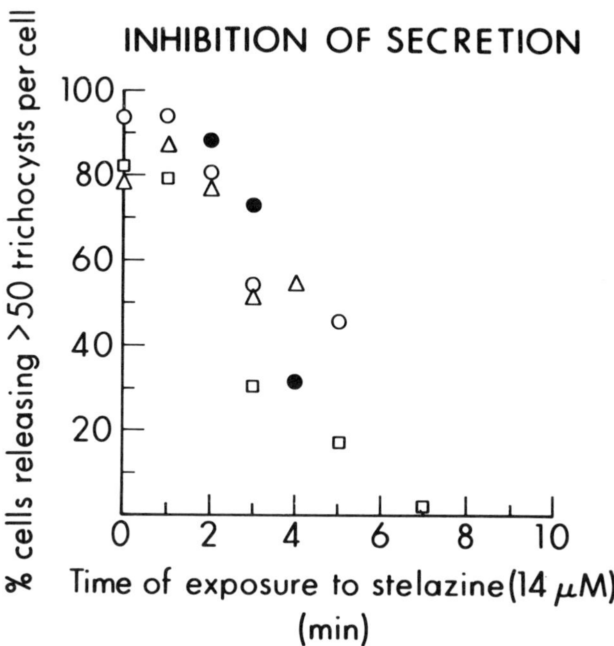


FIGURE 4. Pooled results from four different experiments in which cells were exposed to 14 $\mu$M trifluoperazine (stelazine) prior to testing for secretory response with picric acid. No effect is apparent until 2 minutes of exposure, then a rapid inhibition of release takes place with complete inhibition occurring after 7 minutes' exposure.

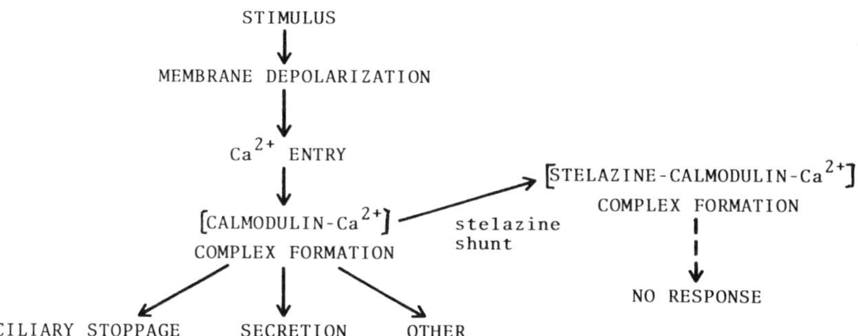


FIGURE 5. The normal chain of events at the cell membrane involves a stimulus-induced membrane depolarization, which may lead to influx of external $Ca^{2+}$. The resulting rise in cytoplasmic $Ca^{2+}$ concentration leads to the formation of a CaM-$Ca^{2+}$ complex, which can now act to regulate a number of $Ca^{2+}$-dependent cellular processes, including secretion and ciliary beat. When trifluoperazine (stelazine) is present, the rise in intracellular $Ca^{2+}$ concentration leads to a $Ca^{2+}$-dependent formation of the stelazine-CaM-$Ca^{2+}$ complex, which prevents the interaction of CaM with its intracellular targets. The "stelazine shunt" is one means of investigating the role of CaM in a variety of physiological processes, and in combination with other types of experimentation, may help clarify the mechanism of CaM action.

trifluoperazine and A23187 might both be expected to have the same effect on release, namely, to stimulate secretion. However, quite the opposite is observed; trifluoperazine leads to an inhibition of secretion when assayed with picric acid. Thus we envision that in the untreated cell, calmodulin acts primarily in secretion subsequent to calcium entry to induce release perhaps by promoting membrane fusion. When trifluoperazine is added, calcium will enter the cell, possibly even at a faster rate than in untreated cells; however, the calcium-calmodulin-trifluoperazine complex that now forms will be inactive (FIGURE 5). The magnesium-induced inhibition of release is not likely to act via the same mechanism, since high $Mg^{2+}$ directly blocks calcium ion influx in *Paramecium*.

The slowing of swimming, contraction, and deciliation induced by trifluoperazine may be effects on $Ca^{2+}$ entry, i.e., an overall increase in cytoplasmic $Ca^{2+}$, perhaps due to a drug effect on membrane pumps? However, if trifluoperazine is properly applied, we might expect a secondary reaction in these situations—inhibition of ciliary reversal or stoppage, inhibition of contraction, and inhibition of deciliation if calmodulin-$Ca^{2+}$ is the activator of these processes. This has proven to be the case for ciliary arrest of metazoan cells.[19] The "stelazine-shunt" of FIGURE 5 may prove a useful tool for deciding whether calmodulin is an intermediate in any given $Ca^{2+}$-mediated response.

## ACKNOWLEDGMENT

The authors thank Dr. R. D. Allen for making his Axiomat available and for his help in using it.

## REFERENCES

1. BEISSON, J., M. LEFORT-TRAN, M. POUPHILE, M. ROSSIGNOL & B. SATIR. 1976. Genetic analysis of membrane differentiation in *Paramecium*. Freeze-fracture study of the trichocyst cycle in wild-type and mutant strains. J. Cell Biol. **69:** 126–143.
2. BROWNING, J. L. & D. L. NELSON. 1976. Biochemical studies of the excitable membrane of *Paramecium aurelia*. I. $^{45}Ca^{2+}$ fluxes across resting and excited membrane. Biochim. Biophys. Acta **448:** 338–351.
3. CHEUNG, W. Y. 1980. Calmodulin plays a pivotal role in cellular regulation. Science **207:** 19–27.
4. DEDMAN, J. R., M. J. WELSH & A. R. MEANS. 1978. $Ca^{2+}$-dependent regulator. Production and characterization of a monospecific antibody. J. Biol. Chem. **253:** 7515–7521.
5. DEL CASTILLO, J. & B. KATZ. 1957. La base "quantale" de la transmission neuromusculaire. *In* Microphysiologie Comparee des Elements excitables. Coll. Internat. C.N.R.S. Paris No. **67:** 245–258.
6. DOUGLAS, W. W. 1968. Stimulus-secretion coupling: the concept and clues from chromaffin and other cells. Br. J. Pharmac. **34:** 451–474.
7. GAROFALO, R. S., J. K. C. KNOWLES & B. H. SATIR. 1978. Restoration of secretory capacity in a non-discharge *Paramecium* mutant by microinjection of wild type cytoplasmic factor(s). J. Cell Biol. **79:** 245a.
8. GILLIGAN, D. M. & B. S. SATIR. 1980. Characterization of components involved in assembly of the intramembrane particle array: rosette, in *Paramecium*. Second International Congress on Cell Biology, Berlin. (In press.)
9. HEILBRUNN, L. V., Ed. 1952. An Outline of General Physiology. Third Edition. W. B. Saunders Co. Philadelphia, Penna.
10. JANISCH, R. 1972. Pellicle of Paramecium caudatum as revealed by freeze etching. J. Protozool. **19:** 470–472.

11. KRETSINGER, R. H. 1977. Evolution of the informational role of calcium in eukaryotes. *In* Calcium-Binding Proteins and Calcium Function. R. H. Wasserman, R. A. Corradino, E. Carafoli, R. H. Kretsinger, D. H. MacLennan & F. L. Siegel, Eds.: 63–72. American Elsevier, Inc. New York, N.Y.
12. LARSEN, F. L. & F. F. VINCENZI. 1979. Calcium transport across the plasma membrane: stimulation by calmodulin. Science **204:** 306–309.
13. LEVIN, R. M. & B. WEISS. 1977. Binding of trifluoperazine to the calcium-dependent activator of cyclic nucleotide phosphodiesterase. Molec. Pharmacol. **13:** 690–697.
14. MAIHLE, N. J. & B. H. SATIR. 1980. Calmodulin in the ciliates *Paramecium tetraurelia* and *Tetrahymena thermophila*. This volume.
15. MAIHLE, N. J., J. R. DEDMAN, A. R. MEANS & B. H. SATIR. 1980. Biochemical characterization and indirect immunofluorescent localization of calmodulin in *Paramecium tetraurelia*. (In preparation.)
16. MEANS, A. R. & J. R. DEDMAN. 1980. Calmodulin—an intracellular calcium receptor. Nature **285:** 73–77.
17. MEYER, W. L., W. H. FISCHER & E. G. KREBS. 1964. Activation of skeletal muscle phosphorylase b kinase by Ca. Biochemistry **3:** 1033–1039.
18. NAITOH, Y. & H. KANEKO. 1972. Reactivated triton-extracted models of *Paramecium:* modification of ciliary movement by calcium ions. Science **176:** 523–524.
19. REED, W. & P. SATIR, 1980. This volume.
20. RINGER, S. 1880–1882. Concerning the influence exerted by each of the constituents of the blood on the contraction of the ventricle. J. Physiol. III:380–393.
21. ROSE, B. & W. R. LOEWENSTEIN. 1975. Calcium ion distribution in cytoplasm visualized by aquorin: diffusion in the cytosol is restricted due to energized sequestering. Science **190:** 1204–1206.
22. SATIR, B. 1974. Membrane events during the secretory process. Symp. for Soc. Exp. Biol. **28:** 399–418.
23. SATIR, B., W. S. SALE & P. SATIR. 1976. Membrane renewal after dibucaine deciliation of *Tetrahymena*. Freeze-fracture technique, cilia, membrane structure. Exp. Cell Res. **97:** 83–91.
24. SATIR, B., C. SCHOOLEY & P. SATIR. 1972. Membrane reorganization during secretion in *Tetrahymena*. Nature **235:** 53–54.
25. SATIR, B., C. SCHOOLEY & P. SATIR. 1973. Membrane fusion in a model system. Mucocyst secretion in *Tetrahymena*. J. Cell Biol. **56:** 153–176.
26. SATIR, B. H. 1978. Membrane fusion. *In* Transport of Macromolecules in Cellular Systems. S. C. Silverstein, Ed.: 431–444. Abakon Verlags. Berlin.
27. SATIR, B. H. 1980. The role of local design in membranes. *In* Membrane-Membrane Interactions. N. B. Gilula, Ed.: 45–58. Raven Press, Inc. New York, N.Y.
28. SATIR, B. H. & S. G. OBERG. 1978. *Paramecium* fusion rosettes: possible function as $Ca^{2+}$-gates. Science **199:** 536–538.
29. SPETH, V. & F. WUNDERLICH. 1972. Evidence for different dispositions of particles associated with freeze-etched membranes. Brief report. Protoplasma **75:** 341–344.

# ROLE OF CALMODULIN IN NEUROTRANSMITTER RELEASE AND SYNAPTIC FUNCTION*

Robert J. DeLorenzo

*Department of Neurology*
*Yale University School of Medicine*
*New Haven, Connecticut 06510*

An understanding of the molecular mechanisms underlying calcium-dependent neurotransmitter release and other calcium-modulated synaptic functions would greatly enhance our knowledge of synaptic transmission and the action of specific neuropharmacologic agents and possibly provide new insights into human disease processes involving synaptic modulation. Although the role of calcium in synaptic function has been of great interest, little is known about the molecular mechanisms of calcium in stimulating neurotransmitter release or its other physiological functions in the nerve terminal.

It is generally accepted that many of calcium's effects on living tissue are mediated by heat-stable, calcium binding regulator proteins. Calmodulin is a major calcium receptor protein in brain and other tissues that has been well characterized and shown to affect several important enzyme systems.[1] Thus, my laboratory has been investigating the possible role of calmodulin in modulating the effects of calcium or neurotransmitter releases and synaptic function.[2-5]

In this report additional evidence is presented to further establish the role of calmodulin in neurotransmitter release and synaptic function. Calmodulin is isolated from synaptosome cytosol and highly enriched synaptic vesicle preparations. Calmodulin and calcium are demonstrated to stimulate the release of acetylcholine and norepinephrine from synaptic vesicles while simultaneously activating synaptic vesicle protein phosphorylation. Depolarization-dependent calcium uptake into synaptosomes is shown to simultaneously stimulate calmodulin modulated vesicle protein phosphorylation and release of acetylcholine and norepinephrine. Calmodulin is also shown to modulate the effects of $Ca^{2+}$ on protein phosphorylation in preparations of synaptic junctional complexes and postsynaptic densities. In an attempt to investigate the mechanism of calcium and calmodulin in stimulating the release of transmitter from vesicles, a morphological study was undertaken to determine if synaptic vesicles change shape or configuration in the presence of calcium and calmodulin. The results suggest that calcium and calmodulin initiate synaptic vesicle and synaptic membrane interactions that may play a role in mediating the effects of these agents on synaptic function. The results presented in this report provide evidence that calmodulin mediates several synaptic events and are consistent with and expand the initial hypothesis from this laboratory that calmodulin plays a major role in modulating the effects of $Ca^{2+}$ on synaptic function and neurotransmitter release.[2,3]

## Materials and Methods

Calmodulin-containing and calmodulin-depleted synaptic vesicle preparations and synaptosome fractions were isolated by differential centrifugation from

*Supported by U.S. Public Health Service (Grant NS 13632) and by Research Career Development Award NS-EA 1 K04 NS 245 to R.J.D.



0077-8923/80/0356-0092 $01.75/0 

rat brain homogenates as described previously.[2] It is important to emphasize that in these experiments each rat brain was removed from the skull and dispersed in ice cold 0.32 M sucrose with 1–2 homogenization strokes within 10–15 seconds following decapitation. After each brain was removed and briefly homogenized to insure rapid cooling to 4°C, all brain homogenates were combined (usually 4–5 brains per preparation) and homogenized with 12 strokes and subjected to our standard rapid subfractionation procedures.[2] $Ca^{2+}$ and calmodulin stimulated protein phosphorylation, neurotransmitter release, and morphological changes were significantly diminished if brains were not cooled to 4°C within 10–15 seconds following decapitation. Employing a protease inhibitor (phenylmethylsulfonylfluoride, 0.3 mM, Sigma, St. Louis, Mo.) in the homogenization buffer also helped prevent loss of these $Ca^{2+}$ and calmodulin-activated systems, suggesting that some of the enzymes stimulated by $Ca^{2+}$ and calmodulin are rapidly broken down postmortem by endogenous protease activity.

### *Synaptic Vesicle Preparations*

The standard reaction mixture for simultaneously studying protein phosphorylation, neurotransmitter release, and morphological changes contained 24 $\mu$M unlabeled ATP or $\gamma$-$^{32}$P-ATP, 4 mM $MgCl_2$, 80 mM KCl, 2.5 mM NaCl, 260 $\mu$M pargyline, 10 mM Tris maleate (pH 6.8) buffer, and approximately 1 mg/ml of vesicle protein. In reactions used to study the interactions of vesicles with synaptic membrane, each reaction mixture contained 0.4 mg/ml vesicle protein and 1 mg/ml synaptic membrane isolated by the procedure of Cotman and Taylor.[6] Other additions or conditions are given in the figure legends. Reactions were initiated by the addition of $Ca^{2+}$ or balance solution[4] and standard reactions were incubated at 37°C for 1 minute. Following incubation, protein phosphorylation was quantitated as described previously.[7] For neurotransmitter studies, vesicles were rapidly isolated following incubation by centrifugation at 4°C[4] and assayed spectrophotometrically for norepinephrine[2,4] and by bioassay for acetylcholine content.[8] Morphological studies were performed as described below.

### *Intact Synaptosomes*

Synaptosomal $^{45}Ca^{2+}$ uptake, neurotransmitter release, and protein phosphorylation were performed by our previously described procedures.[2] Norepinephrine and acetylcholine were assayed as described above. [$^{32}$P]phosphate incorporation into synaptic vesicle, synaptic membrane, and synaptic junctional-complex fractions isolated from intact synaptosomes incubated under various conditions with $^{32}$P was measured by terminating each reaction by osmotically shocking the incubated synaptosome preparation at 4°C and then rapidly isolating the synaptic vesicle,[2] synaptic membrane,[6] and synaptic junctional-complex fractions[6] from each reaction tube by standard procedures. Control synaptosome preparations were also osmotically shocked and stored at 4°C to compare to each subfraction. The isolation procedure was performed in less than 4 hrs. At the end of subfractionation, each fraction was simultaneously solubilized with SDS-stop solution and quantitated for [$^{32}$P]phosphate incorporation[2] (TABLE 4). Rapid isolation of brain synaptosomes as described above was essential in retaining activity in this intact synaptosome preparation.

### *Morphological Studies*

Synaptic vesicles or synaptic vesicles plus synaptic membrane were incubated under standard conditions as described above, and studies by electron microscopy to determine if $Ca^{2+}$ and/or calmodulin could induce vesicle-membrane interactions or vesicle morphological alterations. Following incubation, the reactions were stopped by 2 percent glutaraldehyde fixation at 4°C for 1 hour. The fixed samples were pelleted by centrifugation, postfixed with 1 percent osmium, dehydrated, and embedded in epon.[9] Ultrathin sections of the pelleted samples were prepared by the quantitative procedure of Cotman and Flansburg[10] so that the sample could be examined from the top to the bottom of the entire pellet in a quantitative fashion. For negative staining, samples of fixed or unfixed reaction mixtures were placed on formvar coated grids, stained with uranyl acetate, and blotted dry.[11] Quantitation of vesicle diameter and percent free vesicles were performed by standard morphometric and quantitative procedures on electron micrographs obtained by random sampling of each grid.[14] Special attention was given to the orientation of the vesicle pellet so that random sampling from the top, middle, and bottom of each pellet was obtained.[10]

To study vesicle and membrane interactions, synaptic vesicle and synaptic membrane fractions were prepared and incubated as described above. Following incubation, the membrane was isolated by centrifugation at 25,000 × g for 8 minutes and the resulting membrane pellet was fixed in glutaraldehyde and prepared for electron microscopy as described above. The pellet was oriented so that random electron micrographs could be obtained from the entire pellet[10] and methods of quantitation are described in the legend of TABLE 7.

### *Other Methods*

Calmodulin was isolated from rat brain cytosol, synaptic vesicles, and from synaptosome cytosol by a modification of the method of Lin *et al.*,[15] as described previously.[2] Cyclic nucleotide-dependent phosphodiesterase activity and the ability of calmodulin to stimulate phosphodiesterase activity were determined by the two-step assay method.[15] Amino acid analyses[16] and determinations of isoelectric points by acrylamide tube gel electrophoresis[17] were performed by standard procedures. Postsynaptic density[18] and synaptic junctional complex[6] fractions were prepared by the methods of Cotman's group.

## RESULTS

### *Isolation of Calmodulin from Highly Enriched Synaptic Vesicle Preparations*

Highly enriched preparations of synaptic vesicles isolated under conditions that simulate the intracellular environment following the release of the vesicles from the nerve terminals by osmotic shock demonstrate $Ca^{2+}$ stimulated protein phosphorylation and norepinephrine release when incubated under standard conditions.[2-4] Treatment of these vesicles with EGTA and EDTA has been shown to remove the ability of $Ca^{2+}$ to stimulate phosphorylation and norepinephrine release.[2] The ability of $Ca^{2+}$ to stimulate these events was restored by adding back the synaptic vesicle extract (SVE). Calmodulin was also found to restore the effects of calcium on the treated vesicle preparations.[2] These results suggested that the synaptic vesicle preparations contained calmodulin or a calmodulin-like protein.

To further determine if calmodulin could be isolated from preparations of nerve terminals, attempts were made to isolate calmodulin by standard procedures (METHODS) from synaptic vesicles and from cytosol obtained from osmotically lysed synaptosome preparations. A heat-stable protein was isolated from both of these preparations that has an identical molecular weight and comigrated

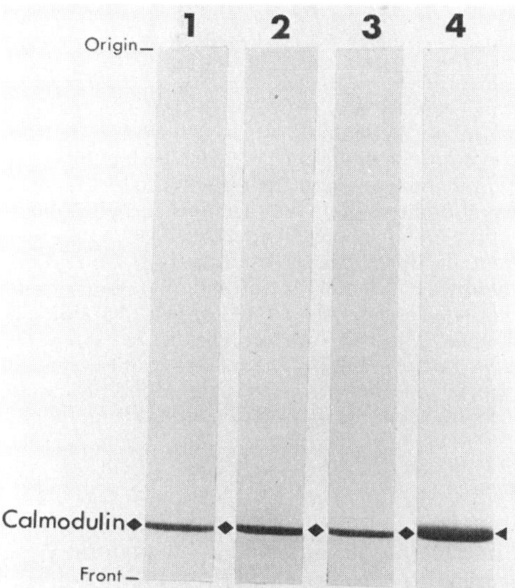


FIGURE 1. Electrophoretic protein patterns on SDS-polyacrylamide gel of calmodulin isolated from whole rat brain cytosol, synaptosome cytosol, and synaptic vesicles, and a mixture of all three calmodulin preparations as shown in channels 1–4, respectively. Whole brain cytosol,[15] synaptosome cytosol,[3] and synaptic vesicle[2] preparations were prepared by standard procedures. Each fraction was then subjected to heat treatment, column chromatography, and preparative gel electrophoresis to obtain a highly enriched preparation of calmodulin as described by Lin *et al*[15] Calmodulin isolated from synaptosome cytosol and synaptic vesicles behaved identically during the isolation procedure to calmodulin isolated from whole brain. Calmodulin represented approximately 0.96%, 0.71%, and 0.92% of the total protein in brain cytosol, synaptosome cytosol, and synaptic vesicle preparation, respectively. Protein was stained with Coomassie brilliant blue.

on SDS-gel electrophoresis with calmodulin isolated from whole rat brain (FIGURE 1). The isoelectric points and amino acid compositions of these proteins were essentially identical to calmodulin from whole rat brain (data not shown). These proteins were also shown to be functionally equivalent to calmodulin, since they also could modulate the activation by $Ca^{2+}$ of phosphodiesterase and protein kinase (TABLE 1). Thus, rat brain synaptosome cytosol and synaptic vesicle preparations contain a heat-stable protein that appears identical to calmodulin based on several physical and functional properties.

TABLE 1

STIMULATION OF ACTIVATOR-DEPLETED PHOSPHODIESTERASE AND SYNAPTIC VESICLE $Ca^{2+}$-PROTEIN KINASE ACTIVITY BY CALMODULIN ISOLATED FROM WHOLE RAT BRAIN, SYNAPTOSOME CYTOSOL, AND SYNAPTIC VESICLE PREPARATIONS

| Activator Added | Phosphodiesterase Activity $A_{660}$ | Phosphodiesterase Activity Activation (Fold) | Protein Kinase Activity cpm | Protein Kinase Activity Activation (Fold) |
|---|---|---|---|---|
| None | 0.17 ± 0.06 | — | 311 ± 21 | — |
| Calmodulin From: | | | | |
| Whole rat brain | 0.96 ± 0.09 | 5.65 | 1580 ± 89 | 5.08 |
| Synaptosome cytosol | 0.89 ± 0.12 | 5.24 | 1632 ± 110 | 5.25 |
| Synaptic vesicles | 0.98 ± 0.07 | 5.76 | 1575 ± 101 | 5.06 |

Phosphodiesterase was prepared and reactions were performed under the conditions of Lin *et al.*[15] The reaction mixture contained 50 μg of activator-depleted phosphodiesterase and 0.1 mM $CaCl_2$ in the presence or absence of 7 μg of calmodulin isolated from whole rat brain, synaptosome cytosol fractions, and highly enriched synaptic vesicle preparations. Cyclic AMP (2 mM) was added to initiate the reactions and phosphodiesterase activity was determined by converting the reaction product, 5-AMP, into adenosine and inorganic phosphate by a two-stage reaction procedure and then the amount of inorganic phosphate released was determined by a colorimetric assay and quantitated spectrophotometrically at 660 nM. Activity is expressed as absorbance (A) at 660 nM and represents the mean ± S.E.M. of 6 determinations. The data are representative of five separate experiments. Protein kinase activity was determined by incubating calmodulin-depleted synaptic vesicles in the absence (none) or presence of various calmodulin preparations (5 μg). Phosphorylation reactions were conducted under standard conditions in the presence of 10 μM $Ca^{2+}$ and quantitated for [$^{32}$P]phosphate incorporation into protein DPH-M (MATERIALS AND METHODS).

### *$Ca^{2+}$ and Calmodulin Stimulated Synaptic Vesicle Neurotransmitter Release and Protein Phosphorylation*

Calmodulin has been shown to mediate the effect of $Ca^{2+}$ on synaptic vesicle norepinephrine release and protein phosphorylation.[2] To determine if other neurotransmitter substances are released from synaptic vesicles under the same conditions as norepinephrine, calmodulin-depleted synaptic vesicles (treated vesicles) were incubated under standard conditions in the presence or absence of $Ca^{2+}$ and/or calmodulin and the effects of these agents on vesicle acetycholine and norepinephrine release were studied. Calmodulin and $Ca^{2+}$ caused a significant increase in the release of both acetylcholine and norepinephrine from the vesicles (TABLE 2). The concentrations of calmodulin required to produce a half maximal increase in release of acetylcholine and norepinephrine were 0.3 and 0.2 μg respectively. These results indicate that the $Ca^{2+}$ stimulated release of two major neurotransmitter substances from isolated synaptic vesicles is modulated by calmodulin. The concentration of free calcium producing a half maximal stimulation of acetylcholine release and norepinephrine release were 0.5 and 0.7 μM, respectively.

Calmodulin and $Ca^{2+}$ also caused a significant increase in the phosphorylation of vesicle proteins while simultaneously increasing neurotransmitter release from the vesicles (FIGURE 2 and TABLE 2). Protein DPH-M[2] (51,000–53,000 daltons) was one of the major phosphoproteins in the vesicle preparation and the effects of $Ca^{2+}$ and calmodulin in this vesicle protein were representative of several other vesicle proteins (FIGURE 2). The concentrations of calmodulin and free $Ca^{2+}$ required to produce a half maximal stimulation of protein DPH-M phosphorylation were 0.5 $\mu$g and 0.6 $\mu$M, respectively.

These results demonstrate that calmodulin modulates the effects of $Ca^{2+}$ on two potentially significant synaptic vesicle processes. Since synaptic vesicles are localized within the presynaptic nerve terminal, these observations indicate that calmodulin may regulate calcium stimulated protein phosphorylation and neurotransmitter release within the nerve terminal.

### *Depolarization-Dependent Neurotransmitter Release and Calmodulin and $Ca^{2+}$ Stimulated Protein Phosphorylation in Intact Synaptosomes*

Although $Ca^{2+}$ and calmodulin stimulate neurotransmitter release and protein phosphorylation in isolated vesicle preparations, it is important to

TABLE 2

EFFECTS OF CALMODULIN AND $Ca^{2+}$ ON SYNAPTIC VESICLE NEUROTRANSMITTER RELEASE AND PROTEIN PHOSPHORYLATION

| Condition | Neurotransmitter Release: Norephinephrine (ng) | Neurotransmitter Release: Acetylcholine (ng) | Protein DPH-M Phosphorylation (cpm/mg) |
|---|---|---|---|
| Plain synaptic vesicles: | | | |
| control | 10 | 15 | 621 |
| $Ca^{2+}$ | 25* | 33* | 1723* |
| calmodulin | 12 | 14 | 618 |
| $Ca^{2+}$ + calmodulin | 27* | 36* | 1801* |
| Calmodulin-depleted synaptic vesicles: | | | |
| control | 6 | 13 | 520 |
| $Ca^{2+}$ | 14 | 17 | 660 |
| calmodulin | 9 | 12 | 524 |
| $Ca^{2+}$ + calmodulin | 25* | 31* | 2600* |

Plain and calmodulin-depleted synaptic vesicles were incubated for 45 seconds under standard conditions to study neurotransmitter release (5 mg vesicle protein) and protein phosphorylation (0.5 mg vesicle protein) in the presence and/or absence of $Ca^{2+}$ (free $[Ca^{2+}]$ 10 $\mu$M) and calmodulin (5 $\mu$g). Reactions were terminated and vesicle norephinephrine and acetylcholine release and protein DPH-M phosphorylation were measured (MATERIALS AND METHODS). Norephinephrine and acetylcholine release are expressed as the amount of neurotransmitter (ng) released from the vesicles into the supernatant. Neurotransmitter data give the mean values for 5 determinations and are representative of 3 separate experiments. Protein DPH-M phosphorylation is expressed as cpm of $[^{32}P]$phosphate incorporated into protein DPH-M per mg of reaction protein. Protein DPH-M (52–53,000 daltons) is a major substrate for the $Ca^{2+}$-calmodulin activated protein kinase system in synaptic vesicles[2] and is designated on the autoradiographs shown in FIGURE 2. Phosphorylation data represent the means of 5 determinations and are representative of 4 separate experiments. The largest SEM values for release and phosphorylation were 2.6 ng and 188 cpm/mg, respectively and were omitted for clarity.

*Designates $P < 0.001$ in comparison to control values.

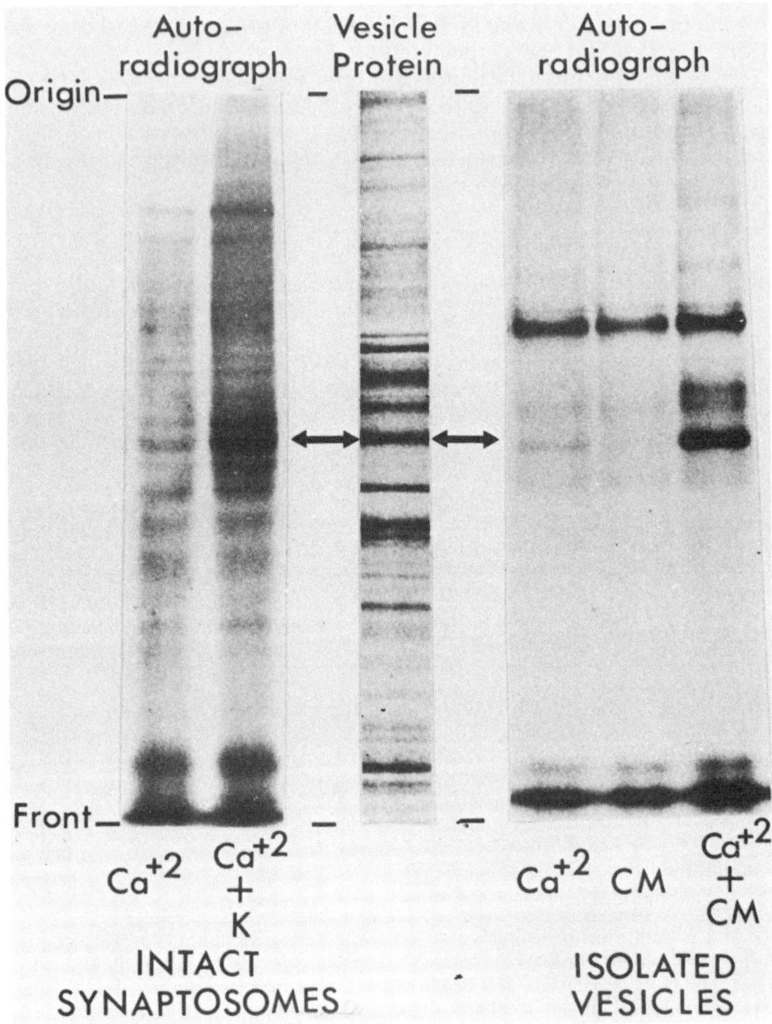


FIGURE 2. Phosphorylation of vesicle protein. Effects of calmodulin (CM) and $Ca^{2+}$ on protein phosphorylation in isolated calmodulin-depleted synaptic vesicles (right) and of depolarization dependent $Ca^{2+}$ uptake on protein phosphorylation of synaptic vesicles isolated from $^{32}P$-labeled intact synaptosomes (left). For experiments with isolated vesicles, $\gamma$-$^{32}P$-ATP was added to the reaction mixture and incubated for one minute in the presence and/or absence of $Ca^{2+}$ (free [$Ca^{2+}$] 10 $\mu$M) and CM (5 $\mu$g). For experiments with intact synaptosomes, synaptosomes were preincubated with $^{32}P$ and then incubated with $Ca^{2+}$ (1 mM) or $Ca^{2+}$ (1 mM) plus $K^{+}$ (65 mM). Following incubation, synaptic vesicles were rapidly isolated from each incubated synaptosome reaction and analyzed for vesicle protein phosphorylations (MATERIALS AND METHODS). Protein DPH-M is designated by arrows.

determine the relationship of these effects to the actions of $Ca^{2+}$ in the nerve terminal. Depolarizing conditions in the presence of $Ca^{2+}$ have been shown by several laboratories to stimulate the phosphorylation of proteins in synaptosome fractions.[2,19-21] Thus, it would be significant to demonstrate that neurotransmitter release is also simultaneously stimulated by $Ca^{2+}$ uptake. Experiments from this laboratory demonstrated that norepinephrine release and protein phosphorylation were simultaneously stimulated in intact synaptosomes by depolarizing conditions that also increased $Ca^{2+}$ uptake.[2] To further establish this observation, the release of both acetylcholine and norepinephrine were studied in intact synaptosome preparations. The results demonstrate that depolarization of the synaptosome membrane by several different conditions simultaneously stimu-

TABLE 3

EFFECTS OF HIGH $K^+$ AND VERATRIDINE (V) IN THE PRESENCE AND ABSENCE OF TETRODOTOXIN (TTX) ON INTACT SYNAPTOSOMAL $Ca^{2+}$ UPTAKE, NEUROTRANSMITTER RELEASE, AND PROTEIN PHOSPHORYLATION

| Condition | $Ca^{2+}$ Uptake (%) | Neurotransmitter Release (%) | | Protein Phosphorylation (%) |
|---|---|---|---|---|
| | | Acetylcholine | Norepinephrine | |
| Control | — | 45 | 52 | 53 |
| $Ca^{2+}$ | 51 | 51 | 56 | 58 |
| $Ca^{2+}$, $K^+$ | 100* | 100* | 100* | 100* |
| $Ca^{2+}$, $K^+$, TTX | 97* | 95* | 99* | 98* |
| $Ca^{2+}$, V | 92* | 91* | 88* | 92* |
| $Ca^{2+}$, V, TTX | 59 | 56 | 58 | 61 |
| $K^+$ | — | 54 | 59 | 61 |
| V | — | 51 | 55 | 57 |

Experimental conditions and procedures were performed as described previously (MATERIALS AND METHODS).[2] Reactions were performed with 1.2 mM $Ca^{2+}$, 60 mM $K^+$, 75 μM veratridine (V), and 0.4 μM TTX. Data are means of six determinations, representative of four experiments, and are expressed as percentage of the maximally stimulated condition. The largest SEM was 8% and were omitted for clarity. The mean values for maximal stimulation (100%) for $Ca^{2+}$ uptake, acetylcholine release, norepinephrine release, and protein phosphorylation were 11.6 nmol of $Ca^{2+}$ per mg of protein, 1.3 ng of acetylcholine released per mg protein, 0.9 ng of norepinephrine released per mg protein, and 801 cpm/500 μg protein.

*Designates $P < 0.001$ in comparison to control conditions.

lates $Ca^{2+}$ uptake, norepinephrine and acetylcholine release, and synaptic vesicle protein phosphorylation (TABLE 3). The time course of phosphorylation demonstrated that it coincided with the time course for neurotransmitter release and $Ca^{2+}$ uptake (data not shown).

Since even the best synaptosome preparations are highly contaminated with other subcellular components, it is necessary to demonstrate that these effects of $Ca^{2+}$ on protein phosphorylation in synaptosome preparations are occurring within the synaptosome and not in the contaminating membrane, mitochondria, or other components in these preparations. To address this problem, we rapidly isolated highly enriched synaptic vesicles at 4°C from intact synaptosome preparations that had been phosphorylated with $^{32}$Pi in the presence of $Ca^{2+}$, but in the presence or absence of depolarizing agents, such as high $K^+$, veratridine (V), or scorpion venom (SV).[2] The results demonstrated that depolarization of the synaptosomes caused both $Ca^{2+}$ uptake and stimulation of the phosphorylation of specific synaptic vesicle proteins within the synaptosome (FIGURE 2 and TABLE 4). The effects of depolarization on the phosphorylation of vesicle proteins was also dependent upon the presence of $Ca^{2+}$ in the media (TABLE 4).

These results indicate that in the intact synaptosome the levels of phosphorylation of several vesicle proteins are regulated by depolarization-dependent $Ca^{2+}$ uptake. The pattern of protein phosphorylation for isolated vesicles from intact synaptosomes (FIGURE 2) demonstrate several differences, but protein DPH-M is clearly phosphorylated in both preparations. Since it has been shown that the $Ca^{2+}$ stimulated phosphorylation of protein DPH-M is modulated by calmodulin in the isolated vesicle system, these findings indicate that this vesicle calmodulin-$Ca^{2+}$ phosphorylating system is activated by the depolarization-dependent entry of $Ca^{2+}$ into the presynaptic nerve terminal. Thus, the physiological $Ca^{2+}$ fluxes activated by membrane depolarization are sufficient to activate the synaptic vesicle calmodulin-protein kinase system within the intact synaptosome.

During the isolation of synaptic vesicles from the intact synaptosome, synaptosome membrane and synaptic junctional complexes were also isolated from synaptosomes incubated under various conditions in the presence of $^{32}$Pi (MATERIALS AND METHODS). The effects of depolarizing conditions on the level of phosphorylation of protein DPH-M in whole synaptosomes, synaptic membrane, synaptic junctional complexes, and synaptic vesicles were examined (TABLE 4). The results demonstrate that the depolarization-dependent stimulation of protein DPH-M phosphorylation observed in intact synaptosomes was enriched in synaptic vesicle and synaptic junctional complex preparations. The vesicle and junction fractions showed a significantly higher specific activity for protein

TABLE 4
EFFECTS OF HIGH $K^+$ AND VERATRIDINE (V) IN THE PRESENCE AND ABSENCE OF TETRODOTOXIN (TTX) ON INTACT SYNAPTOSOME $Ca^{2+}$ UPTAKE AND PROTEIN PHOSPHORYLATION

| Condition | $Ca^{2+}$ Uptake (%) | Protein Phosphorylation (cpm) | | | |
|---|---|---|---|---|---|
| | | Whole Synaptosome | Synaptic Vesicles | Synaptic Membrane | Synaptic Junctional Complex |
| Control | — | 381 | 858 | 419 | 782 |
| $Ca^{2+}$ | 51 | 432 | 881 | 443 | 751 |
| $Ca^{2+}$, $K^+$ | 100* | 721* | 2310* | 507* | 1935* |
| $Ca^{2+}$, $K^+$, TTX | 99* | 706* | 2166* | 516* | 1611* |
| $Ca^{2+}$, V | 87* | 698* | 1967* | 512* | 1431* |
| $Ca^{2+}$, V, TTX | 59 | 427 | 878 | 408 | 710 |

Synaptosomes were incubated under various conditions after preincubation with $^{32}$P.[2] Following the reactions, synaptic vesicles, synaptic membrane, and synaptic junctional complexes were isolated from each reaction mixture (MATERIALS AND METHODS). Data give the mean values of 4 determinations and are representative of three separate experiments. [$^{32}$P]phosphate incorporation is quantitated for protein DPH-M (cpm per 500 $\mu$g protein) and was representative of several other synaptosomal phosphoproteins. $Ca^{2+}$ uptake was determined as described in TABLE 4 and is presented for comparison.

$^*P < 0.001$.

TABLE 5

EFFECTS OF CALMODULIN AND $Ca^{2+}$ ON ENDOGENOUS PROTEIN PHOSPHORYLATION IN SYNAPTIC JUNCTIONAL COMPLEXES AND POSTSYNAPTIC DENSITY PREPARATIONS

| Condition | Protein Phosphorylation: Synaptic Junctional Complex (cpm/mg) | Protein Phosphorylation: Postsynaptic Density (cpm/mg) |
|---|---|---|
| Control | 753 | 368 |
| $Ca^{2+}$ | 881 | 413 |
| Calmodulin | 756 | 373 |
| $Ca^{2+}$ + Calmodulin | 4,586* | 1,063* |

Reactions were performed as described (MATERIALS AND METHODS). The data represent the mean of ten determinations and are representative of four separate experiments. The largest SEM was 161 cpm/mg. Protein phosphorylation is exposed as [$^{32}$P]phosphate incorporated into protein DPH-M (51–53,000 daltons) which is a major representative phosphoprotein in these preparations (FIGURE 2).

*$P < 0.001$ in comparison to control conditions.

phosphorylation than either the whole synaptosome preparation or the plain synaptic membrane.

### *$Ca^{2+}$ and Calmodulin Stimulated Endogenous Protein Phosphorylation in Postsynaptic Density and Synaptic Junctional Complex Preparations*

Previous work in this laboratory has demonstrated that $Ca^{2+}$ and calmodulin stimulate the endogenous phosphorylation of several proteins in postsynaptic density (PSD) preparations.[5] Proteins with identical molecular weights to vesicle proteins DPH-L and DPH-M were prominent phosphoproteins in the PSD fractions. The effect of $Ca^{2+}$ and calmodulin on protein DPH-M is shown in TABLE 5. Independent studies by H. Mahler[22] and P. Siekevitz[23] have also shown that $Ca^{2+}$ and calmodulin stimulate the phosphorylation of several PSD proteins. These results further suggest that calmodulin may play an important role in modulating some of calcium's effects on postsynaptic activity.

It would be important to determine, if the high level of [$^{32}$P]phosphate incorporation into synaptic junctional complexes in intact synaptosome preparations was the result of phosphorylation of exclusively the PSD proteins or a combination of PSD and presynaptic junction protein phosphorylation. To test this possibility, the endogenous phosphorylation of synaptic junctional complexes was investigated. The specific activity (cpm/mg) of calmodulin and $Ca^{2+}$ stimulated protein phosphorylation was over 4-fold higher in the whole synaptic junctional complex in comparison to PSD preparations (TABLE 5). These results might suggest that presynaptic elements in the junction may be phosphorylating in addition to the PSD. Mahler's group has performed several elegant experiments that provide evidence that isolated presynaptic junctional membrane demonstrates $Ca^{2+}$ and calmodulin stimulated protein phosphorylation.[23]

These observations suggest that $Ca^{2+}$ and calmodulin-stimulated protein phosphorylation may play a role in regulating interactions between presynaptic

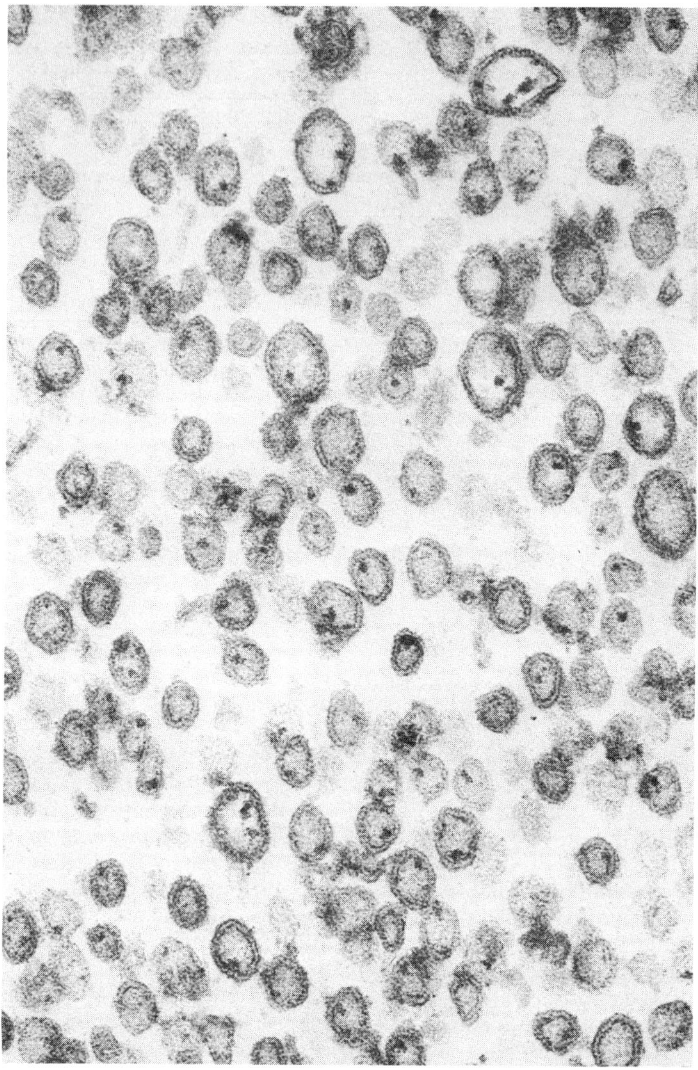

FIGURE 3. Morphology of synaptic vesicles. Synaptic vesicles were incubated under control conditions as described in TABLE 2, pelleted by centrifugation, subjected to fixation, dehydration, and embedding for thin sectioning, and examined by electronmicroscopy (MATERIALS AND METHODS). This electromicrograph is representative of the control condition (×200,000).

and postsynaptic events. The transynaptic location of this important calmodulin regulated system suggests an important function for some of these synaptic phosphoproteins in modulating synaptic transmission. Because of the similarities between synaptic vesicle and synaptic junctional complex calmodulin and $Ca^{2+}$ stimulated protein phosphorylation, experiments were initiated in this laboratory to study the possible role of this system in mediating synaptic vesicle and synaptic membrane interactions.[24]

### *Calmodulin and $Ca^{2+}$ Modulated Synaptic Vesicle and Synaptic Membrane Interactions*

It was observed in this laboratory that suspensions of calmodulin-depleted synaptic vesicle preparations became turbid when incubated in the presence of $Ca^{2+}$ and calmodulin, under conditions that simultaneously stimulated vesicle protein phosphorylation and neurotransmitter release.[24] $Ca^{2+}$ or calmodulin alone were much less effective in causing the turbidity changes than the two agents together. These results suggested that synaptic vesicles might be aggregating or undergoing morphological changes in the presence of $Ca^{2+}$ and calmodulin. To test this possibility, synaptic vesicles were incubated in the presence and/or absence of calmodulin and examined by electron microscopy (MATERIALS AND METHODS).

Thin sectioned specimens of control vesicle preparations (MATERIALS AND METHODS) revealed a vesicle population composed of free and grouped vesicles with a mean vesicle diameter of 410 Å. The number of free vesicles in the control preparation (FIGURE 3) was determined to be 51.3% (TABLE 6). The vesicles

TABLE 6

EFFECTS OF CALMODULIN AND $Ca^{2+}$ ON SYNAPTIC VESICLE INTERACTIONS

| Condition | Free Vesicles (%) |
|---|---|
| Plain Synaptic Vesicles | |
| Control | 51.3 ± 3.1 |
| $Ca^{2+}$ | 11.6* ± 1.3 |
| Calmodulin | 52.8 ± 2.6 |
| $Ca^{2+}$ + Calmodulin | 10.7* ± 0.9 |
| Calmodulin-Depleted Synaptic Vesicles | |
| Control | 48.6 ± 1.8 |
| $Ca^{2+}$ | 31.2* ± 3.1 |
| Calmodulin | 46.5 ± 2.7 |
| $Ca^{2+}$ + Calmodulin | 13.4*† ± 0.8 |

Plain and calmodulin-depleted synaptic vesicles were incubated for one minute under standard conditions in the presence or absence of $Ca^{2+}$ and/or calmodulin, isolated by centrifugation (100,000 × g) and prepared for electron microscopy (MATERIALS AND METHODS). Free vesicles were defined as vesicles that did not touch other vesicles. Free vesicles are expressed as the percent of free vesicles in the preparation. The percent of free vesicles were determined in 500 representative electron micrographs (each containing approximately 150 vesicle profiles) and the data give the mean value and SEM of free vesicles from these 500 determinations. The results shown were representative of 3 separate experiments. The percent free vesicles in the control and stimulated conditions was inversely affected by the force of centrifugation. However, the effects of $Ca^{2+}$ and calmodulin on percent free vesicles were still statistically significant at 4 other speeds of centrifugation. Electron micrographs were randomly selected from the top, middle, and bottom of the vesicle pellet (MATERIALS AND METHODS). The effects of $Ca^{2+}$ and calmodulin on vesicle aggregation observed in thin sections were also seen in negatively stained and freeze dried preparations.

*$P < 0.001$ in comparison to control conditions.

†$P < 0.001$ in comparison to $Ca^{2+}$ conditions.

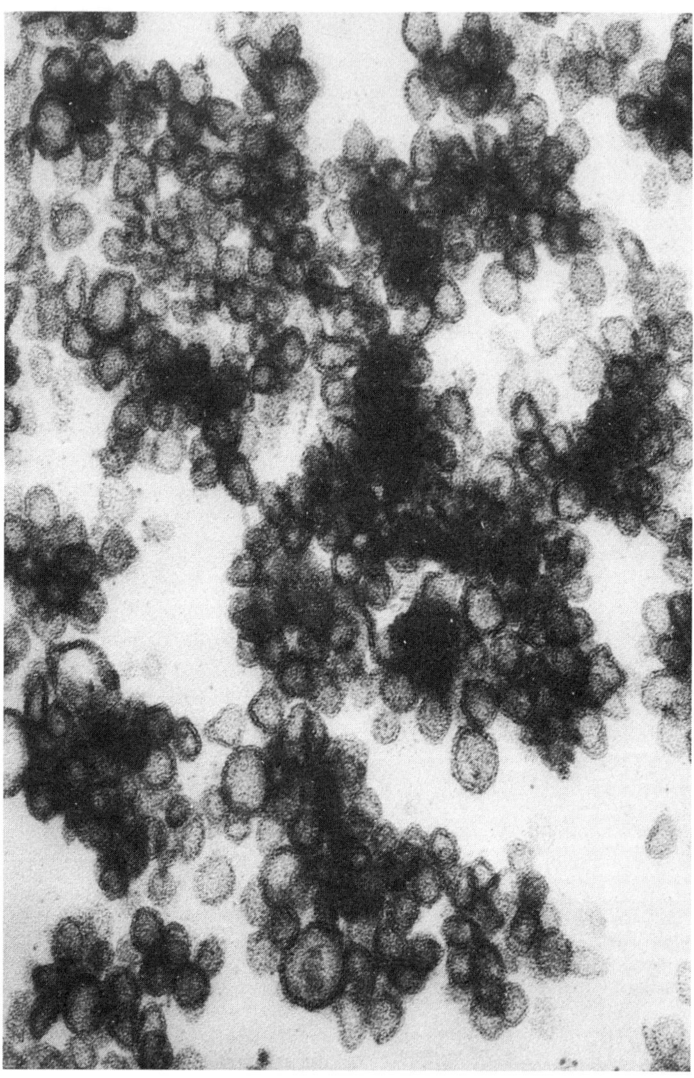

FIGURE 4. Effects of $Ca^{2+}$ and calmodulin on synaptic vesicle morphology. Synaptic vesicles were incubated in the presence of $Ca^{2+}$ and calmodulin as described in TABLE 2 and prepared for examination for electron microscopy (×200,000). This electron micrograph is representative of the effects of $Ca^{2+}$ on synaptic vesicle morphology. Vesicle aggregation and distortion are evident. The effects of $Ca^{2+}$ and calmodulin presented in TABLE 6 for thin sectioned specimens were also seen in negatively stained and freeze-dried preparations (data not shown).

appeared round in shape (FIGURE 3) and the frequency distribution of vesicle diameters revealed a mono-moidle population.

$Ca^{2+}$ and calmodulin produced a dramatic aggregation of synaptic vesicles (FIGURE 4) with a marked decrease in the percent of free vesicles in the preparation (TABLE 6). $Ca^{2+}$ alone had some effect on percent free vesicles, but the effect of $Ca^{2+}$ plus calmodulin produced a statistically significant difference from $Ca^{2+}$ alone (TABLE 6). Calmodulin alone had no significant effect on vesicle interactions. Vesicles were occasionally observed to fuse with each other, form linear membrane structures and have a decreased diameter in the presence of $Ca^{2+}$ and calmodulin. Examination of negatively stained, fixed and unfixed vesicle preparations incubated in the presence or absence of $Ca^{2+}$ and/or calmodulin also demonstrated that $Ca^{2+}$ and calmodulin caused a marked aggregation of synaptic vesicles into tight clusters with apparent vesicle fusion, rupture, and decrease in vesicle diameter. Both the fixed and unfixed vesicles preparations gave similar results after negative staining. Preliminary results with freeze-dried samples also demonstrated that $Ca^{2+}$ and calmodulin stimulated vesicle aggregation.

Although vesicle-vesicle interactions have been observed in intact nerve terminals, synaptic vesicle, and synaptic membrane interactions have been implicated in neurotransmitter release through the process of exocytosis.[25] If $Ca^{2+}$ and calmodulin altered the vesicle membrane causing vesicles to interact, it might be expected that vesicles would also interact with synaptic plasma membrane. To test this possibility, synaptic vesicles and synaptic membrane were incubated under standard conditions in the presence and/or absence of calmodulin. Synaptic membrane was then isolated by centrifugation and examined by electron microscopy (MATERIALS AND METHODS). $Ca^{2+}$ and calmodulin caused a significant increase in the number of synaptic vesicles attached to the synaptic membrane (TABLE 7). $Ca^{2+}$ or calmodulin alone were not as effective as these agents together in causing synaptic membrane and synaptic vesicle interactions. Preliminary results in this laboratory have indicated that the $Ca^{2+}$ and calmodulin stimulated attachment of synaptic vesicles to synaptic membranes is significantly enriched at the synaptic junctional complex.

These results suggest that calmodulin modulates the effects of $Ca^{2+}$ on synaptic membrane and synaptic vesicles interactions. The functional significance of these interactions remains to be elucidated. However, the requirement of calmodulin in modulating $Ca^{2+}$-dependent membrane interactions suggests that these effects may play an important role in vesicle function and possibly in initiating the $Ca^{2+}$ dependent process of exocytosis or other aspects of vesicle and membrane interactions in the nerve terminal.

### *Calmodulin's Role in Modulating the Effects of $Ca^{2+}$ on Synaptic Function*

The research from this laboratory has provided evidence for a role of calmodulin in modulating the effects of $Ca^{2+}$ on synaptic function.[2–5] Calmodulin was isolated from synaptosomal soluble fractions and from highly enriched preparations of synaptic vesicles. These results indicate that calmodulin is present in the nerve terminal and is bound to synaptic vesicles at physiological intracellular conditions. Thus, vesicle-bound calmodulin would be located in close proximity to the presynaptic nerve terminal. Because of the rapid time constants for synaptic transmission, it has been estimated that the influx of $Ca^{2+}$ during the depolarization of the presynaptic nerve terminal could only affect

TABLE 7

EFFECTS OF CALMODULIN AND $Ca^{2+}$ ON SYNAPTIC VESICLE AND SYNAPTIC MEMBRANE INTERACTIONS

| Conditions | Synaptic Vesicle and Synaptic Membrane Interactions (Number Vesicles/$\mu$ Membrane) |
|---|---|
| Plain Synaptic Vesicles | |
| Control | 3.12 ± 0.41 |
| $Ca^{2+}$ | 14.33* ± 0.98 |
| Calmodulin | 4.21 ± 0.29 |
| $Ca^{2+}$ + Calmodulin | 15.26* ± 1.01 |
| Calmodulin-Depleted Synaptic Vesicles | |
| Control | 2.77 ± 0.25 |
| $Ca^{2+}$ | 5.16* ± 0.41 |
| Calmodulin | 3.16 ± 0.12 |
| $Ca^{2+}$ + Calmodulin | 10.21*† ± 0.83 |

Plain or calmodulin-depleted synaptic vesicles were incubated with synaptic membrane for 1 minute under standard conditions in the presence or absence of $Ca^{2+}$ and/or calmodulin. Following incubation, synaptic membranes were immediately isolated from each reaction mixture by centrifugation at 25,000 × g for 10 minutes at 4°C. The isolated membrane pellets were then fixed and prepared for electron microscopy (MATERIALS AND METHODS). Random electron micrography were taken from the top to the bottom of the pellet and the number of synaptic vesicles on each membrane fragment was quantitated. The number of vesicles attached to synaptic membrane is expressed as vesicles per micron ($\mu$) of membrane. The data give the mean values of 600 determinations and are representative of 3 separate experiments.

*$P < 0.001$ in comparison to control conditions.

†$P < 0.001$ in comparison to $Ca^{2+}$ conditions.

structures or enzyme systems in close proximation to the presynaptic membrane, approximately within 200 Å of the presynaptic membrane.[26] The first row of synaptic vesicles in the nerve terminal are directly adjacent to the synaptic junction and are within 200 Å from the presynaptic membrane. Thus, vesicle-bound calmodulin is close enough to be affected in the time range of synaptic transmission by inward diffusion of $Ca^{2+}$ ions during depolarization of the nerve terminal.

Although these results indicate that calmodulin is present within the nerve terminal and thus could modulate some of the effects of $Ca^{2+}$ on synaptic function, it is also necessary to demonstrate that calmodulin mediates the effects of $Ca^{2+}$ on various synaptic enzyme systems and physiological processes such as neurotransmitter release. The results in this paper demonstrate and expand the previous work from my laboratory indicating that several $Ca^{2+}$ stimulated processes associated with synaptic vesicles and synaptic membrane fractions are modulated by calmodulin: 1) $Ca^{2+}$ and calmodulin activated a vesicle-bound protein kinase system that phosphorylates several synaptic vesicle proteins; 2) calmodulin modulates the $Ca^{2+}$ stimulated release of norepinephrine and acetylcholine from isolated synaptic vesicles; 3) calmodulin mediates the $Ca^{2+}$ activation of an endogenous protein kinase system in highly enriched synaptic junctional complex and postsynaptic density fractions and 4) calmodulin modulates the effects of $Ca^{2+}$ on synaptic vesicles and synaptic membrane interactions.

It is also important to demonstrate that the effects of calcium and calmodulin on isolated vesicle and membrane preparations are also occurring in the intact nerve terminal under conditions that stimulate neurotransmitter release. The results in this paper and from previous studies[2] demonstrate that the $Ca^{2+}$ and calmodulin modulated protein kinase system in synaptic vesicles and synaptic junctional complexes is activated by depolarization-dependent $Ca^{2+}$ uptake in intact synaptosomes under conditions that simultaneously stimulate the release of norepinephrine and acetylcholine from the nerve terminal. The activation of this vesicle kinase system within intact synaptosomes by physiological $Ca^{2+}$ fluxes across the synaptosome membrane provides evidence that calmodulin is involved in modulating the effects of $Ca^{2+}$ during the depolarization-dependent entry of calcium into the nerve terminal. When $Ca^{2+}$ and calmodulin activate the vesicle-bound protein kinase system in the isolated vesicle system, these agents simultaneously stimulate release of neurotransmitter substances from the vesicles and activate synaptic vesicle and synaptic membrane interactions. Thus, it is

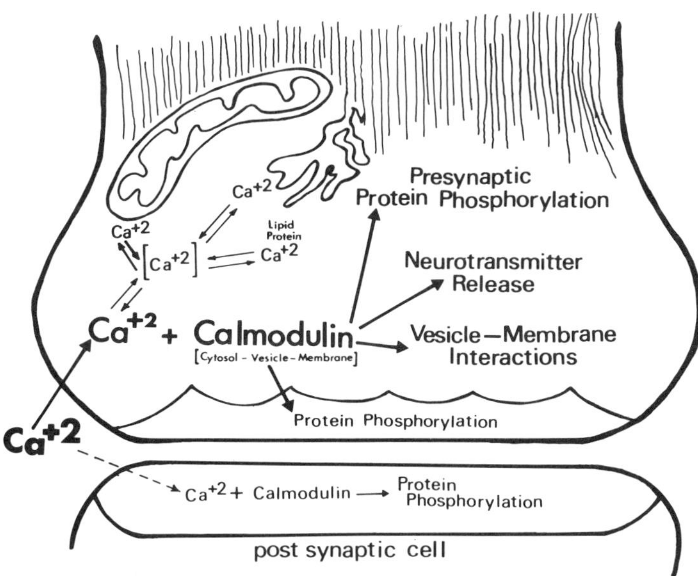


FIGURE 5. Schematic model of the possible role of calmodulin in mediating several of the effects of $Ca^{2+}$ on synaptic function. Calmodulin bound to synaptic vesicles serves as a $Ca^{2+}$ receptor in close proximity to the synaptic junction. Calmodulin may also be present in the synaptosome cytosol and synaptic membrane. Following the depolarization-dependent entry of $Ca^{2+}$ into the nerve terminal, $Ca^{2+}$ is immediately bound to calmodulin near the membrane. The binding of $Ca^{2+}$ to calmodulin in the presynaptic terminal could then initiate several processes: presynaptic protein phosphorylation, neurotransmitter release, vesicle membrane interactions, and other $Ca^{2+}$ regulated synaptic functions. This model serves as framework for initiating further experimentation to more clearly delineate the full functional significance of calmodulin in neurotransmitter release and synaptic function. The transsynaptic localization of the $Ca^{2+}$ and calmodulin protein kinase system suggests that this $Ca^{2+}$ regulated biochemical process may play an important role in modulating synaptic function.

possible that calmodulin is also modulating the effects of $Ca^{2+}$ on neurotransmitter release and vesicle-membrane interactions in the intact synaptosome during the depolarization-dependent uptake of $Ca^{2+}$ into the intact nerve terminal.

The model shown in FIGURE 5 provides a schematic presentation of the proposed role of calmodulin in synaptic function. Calmodulin may play an important role in modulating the effects of $Ca^{2+}$ on synaptic protein phosphorylation, neurotransmitter release, and vesicle-membrane interactions. The transsynaptic localization of the calmodulin-protein kinase system suggests that this biochemical process plays an important role in synaptic function. Studies with the anticonvulsant phenytoin also suggest that this phosphorylation system is involved in synaptic activity.[3] However, many other $Ca^{2+}$-regulated synaptic processes are probably modulated by calmodulin. Ongoing experiments in this laboratory are attempting to further establish the role of calmodulin in neurotransmitter release and synaptic regulation. It is hoped that these studies will provide a greater insight into the molecular mechanisms underlying the effects of $Ca^{2+}$ on synaptic function and stimulate other laboratories to investigate the role of calmodulin in synaptic modulation.

## ACKNOWLEDGMENTS

The suggestions and discussions of G. H. Glaser, C. Cotman, S. Puszkin, and A. Kleinhaus were greatly appreciated.

## REFERENCES

1. CHEUNG, W. Y. 1980. Science **207:** 19–27.
2. DELORENZO, R. J., S. D. FREEDMAN, W. B. YOHE & S. C. MAURER. 1979. Proc. Natl. Acad. Sci. USA **76:** 1838–1842.
3. DELORENZO, R. J. 1980. *In* Antiepileptic Drugs Mechanisms of Action: Advances in Neurology. G. H. Glaser, J. K. Penry & D. M. Woodbury, Eds. Vol. **27:** 399–414. Raven Press, Inc. New York, N.Y.
4. DELORENZO, R. J. & S. D. FREEDMAN. 1978. Biochem. Biophys. Res. Commun. **80:** 183-192.
5. DELORENZO, R. J. 1980. Trans. Am. Soc. Neurchem. **11:** 81.
6. COTMAN, C. W. & D. TAYLOR. 1972. J. Cell Biol. **55:** 696-711.
7. DELORENZO, R. J. 1977. Brain Res. **134:** 125–138.
8. WHITTAKER, V. P. 1969. *In* Handbook of Neurochemistry. A. Lajtha, Ed: 327–364. Plenum Publishing Corp. New York, N.Y.
9. DELORENZO, R. J. & S. D. FREEDMAN. 1977. Biochem. Biophys. Res. Commun. **77:** 1036–1043.
10. COTMAN, C. W. & D. A. FLANSBURG. 1970. Brain Res. **22:** 152–156.
11. BRACKER, C. E., J. RUIZ-HERRERA, S. S. BARTNICKI-GARCIA. 1976. Proc. Natl. Acad. Sci. USA **73:** 4570–4574.
12. MATTHEWS, E. K. & J. J. NORDMANN. 1975. Mol. Pharm. **12:** 778–788.
13. WEIBEL, E. R. & H. ELIAS. 1967. Quantitative Methods in Morphology. Springer-Verlag. Berlin, Germany.
14. WEIBEL, E. R., S. G. KISTLER & W. F. SCHERLE. 1966. J. Cell Biol. **30:** 23–37.
15. LIN, Y. M., Y. P. LIN & W. Y. CHEUNG. 1974. J. Biol. Chem. **249:** 4180–4185.
16. MOORE, S. & W. H. STEIN. 1963. Methods Enzymol. **6:** 819–831.
17. CATSIMPOLLAS, N. 1968. Anal. Biochem. **26:** 480–482.
18. COTMAN, C. W., G. BANKER, L. CHURCHILL & D. TAYLOR. 1974. J. Cell Biol. **63:** 441–455.
19. BERMAN, R. F., J. P. HULLIHAN, W. J. KINNIER & J. E. WILSON. 1980. J. Neurochem. **34:** 431–437.

20. KRUEGER, B. K., J. FORN & P. GREENGARD. 1977. J. Biol. Chem. **252:** 2764–2773.
21. MICHAELSON, D. M. & S. AVISSAR. 1979. J. Biol. Chem. **254:** 12452–12546.
22. MAHLER, H. R. & R. SORENSEN. 1980. Trans. Am. Soc. Neurochem. **11:** 237.
23. GRAB, D., R. CARLIN & P. SIEKEVITZ. 1980. This volume.
24. DELORENZO, R. J. 1979. Trans. Int. Soc. Neurochem. **7:** 68.
25. HEUSER, J. D. & T. S. REESE. 1973. J. Cell Biol. **47:** 315–344.
26. LLINAS, R., I. Z. STEINBERG & K. WALTON. 1976. Proc. Natl. Acad. Sci. USA **73:** 2918–2922.

## DISCUSSION OF THE PAPER

QUESTION: Do you know if the calcium-calmodulin dependent aggregation of vesicles is dependent on ATP?

DR. R. J. DELORENZO (*Yale University, New Haven, CT*): Yes, it is dependent on ATP in our preparations.

DR. B. H. SATIR: Why don't you ever see these vesicles aggregated in normal synapses?

DR. DELORENZO: Fusion or aggregation of vesicles has been cited in intact secretory terminals in the literature (C. Kreutz. 1978. J. Biol. Chem. **253**:2858 and D. Lagunoff. 1973. J. Cell Biol. **57**:252). Our results suggest that calmodulin may mediate both vesicle-vesicle and vesicle-membrane interactions.

DR. SCHUBERT: Did you do controls with other acidic, calcium-modulated proteins?

DR. DELORENZO: Only with troponin C and it did not mimic calmodulin.

# ISOLATION OF THE STRUCTURAL GENE FOR CALMODULIN*

R. P. Munjaal, J. R. Dedman, and A. R. Means

*Department of Cell Biology*
*Baylor College of Medicine*
*Houston, Texas 77030*

Calmodulin levels are presently measured either by the ability of a cell extract to stimulate a calmodulin-dependent enzyme or by radioimmunoassay. Differences have been revealed in the values obtained by the two types of methods in that the enzymatic assays frequently underestimate total calmodulin levels.[1] However, all assays seem to yield similar qualitative values. As an example Watterson *et al.*,[2] using a densitometric procedure, and LaPorte *et al.*,[3] using a phosphodiesterase assay, have reported that transformation of chick embryo fibroblasts by Rous sarcoma virus results in a two-fold elevation in the specific activity of calmodulin. Chafouleas *et al.*[4] have extended these studies to other cell systems and shown that increased calmodulin levels are a general response of mammalian cells transformed by oncogenic viruses. Moreover, the elevated levels of calmodulin were specific for that protein and shown to represent an increase in the number of molecules per cell. The regulation was at the level of synthesis and not due to alterations in protein degradation. These data suggest that viral transformation might result in an increased number of calmodulin mRNA molecules and reveal that the cellular regulation of calmodulin may occur at a number of levels. These levels include the synthesis, processing, translation, and degradation of the mRNA. For these reasons we have partially purified calmodulin mRNA and used this material to isolate a structural gene by molecular cloning techniques.

## Methods and Results

### *Isolation of Calmodulin mRNA*

In most mammalian tissues the concentration of calmodulin is much less than 1% of the total protein. Should the level of calmodulin mRNA exist at a similar ratio to total messenger RNA its purification would be a formidable project. However, the electroplax of the eel, *Electrophorus electricus*, has been shown to contain exceptionally high levels of calmodulin by Childers & Siegel.[5] We have determined, by radioimmunoassay, this tissue to contain 1.0–1.5 grams calmodulin per kilogram wet-weight, which represents approximately 5% of the total protein. It thus appeared that the eel electric organ represented a good source for the isolation of calmodulin mRNA. Total undegraded, electroplax RNA was extracted using a phenol/SDS/chloroform procedure[6] and the poly(A)-containing RNA isolated using oligo(dT)-cellulose chromatography.[7] This RNA fraction was then translated in a cell-free system derived from rabbit reticulocyte lysate. Chafouleas *et al.*[1] demonstrated that the [$^{35}$S]methionine-labeled calmodulin

*See Acknowledgement Section for support.



0077-8923/80/0356-0110 $01.75/0 © 1980, NYAS

synthesized *in vitro* was specifically precipitated using purified sheep anti-calmodulin. The immunoprecipitate represented 10% of the total poly(A) RNA directed protein synthesis. Furthermore, the cell-free synthesized protein migrated as a single band on SDS-polyacrylamide gels, coincident with authentic calmodulin. This observation suggests that there is no precursor form of the protein.

The poly(A) RNA was further purified on 13–25% sucrose gradients in the presence of SDS, EDTA, and sodium acetate (pH 7.0).[8] RNA, which directed calmodulin synthesis, sedimented at an S value of 9–10. The RNA from this fraction, when subjected to agarose-gel electophoresis, migrated as a single band. Approximately 36% of the protein synthesized in a cell-free translation system containing this RNA was calmodulin. This purified calmodulin mRNA fraction was used to direct the synthesis of [$^3$H]cDNA in the presence of reverse transcriptase as shown in FIGURE 1. Following alkaline hydrolysis this radiolabeled single-stranded cDNA was hybridized to an excess of the mRNA fraction. Analysis of the data obtained from the hybridization reaction revealed that approximately 39% of the mRNA fraction was a single species. The remainder of the mRNA was comprised of multiple mRNA species, each represented by only a few copies. The hybridization data agreed with the information obtained from cell-free translation and suggested that 39% of the RNA fraction was calmodulin mRNA.

### *Preparation and Molecular Cloning of Calmodulin ds-cDNA*

Reverse transcriptase was again used to direct synthesis of a $^{32}$P-labeled double-stranded (ds) cDNA from the single-stranded material (FIGURE 1). Following treatment with $S_1$ nuclease, the ds-cDNA was analyzed by gel electrophoresis and radioautography.[8] The major species present (75%) was a ds-cDNA 750 nucleotides in length. A partial restriction map was obtained by subjecting the ds-cDNA to a variety of restriction endonucleases. Treated samples were again analyzed by gel electrophoresis and radioautography. Several enzymes, including *Hae*III and*Hinf*I, cut the ds-cDNA into two or more fragments. However, *Pst*I and *Hha*I did not cut the ds-cDNA. Therefore, 9 cytosine residues were added to the 3′-ends of $S_1$-treated ds-cDNA by terminal deoxynucleotidyl transferase (FIGURE 2). The vector pBR322 was cleaved with *Pst*I, which only cuts this DNA once, within the region of the ampicillin-resistant gene.[9] The linearized plasmid was tailed with 11 guanosine residues again using terminal transferase. The two end-labeled species were mixed in a 1:1 molar ratio and chimeric plasmids were formed by annealing at 70°C, heating at 42°C for 2 hr and then slowly cooled to room temperature overnight. Ligation of the hybrid molecules as well as repair of any gaps that might occur due to the difference in length of the poly(dC) and poly(dG) regions occurred *in vivo* after the subsequent transformation of bacteria.[10] These procedures reconstitute a *Pst*I site on either side of the foreign DNA insert.[8]

The recombinant plasmid mixture was added to two volumes of $CaCl_2$-treated recipient cells which were *E. coli* K strain RRI.[11] After incubation at 4°C for 60 min, aliquots were spread onto L-agar plates containing tetracycline. Tetracycline-resistant clones were selected, transferred to nitrocellulose filters and lysed *in situ* by the method of Grunstein and Hogness.[12] Filters were treated with Denhardt's solution and the bound DNA was hybridized in the presence of

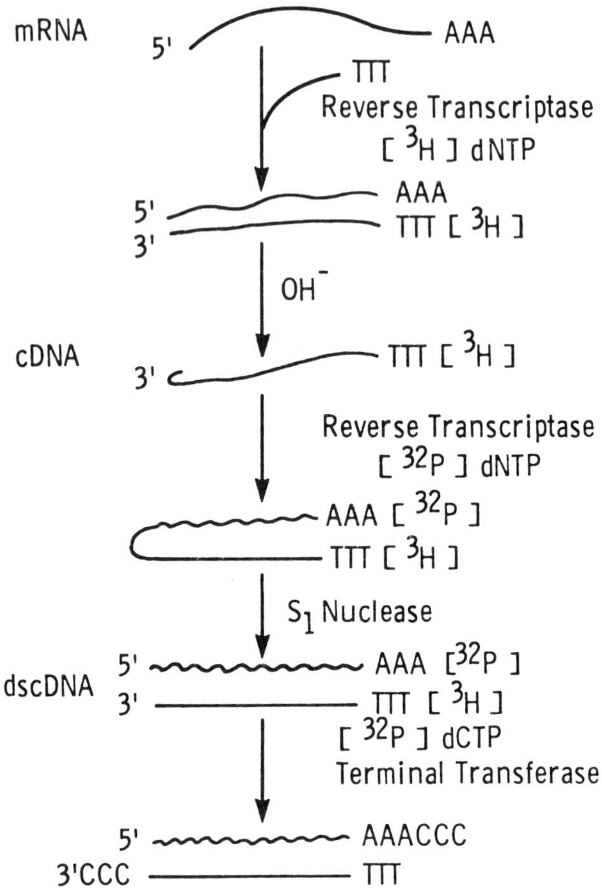


FIGURE 1. Schematic representation of the synthesis of double-stranded calmodulin cDNA from an enriched mRNA preparation.

[$^{32}$P]cDNA synthesized from the partially purified calmodulin mRNA.[9] A replica of each filter was kept by incubating the original tetracycline-resistant plate at 37°C for 48 hr. This technique was described by Stein *et al.*[8] and saves the time required for the tedious and time-consuming transfer of individual colonies to the filters. Twenty-seven of the tetracyline-resistant clones also contained inserted DNA complementary to calmodulin-enriched mRNA as determined by hybridization analysis.

Hybridization-positive colonies were amplified in RRI and recombinant plasmid DNA was purified by the procedures of Katz *et al.*[13] The plasmid DNA was digested with *Pst*I and the fragments were separated by electrophoresis on polyacrylamide slab gels. In all cases only two fragments were observed, an invariant large fragment representing the pBR322 DNA, and a smaller fragment of variable size representing the DNA insert. Of the original 27 positive colonies,

7 were further analyzed and shown to contain inserts between 200 and 400 nucleotides in length and were analyzed further to confirm the presence of calmodulin DNA sequences.

### *Positive Identification of the DNA Insert of pCM9 as Calmodulin*

Many investigators have utilized hybrid-arrested translation to determine the presence of a particular gene sequence. The principle is that the plasmid DNA insert will hybridize with a specific mRNA and prevent its translation. Therefore, one would observe a decrease in the intensity of the newly synthesized

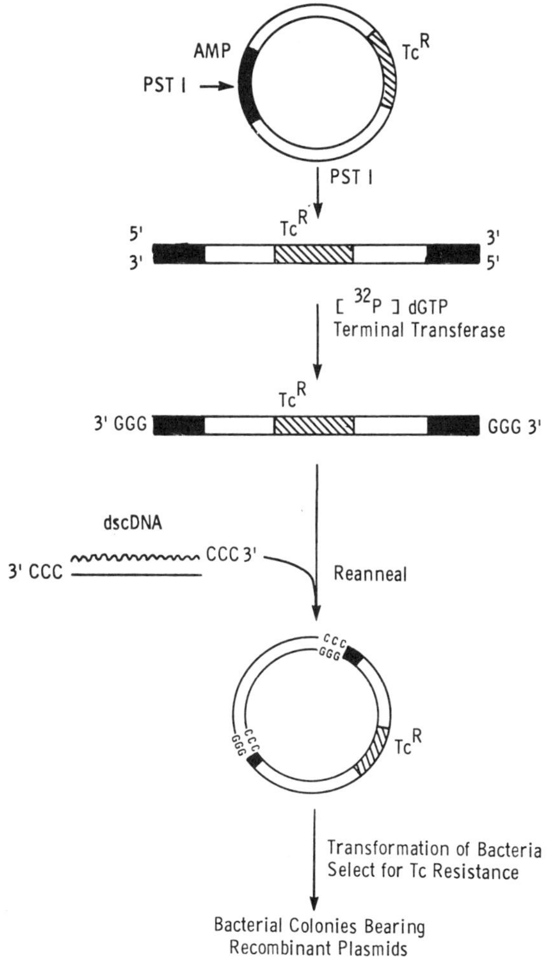


FIGURE 2. Diagram of the strategy used in cloning of the calmodulin structural gene in bacterial plasmid pBR322.

protein band when translation products are analyzed by gel electrophoresis and fluorography. The difficulty with this procedure is that the results are seldom all-or-none. Therefore, we chose to use a positive translation system where the assay is synthesis of calmodulin from purified calmodulin mRNA. The procedure was developed by Noyes and Stark[14] who coupled single-stranded DNA to finely divided cellulose that had been derivitized with diazobenzyloxymethyl. This material was then used as an insoluble support to isolate mRNA by hybridization. For our experiments plasmid calmodulin clone pCM9 was digested with *HhaI* which produces multiple fragments of the parent plasmid pBR322 but does not cut the ds-cDNA insert. The digested DNA was then covalently coupled to the DBM-cellulose and treated with Denhardt's solution containing 2% glycine in order to hydrolyze any remaining diazo groups and block any sites on the cellulose that might nonspecifically bind RNA. Poly(A) RNA from eel electoplax was hybridized to the DNA-cellulose in 50% formamide at 37°C for 2 hr. Following several washings with hybridization buffer (minus RNA), mRNA was eluted in formamide at 80°C. The suspension was centrifuged to remove the cellulose and RNA was precipitated from the supernatant fluid following addition of an equal volume of water and adjusting the salt concentration to 0.5 M, with 2.5 vol of ethanol. RNA was collected by centrifugation and dissolved in water. About 0.5 $\mu$g was translated using a rabbit reticulocyte lysate cell-free translation system. Following translation in the presence of [$^{35}$S]met the products were analyzed by gel electrophoresis. Authentic calmodulin, also labeled with $^{35}$S, was added as a marker. All samples were electophoresed in the presence and absence of $Ca^{2+}$ since this is known to change the electrophoretic mobility of calmodulin even in the presence of SDS. Radioactivity was detected by fluorography and the only band detected had the same $R_f \pm Ca^{2+}$ as did the calmodulin standard. These data suggested that pCM9 contained a DNA insert with sequences complementary to calmodulin mRNA.

The insert of pCM9 was excised with *Pst*I and recovered following electrophoresis on preparative slab gels. The *Pst*I fragment was 350 nucleotide base pairs long. Calmodulin is comprised of 148 amino acids[15] that would require 444

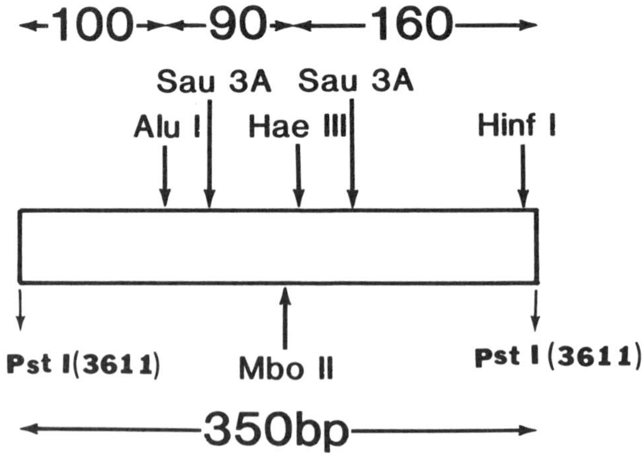


FIGURE 3. A partial restriction endonuclease map of the 350 bp cDNA insert in pCM9.

| 92 GTT | GAC | AAG | GAC | GGT | AAT | GGC | TAC | 100 ATC | AGT | GCA |
|---|---|---|---|---|---|---|---|---|---|---|
| PHE | ASP | LYS | ASP | GLY | ASN | GLY | TYR | ILEU | SER | ALA |
| GCC | GAG | TTG | CGA | CAT | GTC | ATG | 110 ACT | AAC | TTG | GCC |
| ALA | GLU | LEU | ARG | HIS | VAL | METH | THR | ASN | LEU | GLY |

FIGURE 4. Partial DNA sequence of pCM9. The sequence was determined by the methods of Maxam and Gilbert.[19] Note that the sequence is taken from the complementary sequence in FIGURE 3. The sequence was read from the *Hae*III site toward the 3′-terminus. The first codon that could be unambiguously identified from the sequence gel was GTT, which corresponded to amino acid 92 or Phe.[16,17]

nucleotides to be present in the coding sequence of the structural gene. Therefore, pCM9 can maximally contain 80% of the calmodulin cDNA. Since most eukaryotic structural genes contain a noncoding region on the 3′-end, the 80% value is likely to be an overestimation. The *Pst*I fragment was digested with a variety of restriction endonucleases to generate the partial map shown in FIGURE 3. The map shows the inserted calmodulin DNA fragment on an expanded scale. The approximate cleavage sites for the restriction endonucleases *Sau*IIIA, *Mbo*II, *Hae*III, *Alu*I, and *Hin*fI are indicated. Calmodulin DNA was not cleaved by *Pst*I, *Hha*I, or *Hpa*II.

In order to confirm the identity of the cloned DNA of pCM9 as calmodulin, a partial DNA sequence was determined. The complete amino acid sequence of both rat[16] and bovine[17] calmodulins have been published. Recombinant plasmid pCM9 was cleaved with *Pst*I and labeled at the 5′-ends with [$^{32}$P]ATP using polynucleotide kinase after first removing unlabeled phosphate groups with calf intestine alkaline phosphatase.[18,19] The 350 base pair (bp) *Pst*I fragment was isolated on a 8% polyacrylamide slab gel and eluted from the gel with 90% efficiency.[8] This purified fragment was incubated with *Hae*III which cleaves it into two fragments 170 and 180 bp in length. DNA was then sequenced as described by Maxam and Gilbert[19] following separation of the fragments by gel electophoresis. The DNA sequence of a 69 bp fragment codes for a portion of the calmodulin protein corresponding to amino acid residues 96 to 118 as shown in FIGURE 4. This sequence match is well beyond random probability and proves that the inserted DNA of pCM9 contained calmodulin sequences.

## DISCUSSION

The results presented in this paper demonstrate the isolation and molecular cloning of a portion of the calmodulin gene. The cloned DNA fragment will be $^{32}$P-labeled by nick translation as described by Maniatis *et al.*[20] This probe will be utilized to screen the remaining colonies known to contain DNA inserts in order to select one that contains the entire calmodulin structural gene. The nucleotide sequence of this DNA will be determined[19] and should rectify the ambiguities that presently exist concerning the amino acid sequence of the protein. The [$^{32}$P]cDNA will also be employed as an RNA hybridization probe to measure the levels and turnover of calmodulin mRNA in normal and transformed cell lines.[21] Experiments will also be carried out to examine the processing of the primary transcript. In this way it will be possible to determine the mechanism responsible for the elevated levels of calmodulin in cells transformed by oncogenic viruses.[2-4] A third use of the [$^{32}$P]cDNA is to examine the presence of complementary sequences in total DNA from a number of evolutionarily distinct species. This is accomplished by isolating DNA, cleaving with one restriction endonuclease, such as *Eco*RI or *Hin*dIII, and separating the fragments by slab gel electrophoresis. The DNA is then transferred to nitrocellulose filters by the method of Southern,[23] hybridized to the calmodulin [$^{32}$P]cDNA and developed by autoradiography. Positive DNA bands in all samples will indicate that the nucleic acid sequences are evolutionarily conserved as has been shown to be the case for the amino acid sequence. Finally the $^{32}$P-labeled probe can be used to determine the number of calmodulin genes in the genomes of the organisms with which it cross-reacts. The presence of free calmodulin and calmodulin associated in an integral way with enzymes such as phosphorylase kinase suggests the possibility of multiple gene copies.

The complete structural gene will be mapped with respect to restriction endonuclease cleavage sites. With this information in hand we will use appropriate radiolabeled fragments to isolate the natural gene from gene banks prepared from the DNA of both chickens and rats.[23] Several reasons should be listed to justify the isolation of yet another gene from eukaryotic cells. One of the primary goals will be to determine the molecular organization of the entire gene including sequences flanking both 5′ and 3′-ends. This will be accomplished by a combination of techniques including restriction enzyme mapping and heteroduplex analysis with the electron microscope.[23] Together this information will allow the identification of intervening sequences and how they are positioned in the gene relative to structural sequences. On the one hand, the remarkable conservation of calmodulin during evolution would suggest that intervening sequences might be absent in analogy to the histone genes. On the other hand, the presence of such sequences could be predicted due to the fact that the protein can readily be divided into four $Ca^{2+}$-binding domains. It is likely that calmodulin has arisen from a primitive peptide containing one $Ca^{2+}$-binding site during the course of evolution. One can readily envision the use of intervening sequences to divide the $Ca^{2+}$-binding domains as we have recently shown to be the case for the three domains of chicken ovomucoid.[24]

Calmodulin is not regulated by steroid hormones in the chicken oviduct. In fact the number of calmodulin molecules per cell remains constant during estrogen-mediated tissue differentiation, hormone withdrawal, and restimulation.[25] These observations suggest constitutive expression of this protein in oviduct. We would like to compare the sequence of the 5′ and 3′-flanking regions of the natural calmodulin gene with those determined for the hormone-inducible oviduct genes that code for ovalbumin and ovomucoid.[23] Such analyses may provide clues to structural regions that differentiate constitutively expressed genes from the hormone-inducible variety. Finally, as mentioned previously, it is entirely possible that calmodulin arose by gene duplication. It is equally possible that troponin C arose from calmodulin during evolution. Judicious selection of $^{32}$P-labeled probes representing each of the $Ca^{2+}$-binding domains will allow us to perform experiments to answer these questions. In addition, such experiments should aid in the understanding of whether calmodulin represents one of a cluster of genes coding for the several calcium binding proteins known to exist in eukaryotic cells.

## Acknowledgments

The authors would like to express sincere gratitude to Drs. Savio Woo and Chandra Thirumalachary for help with the details of the molecular engineering experiments. This work was supported in part by National Institutes of Health grants HS-07503 (A.R.M.) and GM-25557 (J.R.D.) as well as by American Cancer Society grant NP-3261 and grant Q-611 from the Robert A. Welch Foundation. J.R.D. is the recipient of a Research Career Development Award from the National Institutes of Health and R.P.M. holds a postdoctoral fellowship award from the Muscular Dystrophy Association.

## References

1. Chafouleas, J. G., R. P. Munjaal, J. R. Dedman & A. R. Means. 1979. J. Biol. Chem. **254:** 10262–10267.
2. Watterson, D. M., L. J. Van Eldik, R. E. Smith & T. C. Vanaman. 1976. Proc. Natl. Acad. Sci. USA **73:** 2711–2715.
3. LaPorte, D. C., S. Gidwitz, M. J. Weber & D. R. Storm. 1979. Biochem. Biophys. Res. Commun. **86:** 1169–1177.
4. Chafouleas, J. G., R. L. Pardue, B. R. Brinkley, J. R. Dedman & A. R. Means 1980. Proc. Natl. Acad. Sci. USA (In press).
5. Childers, S. R. & F. L. Siegel. 1975. Biochem. Biophys. Acta **405:** 99–108.
6. Rosen, J. M., S. L. C. Woo, J. W. Holder, A. R. Means & B. W. O'Malley. 1975. Biochemistry **14:** 69–78.
7. Aviv, H. & P. Leder. 1972. Proc. Natl. Acad. Sci. USA **69:** 1408–1412.
8. Stein, J. P., J. F. Catterall S. L. C. Woo, A. R. Means & B. W. O'Malley. 1978. Biochemistry **17:** 5763–5772.
9. Botchan, M., W. Topp & J. Sombrook, 1976. Cell **9:** 269–287.
10. Chang, S. & S. N. Cohen. 1977. Proc. Natl. Acad. Sci. USA **74:** 4811-4815.
11. Bolivar, L., R. L. Rodriquez, P. O. Green, M. D. Bettach, H. L. Heynecker, H. W. Boyer, J. H. Crosa & S. Falkow. 1977. Gene **2:** 95–113.
12. Grunstein, M. & D. S. Hogness. 1975. Proc. Natl. Acad. Sci. USA **72:** 3961–3965.
13. Katz, L., P. H. Williams, S. Sato, R. W. Leavitt & D. R. Helsinki. 1977. Biochemistry **16:** 1677–1683.
14. Noyes, B. E. & G. R. Stark. 1975. Cell **5:** 301–310.
15. Dedman, J. R., J. D. Potter, R. L. Jackson, J. D. Johnson & A. R. Means. 1977. J. Biol. Chem. **252:** 8415–8422.
16. Dedman, J. R., R. L. Jackson, W. E. Schreiber & A. R. Means. 1978. J. Biol. Chem. **253:** 343–346.
17. Watterson, D. M., F. Sharief & T. C. Vanaman. 1980. J. Biol. Chem. **255:** 962–975.
18. McReynolds, L. A., J. J. Monahan, D. W. Bendure, S. L. C. Woo, G. V. Paodoch, W. Salser, J. Dorsen, R. E. Moses & B. W. O'Malley. 1976. J. Biol. Chem. **252:** 1840–1843.
19. Maxam, A. & W. Gilbert. 1977. Proc. Natl. Acad. Sci. USA **75:** 560–564.
20. Maniatis, T., A. Jeffrey & D. G. Kleid. 1975. Proc. Natl. Acad. Sci. USA **72:** 1184–1188.
21. Tsai, S. Y., D. R. Roop, M. J. Tsai, J. P. Stein, A. R. Means & B. W. O'Malley. 1978. Biochemistry **17:** 5773–5780.
22. Southern, E. 1975. J. Mol. Biol. **98:** 503–518.
23. Lai, E. C., J. P. Stein, J. F. Catterall, S. L. C. Woo, M. L. Mace, A. R. Means & B. W. O'Malley. 1979. Cell **18:** 829–842.
24. Stein J. P., J. F. Catterall, P. Kristo, A. C. Ting, A. R. Means & B. W. O'Malley. 1980. Cell (In press.)
25. Means, A. R. & J. R. Dedman. 1980. Nature **285:** 73–77.

## Discussion of the Paper

Dr. Schneider: Have you found any preference for certain triplets for amino acids in your system?

Dr. A. R. Means: That's something that we really haven't examined.

# PLANT CALMODULIN AND THE REGULATION OF NAD KINASE*

Harry W. Jarrett, Harry Charbonneau, James M. Anderson,†
Richard O. McCann, and Milton J. Cormier‡

*Bioluminescence Laboratory*
*Department of Biochemistry*
*University of Georgia*
*Athens, Georgia 30602*

Calmodulin was first discovered in extracts of brain tissue as an activator of cyclic nucleotide phosphodiesterase.[1,2] Since that time the protein has been isolated from a wide variety of animal tissues including protozoans, coelenterates, annelids, and mammals.[3-17]

The first hint that calmodulin may exist in plants was obtained when Anderson and Cormier[18] and Waisman *et al.*[11] independently observed calmodulin-like activity in extracts of tissues from a variety of higher plants. Furthermore, Anderson and Cormier[18] showed that a $Ca^{2+}$-dependent protein activator obtained from plants would not only activate porcine brain phosphodiesterase but also activated plant NAD kinase. They also showed that bovine brain calmodulin would replace the plant protein activator in the $Ca^{2+}$-dependent activation of NAD kinase.

These observations provided the first evidence that calmodulin existed in plants and that it served at least one important function in plants, i.e., the activation of NAD kinase. Plant calmodulin, obtained from peanuts (*Arachis hypogea*) and pea seedlings (*Pisum sativum* L. Cultivar Willet Wonder), was subsequently purified to homogeneity by the use of fluphenazine-sepharose affinity chromatography.[19] This was followed by a detailed characterization of plant calmodulin showing that the chemical, physical, and biological properties of plant and animal calmodulin were remarkably similar.[20] On the other hand, Grand *et al.*[42] have reported that the properties of a plant calmodulin (barley) differ considerably from those of mammalian calmodulin. The reported differences include the lack of trimethyllysine, a significantly different amino acid composition, and the preparation was found to be only one-tenth as active as bovine brain calmodulin. The reasons for these apparent differences are not known. However, it is clear that the properties of the preparation of barley calmodulin differ significantly from those of peanut calmodulin obtained in our laboratory. For example, we find that calmodulin isolated from peanut seeds and bovine brain are similar in their molecular weight, Stokes' radii, amino acid composition including trimethyllysine content, $Ca^{2+}$-dependent enhancement of tyrosine fluorescence, $Ca^{2+}$-dependent interaction with troponin I, equal abilities to activate cyclic nucleotide phosphodiesterase, $Ca^{2+}$-dependent inhibition of calmodulin action by the phenothiazine drugs, electrophoretic mobility and immunological cross-reactivity. Furthermore, evidence has been provided

*Supported by the National Science Foundation (Grant PCM 79-05043).

†Present Address: Crop Sciences Department, North Carolina State University, Raleigh, N.C. 27609

‡To whom correspondence should be addressed.



0077-8923/80/0356-0119 $01.75/0 © 1980, NYAS

recently that calmodulin can be isolated from spinach leaf messenger RNA translation products.[21] These authors have also shown that spinach leaf and bovine brain calmodulin are equally effective in their abilities to stimulate the activity of phosphodiesterase.[21] Taken together, these experiments clearly show that calmodulin is a $Ca^{2+}$-dependent regulatory protein occurring in both plants and animals and may be a characteristic feature of eukaryotes.

In this report we examine some of the similarities and differences between plant and animal calmodulin. We also examine the calmodulin-dependent activation of NAD kinase and provide evidence for the formation of a $Ca^{2+}$-dependent calmodulin NAD kinase complex.

## Methods

Plant and bovine brain calmodulins were purified by use of fluphenazine-sepharose affinity chromatography[19] while erythrocyte calmodulin was purified as previously described.[17] Fluphenazine was a gift from the Squibb Institute, while trifluoperazine was a gift from Smith Kline and French Laboratories.

Assays for NAD kinase were conducted as previously described.[20]

Analytical SDS-PAGE was performed according to the procedure of Laemmli[22] on either 7.5 or 10% polyacrylamide slab gels in the presence of 1.5 mM EDTA. Amino acid analyses were carried out as previously described[20] on a Beckman model 119 C amino acid analyzer.

### *CNBr Cleavage and Peptide Mapping*

An aliquot containing 1 mg of desalted calmodulin in water was lyophilized onto the bottom of a 6 × 50 mm test tube. The lyophilized powder was dissolved in 0.1 ml of freshly prepared 3 mg/ml CNBr in 70% formic acid. Digestion was for 48 hrs at room temperature. Digestion was terminated by freezing and lyophilization. Amino acid analysis of 24 hr acid hydrolysates (6 N HCl, 110°C *in vacuo*) demonstrated that no detectable methionine remained.

The digests were prepared and electrophoresed as described by Swank and Munkries[23] on 12.5% acrylamide, 1.25% bisacrylamide gels except that electrophoresis was at 125 volts for 18 hrs. The gel apparatus was cooled with tap water. Under these conditions, melittin (MW 2846) migrated about ⅔ the length of the gel. Electrophoresis for shorter times showed that no peptides were lost from the gels at the longer times routinely used.

### *Purification of NAD Kinase*

Fourteen day old pea seedlings, grown as previously described,[20] were cut at the soil line. 500 gms were homogenized for 20 sec using a polytron in 2 liters of 50 mM $(NH_4)_2SO_4$, 50 mM Tris, 1 mM EDTA, 0.5 mM phenylmethylsulfonylfluoride, 0.5% polyvinylpyrrolidone, pH 8. The resulting suspension was filtered through four layers of cheesecloth and centrifuged at 48,000 × g for 30 min. The pellet was discarded and 144 g $(NH_4)_2SO_4$ per liter (25% saturated) was added to the supernatant. This was stirred at 4°C for 30 min and centrifuged at 10,000 × g for 10 min. The pellet was discarded and 158 g $(NH_4)_2SO_4$ was added per liter of the supernatant (50% saturated). This was again stirred and centrifuged as above.

The supernatant was discarded and the pellet was suspended in 500 ml of 50 mM $(NH_4)_2SO_4$, 50 mM Tris, 1 mM EDTA pH 8, dialyzed exhaustively against this buffer, and applied to a 3.2 × 40 cm column of DEAE-cellulose equilibrated in the same buffer. The NAD kinase activity was not bound by the column and was collected in the flow-through. To the flow-through was added 390 g $(NH_4)_2SO_4$ per liter (60% saturated) and the precipitated protein was collected as above, dissolved in 50 ml of 50 mM $(NH_4)_2SO_4$, 50 mM Tris, 50 mM KCl, 0.1 mM $CaCl_2$, 10 mM $MgCl_2$ (buffer A) and dialyzed exhaustively versus this buffer. The dialyzed material was clarified by centrifugation at 10,000 × g for 10 min and was applied to a 1.6 × 40 cm calmodulin-sepharose column (0.4 mg calmodulin/ml resin), prepared as described by Klee and Krinks,[25] and preequilibrated with buffer A. After most of the protein was eluted from the column, buffer A was made 0.2 M in KCl in order to remove protein adsorbed to the column by nonspecific interaction. Buffer A was then modified by replacing the $CaCl_2$ with 0.1 mM EGTA (Buffer B). The column was eluted with buffer B in order to elute proteins bound to the column via a $Ca^{2+}$-dependent interaction.

TABLE 1

EFFECTS OF CALMODULIN AND TRIFLUOPERAZINE ON PLANT NAD KINASE ACTIVITY

| Additions or Deletions | NAD Kinase Activity (pmoles $min^{-1}$ $mg^{-1}$) |
|---|---|
| None | 153.6 ± 8.1 |
| − ATP | 2.2 ± 1.5 |
| − $NAD^+$ | 2.9 ± 2.2 |
| − $Ca^{2+}$ | 1.6 ± 1.8 |
| − Calmodulin | 1.2 ± 2.7 |
| + Trifluoperazine (50 $\mu$m, Final Concentration) | 7.2 ± 6.7 |

The complete assay medium contained 50 mM Tris, 50 mM KCl, 10 mM $MgCl_2$, 2 mM NAD, 3 mM ATP, 0.2 mM $CaCl_2$, 20 $\mu$g/ml bovine brain calmodulin, and 0.2 ml of calmodulin-sepharose purified NAD kinase. Additions and deletions from this standard assay mixture are as noted. When $CaCl_2$ was deleted it was replaced by 0.2 mM EGTA.

## RESULTS

We previously reported that the activity of relatively crude preparations of plant NAD kinase was stimulated approximately five-fold upon the addition of $Ca^{2+}$ and calmodulin.[18] Thus far this is the only plant enzyme known to be activated by calmodulin. When the enzyme is further purified, as outlined under METHODS, NAD kinase is apparently inactive in the absence of $Ca^{2+}$ and calmodulin (TABLE 1). Also as shown in TABLE 1, low concentrations of trifluoperazine inhibits the Ca-calmodulin-dependent activation of NAD kinase as has been observed for other calmodulin-activated enzymes (see Cheung[24] for a review).

The activation of NAD kinase by calmodulin apparently involves the $Ca^{2+}$-dependent formation of a Ca-calmodulin-NAD kinase complex. As illustrated in FIGURE 1, NAD kinase is adsorbed to a calmodulin-sepharose column in the presence of $Ca^{2+}$. Most of the protein, but not NAD kinase activity, is eluted from the column by washing with buffer followed by buffer plus 0.2 M KCl (see METHODS). NAD kinase was then eluted by removal of free $Ca^{2+}$ from the buffer, i.e., by inclusion of EGTA in the column buffer. FIGURE 2 is an SDS-PAGE profile of the proteins eluted by the EGTA pulse. The absorbance at 280 nm was measured for each column fraction examined. These values are given in the bar graphs associated with each well (FIGURE 2). Note that the major fractions have four major protein species, including one observed at the dye front, and additional minor ones. The EGTA pulse, containing all of the NAD kinase activity, was applied a second time to the calmodulin-sepharose column. The column was washed with buffer plus 0.2 M KCl until the absorbance at 280 nm reached baseline. EGTA was added to the buffer and the NAD kinase was again eluted. FIGURE 3 shows an SDS-PAGE profile of the combined NAD kinase-containing fractions. Note the presence of four major protein species and several

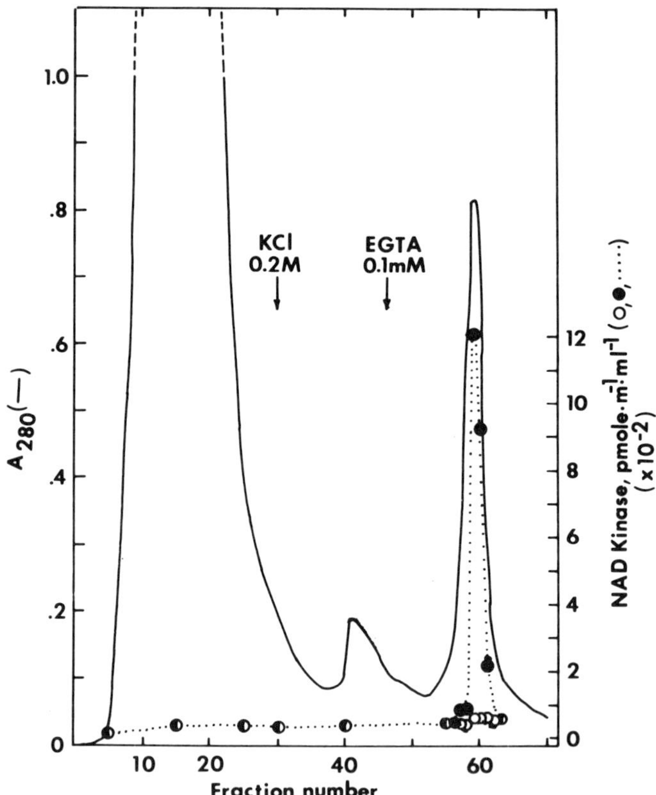


FIGURE 1. Calmodulin-sepharose purification of NAD Kinase. A 50 ml volume of partially purified NAD kinase was applied to a calmodulin-sepharose column (see METHODS) equilibrated with buffer A (50 mM Tris, 50 mM KCl, 10 mM $MgCl_2$, 0.1 mM $CaCl_2$ at pH 8.0). The arrows indicate the points at which buffer A was modified to include 0.2 M KCl (buffer B) and finally where the $CaCl_2$ in buffer B was replaced by 0.1 mM EGTA. Fractions were 6.9 ml of which duplicate 0.1 ml aliquots were used to assay NAD kinase activity for the fractions indicated. The open circles (-O-) represent the activity without and the filled circles (-●-) the activity with 20 μg of calmodulin added per ml of the assay mixture.

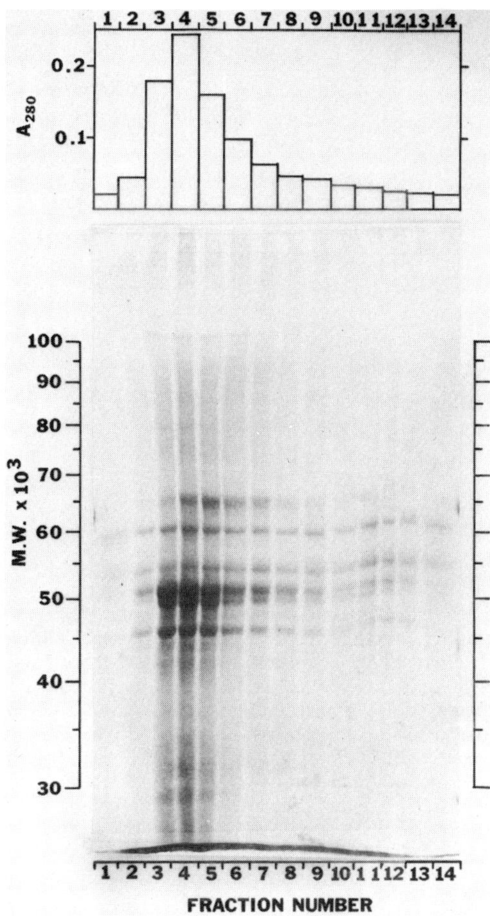


FIGURE 2. SDS gel electrophoresis of plant proteins eluted from calmodulin-sepharose by EGTA—50 ml of partially purified NAD kinase was applied to the calmodulin-sepharose and eluted as described for FIGURE 1. After the $A_{280}$ returned to baseline, following the 0.2 M KCl wash, the column was eluted with EGTA and 6.3 ml fractions were collected. One ml of each fraction with $A_{280} > 0$ (14 fractions) through the peak was dialyzed versus 10 mM $NH_4HCO_3$, lyophilized and prepared for electrophoresis by the method of Laemmli[22] in a total volume of 50 $\mu$l, all of which was applied to a well. Molecular weight standards (bovine serum albumin, ovalbumin, carbonic anhydrase, and lysozyme) were run on adjacent wells (not shown) on the 10% polyacrylamide gel.

minor ones, one of which again appears at the dye front. The electrophoretic mobility of this species was compared to calmodulin on 15% SDS-PAGE gels (data not shown) and was found to have significantly greater mobility. Thus this species does not appear to be calmodulin and thus could not have been derived by incomplete coupling of calmodulin to the sepharose.

Although the biological properties of plant and mammalian calmodulin are indistinguishable, based on their abilities to activate porcine brain phosphodies-

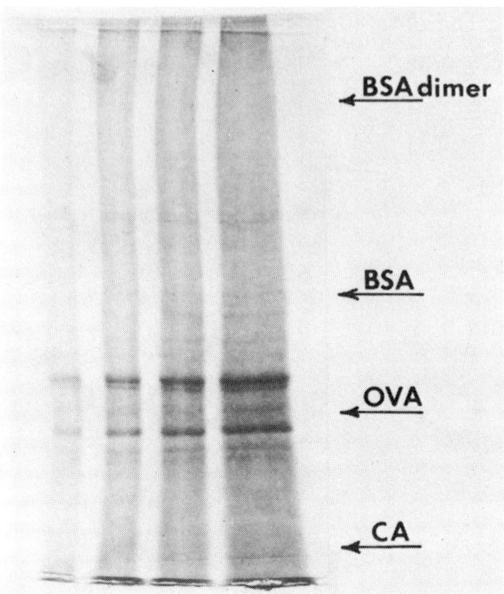


FIGURE 3. SDS gel electrophoresis of protein species bound by calmodulin-sepharose on repeated chromatography. Fractions 59 and 60 from the experiment in FIGURE 1 were pooled and dialyzed exhaustively against Buffer A (see METHODS) and reapplied to the calmodulin-sepharose column and eluted as described for FIGURE 1. The peak fractions of NAD kinase activity were pooled (four fractions, 25 ml) dialyzed versus 10 mM $NH_4HCO_3$, lyophilized and prepared for electrophoresis on 7.5% acrylamide gels. The sample wells contained a volume (from left to right) equivalent to 0.1, 0.25, 0.5, or 1.0 ml of the pool, respectively. The position of bovine serum albumin (BSA) and its dimer, ovalbumin (OVA), and carbonic anhydrase (CA), run an adjacent wells is indicated.

terase and NAD kinase, a comparison of peptides derived from CNBr digestion shows that some differences are to be expected in their primary structures. Of the three calmodulins compared in FIGURE 4, only the one from bovine brain has a known primary structure.[15] From this sequence, it can be shown that cleavage at the nine methionine residues would produce ten peptides of which only six would be of molecular weight 1,000 or greater and should be resolved by the gel system used; two other peptides are of molecular weight ~500 and may be resolved (Swank & Munkries[23]). The peptide maps presented here show 6 major and 2 minor peptides for bovine brain calmodulin and are thus reasonably consistent with the sequence data. That the human erythrocyte protein gives precisely the same banding pattern in the present experiment, and has also been shown to have nearly identical tryptic peptide maps,[43] demonstrates further that these mammalian calmodulins are similar if not identical proteins. However, the present experiment clearly shows that there are obvious differences between plant and mammalian calmodulin. The plant and the mammalian calmodulins do not differ significantly in methionine content[15,20] and should give approximately the same number of peptides. This appears to be the case; six major and four minor bands were found. Of the six major species, at least five and perhaps all six differ in mobility from their apparent counterparts in the mammalian calmodulin digest. The differences in mobility are small but have been consistently observed in five separate experiments. These observations are indicative of some differences in the primary structure of the plant and mammalian calmodulins which is perhaps not surprising in light of the evolutionary separation of the organisms involved.

## DISCUSSION

The results presented in TABLE 1 show that the NAD kinase in higher plants is completely dependent upon both calmodulin and $Ca^{2+}$ for activity. This complete dependency for enzymic activity has only been observed for one other calmodulin-activated enzyme, i.e., the myosin light chain kinase from smooth muscle.[9,10]

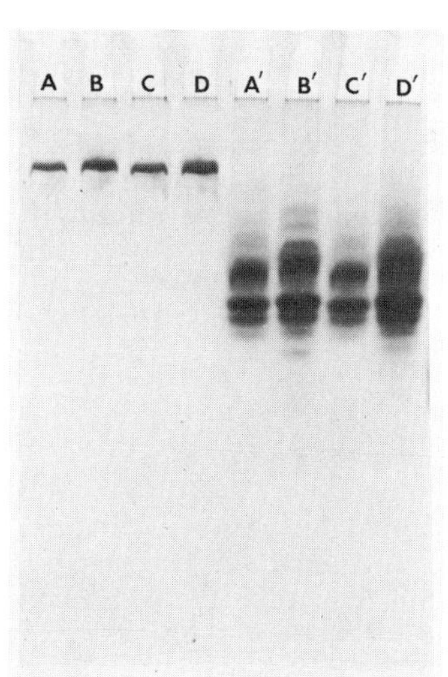


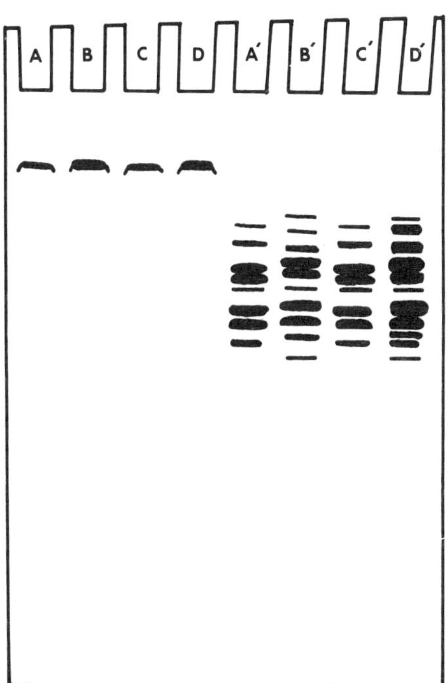


FIGURE 4. (Left) Gel electrophoresis of plant and mammalian calmodulin CNBr digests. From left to right: 10 μg each of A, human erythrocyte; B, Peanut seed; C, Bovine brain calmodulins; and D, mixture of 5 μg each of the above calmodulins. A′, B′ and C′ represent 200 μg each of the 48 hr CNBr digests of the respective calmodulins while D′ is a mixture of 100 μg of each digest. Notice that no undigested calmodulin is apparent in the digests. The gels were by the method of Swank and Munkries[23] (see METHODS). (Right) Graphic representation of the CNBr peptide maps. 50, 100, 200, and 400 μg samples of each CNBr digest shown on the left were run on the same gel system and these data were used to construct a drawing to the scale of the gel shown on the left. At the lower and higher concentrations, additional details concerning the number of major and minor peptides that are poorly resolved at any one concentration alone were obtained and this information is incorporated into the drawing.

The results presented in FIGURE 1 demonstrate that the interaction between calmodulin and NAD kinase is a direct, $Ca^{2+}$-dependent one. Taken together, these results conclusively rule out the possibility that calmodulin activation of NAD kinase occurs in some indirect fashion; rather, the activation is the result of a direct, noncovalent association of the activator and enzyme molecules. It appears likely that the dependency on $Ca^{2+}$ for activity and for binding to calmodulin-sepharose is due to $Ca^{2+}$ binding to calmodulin rather than $Ca^{2+}$ binding to the enzyme (although this latter possibility can not be ruled out at this time). A variety of physical measurements on both mammalian[8,26-31] and plant[20] calmodulins have shown a $Ca^{2+}$-dependent conformation change in calmodulin that coincides with the $Ca^{2+}$-dependence for calmodulin activation of its target enzymes. For calmodulin activation of NAD kinase these relationships can be illustrated as follows:

$$Ca^{2+} + \text{calmodulin} \rightleftharpoons Ca^{2+}\text{-calmodulin} \rightleftharpoons Ca^{2+}\text{-calmodulin}^*$$

$$Ca^{2+}\text{-calmodulin}^* + \underset{\text{(inactive)}}{\text{NAD kinase}} \rightleftharpoons \underset{\text{(active)}}{Ca^{2+}\text{-calmodulin-NAD kinase}}$$

where $Ca^{2+}$-calmodulin* indicates the active conformation of calmodulin. This mechanism is analogous to the calmodulin-dependent activation of several enzymes of mammalian origin (see Cheung[24] for review); however, in the present case the NAD kinase is inactive in the absence of calmodulin, making this an unusual though not unique (myosin light chain kinase also shows this behavior) type of calmodulin regulation.

The question of whether plant cells contain enzymes other than NAD kinase that are regulated by $Ca^{2+}$ and calmodulin is raised by the polyacrylamide slab gels shown in FIGURES 2 and 3. Although plant NAD kinase may be a multisubunit protein, the number of protein species observed on the gels in FIGURES 2 and 3, and the observation that these species appear not to be present in stoichiometric amounts, leave open the possibility that some of them may be calmodulin-dependent enzymes in plants as yet unidentified.

### *Significance of Calmodulin in Plants*

The recent demonstration that calmodulin exists in higher plants,[18-20] and that its $Ca^{2+}$-binding properties are similar to mammalian calmodulin,[20] suggest that plant cells undergo $Ca^{2+}$-dependent regulatory events that are mediated through calmodulin. This implies a second messenger role for $Ca^{2+}$ in plant cells with calmodulin being a primary target for free $Ca^{2+}$.

In recent years there has been increasing evidence for a role for $Ca^{2+}$ in the regulation of a number of physiological processes in plants. The observations include: (1) a light-induced uptake of extracellular calcium,[32] (2) a $Ca^{2+}$-dependent stimulation of chloroplast rotation in *Mougeotia*,[33] (3) a $Ca^{2+}$ and phytochrome-dependent depolarization of *Nitella* cells,[34] (4) a $Ca^{2+}$-dependent inhibition of cytoplasmic streaming in *Nitella*,[35] and (5) a $Ca^{2+}$-dependent regulation of directional growth that involves the formation of an intracellular $Ca^{2+}$ gradient.[36-39] In this latter series of observations, the authors have shown that the growing tips of certain plant cells contain elevated levels of $Ca^{2+}$ relative to other regions of the cell and that disruption of the $Ca^{2+}$ gradient causes tip growth to cease.

In this report we have presented evidence that a specific enzyme in plants, i.e. NAD kinase, is converted from an inactive to an active form of the enzyme in the presence of $Ca^{2+}$ and calmodulin. This may be important from the point of view that NADP is the receptor of the reducing power derived from water during photosynthesis and is involved in numerous metabolic reactions associated with photosynthesis.

Earlier observations have shown that illumination of green leaves[40] or of *Chlorella*[41] causes an increase in the NADP/NAD ratio. One way in which these observations may be interrelated is illustrated by FIGURE 5. In animal cells, calmodulin is presumably a target for stimulus-induced changes in intracellular free $Ca^{2+}$ of the magnitudes shown in FIGURE 5. We suggest that something

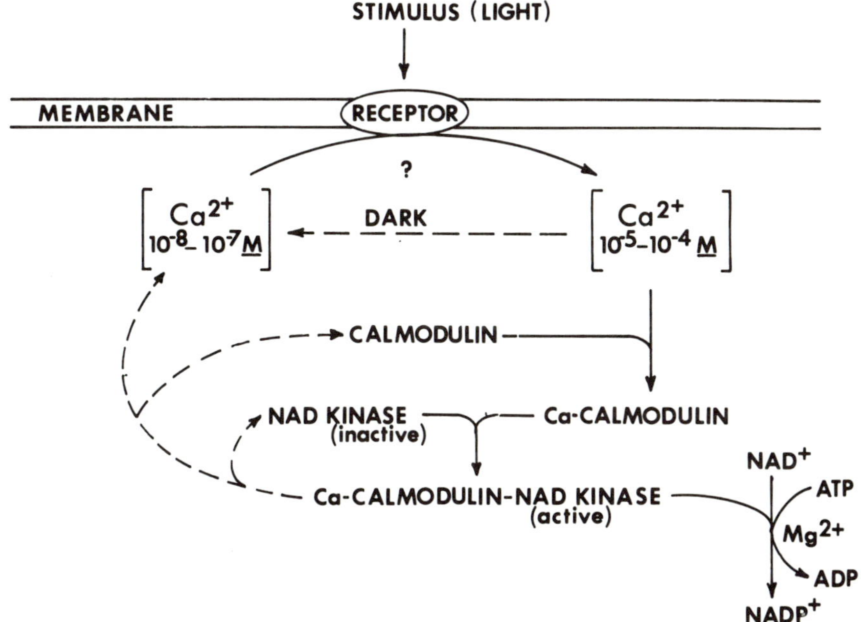


FIGURE 5. Model for the involvement of $Ca^{2+}$ and calmodulin during stimulus-response coupling in plants.

similar happens in plants except that in plants the stimulus may be sustained rather than transient. For example, sustained increases in intracellular $Ca^{2+}$ may occur as a result of illumination. Calmodulin could then function as the target for these localized changes in $Ca^{2+}$ levels, resulting in the activation of NAD kinase and perhaps other calmodulin-dependent enzymes.

The scheme shown in FIGURE 5 is a proposed model in which the stimulus (light) is translated into a response (increase in NADP/NAD ratio) via $Ca^{2+}$ and calmodulin. Perhaps other $Ca^{2+}$-dependent stimulus-response processes in plants operate in a similar manner.

## ACKNOWLEDGMENT

Dr. Jarrett is a recipient of an Individual Fellowship Award HL-05698 from the National Institutes of Health.


## REFERENCES


1. CHEUNG, W. Y. 1970. Biochim. Biophys. Res. Commun. **38:** 533–538.
2. KAKIUCHI, S., R. YAMAZAKI & H. NABAJIMA. 1970. Proc. Jpn. Acad. **46:** 587–592.
3. TEO, T. S., T. H. WANG & J. H. WANG. 1973. J. Biol. Chem. **248:** 588–595.
4. TEO, T. S. & J. H. WANG. 1973. J. Biol. Chem. **248:** 5950–5955.
5. LIN, Y. M., Y. P. LIN & W. Y. CHEUNG. 1974. J. Biol. Chem. **249:** 4943–4954.
6. WOLFF, D. J. & C. O. BROSTROM. 1974. Arch. Biochem. Biophys. **163:** 349–358.
7. CHILDERS, S. R. & F. L. SIEGEL. 1975. Biochim. Biophys. Acta **405:** 99–108.
8. DEDMAN, J. R., J. D. POTTER, R. L. JACKSON, J. D. JOHNSON & A. R. MEANS. 1977. J. Biol. Chem. **252:** 8415–8422.
9. YAGI, K., M. YAZAWA, S. KAKIUCHI, M. OHSHIMA & K. VENISHI. 1978. J. Biol. Chem. **253:** 1338–1340.
10. DABROWSKA, R., J. M. SHERRY, D. K. AROMATORIO & D. J. HARTSHORNE. 1978. Biochemistry **17:** 253–258.
11. WAISMAN, D. M., F. C. STEVENS & J. H. WANG. 1978. J. Biol. Chem. **253:** 1106–1113.
12. JONES, H. P., J. C. MATTHEWS & M. J. CORMIER. 1979. Biochemistry **18:** 55–60.
13. HEAD, J. F., S. MADER & B. KAMINER. 1979. J. Cell. Biol. **80:** 211–218.
14. JAMIESON, G. A., T. C. VANAMAN & J. J. BLUM. 1979. Proc. Natl. Acad. Sci. USA **76:** 6471–6475.
15. VANAMAN, T. C., F. SHARIEF & D. M. WATTERSON. 1977. *In* Calcium Binding Proteins and Calcium Function. R. H. Wasserman, R. A. Corradino, E. Carafoli, R. H. Kretsinger, D. A. MacClennan & F. L. Siegel, Eds.: 107–116. Elsevier/North-Holland. New York, N.Y.
16. GRAND, R. J. A. & S. V. PERRY. 1978. FEBS Lett. **92:** 137–142.
17. JARRETT, H. W. & J. T. PENNISTON. 1978. J. Biol. Chem. **253:** 4676–4682.
18. ANDERSON, J. M. & M. J. CORMIER. 1978. Biochem. Biophys. Res. Commun. **84:** 595–602.
19. CHARBONNEAU, H. & M. J. CORMIER. 1979. Biochem. Biophys. Res. Commun. **90:** 1039–1047.
20. ANDERSON, J. M., H. CHARBONNEAU, H. P. JONES, R. O. MCCANN & M. J. CORMIER. 1980. Biochemistry **19:** 3113–3120.
21. VAN ELDIK, L. J., A. R. GROSSMAN, D. B. IVERSON & D. M. WATTERSON. 1980. Proc. Natl. Acad. Sci. USA **77:** 1912–1916.
22. LAEMMLI, V. K. 1970. Nature **227:** 680–685.
23. SWANK, R. T. & MUNKRES, K. D. 1971. Anal. Biochem. **39:** 462–477.
24. CHEUNG, W. Y. 1980. Science **207:** 19–27.
25. KLEE, C. B. & M. H. KRINKS. 1978. Biochemistry **17:** 120–126.
26. WANG, J. H., T. S. TEO, H. C. HO & F. C. STEVENS. 1975. Adv. Cyclic Nucleotide Res. **5:** 179–194.
27. LIU, Y. P. & W. Y. CHEUNG. 1976. J. Biol. Chem. **251:** 4193–4198.
28. KLEE, C. B. 1977. Biochemistry **16:** 1017–1024.
29. WOLFF, D. J., P. G. POIRIER, C. O. BROSTROM & M. A. BROSTROM. 1977. J. Biol. Chem. **252:** 4108–4117.
30. RICHMAN, P. G. & C. B. KLEE. 1979. J. Biol. Chem. **254:** 5372–5376.
31. SEAMON, K. B. 1980. Biochemistry **19:** 207–215.
32. DREYER, E. M. & M. H. WEISENSEEL. 1979. Planta **146:** 31–39.
33. HAUPT, W. 1959. Planta **53:** 484–501.
34. WEISENSEEL, M. H. & H. K. RUPPERT. 1977. Planta **137:** 225–229.
35. HAYAMA, T., T. SHIMMEN & M. TAZAWA. 1979. Protoplasma **99:** 305–321.

36. JAFFE, L. A., M. H. WEISENSEEL & L. F. JAFFE. 1975. J. Cell Biol. **67:** 488–492.
37. HERTH, W. 1978. Protoplasma **96:** 275–282.
38. REISS, H. D. & W. HERTH. 1978. Protoplasma **97:** 373–377.
39. REISS, H. D. & W. HERTH. 1979. Planta **145:** 225–232.
40. ORGEN, W. L. & D. W. KROGMANN. 1965. J. Biol. Chem. **240:** 4603–4608.
41. OH-HAMA, T. & S. MIYACHI. 1959. Biochim. Biophys. Acta **34:** 202–210.
42. GRAND, R. J. A., A. C. NAIRN & S. V. PERRY. 1980. Biochem. J. **185:** 755–760.
43. JARRETT, H. W. & J. KYTE. 1979. J. Biol. Chem. **254:** 8237–8244.

## DISCUSSION OF THE PAPER

DR. J. D. PUETT (*Vanderbilt University, Nashville, TN*): Where does phytochrome fit into your scheme?

DR. M. J. CORMIER: We don't know. However, phytochrome may function by affecting cytosolic $Ca^{2+}$ levels and thus activating NAD kinase via calmodulin.

DR. T. C. VANAMAN: Have you looked to see if there might be other calmodulin regulated enzymes that use NADP?

DR. CORMIER: Yes. The data are too premature to really say anything.

DR. S. J. ROUX (*University of Texas, Austin, TX*): Phytochrome apparently does activate a calcium ATPase on the plasma membrane of plant cells. It apparently does so by causing a transient increase of calcium concentration in the plant cytoplasm which in turn leads to the activation of the calcium ATPase. Some of this work is still in progress but there are some publications already out.

DR. CORMIER: How are you measuring calcium levels?

DR. ROUX: With calcium-45 and with murexide.

# PHOSPHORYLATION OF MYOSIN AS A REGULATORY COMPONENT IN SMOOTH MUSCLE*

D. J. Hartshorne and A. J. Persechini

*Departments of Biochemistry and Nutrition and Food Science*
*College of Agriculture*
*University of Arizona*
*Tucson, Arizona 85721*

Over the last few years our knowledge of the biochemistry of smooth muscle has increased considerably. It is assumed that the basic mechanism of contraction in smooth muscle is similar to that in skeletal muscle and that length changes occur as a result of interactions between the thick and thin filaments. The major components of these filaments, i.e. myosin, actin, and tropomyosin, are similar to their skeletal muscle counterparts with respect to gross physical properties and several biochemical characteristics. It is clear, however, that the proteins are not identical and each component is characteristic of the muscle type from which it is isolated. Of the three proteins listed above myosin is the most distinctive and myosins from skeletal and smooth muscles show differences in light chain composition and in their basic ATPase profiles.[1] The contractile activity of smooth muscle, in common with that of skeletal and cardiac muscle, is initiated by an increase in the intracellular concentration of $Ca^{2+}$. The proteins of the contractile apparatus that recognize the changes in the $Ca^{2+}$ concentration and subsequently modify the actin-myosin interactions are termed regulatory proteins, and herein lies a fundamental difference between smooth and striated muscles.

## Regulatory Mechanism in Smooth Muscle

In the absence of the regulatory proteins (i.e., troponin and tropomyosin) myosin and actin from skeletal muscle form an unregulated but active complex. The function of the regulatory system is to inhibit ATPase activity, or reduce the number of cross-bridge contacts with actin when $Ca^{2+}$ is removed. Regulation in skeletal muscle therefore is achieved by inhibiting an active state, and the inhibition results from the troponin-tropomyosin-actin interactions. However, actomyosin from smooth muscle in the absence of regulatory proteins is dormant and the function of the regulatory system is to activate ATPase activity but only in the presence of $Ca^{2+}$. This fundamental difference in the requirements for regulation in the two muscle types is generally accepted and the controversy that exists is centered on the nature of the activating factor in smooth muscle. The most popular theory is that actin-activation of the $Mg^{2+}$-ATPase activity of myosin is achieved only when the myosin is phosphorylated. The sites of phosphorylation being the 20,000 $M_r$ light chains. Thus in this mechanism the regulatory proteins would be a myosin light chain kinase (MLCK) which activates the system and a myosin light chain phosphatase (MLCP) which returns the

*Supported by the National Institutes of Health (Grant HL 23615).



0077-8923/80/0356-0130 $01.75/0 © 1980, NYAS

myosin to its dormant (or relaxed) state. The alternate viewpoint is that regulation is achieved without phosphorylation of myosin[2] and is under the control of a system termed leiotonin.[3,4] At this time there is no adequate explanation for the two seemingly incompatible viewpoints and the evaluation of each system must be a priority for future research. However, it is possible to analyze the phosphorylation scheme and specifically to ask whether or not the state of myosin phosphorylation can account fully for the observed biochemical properties.

## *Phosphorylation Scheme*

The basic tenets of the phosphorylation hypothesis, as established by the work of several investigators,[5-8] can be categorized as follows: 1) The two myosin light chains of $M_r$ 20,000 are phosphorylated by the MLCK in the presence of $Ca^{2+}$. Two moles of phosphate are incorporated per mole of myosin. 2) The $Ca^{2+}$ concentration at which phosphorylation is initiated is the same as that required to activate the $Mg^{2+}$-ATPase activity of actomyosin. 3) The event of phosphorylation allows the actin activation of the $Mg^{2+}$-ATPase activity of myosin; and 4) when the $Ca^{2+}$ concentration is reduced the MLCK is inactivated and the phosphate groups are removed from the light chains by the MLCP and actin activation is lost.

This sequence of events is summarized in FIGURE 1. The signal for the onset of contraction is an increase in the intracellular $Ca^{2+}$ concentration. The $Ca^{2+}$-calmodulin complex is formed and this in turn interacts with the larger kinase subunit to form the active ternary MLCK complex. Phosphorylation of myosin then results and it is assumed that this step is obligatory to the onset of contraction. The basis for this assumption is that actin activation of $Mg^{2+}$-ATPase activity can occur only when myosin is phosphorylated. As long as $Ca^{2+}$ is present, the steady-state ATP hydrolysis, or cross-bridge cycling will continue. The kinetics of this phase are not established although it has been suggested[9] that the rate limiting step is a conformational change occurring in the actin-myosin-products complex and this is indicated in FIGURE 1. When the $Ca^{2+}$ level within the cell is reduced the MLCK is inactivated and the MLCP removes the phosphate groups from the myosin light chains. Actin-activation is lost and relaxation follows.

## *Properties of the Myosin Light Chain Kinase*

It is assumed that the MLCK initiates the contractile process by phosphorylating the myosin light chains. Cast in this role it is obviously a critical factor involved in muscle shortening or the development of tension, and there has been some effort expended to characterize the enzyme and establish various kinetic parameters.

Initially, it was found that the MLCK from chicken gizzard is composed of two distinct components.[10] One of these is a subunit of $M_r$ 105,000 and the other was identified as calmodulin.[11] Neither component alone possessed MLCK activity. This basic subunit composition was later confirmed for the MLCK from turkey gizzard[12] although an $M_r$ of about 125,000 was reported for the larger subunit. The MLCKs from skeletal[13,14] and cardiac muscle[15] are also dependent on calmodulin, although in these two muscle types there is no clear indication of a functional role for the MLCK.

It is assumed, based on the analogy with phosphodiesterase, that the $Ca^{2+}$-calmodulin complex interacts with the larger kinase subunit to form the active ternary complex of the MLCK. The stoichiometry of the two protein components was determined to be unity,[16] and this was done using two independent techniques. The $K_m$ of the kinase for ATP is about 65 $\mu$M,[16] which is of a slightly higher affinity than the values reported for skeletal muscle[14,17] (200–400 $\mu$M), cardiac muscle[15] (175 $\mu$M), and the MLCK from blood platelets[18] (121 $\mu$M). The $V_{max}$ of the MLCK from chicken gizzard ranged from 5–13 $\mu$moles $min^{-1}mg^{-1}$ kinase,[16] which is similar to values obtained from MLCKs from turkey gizzard[12] and skeletal muscle,[14,17] but considerably higher than the activity obtained from cardiac muscle[15] and rat myoblasts.[19] Under the conditions used to determine the

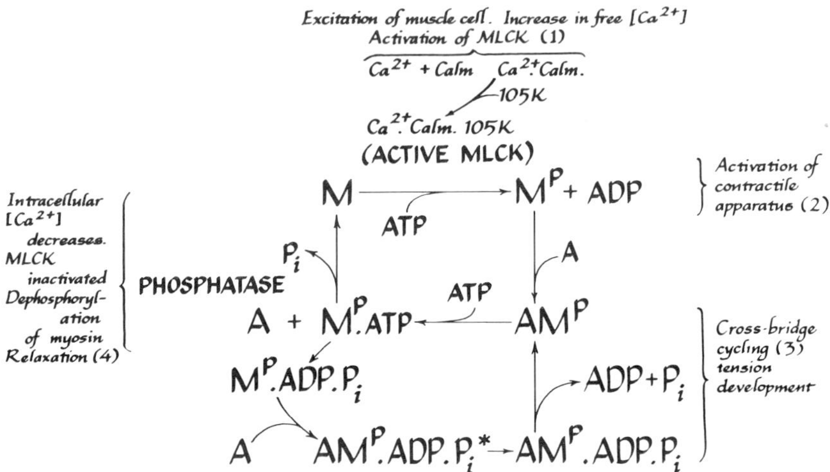


FIGURE 1. Scheme to illustrate the role of phosphorylation of myosin as a regulatory mechanism in smooth muscle. The abbreviations used are: M, myosin; $M^P$, phosphorylated myosin; A, actin; calm, calmodulin; 105 K, the larger myosin light chain kinase subunit. The sequence of events going from excitation to contraction to relaxation are indicated 1 through 4. Some of the intermediate stages in the steady-state ATPase activity are shown. Two intermediates for the actin-myosin-products complex are indicated, and the transition between the two is thought to be a rate-limiting step in the hydrolysis of ATP.[9]

$V_{max}$ for the chicken gizzard MLCK, the molar ratio of the 20,000 $M_r$ light chain to MLCK was in excess of 4000:1. The $Q_{10}$ for the MLCK from gizzard is close to 2 and optimal activity was found in the neutral pH range,[20] which is distinct from the acidic pH optimum reported for the skeletal muscle MLCK.[21] Unlike the $Mg^{2+}$-ATPase activity of smooth muscle actomyosin, the MLCK does not have a requirement[16] for free $Mg^{2+}$. Using the purified MLCK and isolated myosin light chains as the substrate, it was found that neither cAMP nor cGMP affected the enzyme activity,[16] although it should be added that in the presence of the cAMP-dependent protein kinase an inhibition of the MLCK activity was observed.[12] This effect was traced to the phosphorylation of the larger MLCK subunit which resulted in a decreased affinity for calmodulin and subsequent inhibition of MLCK activity.

One of the reasons for measuring the $V_{max}$ of the isolated MLCK was to question whether or not the phosphorylation of myosin could be rate-limiting in the activation of $Mg^{2+}$-ATPase activity or the onset of contraction. If a representative value of 10 $\mu$moles $P_i$ min/mg kinase is taken as the $V_{max}$ of the MLCK (determined using isolated myosin light chains as the phosphate acceptor) a turnover number of about 17 $s^{-1}$ can be calculated. A corresponding value for the $Mg^{2+}$-ATPase activity of actomyosin would give a specific activity of over 2 $\mu$moles $P_i$ liberated $min^{-1}$ $mg^{1}$ myosin, which is considerably higher than the values reported (i.e., 10 to 300 nmoles $P_i$/min/mg, see review[1]). Thus, it would appear that the phosphorylation of myosin is considerably faster than the rate of cross-bridge turnover and is unlikely to be a rate-limiting step in the contractile process. However, this conclusion should be regarded as tentative since preliminary experiments[20] have indicated that the rate of phosphorylation of the isolated light chains is faster than that for whole myosin.

## *Evidence to Support the Phosphorylation Scheme*

The original observation that led to the suggestion that phosphorylation of myosin plays a regulatory role was that as the extent of phosphorylation increased the actin-activated ATPase activity of smooth muscle myosin also increased.[5] This general relationship has subsequently been confirmed by several investigators.[7,8,23,24] Using myofibrils and actomyosin from chicken gizzard, a linear relationship between the extent of phosphorylation and ATPase activity was reported.[25] At saturating levels of phosphorylation a specific activity of about 300 nmoles $P_i$/min/mg myosin was obtained. Experiments carried out with reconstituted actomyosins also indicate a dependence of ATPase activity on the level of myosin phosphorylation. With purified MLCK, partially purified gizzard myosin and actin, and tropomyosin from skeletal muscle, a specific activity of about 60 nmoles $P_i$/min/mg myosin was obtained for the fully phosphorylated myosin.[10] A similar activity was observed by Aiba *et al.*[26] using column-purified phosphorylated gizzard myosin and skeletal muscle actin. In studies using intact muscle,[27,28] or various types of skinned fiber[29,30] it was found that contraction is associated with the phosphorylation of myosin. All of this evidence is consistent with the hypothesis that phosphorylation of myosin is a prerequisite for actin-activation of the $Mg^{2+}$-ATPase activity. However, it does not prove such a relationship, and it is possible that other $Ca^{2+}$-dependent mechanisms are involved and the phosphorylation of myosin could be coincidental with another process. One approach to identifying the critical process would be to use protein components of established homogeneity, and here myosin presents the major technical problem.

This is obviously one of the priorities for the future, but until this is achieved there are other experimental approaches that suggest that myosin phosphorylation is linked to the regulation of smooth muscle activity. The use of the analog adenosine 5′-O-(3-thiotriphosphate), ATP$\gamma$S, provides one example. This compound acts as a thiophosphate donor for the MLCK but the resultant thiophosphorylated light chain is resistant to hydrolysis by the phosphatase. The net result is that myosin becomes trapped in the phosphorylated state, and the effect of $Ca^{2+}$ can be tested without complications arising due to MLCP activity. Following preincubation of ATP$\gamma$S with gizzard myosin it was shown that as the level of thiophosphorylation increased the actin-activated $Mg^{2+}$-ATPase activity of the myosin in the absence of $Ca^{2+}$ also increased until $Ca^{2+}$ sensitivity was lost.[31] This experiment is important for two reasons: it suggests that $Ca^{2+}$-binding by the myosin light chains does not form part of the regulatory process, and it also provides strong support for the phosphorylation theory of regulation. The latter is valid because ATP$\gamma$S uncouples the $Ca^{2+}$-dependence of the system, and it is unlikely that an unknown $Ca^{2+}$-dependence process would show the same response to ATP$\gamma$S. The $Ca^{2+}$ sensitivity of tension development was also lost on preincubation with ATP$\gamma$S using mechanically disrupted chicken gizzard fibers[29] and functionally skinned rabbit ileum strips.[32] Thus an advantage associated with the use of ATP$\gamma$S is that it perturbed the normal response and this perturbation could be correlated to the thiophosphorylation of myosin.

Another approach that is supportive of the phosphorylation scheme is the use of various phenothiazine derivatives. It was found[33] that trifluoperazine binds to calmodulin with a high affinity and inhibits calmodulin-dependent processes, including the MLCK.[34] The phenothiazines inhibited tension development when applied to strips of skinned intestinal and arterial muscle.[30] This suggests that a calmodulin-dependent process is implicated in the contraction of smooth muscle and it is reasonable to assume that the MLCK is involved. Thus there is substantial evidence to suggest that the phosphorylation of myosin is an integral component of the regulatory mechanism in smooth muscle. On the other hand, from the available evidence it cannot be claimed that it is the sole component of the control system and it is possible that other mechanism(s) also are implicated. One of our objectives is to identify the role of myosin phosphorylation in the regulatory process and an approach that we have initiated is to correlate the rates and extent of myosin phosphorylation with the ATPase characteristics. Some of these results are presented below.

## *Correlation of Phosphorylation and ATPase Activity*

If it is assumed that the only factor that influences the actin-activated $Mg^{2+}$-ATPase activity of myosin is the state of phosphorylation of the myosin light chains (under constant assay conditions) then it is possible to construct a theoretical time course of phosphate release (i.e., ATPase activity) based on a given extent of phosphorylation. The assumption that is made is that myosin which is fully phosphorylated (2 moles $P_i$/mole myosin) represents the optimal specific ATPase activity and that at lower levels of phosphorylation the specific activity is directly proportional to the extent of phosphorylation. An added assumption is that the two myosin heads are independent, both with respect to phosphorylation and ATPase activity.

In FIGURE 2 the phosphorylation of myosin was measured and based on this curve the ATPase data were constructed. The specific $Mg^{2+}$-ATPase activity of the actomyosin was assumed to be 135 nmoles $P_i$/min/mg myosin for fully phosphorylated myosin. At lower levels of phosphorylation, taken from the measured values, the ATPase activities were calculated and summed over 0.1 min intervals to give the total $P_i$ accumulation. This was done assuming that the myosin was dephosphorylated at the start of the experiment, and the results are shown in the lower curve of FIGURE 2. A predicted curve was also calculated assuming that the myosin was 50% prephosphorylated and this is shown as the upper dashed line in FIGURE 2. The general feature of both curves is that they are nonlinear and show an increasing rate with respect to time. This is more dramatic for the 0% prephosphorylated sample and in this case a lag phase is clearly observed. This type of behavior for the ATPase activity has been reported for gizzard actomyosin[25] and for acto-gizzard heavy meromyosin.[35] An example is also shown in FIGURE 3 for a mixture of gizzard myosin, skeletal actin, and a crude MLCK preparation. The phosphate incorporation was nonlinear, showing

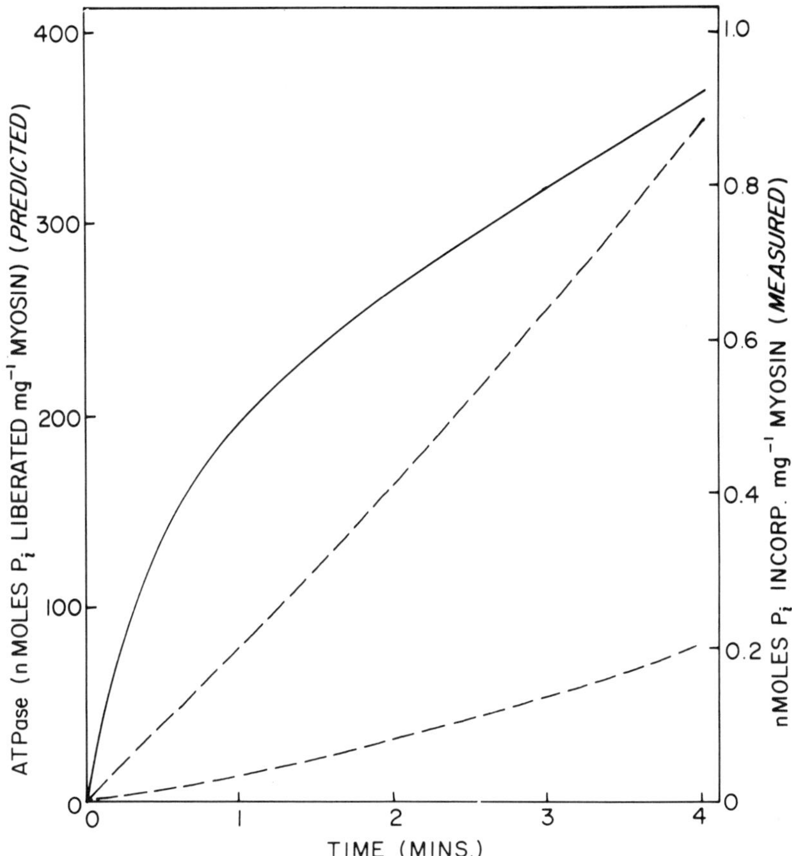


FIGURE 2. The phosphorylation of gizzard myosin and the predicted patterns of ATPase activity. Myosin phosphorylation is indicated by the solid line. The ATPase curves were calculated from the measured phosphorylation data (taken from FIGURE 3) assuming a specific $Mg^{2+}$ ATPase activity of 135 nmoles $P_i$ $min^{-1}$ $mg^{-1}$ myosin for fully phosphorylated myosin and assuming that the myosin was not prephosphorylated (lower dashed curve) or was 50% prephosphorylated (upper dashed curve).

a rapid initial phase and a slower subsequent phase. After 4 min the phosphorylation was not complete and approximately 22% of the total 20,000 $M_r$ light chains were phosphorylated. The ATPase curve initially showed a lag phase followed by increasing ATPase activity. The measured ATPase activity was approximated by a predicted curve based on the measured phosphorylation data

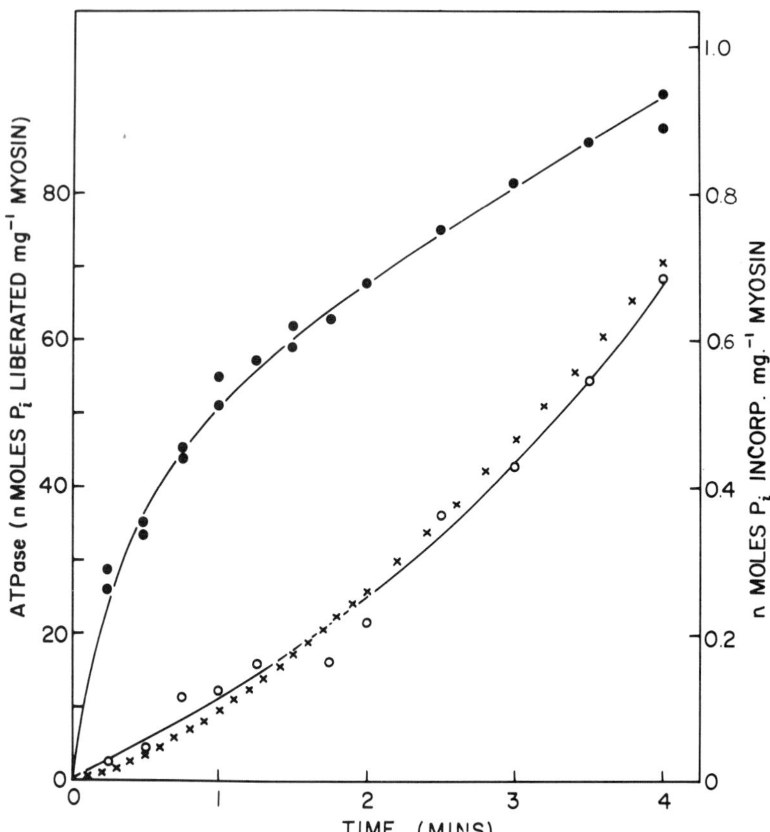


FIGURE 3. The phosphorylation of gizzard myosin and the actin-activated $Mg^{2+}$ ATPase activity. ATPase assays (O); phosphorylation assays (●). Conditions of ATPase and phosphorylation assays: 1 mM $\gamma$-labeled [$^{32}$P]ATP at approx. 5000 cpm per nmole; 4 mM $MgCl_2$; 60 mM KCl; 20 mM Tris-HCl (pH 7.6); 25°C; myosin, 0.69 mg/ml; skeletal muscle actin, 0.43 mg/ml; crude kinase, 14 $\mu$g/ml. The protein preparations and assay techniques were as described earlier: myosin;[36] skeletal muscle actin;[37] crude kinase;[20] $^{32}$P incorporation;[20] ATPase assays.[38] The predicted ATPase activity (X) was based on the measured $^{32}$P incorporation assuming a specific $Mg^{2+}$ ATPase activity of 115 nmoles $P_i$ $min^{-1}$ $mg^{-1}$ myosin for the fully phosphorylated myosin. The latter value was estimated from the rate of $P_i$ release over the last 30 seconds which was then corrected to represent complete phosphorylation.

and assuming a specific activity of 115 nmoles/min/mg for the fully phosphorylated myosin. In this example the observed ATPase activity was similar to that which was predicted from the phosphorylation data. However this pattern of behavior is not always shown and is more frequently encountered at relatively low concentrations of kinase.

Initially our objective was to analyze the phosphorylation and ATPase data using a system that might approximate the physiological state. For this we chose actomyosin, which was extracted as a complex. These preparations contained endogenous MLCK and MLCP. Both fresh and frozen gizzards were used as a source of the actomyosin, although in terms of the phosphorylation and ATPase behavior both preparations were similar. Using a reconstituted system our concern was that certain components might be absent and consequently that the biochemical characteristics may not be representative of those occurring *in vivo*.

The time course of myosin phosphorylation and of $Mg^{2+}$-ATPase activity of actomyosin are shown in FIGURE 4. The phosphorylation of myosin is similar to

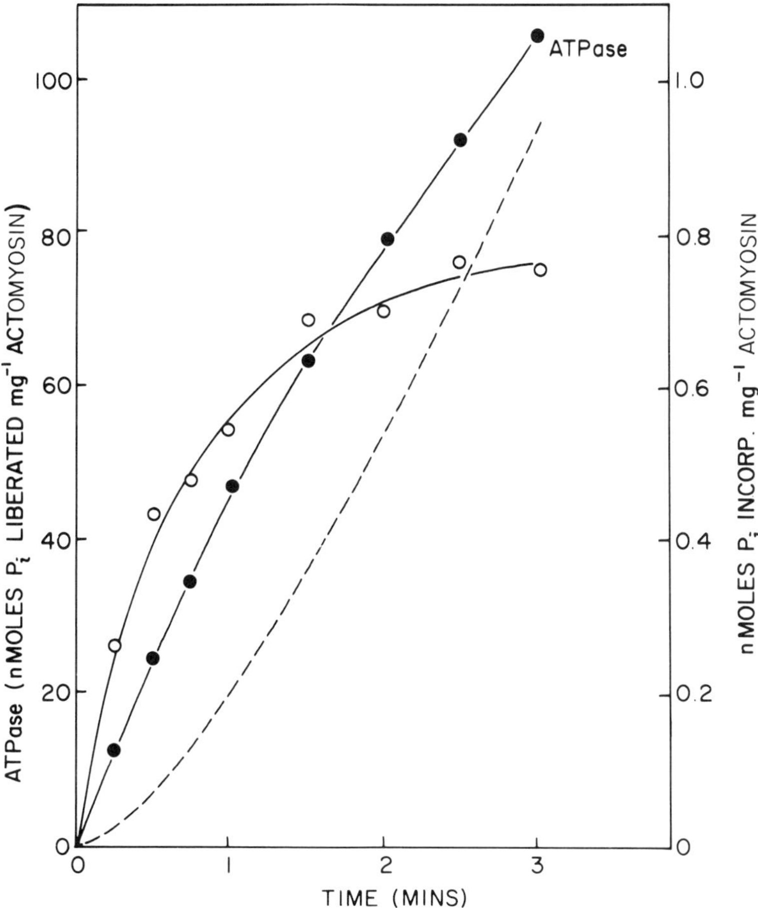


FIGURE 4. Phosphorylation and $Mg^{2+}$ ATPase activity of gizzard actomyosin. ATPase assays (●); phosphorylation assays (○). Assay conditions were as given in FIGURE 3, except that actomyosin 1 mg/ml was used. Actomyosin was prepared essentially as described earlier[36] except that 20 mM imidazole (pH 7.2) was used. The predicted ATPase activity (— —) was based on the measured $^{32}P$ incorporation assuming a specific $Mg^{2+}$ ATPase activity of 150 nmoles $P_i$ $min^{-1}$ $mg^{-1}$ actomyosin for the fully phosphorylated myosin. This relatively high specific activity was chosen in an attempt to match the observed ATPase data and to emphasize the difference between the predicted and measured curves.

that shown earlier (FIGURE 3) using the reconstituted system. The content of myosin in the actomyosin preparations was determined by estimating the area of the myosin heavy chains on sodium dodecyl sulfate polyacrylamide gels and found to be 67.5% ± 2% (6 determinations). The content of the myosin light chains of $M_r$ 20,000 was therefore approximately 2.9 nmoles/mg actomyosin. The extent of $^{32}P$ incorporation after 3 min was about 26% of the available light chains. Despite the close similarity of the time course of phosphorylation to that shown in FIGURE 3, the ATPase kinetics differed considerably. The latter was characterized by a rapid initial phase (i.e., no lag period) and a slightly slower second phase. Assuming that the ATPase profile could be approximated by two linear components, the rates for the two phases were estimated to be about 45 and 28 nmoles $P_i$/min/mg actomyosin (or, 67 and 41 nmoles $P_i$/min/mg myosin, respectively). The slower rate is similar to the values which are usually reported for the steady-state $Mg^{2+}$-ATPase activity of smooth muscle actomyosins.[1] A predicted ATPase curve is also shown in FIGURE 4. This curve was based on the observed phosphorylation data and a specific activity of 150 nmoles $P_i$/min/mg actomyosin was chosen for the fully phosphorylated myosin. The predicted and measured ATPase curves were quite distinct in that the latter did not exhibit a lag phase and with increasing time there was a tendency toward decreasing ATPase rates.

The relationship between ATPase activity and myosin phosphorylation was also examined under conditions where the ATPase activity was slightly reduced. This was achieved by decreasing the $Mg^{2+}$ concentration in the assay to 1 mM (i.e., equimolar with the ATP). The results are shown in FIGURE 5. The phosphorylation of myosin was similar to that shown above and it appears that the MLCK activity is not reduced under these assay conditions. This is consistent with an earlier finding[16] that the MLCK does not show a requirement for free $Mg^{2+}$. The ATPase data showed a marked deviation from linearity. The initial ATPase rate was quite high, estimated over the first 30 seconds as 66 nmoles $P_i$/min/mg actomyosin, and this was followed by a slower phase of approximately 17 nmoles $P_i$/min/mg actomyosin. Once again the ATPase profile did not show a lag period and the ATPase rates decreased as a function of time.

The examples presented above indicated that the phosphorylation of myosin was relatively consistent, and showed a nonlinear time dependence. The extent of phosphorylation was never complete over this time course and approached 30% of the total light chain sites. The pattern of the phosphorylation using actomyosin resembles that found earlier[20] for myosin alone (i.e., in the absence of actin and tropomyosin) and therefore the limited extent of phosphorylation is unlikely to be affected by the superprecipitation of actomyosin. One factor that has not been investigated systematically and that could influence the shape of the phosphorylation curve is the level of prephosphorylation that is present in the actomyosin preparations. If this is relatively high, approaching 50% of the available sites, then the observed nonlinearity could merely be a consequence of substrate limitation. This possibility must be explored in the future, but even allowing that this phenomenon occurs the significance of the shape of the ATPase curve is not diminished. Reference to FIGURE 2 shows that even if 50% prephosphorylation is assumed the predicted ATPase curve still does not resemble the observed ATPase data.

The reason underlying the ATPase response is not understood. The influence that superprecipitation has on the ATPase activity of actomyosin remains to be tested. It was suggested[39] that during superprecipitation the ATPase activity is higher than after superprecipitation and thus the decline in the ATPase activity

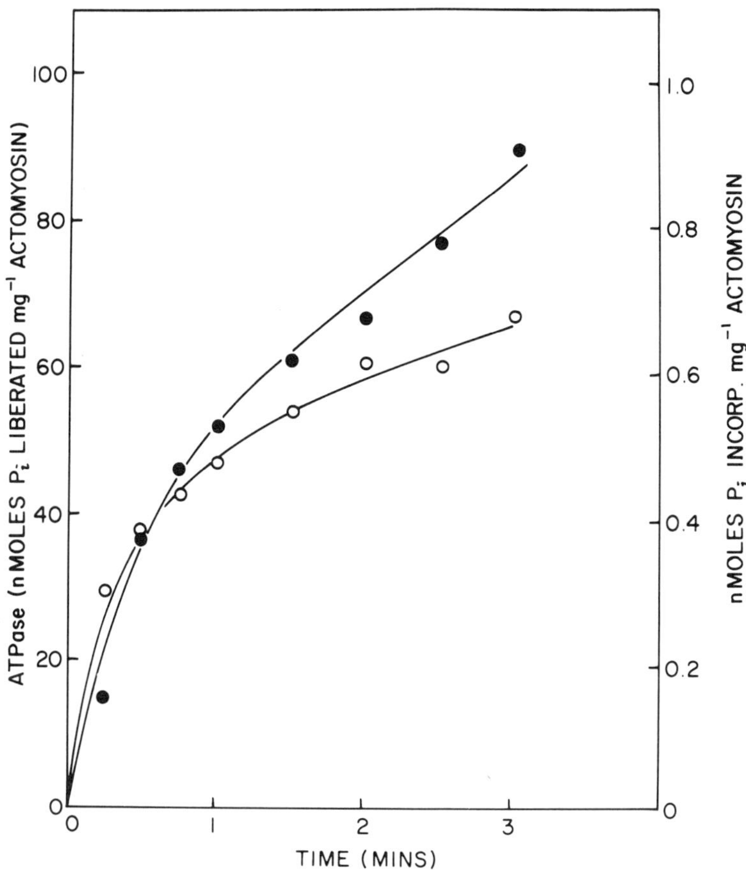


FIGURE 5. Phosphorylation and $Mg^{2+}$ ATPase activity of gizzard actomyosin. ATPase assays (●); phosphorylation assays (○). Assay conditions were as given in FIGURES 3 and 4, except that 1 mM $MgCl_2$ was used.

that was observed in our studies could be due, at least partly, to a phenomenon associated with superprecipitation. However, even if this proves to be true it is clear that activation of ATPase activity does not depend simply on the state of myosin phosphorylation, other factors are involved. This does not necessarily mean that additional protein components are implicated in the regulatory mechanism of smooth muscle, but only that the translation from myosin phosphorylation to ATPase activity is more complex than the simplest model predicts.

## CONCLUSIONS

The role of myosin phosphorylation in the regulatory mechanism of smooth muscle is discussed. The evidence in favor of this hypothesis is presented and the conclusion derived that phosphorylation is at least a component of the regulatory mechanism. There is not sufficient evidence to conclude that myosin phosphorylation is the only component of the regulatory scheme and some preliminary data are presented that suggest that our current conception of the phosphorylation theory is probably oversimplified.

## References


1. Hartshorne, D. J. & A. Gorecka. 1980. The biochemistry of the contractile proteins of smooth muscle. *In* Handbook of Physiology. R. M. Berne, A. P. Somlyo & H. V. Sparks, Eds. Vol. **2:** 93–120. American Physiology Society. Bethesda, Md.
2. Mikawa, T., Y. Nonomura & S. Ebashi. 1977. J. Biochem. (Tokyo) **82:** 1789–1791.
3. Ebashi, S., T. Mikawa, M. Hirata, T. Toyo-oka & Y. Nonomura. 1977. Regulatory proteins of smooth muscle. *In* Excitation-Contraction Coupling in Smooth Muscle. R. Casteels, T. Godfraind & J. C. Rüegg, Eds.:325–334. Elsevier/North Holland Biomedical Press. Amsterdam, Netherlands.
4. Mikawa, T., Y. Nonomura, M. Hirata, S. Ebashi & S. Kakiuchi. 1978. J. Biochem. (Tokyo) **86:** 1633–1636.
5. Sobieszek, A. 1977. Vertebrate smooth muscle myosin. Enzymatic and structural properties. *In* The Biochemistry of Smooth Muscle. N. L. Stephens, Ed.:413–443. University Park Press. Baltimore, Md.
6. Aksoy, M. O., D. Williams, E. M. Sharkey & D. J. Hartshorne. 1976. Biochem. Biophys. Res. Commun. **69:** 35–41.
7. Gorecka, A., M. O. Aksoy & D. J. Hartshorne. 1976. Biochem. Biophys. Res. Commun. **71:** 325–331.
8. Chacko, S., M. A. Conti & R. S. Adelstein. 1977. Proc. Natl. Acad. Sci. USA **74:** 129–133.
9. Marston, S. B. & E. W. Taylor. 1978. FEBS Lett. 86:167–170.
10. Dabrowska, R., D. Aromatorio, J. M. F. Sherry & D. J. Hartshorne. 1977. Biochem. Biophys. Res. Commun. **78:** 1263–1272.
11. Dabrowska, R., J. M. F. Sherry, D. K. Aromatorio & D. J. Hartshorne. 1978. Biochemistry **17:** 253–258.
12. Adelstein, R. S., M. A. Conti, D. R. Hathaway & C. B. Klee. 1978. J. Biol. Chem. **253:** 8347–8350.
13. Yazawa, M., H. Kuwayama & K. Yagi. 1978. J. Biochem. (Tokyo) **84:** 1253–1258.
14. Nairn, A. C. & S. V. Perry. 1979. Biochem. J. **179:** 89–97.
15. Walsh, M. P., B. Vallet, F. Autric, & J. G. Demaille. 1979. J. Biol. Chem. **254:** 12136–12144.
16. Hartshorne, D. J., R. F. Siemankowski & M. O. Aksoy. 1980. Ca regulation in smooth muscle and phosphorylation: some properties of the myosin light chain kinase. *In* Regulatory Mechanisms of Muscle Contraction. S. Ebashi, K. Maruyama & M. Endo, Eds. Tokyo, Japan. Sci. Soc. Press. Springer-Verlag. Berlin, Germany. (In press.)
17. Yazawa, M. & K. Yagi. 1978. J. Biochem. (Tokyo) **84:** 1259–1265.
18. Hathaway, D. R. & R. S. Adelstein. 1979. Proc. Natl. Acad. Sci. USA **76:** 1653–1657.
19. Scordilis, S. P. & R. S. Adelstein. 1978. J. Biol. Chem. **253:** 9041–9048.
20. Mrwa, U. & D. J. Hartshorne. 1980. Fed. Proc. **39:** 1564–1568.
21. Pires, E. M. V. & S. V. Perry. 1977. Biochem. J. **167:** 137–146.
22. Conti, M. A. & R. S. Adelstein. 1980. Fed. Proc. **39:** 1569–1573.
23. Small, J. V. & A. Sobieszek. 1977. Eur. J. Biochem. **76:** 521–530.
24. Ikebe, M., T. Aiba, H. Onishi & S. Watanabe. 1978. J. Biochem (Tokyo) **83:** 1643–1655.
25. Sobieszek, A. 1977. Eur. J. Biochem. **73:** 477–483.
26. Aiba, T., I. Ohtsuka, H. Onishi & S. Watanabe. 1979. J. Biochem. (Tokyo) **86:** 1275–1282.
27. Barron, J. T., M. Bárány & K. Bárány. 1979. J. Biol. Chem. **254:** 4954–4956.
28. Janis, R. A. & R. T. Gualteri. 1978. The Physiologist **21:** 59.
29. Hoar, P. E., W. G. L. Kerrick & P. S. Cassidy. 1979. Science **204:** 503–506.

30. KERRICK, W. G. L., P. E. HOAR & P. S. CASSIDY. 1980. Fed. Proc. **39:** 1558–1563.
31. SHERRY, J. M. F., A. GORECKA, M. O. AKSOY, R. DABROWSKA & D. J. HARTSHORNE. 1978. Biochemistry **17:** 4411–4418.
32. CASSIDY, P. S., P. E. HOAR & W. G. L. KERRICK. 1979. J. Biol. Chem. **254:** 11148–11153.
33. LEVIN, R. M. & B. WEISS. 1977. Mol. Pharmacol. **13:** 690–697.
34. HIDAKA, H., M. NAKA & T. YAMAKI. 1979. Biochem. Biophys. Res. Commun. **90:** 694–699.
35. ONISHI, H. & S. WATANABE. 1979. J. Biochem (Tokyo) **85:** 457–472.
36. HARTSHORNE, D. J., A. GORECKA & M. O. ASKOY. 1977. Aspects of the regulatory mechanism in smooth muscle. *In* Excitation-Contraction Coupling in Smooth Muscle. R. Casteels, T. Godfraind & J. C. Rüegg, Eds.:377–384. Elsevier/North-Holland Biomedical Press. Amsterdam, Netherlands.
37. DRISKA, S. & D. J. HARTSHORNE. 1975. Arch. Biochem. Biophys. **167:** 203–212.
38. FERENCZI, M. A., E. HOMSHER, D. R. TRENTHAM & A. G. WEEDS. 1978. Biochem. J. **171:** 155–163.
39. IKEBE, M., T. AIBA, H. ONISHI & S. WATANABE. 1978. J. Biochem. (Tokyo) **83:** 1643–1655.


## DISCUSSION OF THE PAPER

DR. R. H. KRETSINGER: Have you evidence as to whether both light chains of the myosin hexamer need to be phosphorylated? Second, have you done reconstitution experiments in which you've used light chains from skeletal or cardiac muscle? Third, I infer from what you said that there are light chain kinases in skeletal and cardiac muscle. Have you crossed them with the obvious combinations?

DR. D. HARTSHORNE: The evidence that does exist suggests that there is no requirement for both myosin heads to be phosphorylated. This is very preliminary, though. In our hands, it has so far been impossible to get skeletal muscle light chains or cardiac muscle light chains hybridized into the myosin. It is something that people are trying to do. The skeletal muscle light chain kinase is not very effective with smooth muscle myosin.

DR. KRETSINGER: What's it doing to skeletal muscle?

DR. HARTSHORNE: I don't know.

# REGULATION OF MYOSIN LIGHT CHAIN KINASE BY REVERSIBLE PHOSPHORYLATION AND CALCIUM-CALMODULIN

R. S. Adelstein, M. A. Conti, and M. D. Pato

*Section on Molecular Cardiology, Cardiology Branch*
*National Heart, Lung, and Blood Institute*
*National Institutes of Health*
*Bethesda, Maryland 20205*

## INTRODUCTION

The contractile proteins actin and myosin are present in most, if not all, eukaryotic cells. Their role in muscle contraction has been well-documented in all types of muscle cells: skeletal, cardiac, and smooth.[1] Their function in nonmuscle cells is still obscure, but they are thought to play a role in cell migration, cytokinesis, and possibly in cell division. In addition they are also thought to function in specialized cell processes, such as phagocytosis in macrophages and clot retraction in blood platelets.[2,3]

The contractile response in both muscle and nonmuscle cells results from the interaction of the globular portion of bipolar myosin filaments with the filaments formed by actin polymers (FIGURE 1). The myosin "head" attaches to the actin polymer, the angle of attachment is altered in such a manner that the filaments slide past each other, and muscle contraction or its equivalent in nonmuscle cells, takes place.[4,5] The energy for this process is provided by ATP and is released by the interaction of actin with myosin; actin activates the ability of myosin to hydrolyze ATP.

The *in vitro* biochemical correlate of the *in vivo* contractile response is the actin-activated MgATPase activity of myosin. In the laboratory this activity is measured at relatively low ionic strength ($<50$ mM KCl) in the presence of $Mg^{2+}$ and can be quantitated by using purified proteins.

One major difference between myosin purified from skeletal and cardiac muscle and myosin isolated from smooth muscle and nonmuscle cells involves the ability of actin to activate the myosin MgATPase activity. Whereas skeletal and cardiac muscle myosin MgATPase activity can be activated by actin without prior modification of the myosin molecule, myosin isolated from smooth muscle and nonmuscle cells cannot. These myosins require a covalent phosphorylation of the 20,000 dalton light chain of myosin as a prerequisite for activation of their MgATPase activity by actin (FIGURE 2).[6]

Although phosphorylation of skeletal and cardiac muscle myosin has no effect on the steady-state, actin-activated myosin MgATPase activity, both skeletal and cardiac muscles contain the necessary enzymes to carry out myosin phosphorylation and dephosphorylation. Moreover there is evidence from studies with intact skeletal muscle that phosphorylation of skeletal muscle myosin increases with tetanic stimulation.[7,8] Myosin phosphorylation appears to modulate actin-myosin interaction in skeletal and cardiac muscle, in contrast to smooth muscle and nonmuscle cells, where it appears to be the dominant regulatory system.

At present there are three well-characterized regulatory systems controlling



0077-8923/80/0356-0142 $01.75/0 © 1980, NYAS

actin-myosin interactions in muscle and nonmuscle cells: a) the reversible phosphorylation of myosin; b) the direct binding of $Ca^{2+}$ to myosin; and c) the interaction of actin with troponin-tropomyosin.

The first of these is the dominant regulatory system in smooth muscle and nonmuscle cells, and the $Ca^{2+}$-binding protein calmodulin plays a major role in this mechanism.[6,9] The second regulatory mechanism, which exists in invertebrate cells such as scallops, has been extensively described by Szent-Gyorgyi and his co-workers.[10] The third mechanism is the dominant regulatory system in skeletal and cardiac muscle cells. The $Ca^{2+}$-binding protein troponin C, which resembles calmodulin in its structure,[11] plays an important role in this system[6,12] (FIGURE 1).

We have been studying the myosin phosphorylating system in smooth muscle and nonmuscle cells. In this system $Ca^{2+}$ initiates contraction by binding to the ubiquitous calcium-binding protein calmodulin. Calcium-calmodulin then activates an enzyme, myosin light chain kinase which phosphorylates the 20,000

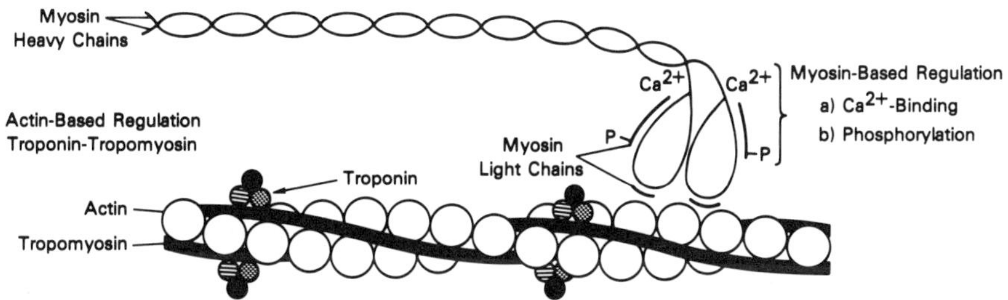


FIGURE 1. Diagrammatic representation of the three major regulatory systems controlling actin-myosin interaction (see text). Each of these regulatory systems plays a dominant role in a different type of muscle. Note that the exact position of the myosin light chains on the myosin head is not known. (Reprinted from Reference 6).

dalton light chain of myosin. Phosphorylated myosin, but not unphosphorylated myosin, can interact with actin and contraction can occur. The system is cyclical, in that a phosphatase can catalyze dephosphorylation of myosin, restoring it to a form that cannot be activated by actin (FIGURE 2).

We have been particularly interested in three aspects of this regulatory process: a) How $Ca^{2+}$-calmodulin regulates the enzyme myosin light chain kinase; b) How phosphorylation of myosin light chain kinase by cAMP-dependent protein kinase modulates the activity of myosin kinase; and c) The purification and characterization of a phosphatase that dephosphorylates myosin kinase as well as the 20,000 dalton light chain of myosin.

## PROPERTIES OF MYOSIN KINASE

Myosin light chain kinase was purified from turkey gizzards[13,14] and human platelets[15] as described previously. The final step of purification was chromatography on an affinity column of calmodulin bound to Sepharose 4B. The purified kinase from smooth muscle has a molecular weight of approximately 130,000 by SDS-polyacrylamide gel electrophoresis. The platelet kinase has a molecular

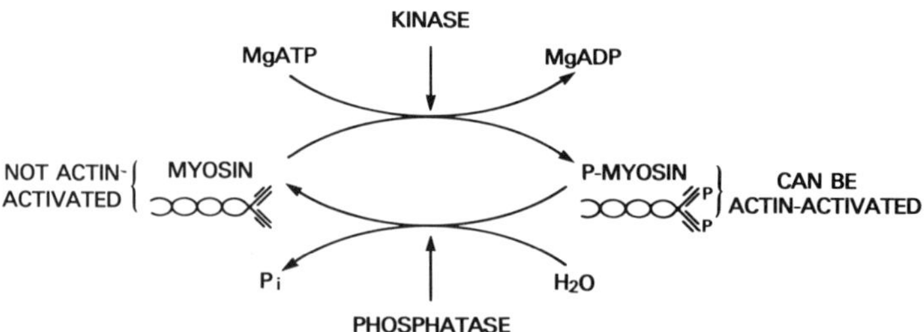


FIGURE 2. Control and effect of myosin phosphorylation. Stick diagram of the myosin molecule illustrates the heavy chains, which are coiled about each other in the rod portion of the molecule, and the two sets of light chains. The larger light chain (20,000 daltons) is phosphorylated.

weight of 105,000. Both kinases have an absolute requirement for calcium and calmodulin for activity.[15,16] TABLE 1 shows the $Ca^{2+}$-calmodulin requirement of the smooth muscle kinase. Note that the enzyme is inhibited in the absence of $Ca^{2+}$ or calmodulin.

This requirement for calmodulin can be removed by digesting smooth muscle myosin kinase briefly with trypsin. Whether there is a concomitant loss of kinase activity depends on whether or not calmodulin is bound to the kinase during digestion. The results of partial proteolysis are consistent with the idea that: 1) Myosin kinase has at least one site, distinct from the active site, which can bind calmodulin. 2) This site or a nearby, interacting site, inhibits the enzyme in the absence of calmodulin. 3) Proteolysis removes this particular part of the molecule, relieving the inhibition, and allowing the kinase to be active in the absence of $Ca^{2+}$-calmodulin.

The stoichiometry of binding of calmodulin to the smooth muscle myosin kinase is one:one.[17] This has been demonstrated by cross-linking the kinase with dimethyl suberimidate in the presence and absence of bound $^{14}C$-labeled cal-

TABLE 1

DEPENDENCE OF SMOOTH MUSCLE MYOSIN LIGHT CHAIN KINASE ON CALCIUM AND CALMODULIN[14]

| Myosin Kinase Assay* | Incorporation into L.C./min. (dpm) | Kinase Specific Activity, ($\mu$mol/mg per min) |
|---|---|---|
| + $Ca^{2+}$ + calmodulin | 278,000 | 3.0 |
| + EGTA + calmodulin | 3,000 | — |
| + $Ca^{2+}$ | 3,900 | — |
| + EGTA | 2,900 | — |

*Myosin light chain kinase activity was assayed in 0.1 ml, 20 mM Tris-HCl, pH 7.3, 10 mM $MgCl_2$, 0.1 mM $AT^{32}P$ (0.5 Ci/mmol), $10^{-8}$M myosin kinase, 0.2 mg/ml smooth muscle myosin light chain at 24°C. Either 0.2 mM $CaCl_2$ (in excess over EGTA) or 2 mM EGTA was present in the assay and $10^{-6}$ M calmodulin was added to the assay as indicated above. Background level of dpm for the Millipore assay performed under the above conditions is 3,000–4,000. Assays in the absence of calmodulin or calcium showed no increase in dpm with time over the background.

modulin. It has also been demonstrated by titrating the enzyme with calmodulin. Full activation of myosin kinase occurs at a molar ratio of one calmodulin per kinase.

## PHOSPHORYLATION OF MYOSIN KINASE

Myosin light chain kinase isolated from both smooth muscle cells[14] and human platelets[18] is a substrate for the catalytic subunit of protein kinase. Phosphorylation of both enzymes, in the absence of calmodulin, results in the

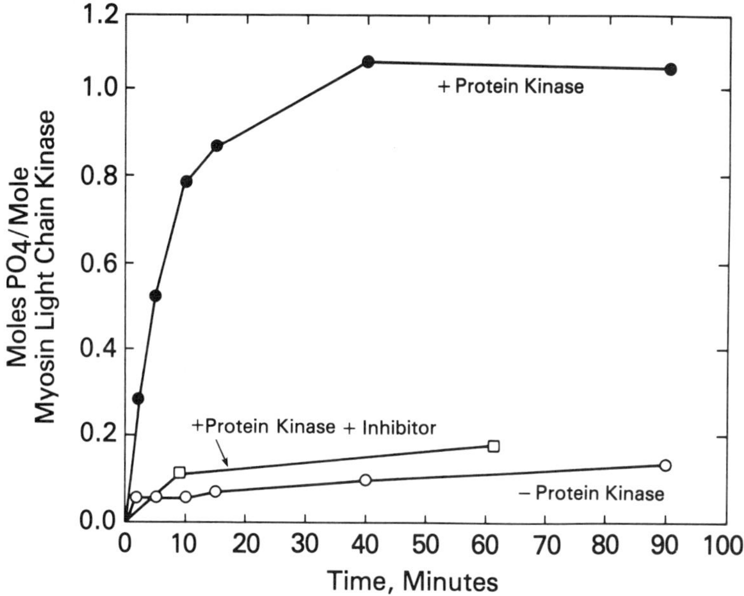


FIGURE 3. Time dependence of incorporation of $^{32}P$ from $\gamma$-$AT^{32}P$ into myosin light chain kinase by the catalytic subunit of cAMP-dependent protein kinase. $^{32}P$ incorporation into myosin kinase was assayed in the presence of added protein kinase (●) or of protein kinase plus the specific inhibitor of protein kinase (□), or in the absence of either protein kinase or inhibitor (○). Assays were performed by Millipore filtration and location of radioactive label in the 130,000 dalton band of the mysoin kinase confirmed by radioautography of slab gels. (Reprinted from Reference 14).

incorporation of approximately 1 mole of phosphate/mole of kinase (FIGURE 3). The effect of phosphorylation is a decrease in the activity of myosin kinase (TABLE 2). In the case of smooth muscle, the decrease in kinase activity can be attributed to a marked decrease in the binding of calmodulin by the phosphorylated enzyme (FIGURE 4). There is only a slight effect of phosphorylation on the maximum velocity of the kinase.[14] The kinase isolated from human platelets also shows a decreased activity upon phosphorylation. However the phosphorylated platelet enzyme shows a decrease in both maximum velocity and the binding of

TABLE 2
THE EFFECT OF PHOSPHORYLATION AND DEPHOSPHORYLATION ON SMOOTH MUSCLE MYOSIN KINASE ACTIVITY[23]

| | Specific Activity* ($\mu$mol/mg/min) |
|---|---|
| Prior to phosphorylation | 4.25 |
| Phosphorylated | 0.42 |
| Following dephosphorylation | 4.12 |

*$6.7 \times 10^{-9}$ M calmodulin and 0.2 mM $CaCl_2$; activity was measured by determining the transfer of $^{32}P$ from $AT^{32}P$ to the 20,000 dalton light chain of smooth muscle myosin.[14]

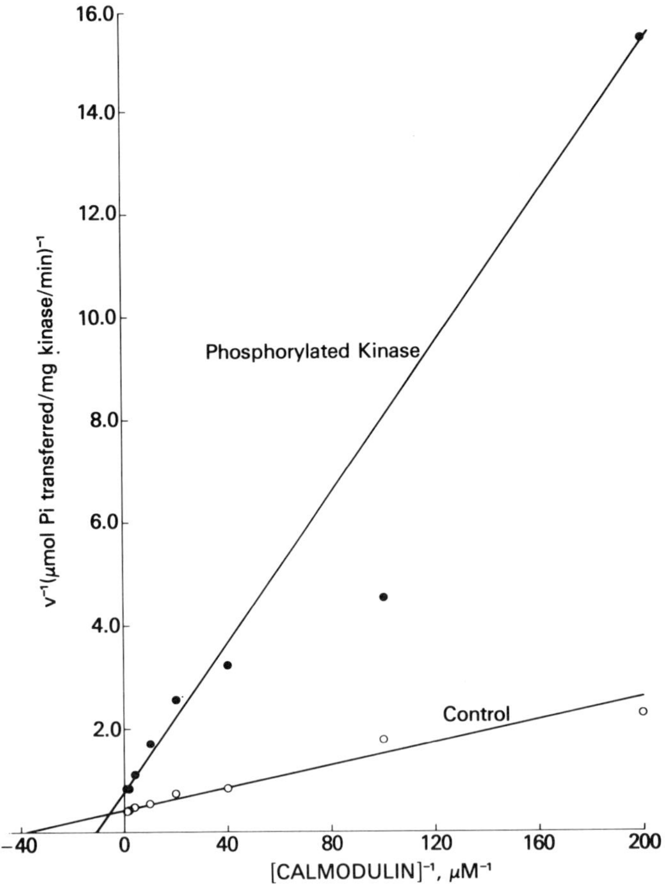


FIGURE 4. Double reciprocal plot of calmodulin concentration vs myosin kinase activity. The data for this plot was obtained from titration curves of the phosphorylated (●) and unphosphorylated (○) kinases with calmodulin. Myosin light chain kinase activity was assayed by the ability to transfer $^{32}P$ from $\gamma$-$AT^{32}P$ to smooth muscle myosin light chains at 24°C. Kinase concentration was $10^{-9}$M and assay time was typically 3–5 min. Incorporation was linear with time under the assay conditions. Calmodulin concentration was determined by absorption at 230 nm ($\epsilon^{1\%}_{230nm} = 25$) and appropriate dilutions made into 1 mg/ml bovine serum albumin monomer. (For details see Conti & Adelstein.[14])

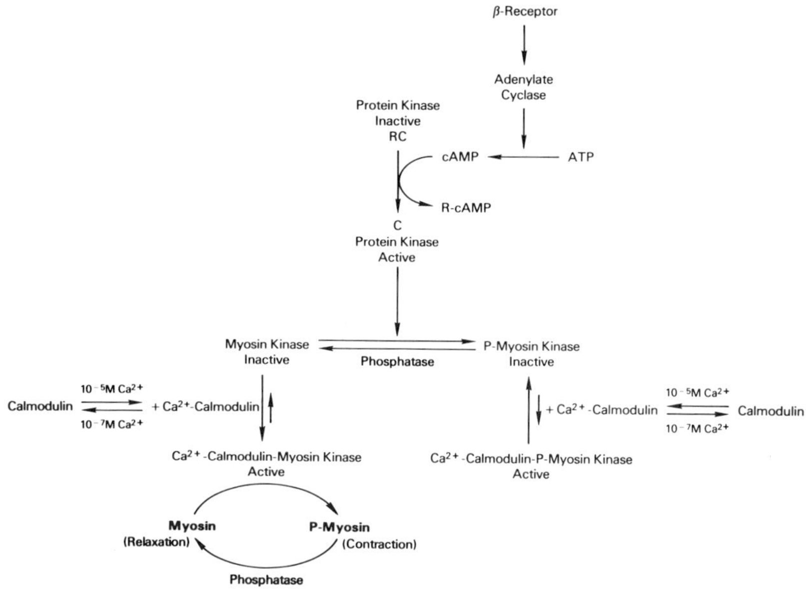


FIGURE 5. A schematic representation of one mechanism by which calcium and cyclic adenosine 3′:5′-monophosphate regulate smooth muscle contraction. Calcium binds to calmodulin (16,500 daltons) which permits binding of calmodulin to myosin light chain kinase. The active calmodulin-myosin kinase complex phosphorylates myosin (P-myosin), which results in contraction. Dephosphorylation of myosin by a phosphatase promotes relaxation. Cyclic adenosine 3′:5′-monophosphate is synthesized from ATP by the enzyme, adenylate cyclase, which is coupled to specific receptors on the outer smooth muscle cell membrane (i.e., the beta-adrenergic receptor). Cyclic adenosine 3′:5′-monophosphate activates protein kinase by liberating a catalytic subunit (C), which can phosphorylate myosin kinase. The phosphorylated kinase binds calmodulin much more weakly than the unphosphorylated kinase and therefore at the same calmodulin concentration the phosphorylated myosin kinase is in the inactive form while the unphosphorylated kinase is in an active form (see also FIGURE 4). This equilibrium is qualitatively reflected in the lengths of the arrows indicating conversion of myosin kinase from inactive to active forms upon binding of calmodulin. The net effect of this sequence is to inhibit phosphorylation of myosin, thus favoring relaxation. (Reprinted from Reference 14).

TABLE 3

PROPERTIES OF MYOSIN PHOSPHATASES I AND II

| | | | Specific Activity (μmol/mg/min) | |
|---|---|---|---|---|
| | Molecular Weight | $Mg^{2+}$-dependent | Myosin Light Chains (20,000 daltons) | Myosin Kinase |
| Phosphatase I | 153,000 (trimer of 60,000 55,000 and 38,000) | No | 0.53 | 0.32 |
| Phosphatase II | 43,000 | Yes | 0.32 | 0.01 |

Substrate concentrations were approximately 2 μM.

calmodulin. (D. R. Hathaway, C. R. Eaton & R. S. Adelstein, unpublished results).

The significance of phosphorylation of myosin kinase is not known at the present time, and it will not be known until it is shown to be an effective modulatory process *in vivo*. The manner in which a muscle responds to an increase in cyclic AMP depends on the muscle type. In smooth muscle, a rise in cyclic AMP is associated with muscle relaxation, whereas in cardiac muscle it is associated with an increased contractile response. The mechanism by which a rise in cyclic AMP results in a decrease in contraction in smooth muscle and its equivalent in platelets might be related to the phosphorylation of myosin kinase. As outlined above (and in FIGURE 5) phosphorylation of smooth muscle myosin is required for smooth muscle contraction. A decrease in the activity of the enzyme that catalyzes myosin phosphorylation would result in an increase in unphosphorylated myosin, which cannot interact with actin. Phosphorylation of myosin kinase decreases the ability of the kinase to bind calmodulin and results in a decreased kinase activity. Recent experiments utilizing a crude fraction of actomyosin isolated from vascular smooth muscle support the idea that an increase in the activity of cyclic AMP-dependent protein kinase is associated with a decrease in myosin light chain phosphorylation.[19] Moreover, utilizing a preparation of skinned smooth muscle fibers Hoar & Kerrick demonstrated that an increase in protein kinase activity resulted in a decrease in tension.[20] However, whether phosphorylation of myosin kinase is the sole mechanism causing this decrease in contractility and how it may integrate with other proposed control mechanisms in smooth muscle is still not known.

## THE ROLE OF PHOSPHATASES

We have also been interested in the enzymes that catalyze the reverse of the phosphorylating steps, namely phosphatases that dephosphorylate myosin as well as myosin kinase (FIGURE 5). To date there has been only one detailed report on myosin light chain phosphatases. Morgan *et al.* isolated a highly specific phosphatase from skeletal muscle.[21] Recently Mary D. Pato in our laboratory isolated two different phosphatases from turkey gizzard smooth muscle.[22]

Both phosphatases were purified using thiophosphorylated light chains coupled to Sepharose 4B. The two phosphatases differed in a number of properties, summarized in TABLE 3. Phosphatase I is a trimer, being composed of three different subunits, 60,000 daltons, 55,000 daltons, and 38,000 daltons. It differs from phosphatase II, which is a monomer with a molecular weight of 43,000 by SDS-polyacrylamide gel electrophoresis, in that it can also dephosphorylate myosin light chain kinase in addition to the phosphorylated light chain of myosin. Moreover phosphatase I has no divalent cation requirement whereas phosphatase II is dependent on $Mg^{2+}$ for activity.

The isolation of a phosphatase from smooth muscle that catalyzed dephosphorylation of myosin light chain kinase allowed us to see if the effect of phosphorylating that enzyme could be reversed. TABLE 2 shows that dephosphorylation of the phosphorylated myosin light chain kinase restores the kinase activity to the level found prior to phosphorylation. The mechanism by which it increases the kinase activity appears to be by restoration of the high affinity of the dephosphorylated myosin kinase for calmodulin (M. A. Conti & R. S. Adelstein, unpublished results).

The factors that regulate myosin phosphatases are not known at present. One possibility is that the light chain phosphatase is continually acting to dephosphorylate myosin, and activation of the kinase by $Ca^{2+}$-calmodulin reverses its effect when $Ca^{2+}$ initiates smooth muscle contraction. However the presence of at least two phosphatases that catalyze dephosphorylation of smooth muscle myosin, one of which can also catalyze the dephosphorylation of myosin kinase (which would favor contraction, rather than relaxation) suggests that the regulation of phosphatases will be both important and complex.

## SUMMARY

1) Myosin light chain kinases from smooth muscle and platelets can be phosphorylated by the catalytic subunit of cAMP-dependent protein kinase. 2) Phosphorylation of both kinases, in the absence of calmodulin, markedly decreases kinase activity. 3) The decrease in smooth muscle myosin kinase activity is due to a decreased affinity of the phosphorylated kinase for calmodulin. 4) Dephosphorylation of the smooth muscle kinase by a phosphatase isolated from smooth muscle restores the affinity of the kinase for calmodulin.

## ACKNOWLEDGMENT

The authors gratefully acknowledge the expert editorial assistance of Mrs. Exa Murray.

## REFERENCES

1. TAYLOR, E. W. 1979. C. R. C. Crit. Rev. Biochem. **6:**103–164.
2. KORN, E. D. 1978. Proc. Natl. Acad. Sci. USA **75:** 588–599.
3. DEDMAN, J. R., B. R. BRINKLEY & A. R. MEANS. 1979. Adv. Cycl. Nucl. Res. **11:** 131–174.
4. HUXLEY, A. F. & R. NIEDERGERKE. 1954. Nature **173:** 971–973.
5. HUXLEY, H. E. & J. HANSON. 1954. Nature **173:** 973–976.
6. ADELSTEIN, R. S. & E. EISENBERG. 1980. Ann. Rev. Biochem. **49:** 921–956.
7. MANNING, D. R. & J. T. STULL. 1979. Biochem. Biophys. Res. Commun. **90:** 164–170.
8. BARANY, K., M. BARANY, J. M. GILLIS & M. J. KUSHMERICK. 1979. J. Biol. Chem. **254:** 3617–3623.
9. DABROWSKA, R., D. AROMATORIO, J. M. F. SHERRY & D. J. HARTSHORNE. 1978. Biochemistry **17:** 253–258.
10. SZENT-GYORGYI, A. G. 1975. Biophysical Journal **15:** 707–723.
11. WATTERSON, D. M., F. SHARIEF, & T. C. VANAMAN. 1980. J. Biol. Chem. **255:** 962–975.
12. WEBER, A. & J. M. MURRAY. 1973. Physiol. Rev. **53:** 612–673.
13. ADELSTEIN, R. S., M. A. CONTI, D. R. HATHAWAY & C. B. KLEE. 1978. J. Biol. Chem. **253:** 8347–8350.
14. CONTI, M. A. & R. S. ADELSTEIN. 1980. Fed. Proc. **39:**1569–1573.
15. HATHAWAY, D. R. & R. S. ADELSTEIN. 1979. Proc. Natl. Acad Sci. USA **76:** 1653–1657.
16. DABROWSKA, R. & D. J. HARTSHORNE. 1978. Biochem. Biophys Res. Commun. **85:** 1352–1359.
17. ADELSTEIN, R. S. & C. B. KLEE. 1980. *In* Calcium and Cell Function. W. Y. Cheung, Ed. Vol. 1. Academic Press, Inc. New York, N.Y. (In Press).
18. HATHWAY, D. R., C. R. EATON & R. S. ADELSTEIN. 1980. *In* The Regulation of Coagulation. K. G. Mann & F. B. Taylor, Eds.: 271–276. Elsevier, New York, N.Y.
19. SILVER, P. J. & J. DiSALVO. 1979. J. Biol. Chem. **254:** 9951–9954.

20. Hoar, P. E. & W. G. L. Kerrick. 1980. Fed Proc. **39:** 1817a.
21. Morgan, M., S. V. Perry & J. Ottaway. 1976. Biochem. J. **157:** 687–697.
22. Pato, M. D. & R. S. Adelstein. 1980. J. Biol. Chem. **255:** 6535–6538.
23. Adelstein, R. S., M. D. Pato & M. A. Conti. 1980. *In* Regulatory Mechanisms of Muscle Contraction. S. Ebashi, *et al.*, Eds. Japan Sci. Soc. Press. Tokyo Springer-Verlag. Berlin. (In press.)

## Discussion of the Paper

Dr. D. M. Watterson: Could you summarize the evidence for autophosphorylation? We have found that cyclic AMP-dependent protein kinase will copurify with calmodulin binding proteins and will actually interact with calmodulin-Sepharose resins. We do not know if this interaction is real or artifactual, but if you put in $^{32}$P-labeled ATP you will readily get radioactivity incorporated into protein.

Dr. R. S. Adelstein: We get very little autophosphorylation either in the smooth muscle enzyme or in the platelet enzyme. We see very rapid autophosphorylation which is completely calmodulin dependent in the cardiac enzyme. We tested for histone phosphorylating activity and found none. The cyclic AMP-dependent protein kinase inhibitor protein does not inhibit the autophosphorylation.

Dr. D. Hartshorne: We found that the autophosphorylation occurred under all conditions as soon as ATP is present. There is no requirement for calcium or calmodulin. The levels of autophosphorylation were less than one mole of isotope incorporated per mole of protein.

Dr. Watterson: What about the addition of protease inhibitors to your preparations?

Dr. Adelstein: We have tried a number of known protease inhibitors but they do not entirely inhibit protease activity. We find that storing the kinase at −70°C in the presence of EGTA will inhibit the proteolysis.

Dr. Hartshorne: There is no effective protease inhibitor that we know about.

Dr. F. F. Vincenzi: How significant, physiologically, is the decrease in affinity for calmodulin after phosphorylation?

Dr. Adelstein: The physiological relevance is unclear.

Dr. Vanaman: What is the relationship between the phosphorylation domain and the calmodulin binding domains on the myosin light chain kinase?

Dr. Adelstein: They are not the same. You can remove the phosphorylated domain with a brief trypsin treatment and still have a calcium sensitive kinase.

Dr. D. Flockhart (*Vanderbilt University, Nashville, TN*): Did you do a phosphate determination on the kinase before you did your phosphorylation experiments?

Dr. Adelstein: No.

# CALCIUM CONTROL OF MUSCLE PHOSPHORYLASE KINASE THROUGH THE COMBINED ACTION OF CALMODULIN AND TROPONIN*

Philip Cohen,† Claude B. Klee,‡ Colin Picton,† and Shirish Shenolikar†

*†Department of Biochemistry*
*University of Dundee*
*Dundee, DD1 4HN*
*Scotland*
*and*
*‡The National Cancer Institute*
*National Institutes of Health*
*Bethesda, Maryland 20014*

## 1. Discovery of the Calcium Control of Phosphorylase Kinase

Phosphorylase kinase catalyzes the formation of active phosphorylase *a* from inactive phosphorylase *b* and thereby stimulates glycogenolysis in skeletal muscle.[1] Since phosphorylase *a* is formed within seconds of the onset of muscle contraction,[2,3] it was realized many years ago that a mechanism must exist for coupling glycogenolysis and contraction, allowing energy production to be linked to the energy requirements of the tissue.

Electrical excitation of muscle causes $Ca^{2+}$ to be released from the sarcoplasmic reticulum leading to elevated levels of $Ca^{2+}$ in the muscle sarcoplasm. The $Ca^{2+}$ bind to troponin C, one of the protein components of the troponin complex, causing it to bind more tightly to the inhibitory component of the complex, troponin I. This prevents troponin I from inhibiting the interaction of actin with myosin and therefore allows contraction to take place.[4]

Soon after the discovery of phosphorylase kinase it became obvious that $Ca^{2+}$ would be a very attractive candidate as an activator of the enzyme, and by 1971 it had been established that phosphorylase kinase was almost completely dependent on $Ca^{2+}$, the activation occurring at concentrations believed to exist in contracting muscle.[5-8] In a particularly elegant experiment, it was shown that the conversion of phosphorylase *b* to *a in vitro* could be arrested instantaneously by the addition of a sarcoplasmic reticulum preparation and restarted by the readdition of $Ca^{2+}$.[8] This made the important demonstration that the naturally occurring calcium chelator could compete successfully for the $Ca^{2+}$ that were tightly bound to phosphorylase kinase, indicating that the process could be reversible *in vivo*.

## 2. Identification of the $\delta$ Subunit of Phosphorylase Kinase as Calmodulin

In 1973 two laboratories reported that phosphorylase kinase contained three types of subunit, termed $\alpha$, $\beta$, and $\gamma$, whose molecular weights were 145,000,

*Supported by research grants from the Medical Research Council, London and the British Diabetic Association.



0077-8923/80/0356-0151 $01.75/0 © 1980, NYAS

128,000 and 45,000 respectively.[9,10] The $\alpha$ and $\beta$ subunits were shown to be the components phosphorylated by cyclic AMP-dependent protein kinase,[10,11] a reaction which enhances the activity considerably (see Section 7). Very recently, evidence has been presented which indicates that the $\gamma$ subunit is likely to be the catalytic subunit of the enzyme.[12]

These findings raised the question of which subunit bound the $Ca^{2+}$ that were essential for activity, since the subunit structure appeared to exclude the involvement of calmodulin, molecular weight 17,000. However, in 1978 we reported that phosphorylase kinase contained a fourth subunit termed the $\delta$ subunit.[13] This component was much smaller than the $\alpha$, $\beta$, and $\gamma$ subunits and due to its acidic nature stained rather poorly with Coomassie blue. It was therefore missed for several years as a faint band migrating with the bromophenol dye-front on the 5% polyacrylamide gels previously used to resolve the $\alpha$, $\beta$, and $\gamma$ subunits.[10,11] This component was however clearly visible on 7.5% or 10% polyacrylamide gels.[13]

The $\delta$ subunit has been shown to be identical to calmodulin. This was initially suggested by its heat stability, behavior on ion exchange chromatography, and mobility on polyacrylamide gels in the presence of sodium dodecyl sulphate, and confirmed by its amino acid composition and ability to activate other calmodulin-dependent enzymes.[13] The amino acid sequence of the $\delta$ subunit has now been completed.[14] It is identical to that reported for bovine brain calmodulin[15] except for two amino acid substitutions; asparagine-24 being changed to aspartic acid, and glutamic acid-135 being changed to glutamine.

## 3. The $\delta$ Subunit is the Component Which Confers Calcium Sensitivity to the Phosphorylase Kinase Reaction

It has been established that the $\delta$ subunit is present in equimolar proportions with the $\alpha$, $\beta$, and $\gamma$ subunits.[13,16] Phosphorylase kinase therefore has the structure $(\alpha\beta\gamma\delta)_4$ and a molecular weight of 1,300,000.[16]

Kilimann and Heilmeyer[17] measured the binding of $Ca^{2+}$ to phosphorylase kinase in 1.0 mM sodium glycerophosphate buffer pH 6.8. Under these conditions each molecule of enzyme (molecular weight 1,300,000) bound 12 $Ca^{2+}$ with a dissociation constant of $1.6 \times 10^{-8}$ M and a further 4 $Ca^{2+}$ with a dissociation constant of $6 \times 10^{-7}$ M. These measurements, which were made before calmodulin was identified as a subunit, are in remarkably good agreement with the known calcium binding properties of calmodulin, which is known to bind 4 molecules of $Ca^{2+}$ per mole. Furthermore, the existence of more than one class of calcium binding site is well documented.[18] The calcium binding studies therefore support the idea that all the high affinity calcium binding sites on phosphorylase kinase are located on the $\delta$ subunit. This in turn suggests that calmodulin is the component which confers calcium sensitivity to the phosphorylase kinase reaction. The recent finding that antibody to rat testis calmodulin[19] or the $\delta$ subunit itself[20] inhibits phosphorylase kinase is also consistent with the idea that calmodulin is associated with the catalytic center of the enzyme.

The calcium binding properties of phosphorylase kinase do however indicate that the interaction of the $\delta$ subunit with the enzyme increases the affinity of calmodulin for $Ca^{2+}$. This would be analogous to the situation with troponin C, where the calcium binding properties of the protein are altered considerably through its association with troponin I in the troponin complex.[21]

## 4. Phosphorylase Kinase Binds a Second Molecule of Calmodulin, Termed the δ′ Subunit, Which Produces Additional Activation of the Enzyme

Although all preparations of phosphorylase kinase contain stoichiometric quantities of calmodulin, the activity of the enzyme is increased by the addition of further calmodulin to the assays.[16] Half-maximal activation is observed at 0.01 μM calmodulin.[16,22]

This additional activation is caused by the binding of a second molecule of calmodulin to phosphorylase kinase. The calmodulin-stimulated activity can be prevented by addition of either the anti-psychotic drug trifluoperazine[22] or the calmodulin binding protein isolated from brain, termed calcineurin.[23,24] In contrast, these compounds have little or no effect on the calcium-dependent activity in the absence of calmodulin.[23,24] It would therefore appear that trifluoperazine and calcineurin differentiate between the tightly bound molecule of calmodulin, the δ subunit, which is an integral component of the enzyme, and a second molecule of calmodulin, the δ′ subunit which stimulates the activity.

The presence of a second calmodulin binding site has been established by direct binding studies using $^{14}C$-labeled calmodulin.[25] If phosphorylase kinase is mixed with [$^{14}C$]calmodulin and subjected to glycerol density gradient centrifugation, 1 molecule of [$^{14}C$]calmodulin per $\alpha\beta\gamma\delta$ unit remains associated with phosphorylase kinase, provided that $Ca^{2+}$ are present in the gradient. If $Ca^{2+}$ are excluded, no [$^{14}C$]calmodulin is bound to the enzyme. This demonstrates that the interaction of the δ′ subunit with phosphorylase kinase requires $Ca^{2+}$, and also that the exchange between [$^{14}C$]calmodulin and the δ subunit is very slow.[25] Indeed this technique has shown that the rate of exchange of [$^{14}C$]calmodulin with the δ subunit is only 15% per week in the absence of $Ca^{2+}$ at pH 7.0 and 0°C.[25]

Phosphorylase kinase binds to calmodulin-Sepharose in the presence of $Ca^{2+}$ and can be displaced if the $Ca^{2+}$ are replaced by EDTA.[16] This also demonstrates that the interaction of phosphorylase kinase with the δ′ subunit is calcium dependent.

## 5. Interaction of Phosphorylase Kinase with the δ and δ′ Subunits, and Identification of the Calmodulin Binding Subunits

The preceding sections have demonstrated that the δ and δ′ subunits can be easily distinguished. The δ subunit remains bound to the enzyme even in the presence of EDTA or EGTA and this explains why it is recovered in stoichiometric amounts with the α, β, and γ subunits in purified phosphorylase kinase.[10] Conversely, the δ′ subunit is only bound in the presence of $Ca^{2+}$ and this explains why it is not found in purified phosphorylase kinase.

Although the δ subunit is firmly bound to phosphorylase kinase in the absence of $Ca^{2+}$, its binding is even stronger in the presence of $Ca^{2+}$. Thus the δ subunit remains associated with the enzyme even in the presence of 8 M urea provided that $Ca^{2+}$ are present.[16] This property has allowed two-dimensional protein separation techniques to be used to demonstrate that the δ subunit is complexed to the γ subunit.[25] This result has been confirmed by cross-linking studies using dimethylsuberimidate that result in the formation of a γδ cross-linked complex of molecular weight 60,000.[25]

If phosphorylase kinase is mixed with [$^{14}$C]calmodulin and immediately cross-linked with dimethylsuberimidate, two $^{14}$C-labeled cross-linked species are formed that have molecular weights of 162,000 and 145,000, respectively.[25] These species are only formed in the presence of $Ca^{2+}$ (unlike the $\gamma\delta$ complex) and therefore represent cross-linked complexes involving the $\delta'$ subunit. The molecular weights indicate that the two species have the structures $\delta'\alpha$ and $\delta'\beta$. This conclusion is supported by experiments in which phosphorylase kinase is subjected to limited proteolysis with trypsin and chymotrypsin.

Limited proteolysis leads to considerable activation of phosphorylase kinase (Section 7). Incubation with trypsin cleaves the $\alpha$ and $\beta$ subunits to species of lower molecular weight, leaving the $\gamma$ and $\delta$ subunits untouched. The trypsin-activated enzyme no longer binds to calmodulin-Sepharose,[25,26] indicating that the ability to interact with the $\delta'$ subunit has been destroyed. Incubation with chymotrypsin leads to a selective cleavage of the $\alpha$ subunit, leaving the $\beta$, $\gamma$, and $\delta$ subunits untouched. The chymotrypsin-activated enzyme shows reduced binding to calmodulin-Sepharose, and the cross-linked species of molecular weight 162,000 is no longer observed after incubation with dimethylsuberimidate.[25] Thus the results of both cross-linking and limited proteolysis demonstrate that the $\delta'$ subunit interacts with both the $\alpha$ and $\beta$ subunits of phosphorylase kinase.

## 6. Troponin C, the Troponin Complex, and Artificial Thin Filaments Can Substitute for the $\delta'$ Subunit in the Activation of Phosphorylase Kinase

Troponin C is 50% homologous to calmodulin in its amino acid sequence, and is present in very high concentrations in skeletal muscle. It has been shown that troponin C, the troponin complex, and even artificial thin filaments made by mixing actin, tropomyosin and the troponin complex at physiological concentrations, can replace the $\delta'$ subunit as the activation of phosphorylase kinase.[22,23] Saturating concentrations of troponin C or the $\delta'$ subunit produce the same degree of activation of phosphorylase kinase, and no further activation is observed if the two protein activators are added together.[22,23]

The $A_{0.5}$ for troponin C or troponin complex is near 1.0 $\mu$M, about 100-fold greater than the $A_{0.5}$ for the $\delta'$ subunit.[22,23] The possibility that the activation by troponin is caused by a 1% contamination with calmodulin has however been eliminated by the use of the calmodulin-binding protein termed calcineurin. This protein blocks the activation by the $\delta'$ subunit, but has no effect on the activation by troponin.[23] The detailed kinetics of activation of phosphorylase kinase by troponin (Section 7) also exclude trace contamination with calmodulin as the cause of the observed activation.

It would be anticipated that troponin C, like the $\delta'$ subunit binds to the $\alpha$ and $\beta$ subunits of phosphorylase kinase, but this demonstration has not yet been made.

## 7. Regulation of the Different Forms of Phosphorylase Kinase by $Ca^{2+}$, Calmodulin, and Troponin[22]

A detailed kinetic analysis of the effects of $Ca^{2+}$, calmodulin ($\delta'$ subunit), and troponin on the activities of the completely undegraded and dephosphorylated *b* form of phosphorylase kinase, the phosphorylated *a* form, and the trypsin-activated *a'* form have been carried out.[22]

### *Phosphorylase Kinase b*[22]

The $A_{0.5}$ for $Ca^{2+}$ in the absence of the $\delta'$ subunit is 23 $\mu$M (at pH 6.8 and 8.57 mM $Mg^{2+}$). In the presence of the $\delta'$ subunit the $A_{0.5}$ for $Ca^{2+}$ is 20 $\mu$M. If these results are interpreted in terms of the known calcium binding properties of calmodulin,[18] it would appear that 3 or 4 molecules of calcium must be bound to either the $\delta$ subunit or the $\delta'$ subunit before activation of phosphorylase kinase *b* can take place. Formation of a calmodulin-${Ca^{2+}}_{3-4}$ complex has also been shown to be essential for the activation of the high $K_m$ cyclic nucleotide phosphodiesterase.[27]

In the presence of the troponin complex, the $A_{0.5}$ for $Ca^{2+}$ was 4 $\mu$M (pH 6.8 and 8.57 mM $Mg^{2+}$). Since phosphorylase kinase *b* has very little activity either in the presence or absence of the $\delta'$ subunit at $Ca^{2+}$ concentrations below 3.0 $\mu$M, the activity is largely dependent on troponin and $Ca^{2+}$ in the $\mu$M range. The activation is 15–25 fold at saturating concentrations of troponin and $Ca^{2+}$ concentrations between 1.0 and 3.0 $\mu$M.

### *Phosphorylase Kinase a and a'*[22]

Phosphorylation of the *b* form by cyclic AMP-dependent protein kinase at low concentrations of $Mg^{2+}$ *in vitro* is accompanied by the phosphorylation of one serine residue on the $\alpha$ subunit and one serine on the $\beta$ subunit.[10,28,29] Both serine residues become phosphorylated *in vivo* in response to adrenaline.[30]

The activity of the phosphorylated *a* form, labeled in both serine residues, is 15-fold higher than that of the *b* form at pH 6.8 and saturating concentrations of $Ca^{2+}$ (80 $\mu$M). However the $A_{0.5}$ for $Ca^{2+}$ is 1.6 $\mu$M, 15-fold lower than the *b* form. Consequently, phosphorylation is accompanied by over 100-fold activation at $Ca^{2+}$ concentrations in the $\mu$M range. The *a* form has less than 5% of its maximum activity at 0.1 $\mu$M $Ca^{2+}$ and very low activity at 0.03 $\mu$M $Ca^{2+}$. Phosphorylase kinase *a* can only be activated slightly by either the $\delta'$ subunit (1.3-fold) or by the troponin complex (1.2-fold).

The conversion of phosphorylase kinase *b* to *a'* by incubation with trypsin is accompanied by the degradation of both the $\alpha$ and $\beta$ subunits,[10] and even more dramatic changes in activity than those observed after phosphorylation. The activity of the *a'* form is 20-fold higher than the *b* form at pH 6.8 and saturating of $Ca^{2+}$. However the $A_{0.5}$ for $Ca^{2+}$ is 0.07 $\mu$M, 300-fold lower than that of the *b* form. Consequently, limited proteolysis is accompanied by several hundred-fold activation at $Ca^{2+}$ concentrations in the $\mu$M range. Proteolysis is also accompanied by a loss of the absolute requirement for $Ca^{2+}$ and the *a'* form shows 25% of its maximum activity even in the presence of 1.0 mM EGTA. The *a'* form is not activated significantly by either the $\delta'$ subunit or by troponin.

Although it is possible that modification of the $\alpha$ and $\beta$ subunits by proteolysis or phosphorylation might alter the binding of $Ca^{2+}$ to the $\delta$ subunit, such large changes are perhaps unlikely, particularly as the $\delta$ subunit is complexed with the $\gamma$ subunit.[25] We therefore propose that covalent modification of the $\alpha$ and $\beta$ subunits not only increases the catalytic activity of the $\gamma$ subunit, but allows it to be activated by calmodulin-${Ca^{2+}}_1$ or calmodulin-${Ca^{2+}}_2$ rather than the calmodulin-${Ca^{2+}}_{3-4}$ required for the activation of the *b* form. This idea is attractive since it suggests a biological role for the different classes of calcium binding sites in calmodulin, and introduces a mechanism for drastically altering the dependence of an enzyme of $Ca^{2+}$, without effecting its calcium binding properties.

The hypothesis is however likely to be oversimplified. Phosphorylase kinase has the structure $(\alpha\beta\gamma\delta)_4$ and therefore contains 4 molecules of calmodulin and not just one, and 16 calcium binding sites and not just 4. Furthermore, as mentioned in the third section, the affinity of at least some of the calcium binding sites appears to be increased by the interaction of the $\delta$ subunit with the $\gamma$ subunit.

## 8. Proportion of the Calmodulin in Skeletal Muscle That Is Bound to Phosphorylase Kinase

The amount of calmodulin in rabbit and rat skeletal muscle has been estimated to be 50 mg and 30 mg per 1,000 g of muscle, respectively.[16,31] Phosphorylase kinase represents 0.8% of the protein in low ionic strength EDTA extracts of rabbit skeletal muscle,[32] and it can therefore be estimated that almost 20 mg of calmodulin per 1,000 g of muscle should be bound to phosphorylase kinase as the $\delta$ subunit.[16]

These calculations have been confirmed using ICR/IAn mice which lack muscle phosphorylase kinase activity, and in which the $\alpha$, $\beta$, and $\gamma$ subunits are completely absent.[33] When muscle extracts from either normal mice or rabbits are fractionated by a 0–30% ammonium sulphate precipitation, all the phosphorylase kinase activity and 35–40% of the calmodulin is recovered in the 0–30% ammonium sulphate precipitate. If this experiment is repeated using muscle extracts prepared from ICR/IAn mice, the 0–30% ammonium sulphate precipitate contains 10–20-fold less calmodulin, while the 30% ammonium sulphate supernatant contains the same amount of calmodulin found in normal mice.[16]

These experiments demonstrate that the $\delta$ subunit accounts for between one-third and one-half of the calmodulin present in skeletal muscle extracts. They also show that the calmodulin in ICR/IAn mice is reduced by 35–40%, i.e., the amount that is bound to phosphorylase kinase. This suggests that calmodulin is synthesized in amounts sufficient to saturate calmodulin binding proteins, and that a mechanism exists for preventing the further accumulation of this protein.[16]

The calmodulin-dependent myosin light chain kinase activity in ICR/IAn mice is the same as in mice with normal phosphorylase kinase activity,[16] confirming that there is not a generalized defect in calmodulin-dependent enzymes in this strain.

It is important to emphasize that phosphorylase kinase is only a major calmodulin binding protein in white, fast-twitch anaerobic muscle fibers. The activity of phosphorylase kinase is 20-fold lower in red, slow-twitch oxidative fibers, 50-fold lower in cardiac muscle, and over 100-fold lower in nonmuscle tissues such as brain.[34] Conversely, the levels of calmodulin are as high or much higher in these other cell types.[18]

## 9. Activation of Phosphorylase Kinase *b* by Troponin May Be the Key Event in Coupling Glycogenolysis and Contraction

The preceding sections have introduced some serious doubts regarding the physiological role of the $\delta'$ subunit. Since 35–40% of the calmodulin in mammalian skeletal muscle is already bound to phosphorylase kinase as the $\delta$ subunit, activation by the $\delta'$ subunit would require most of the remaining calmodulin of the tissue. However much of this may be bound to other calmodulin-dependent proteins such as myosin light chain kinase, whose molar concentration is about half that of phosphorylase kinase.[35] Furthermore, activation of phosphorylase kinase *b* by the δ′ subunit requires very high concentrations of $Ca^{2+}$ ($A_{0.5}$ = 20 μM), while phosphorylase kinase *a* is not activated by the δ′ subunit to a significant extent[22] (Section 7).

These findings raise the question of whether troponin C rather than the δ′ subunit is the physiological activator of phosphorylase kinase. It has been established that the troponin complex, and even artificial thin filaments are just as effective activators of phosphorylase kinase as is troponin C itself. 100-fold higher concentrations of troponin complex (1.0 μM) are required for the activation of phosphorylase kinase as compared to the δ′ subunit, but the average concentration of troponin C in skeletal muscle is 100 μM,[4] which is clearly sufficient for maximal activation. It should also be mentioned that the calmodulin-dependent enzymes, myosin light chain kinase and cyclic nucleotide phosphodiesterase, are not activated by troponin, even at concentrations 1,000-fold

TABLE 1

RELATIVE ACTIVITIES OF PHOSPHORYLASE KINASE *b* AND *a* AT pH 6.8 AND SELECTED CONCENTRATIONS OF CALCIUM IONS[22]

| $Ca^{2+}$ Concentration (μM) | Relative Activity of Phosphorylase Kinase Form | | |
|---|---|---|---|
| | b | b + troponin complex (15 μM) | a |
| 0.01 | 1.0 | 1.0 | 1.2 |
| 0.1 | 1.2 | 1.4 | 50 |
| 1.0 | 5.0 | 75 | 750 |
| 3.0 | 10 | 250 | 1,100 |
| 10 | 30 | 330 | 1,350 |
| 100 | 100 | 500 | 1,500 |

*b*, dephosphorylated form; *a*, phosphorylated form.

greater than those required for the activation by calmodulin[36] (C.B. Klee, unpublished work).

Since the activation by troponin occurs at lower concentrations of $Ca^{2+}$ than the activation by calmodulin, the activity of phosphorylase kinase *b* is almost totally dependent on troponin at concentrations of $Ca^{2+}$ in the μM range. The activation is 25-fold at 3 μM $Ca^{2+}$ which corresponds to 20–25% of the activity of the *a* form at the same concentration of $Ca^{2+}$ (TABLE 1).[22]

In the absence of adrenaline, phosphorylase kinase is essentially in the dephosphorylated *b* form in both resting and contracting muscle.[37,38] It has therefore been suggested that the activation of phosphorylase kinase *b* by $Ca^{2+}$ must be responsible for the coupling between muscle contraction and glycogenolysis.[8] However, the rate of formation of phosphorylase *a* in isolated frog muscle varies by more than 100-fold according to the frequency of electrical stimulation of the muscle, and phosphorylase *a* is formed with a half-time of less than a second following tetanic stimulation of either frog muscle or mouse muscle.[2,3] We have pointed out that the activity of phosphorylase kinase *b* in the presence of $Ca^{2+}$ is insufficient to account for the extremely rapid formation of phosphorylase *a* during a muscle tetanus.[39] The activation of phosphorylase kinase *b* by troponin eliminates this discrepancy and provides a mechanism for achieving more than 100-fold activation of the enzyme in response to $Ca^{2+}$. An increase in $Ca^{2+}$ concentration from $\leq$ 0.1 $\mu$M to 1.0 $\mu$M produces a 5-fold activation in the absence and a 75-fold activation in the presence of troponin, while an increase from $\leq$ 0.1 $\mu$M to 3 $\mu$M produces a 10-fold activation in the absence and 250-fold activation in the presence of troponin. (TABLE 1).

Phosphorylase kinase and the other enzymes of glycogen metabolism are linked together *in vivo* on glycogen particles.[32,40] For the activation of phosphorylase kinase by troponin to be physiologically significant, the glycogen particles would have to be localized at the level of the thin filaments during muscle contraction. Such a localization was established by histochemical techniques several years ago.[41] We have also demonstrated that a major proportion of the enzymes of glycogen metabolism remain associated with the myofibrillar fraction when skeletal muscle is extracted in the presence of $Ca^{2+}$. This does not occur if the muscle is extracted with EDTA.[22,23] The recent demonstration by immunofluorescence techniques that much of the calmodulin in skeletal muscle is localized in the thin filaments[42] is also significant, in view of the finding that phosphorylase kinase is the major calmodulin-binding protein in this tissue. (Section 8).

## 10. Regulation of Phosphorylase Kinase *a* by $Ca^{2+}$

It is now well established that phosphorylase kinase *a* is formed in *resting* muscle in response to adrenaline.[30,37,38] Furthermore, the level of phosphorylase *a* is elevated and reaches values as high or higher than those obtained during tetanic stimulation,[37,38] showing that phosphorylase kinase *a* must be at least partially active under these conditions. Since the concentration of $Ca^{2+}$ in resting muscle are considered to be $\leq$ 0.1 $\mu$M, and the *a* form shows less than 5% of its maximum activity at 0.1 $\mu$M $Ca^{2+}$ and very low activity below 0.03 $\mu$M $Ca^{2+}$ (TABLE 1),[22] the high levels of phosphorylase *a* that are formed in resting muscle in response to adrenaline are still a little difficult to understand. One possibility is that phosphorylase phosphatase becomes inactivated in response to adrenaline, so that the activity of the *a* form at 0.1 $\mu$M $Ca^{2+}$ is sufficient to promote the conversion of phosphorylase *b* to *a*. Skeletal muscle contains a protein termed inhibitor-1, which is a powerful inhibitor of phosphorylase phosphatase *in vitro* after it has been phosphorylated by cyclic AMP-dependent protein kinase.[43,44] This protein has been shown to be phosphorylated *in vivo* and its degree of phosphorylation is increased considerably in response to adrenaline.[45,46] Another possibility is that adrenaline can increase the concentration of $Ca^{2+}$ in the muscle cytoplasm to levels that partially activate phosphorylase kinase, but that are insufficient to initiate contraction (e.g., 0.3–0.5 $\mu$M?).

## 11. Summary

Although it has been believed for several years that $Ca^{2+}$ are the means by which glycogenolysis and muscle contraction are synchronized, it is only over the past two years that this concept has started to be placed on a firm molecular basis.

The current evidence suggests that the regulation of phosphorylase kinase by

$Ca^{2+}$ *in vivo* is achieved through the interaction of this divalent cation with calmodulin (the δ subunit) and troponin C, and that the relative importance of these two calcium binding proteins depends on the state of phosphorylation of the enzyme (FIGURE 1). In the low-activity dephosphorylated *b* form, increasing $Ca^{2+}$ from 0.1 μM to concentrations in the μM range produces a 5–10-fold activation through the binding of $Ca^{2+}$ to the δ subunit, and a further 15–25-fold activation through the binding of $Ca^{2+}$ to troponin C (TABLE 1). Troponin C rather than the δ subunit is therefore the dominant calcium dependent regulator of the *b* form, providing an attractive mechanism for coupling glycogenolysis and muscle contraction. On the other hand, the high-activity phosphorylated *a* form is only

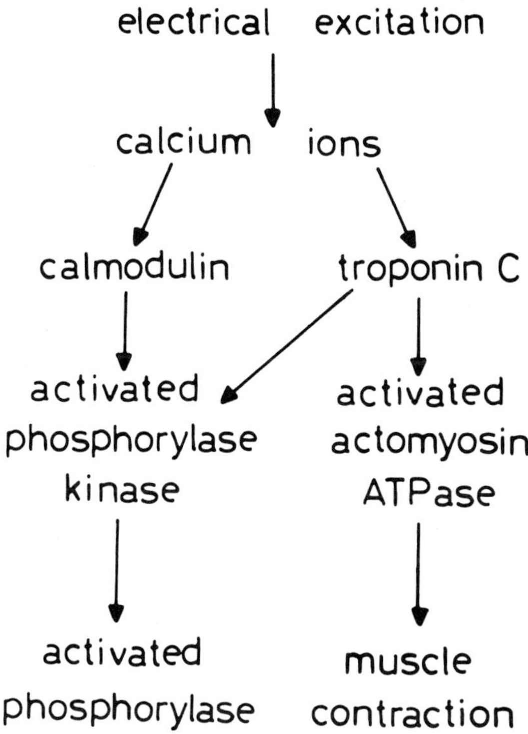


FIGURE 1. Calcium control of phosphorylase kinase through the combined action of calmodulin (the δ subunit) and troponin C.

activated very slightly by troponin (Section 7). The δ subunit is therefore the dominant calcium dependent regulator of the hormonally activated state of the enzyme.

It has recently become clear that phosphorylase kinase not only phosphorylates and activates phosphorylase, but also phosphorylates glycogen synthase, decreasing its activity.[47–51] The regulation of phosphorylase kinase by $Ca^{2+}$ may therefore also provide a mechanism for achieving synchronous control of the pathways of glycogenolysis and glycogen synthesis.

## References


1. Krebs, E. G. & E. H. Fischer. 1956. Biochem. Biophys. Acta **20:** 150–157.
2. Danforth, W. H., E. Helmreich & C. F. Cori. 1962. Proc. Natl. Acad. Sci. USA **48:** 1191–1199.
3. Danforth, W. H. & J. B. Lyon. 1964. J. Biol. Chem. **239:** 4047–4050.
4. Perry, S. V. 1974. Biochem. Soc. Symp. **39:** 115–132.
5. Meyer, W. L., E. H. Fischer & E. G. Krebs. 1964. Biochemistry **3:** 1033–1044.
6. Osawa, E., K. Hosoi & S. Ebashi. 1967. J. Biochem. (Tokyo) **61:** 531–533.
7. Heilmeyer, L. M. G., F. Meyer, R. H. Haschke & E. H. Fischer. 1970. J. Biol. Chem. **245:** 6649–6656.
8. Brostrom, C. O., F. L. Hunkeler & E. G. Krebs. 1971. J. Biol. Chem. **246:** 1961–1967.
9. Hayakawa, T., J. P. Perkins & E. G. Krebs. 1973. Biochemistry **12:** 567–573.
10. Cohen, P. 1973. Eur. J. Biochem. **34:** 1–14.
11. Hayakawa, T., J. P. Perkins & E. G. Krebs. 1973. Biochemistry **12:** 574–580.
12. Skuster, J. R., K. F. Jesse Chan & D. J. Graves. 1980. J. Biol. Chem. **255:** 2203–2210.
13. Cohen, P., A. Burchell, J. G. Foulkes, P. T. W. Cohen, T. C. Vanaman & A. C. Nairn. 1978. FEBS Lett. **92:** 287–293.
14. Grand, R. J., S. Shenolikar & P. Cohen. 1980. Eur. J. Biochem. (submitted.)
15. Watterson, D. M., F. Sharief & T. C. Vanaman. 1980. J. Biol. Chem. **255:** 962–975.
16. Shenolikar, S., P. T. W. Cohen, P. Cohen, A. C. Nairn & S. V. Perry. 1979. Eur. J. Biochem. **100:** 329–337.
17. Kilimann, M. & L. M. G. Heilmeyer. 1977. Eur. J. Biochem. **73:** 191–197.
18. Klee, C. B., T. H. Crouch & P. G. Richman. 1980. Ann. Rev. Biochem. **49** (In press.)
19. Cohen, P. 1980. *In* Calcium Binding Proteins as Cellular Regulators. W. Y. Cheung, Ed. Academic Press, Inc. New York, N. Y. (In press.)
20. Jennissen, H. P., R. W. Veh, J. K. H. Peterson & H. P. Neubauer. 1979. Hoppe-Seyler's Z. Physiol. Chem. **360:** 293.
21. Potter, J. D. & J. Gergely. 1975. J. Biol. Chem. **250:** 4628–4633.
22. Cohen, P. 1980. Eur. J. Biochem. (In press.)
23. Cohen, P., C. Picton & C. B. Klee. 1979. FEBS Lett. **104:** 25–30.
24. Klee, C. B., T. H. Crouch & M. H. Krinks. 1979. Proc. Natl. Acad. Sci. USA **76:** 6270–6273.
25. Picton, C., C. B. Klee & P. Cohen. 1980. Eur. J. Biochem. (In press.)
26. DePaoli-Roach, A. A., J. B. Gibbs & P. J. Roach. 1979. FEBS Lett. **105:** 321–324.
27. Crouch, T. H. & C. B. Klee. 1980. J. Biol. Chem. (In press.)
28. Cohen, P., D. C. Watson & G. H. Dixon. 1975. Eur. J. Biochem. **51:** 79–92.
29. Yeaman, S. J., P. Cohen, D. C. Watson & G. H. Dixon. 1977. Biochem. J. **162:** 411–421.
30. Yeaman, S. J. & P. Cohen. 1975. Eur. J. Biochem. **51:** 93–104.
31. Yagi, K., M. Yazawa, S. Kakiuchi, M. Oshima & K. Uenishi. 1978. J. Biol. Chem. **253:** 1338–1340.
32. Cohen, P. 1978. Curr. Top. Cell. Reg. **14:** 117–196.
33. Cohen, P. T. W., A. Burchell & P. Cohen. 1976. Eur. J. Biochem. **66:** 347–356.
34. Burchell, A., J. G. Foulkes, P. T. W. Cohen, G. D. Condon & P. Cohen. 1978. FEBS Lett. **92:** 68–72.
35. Pires, E. M. V. & S. V. Perry. 1977. Biochem. J. **167:** 137–146.
36. Walsh, M. P., B. Vallet, J. C. Cavadore & J. G. Demaille. 1980. J. Biol. Chem. **255:** 335–337.
37. Posner, J. B. & E. G. Krebs. 1965. J. Biol. Chem. **240:** 982–985.
38. Drummond, G. I., J. P. Harwood & C. A. Powell. 1969. J. Biol. Chem. **244:** 4235–4240.
39. Cohen, P. 1974. Biochem. Soc. Symp. **39:** 51–73.
40. Meyer, F., L. M. G. Heilmeyer, R. H. Haschke & E. H. Fischer. 1970. J. Biol. Chem. **245:** 6642–6648.
41. Sigel, P. & D. Pette. 1969. J. Histochem. Cytochem. **17:** 225–237.
42. Harper, J. F., W. Y. Cheung, R. W. Wallace, H. L. Huang, S. N. Levine & A. L. Steiner. 1980. Proc. Natl. Acad. Sci. USA **77:** 366–370.
43. Huang, F. L. & W. H. Glinsmann. 1976. Eur. J. Biochem. **70:** 419–426.

44. NIMMO, G. A. & P. COHEN. 1978. Eur. J. Biochem. **70:** 353–365.
45. FOULKES, J. G. & P. COHEN. 1979. Eur. J. Biochem. **97:** 251–256.
46. FOULKES, J. G., L. S. JEFFERSON & P. COHEN. 1980. FEBS Lett. **112:** 21–24.
47. ROACH, P. J., A. A. DEPAOLI-ROACH & J. LARNER. 1978. J. Cyc. Nuc. Res. **4:** 245–257.
48. EMBI, N., D. B. RYLATT & P. COHEN. 1979. Eur. J. Biochem. **100:** 339–347.
49. DEPAOLI-ROACH, A. A., P. J. ROACH & J. LARNER. 1979. J. Biol. Chem. **254:** 4212–4219.
50. SODERLING, T. R., A. K. SRIVASTAVA, M. A. BASS & B. S. KHATRA. 1979. Proc. Natl. Acad. Sci. USA **76:** 2536–2540.
51. WALSH, K. Y., D. M. MILLIKIN, K. K. SCHLENDER & E. M. REIMANN. 1979. J. Biol. Chem. **254:** 6611–6616.

## DISCUSSION OF THE PAPER

DR. W. DRABIKOWSKI (*Nencki Institute of Experimental Biology, Warsaw, Poland*): Did you test parvalbumins?

DR. P. COHEN: Yes. We used four different parvalbumins corresponding to both evolutionary lineages of the protein. Parvalbumins had no effect on phosphorylase kinase activity even at concentrations as high as 0.2 mM.

QUESTION: You mentioned that calmodulin antibody will inhibit your enzyme and that this can be reversed with added calmodulin. Can it be reversed with troponin C?

DR. COHEN: Yes, at high concentrations.

# THE EFFECTS OF TRIFLUOPERAZINE ON THE MACROPHAGE-LIKE CELL LINE, J774*

Mark G. Speaker,† Thomas W. Sturgill,† Seth J. Orlow,†
Grace Hua Chia,† Sharon Pifko-Hirst,† and Ora M. Rosen†‡

*†Department of Molecular Pharmacology*
*and*
*‡Department of Medicine*
*Albert Einstein College of Medicine*
*Bronx, New York 10461*

## INTRODUCTION

The macrophage-like cell line J774 was originally established by Ralph[1] from a murine reticulum cell sarcoma. The cloned lines derived from J774, J774.2, and J774.16, maintain a number of differentiated macrophage-like functions *in vitro*.[2] The cells secrete lysozyme and plasminogen activator,[3] migrate, and phagocytize antibody-coated erythrocytes via specific Fc-receptors.[1,4] Stable variants have been selected that are defective in some of these functions. For example, variants defective in Fc-mediated phagocytosis[4] and variants resistant to the growth inhibitory effects of cyclic AMP and cholera toxin, which are defective in cyclic AMP-dependent protein kinase and adenylate cyclase, respectively,[5] have been obtained. Since phagocytosis, enzyme secretion, and migration are macrophage properties that involve fundamental cellular processes, components of which are considered to be $Ca^{2+}$-sensitive, we have begun to study the role of the $Ca^{2+}$-calmodulin (CaM) system in J774.

Phenothiazine antipsychotics,[6,8,9-13] as well as certain sulfonamide derivatives,[7,10,12] inhibit the calcium-dependent activation by calmodulin of a number of enzymes, such as cyclic nucleotide phosphodiesterase,[6,7] adenylate cyclase,[8] membrane-bound $Ca^{2+}$-ATPase,[9-11] myosin light chain kinase and myosin ATPase,[12] and phospholipase $A_2$.[13] Trifluoperazine (Stelazine®, Smith, Kline & French, Philadelphia, Penna.), a phenothiazine, exhibits specific calcium-dependent binding to calmodulin.[14,15] We have used trifluoperazine to probe the role of $Ca^{2+}$-CaM in Fc-mediated phagocytosis and to select variants that may be defective in some component of the $Ca^{2+}$-CaM system.

## MATERIALS

Trifluoperazine dihydrochloride (Tfp) and [$^3$H]trifluoperazine were gifts from Dr. Harry Green of Smith, Kline & French. [$\alpha$-$^{32}$P]ATP (35 Ci/mmole), [$\gamma$-$^{32}$P]ATP (40 Ci/mmole), [$^3$H]cAMP (50 Ci/mmole), [$^3$H]cGMP (10 Ci/mmole) and $^{125}$I-labeled N-succinimidyl-3-(4-hydroxyphenol) propionate (Bolton Hunter reagent, 1500 Ci/mmole) were obtained from New England Nuclear (Boston, Mass.). Soybean trypsin inhibitor, aprotinin, alcohol dehydrogenase, fumarase, and

*Supported by National Institutes of Health Training Grant 5T32GM7288 from the National Institute of General Medical Sciences (M.G.S. and S.J.O.), National Institutes of Health Fellowship IF32AM06071 (T.W.S.), National Institutes of Health Grant AM-09038 (O.M.R.), and American Cancer Society Grant BC-121J (O.M.R.).



0077-8923/80/0356-0162 $1.75/0 © 1980, NYAS

catalase were from Boehringer-Mannheim (Indianapolis, Ind.). Phenylmethylsulfonyl fluoride (PMSF), Tween 80, and ethylmethane sulfonate (EMS) were obtained from Sigma (St. Louis, Mo.). Pentex bovine serum albumin and $\gamma$-globulin were obtained from Miles Laboratories (Elkhart, Ind.). Polyethylenimine (PEI)-cellulose thin layer plates were from Merck. Chicken gizzard myosin light chains were the generous gift of Dr. Robert Adelstein (National Institutes of Health, Bethesda, Md.). Rat embryo fibroblasts used as feeder layers for cloning in soft agar were obtained from Microbiological Associates (Bethesda, Maryland) and polyethylene glycol 6000, from Fisher Scientific (Pittsburgh, Penna.). Thioglycollate-stimulated peritoneal macrophages were provided by Drs. John Loike and Samuel Silverstein.

## METHODS

Cell viability was monitored by trypan blue exclusion. Trifluoperazine was added to cultures after the cells had been allowed to attach to the culture dish. Care was taken to prevent exposure of the Tfp to light.

Fc-mediated phagocytosis was assayed as described by Muschel *et al.*[4] The studies depicted in FIGURE 2 were performed on J774.16, a clone of J774 that exhibits a somewhat higher level of Fc-mediated phagocytosis than clone J774.2. Determinations of cyclic AMP content were performed as previously described[5] and adenylate cyclase activity was assayed according to Salomon *et al.*[16]

### *Preparation of Cell Extracts*

For assays of CaM and adenylate cyclase, the cells ($5 \times 10^7$/ml) were disrupted in a glass homogenizer (Thomas) with a motor driven teflon pestle at 0°C. The homogenization medium contained 20 mM Tris-HCl buffer, pH 7.5, 1 mM EGTA, 2 mM $MgSO_4$, 0.2 mM dithiothreitol, and 10 mM 2-mercaptoethanol. For studies of CaM-binding proteins, cyclic nucleotide phosphodiesterase, and calcium-dependent phosphorylation, extracts were prepared as above, except that 0.34 M sucrose, 0.5 mg/ml soybean trypsin inhibitor, 0.06 mg/ml aprotinin, and 1 mM PMSF were added to the homogenization medium. Homogenates were centrifugated at 4°C for 5 minutes at 500 × g, the pellet was discarded and the supernatant fluid was centrifuged for 15 minutes at 8,700 × g. The supernatant fluid was then centrifuged again for 30 minutes at 50,000 × g and the final supernatant fluid employed in the analyses to be described.

### *Purification of CaM and Cyclic Nucleotide Phosphodiesterase*

Purifications were based on the procedures of Dedman *et al.*[17] for CaM and those of Wang and Desai,[18] and Watterson and Vanaman[19] for cyclic nucleotide phosphodiesterase (PDE). Fresh bovine brains were homogenized in a Waring blender in 3 volumes of 20 mM Tris-HCl buffer, pH 7.5, containing 0.1 mM EGTA, 8 mM 2-mercaptoethanol, and 0.2 mM PMSF. Following homogenization and centrifugation at 10,000 × g for 20 minutes, the supernatant fluid was adsorbed batchwise to DEAE cellulose (100 mg protein/ml resin), previously equilibrated with the homogenization buffer. The resin was washed with 3 volumes of homogenization buffer and then with 3 volumes of the same buffer containing 0.145 M ammonium sulfate. The material eluted in this last wash contained cyclic nucleotide phosphodiesterase activity and was saved for further purification. The DEAE cellulose was next washed with 3 volumes of homogenization buffer containing 0.45 M ammonium sulfate. The protein in this eluate was precipitated with 90% ammonium sulfate, collected by centrifugation, and resuspended by dialysis against 10 mM imidazole buffer, pH 6.1, containing 1 mM EGTA. The solution was then heated at 90°C for 2 minutes in a hot water bath. The precipitate was removed by centrifugation and the supernatant fluid applied to a column of DEAE-cellulose (10 mg protein/ml resin), previously equilibrated with the dialysis buffer. Proteins were eluted with a NaCl gradient (0–500 mM) and fractions containing CaM as assessed by SDS-polyacrylamide gel electrophoresis were pooled, concentrated by negative-pressure dialysis, and stored at −60°C. The purified preparation of CaM exhibited only 1 protein band on SDS-polyacrylamide gel electrophoresis.

Cyclic nucleotide phosphodiesterase was prepared by subjecting the 0.145 M ammonium sulfate eluate from DEAE-cellulose to gel filtration on ACA-34 in the presence of 0.1 mM EGTA. Peak fractions were pooled, concentrated to 5 mg protein/ml, and applied to a 1 ml column of CaM-Sepharose in the presence of

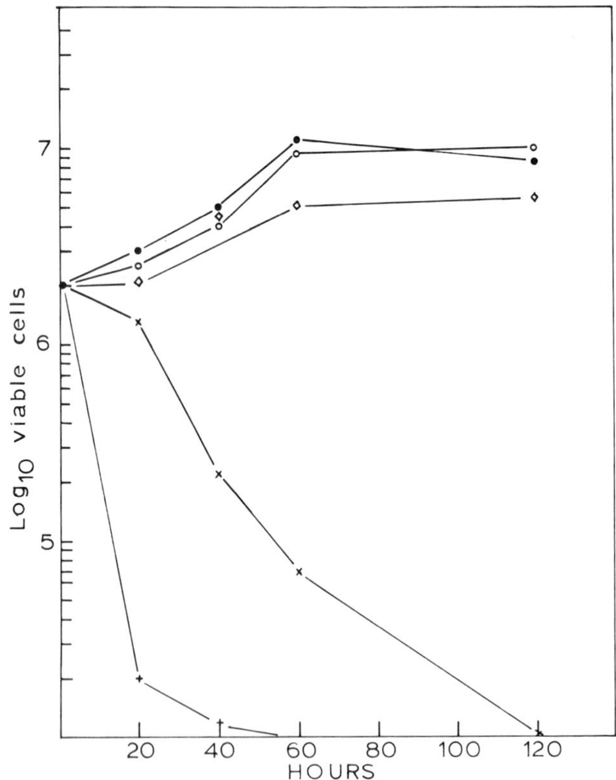


FIGURE 1. Effect of trifluoperazine on the growth of J774.2. Control, (●); Trifluoperazine: 5 $\mu$M (0), 10 $\mu$M (●), 25 $\mu$M (X) and 50 $\mu$M (+).

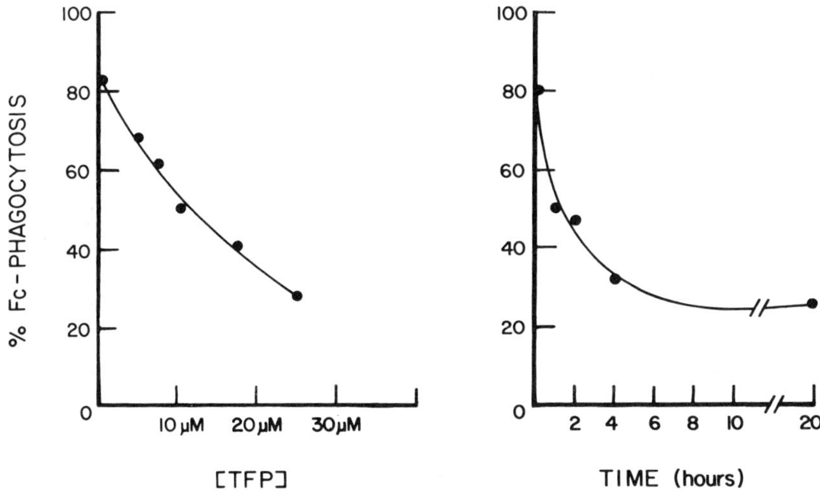


FIGURE 2. Inhibition of Fc-mediated phagocytosis by trifluoperazine. The left panel shows the relationship between inhibition of Fc-mediated phagocytosis and the concentration of Tfp. Cells (J774.16) were incubated with the drug for 4 hours. The panel on the right depicts the time course for incubation at 25 $\mu$M Tfp.

0.5 mM $Ca^{2+}$. The resin was washed with homogenization buffer containing 0.5 mM $Ca^{2+}$ and 100 mM NaCl and then eluted with the same buffer containing 2 mM EGTA. The phosphodiesterase activity eluted by EGTA could be stimulated 10-fold by addition of CaM and $Ca^{2+}$.

## *Iodination of CaM*

CaM was iodinated by the Bolton-Hunter method as described by Chafouleas *et al.*[20]

## *Radioimmunoassay for CaM*

An antiserum to CaM was prepared based on the procedure of Dedman *et al.*[21] A goat was immunized initially with 10 mg of homogeneous bovine brain CaM. The protein was prepared for intramuscular injection in complete Freund's adjuvant. One month later 1 mg CaM was administered intramuscularly and an additional 1.0 mg adsorbed to alumina was administered intravenously. The animal was bled two weeks later. Antibodies were purified by affinity chromatography on CaM-Sepharose as described by Dedman *et al.*[21] Preimmune serum was devoid of activity.

The assay was initiated by incubating the antibody, [$^{125}$I]CaM ($10^5$ cpm), and the sample to be assayed, on ice for 1 hour in a volume of 100 $\mu$l of 20 mM potassium phosphate buffer, pH 7.4, and 0.15 M NaCl. Bovine $\gamma$-globulin, (0.5 mg in 0.5 ml) and 0.5 ml of 18.7% polyethylene glycol were then added and the mixture kept on ice for an additional hour. The resulting precipitate was

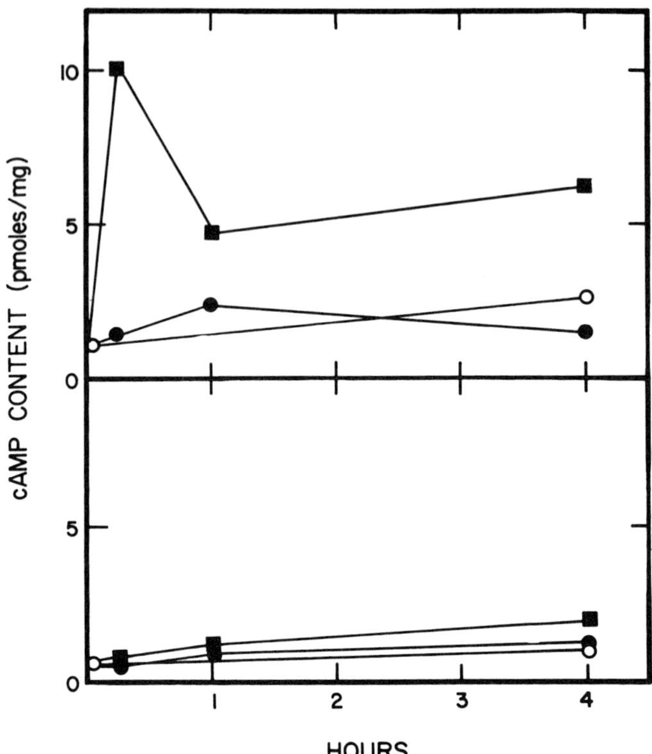


FIGURE 3. Effects of trifluoperazine and methylisobutylxanthine on the cyclic AMP content of wild-type, J774.2 (upper panel), and adenylate cyclase deficient variant, $CT_2$ (lower panel). No addition, (○); 25 μM Tfp, (●); 50 μM methylisobutylxanthine (■).

collected by centrifugation at 3,000 × g for 10 minutes. The supernatant fluid was aspirated and the pellet assayed for $^{125}I$ in a gamma counter. The standard curve for the radioimmunoassay is presented in FIGURE 7. Parvalbumin did not compete for the anti-calmodulin antibody; troponin C was 1% as effective as calmodulin in displacing [$^{125}I$] CaM from the immune complex.

### *Cyclic Nucleotide Phosphodiesterase Assay*

Assays contained 40 μl of reaction mixture consisting of 50 mM Tris-HCl buffer, pH 7.4; 5 mM $MgCl_2$, 200 μM cAMP, 2 μCi [$^3H$]cAMP, 2 mM dithiothreitol, 1 mg/ml bovine serum albumin; 10 μl of either 25 mM $CaCl_2$ or 5 mM EGTA, and 20 μl of sample to be assayed. Incubation was for 10 minutes at 37°C and the reaction was stopped by addition of a solution (20 μl) containing 12.5 mM adenosine, 12.5 mM cAMP, 12.5 mM 5′-AMP, and 0.2 M EDTA. Percent conversion of cAMP to 5′-AMP and adenosine was then determined by spotting 3 μl of the assay mixture on PEI-cellulose thin layer plates as previously described.[22]

### *Assay of $Ca^{2+}$-Dependent Endogenous Phosphorylation and Myosin Light Chain Kinase*

Assays were performed in a volume of 40 μl containing 25 mM Tris-HCl buffer, pH 7.5, 10 mM $MgCl_2$, 0.1 mM ATP, [γ-$^{32}$P]ATP (5–10 μCi), 1 μg of purified brain CaM, 0.25 mg/ml soybean trypsin inhibitor, 0.08 mg/ml aprotinin, 0.25 mM PMSF, and 5 mM EGTA or 0.5 mM $CaCl_2$. For assays of myosin light chain kinase, 6 μg of purified chicken gizzard myosin light chains were included in the reaction mixture. Incubation was carried out for 5 minutes at 30°C for endogenous phosphorylation and for 2 minutes at 25°C for myosin light chain phosphorylation. Reactions were stopped by the addition of 10 μl of a solution containing 50 mM Tris-HCl, pH 6.8, 10% SDS, 50% glycerol, and 1.4 M 2-mercaptoethanol, followed by heating at 100°C for 3 minutes. The samples were then subjected to electrophoresis on 12.5% SDS-polyacrylamide slab gels according to the procedure of Laemmli.[23] Gels were stained in 10% acetic acid, 0.2% Coomassie blue, 50% methanol for 1 hour and then destained with 10% acetic acid, 30% methanol. For radioautography, dried gels were placed at −30°C with either Kodak BB5 medical X-ray film and a DuPont Quanta III intensifying screen for 2 hours or with Kodak AA5 medical X-ray film overnight. Incorporation of $^{32}$P into myosin light chains was quantitated by cutting out the stained light chains from the gels, either before or after drying for radioautogra-

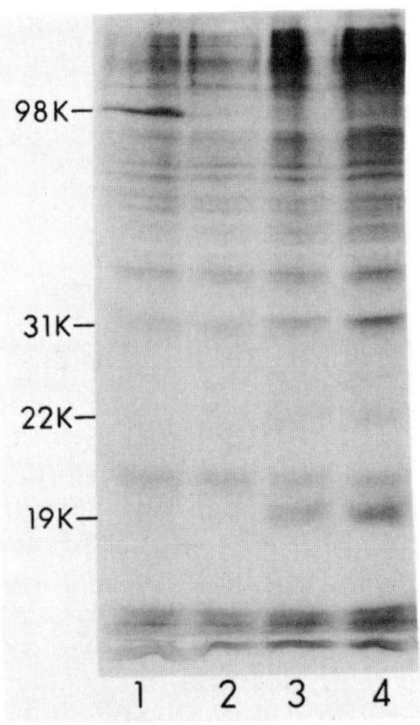


FIGURE 4. $Ca^{2+}$-dependent and cyclic AMP-dependent phosphorylation in cell extracts of J774.2. Additions to the assays were: Lane 1, 0.4 mM $Ca^{2+}$; lane 2, none; lane 3, 0.5 μM cyclic AMP; lane 4, 5 μM cyclic AMP.

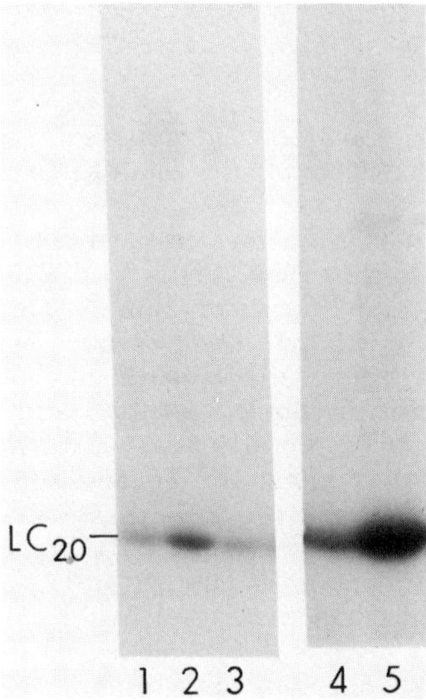


FIGURE 5. Myosin light chain kinase activity in J774.2. The samples were incubated with [$\gamma$-$^{32}$P]ATP and exogenous myosin light chains (see METHODS). Lanes 1–3 demonstrate phosphorylation by cell extracts prepared as described in METHODS. Lanes 4 and 5 show phosphorylation by material that was chromatographed on DEAE-cellulose to remove endogenous CaM, and then applied to and eluted from a CaM-Sepharose column, as described in METHODS for purification of PDE. Shown is a radioautogram of a 12.5% SDS-polyacrylamide gel. $LC_{20}$ indicates the 20,000 dalton myosin light chain. Assays were performed in the presence of: 5 mM EGTA (lanes 1 and 4); 0.4 mM $CaCl_2$ (lanes 2 and 5); 0.4 mM $CaCl_2$ and 25 $\mu$M Tfp (lane 3).

phy, and dissolving the gel slices in 0.5 ml of 30% hydrogen peroxide at 80°C for 3 hours prior to liquid scintillation spectrometry. Incorporation measured in the absence of added enzyme was approximately 50 cpm.

### *Assay for CaM Binding Proteins*

EGTA, to a final concentration of 1 mM, was added to samples containing 10–150 $\mu$g of protein in 20 mM Tris-HCL buffer, pH 7.5. [$^{125}$I]CaM (5 $\times$ $10^5$ cpm) was added to the sample in 10 $\mu$l, the volume was brought up to 150 $\mu$l with the buffer, and the sample thoroughly mixed. Five minutes prior to applying the sample to a gel, 10 $\mu$l of 40 mM $CaCl_2$ was added, followed by 40 $\mu$l of 50% glycerol, 5% 2-mercaptoethanol, and 0.01% bromophenol blue. Samples were applied to nondenaturing 7.5% polyacrylamide gels prepared according to Davis[24] containing 1 mM $CaCl_2$ in all gel and electrode buffers, and 2.5 mM mercaptopropionate in the upper electrode buffer. Electrophoresis was carried out at 4°C on slabs or cylinders of polyacrylamide with 2.5 cm stacking and 8.5 cm resolving gels. Electrophoresis was terminated while the dye front and unbound CaM remained on the gel. The resolving gel was cut into 40 slices and assayed for [$^{125}$I] in a gamma spectrometer. A rapid assay for the $R_f$ 0.5 binding activity (vide infra) was performed by subjecting the sample to electrophoresis on shorter cylindrical gels with a 1.5 cm stacking and 4.5 cm resolving gels. The shorter resolving gel was divided into three equal pieces and counted in a gamma spectrometer. Binding activity was expressed as the percent of total counts bound in a particular peak per 100 μg of protein applied to the gel.

### *Selection of Trifluoperazine Resistant Variants*

Cells ($10^7$) were exposed to the mutagen ethylmethane sulfonate (EMS) (500 μg/ml) for 24 hours in a 150 mm tissue culture dish (Lux). This treatment reduced the cloning efficiency from 85% to 10% as determined by cloning 500 cells in soft agar 24 hours after the end of the EMS treatment. Cloning in soft agar was performed as previously described[5] based on the method of Coffino *et al.*[25]

After exposure to EMS, the cells were cultured for 10 generations before exposure to selective conditions, allowing for phenotypic lag as suggested by Thilly *et al.*[26] The selection was performed by cloning mutagenized cells directly in soft agar containing 30 μM trifluoperazine and 100 μg/ml Tween 80. Tween 80 was included to decrease selection of transport variants as described by Ling *et*

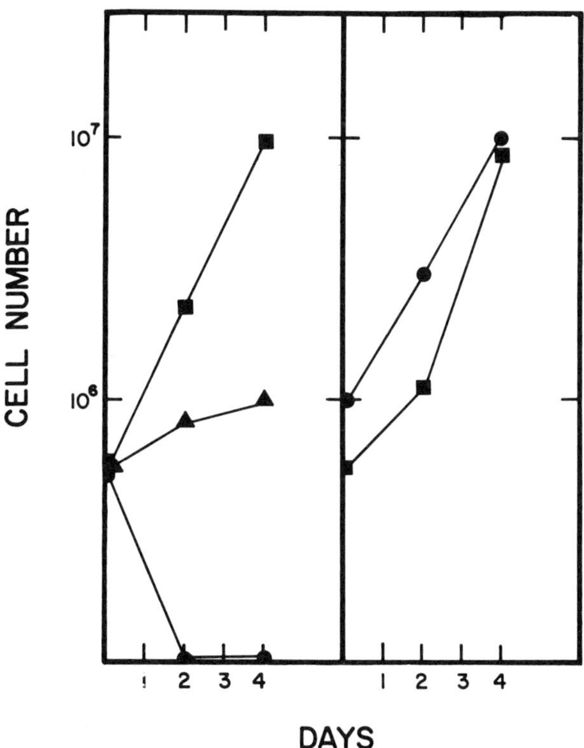


FIGURE 6. Effect of trifluoperazine on the growth of variant, C2 (right panel) compared to parental J774.2 (left panel). No drug, (■), 10 μM Tfp, (▲); 20 μM or 30 μM Tfp, (●).

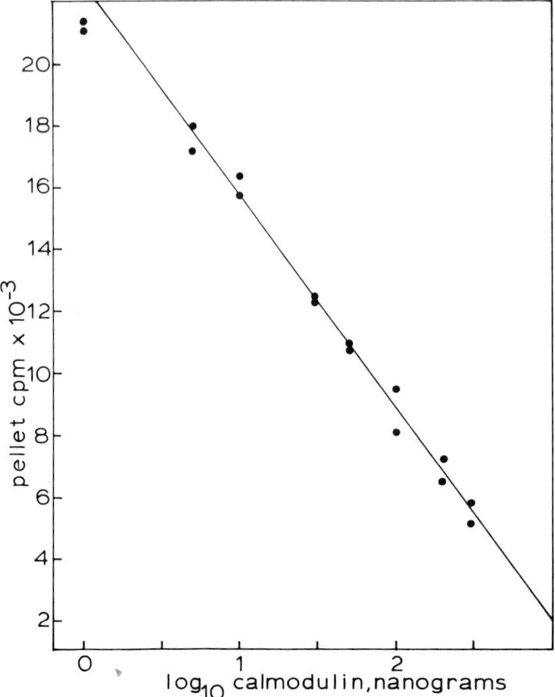


FIGURE 7. Standard curve for the radioimmunoassay of calmodulin.

*al.*[27] Mutagenized cells were cloned at a density of $10^6$ cells per 60 mm dish. It was important that the feeder layer be almost confluent from the beginning of the selection, otherwise it did not remain viable in the presence of the drug. After a week, clones of 50–100 cells were picked, grown in nonselective medium and then tested for drug resistance.

### *Protein Determinations*

Protein was measured by the modified deoxycholate-trichloroacetic acid Lowry procedure of Bensadoun and Weinstein,[28] or with the commercially available BioRad® protein assay, based on the method of Bradford[29] using bovine serum albumin as standard.

### *Determination of the Native Molecular Weight of the $^{125}$I-CaM Complex*

Standard proteins and cell extracts containing [$^{125}$I] CaM were subjected to electrophoresis at 4°C on nondenaturing polyacrylamide slab gels of different porosities. The polyacrylamide concentrations used were 6%, 7.5%, 8.5%, and 10%, and the buffer system used was that of Davis.[24] The $R_f$s of the standard proteins were determined by Coomassie blue staining, and that of the CaM complexes by radioautography. The data was treated according to the method of Hedrick and Smith.[30] The proteins employed as standards were bovine serum albumin, transferrin, alcohol dehydrogenase, bovine heart cyclic AMP-dependent protein kinase, fumarase, and catalase.

## Results and Discussion

### *Effect of Tfp on J774*

The addition of 10–15 μM Tfp to growing cultures of J774.2 inhibited growth (Figures 1 and 6). Higher concentrations (25 μM) led to cell death after 2–3 days of continuous exposure. Removal of the drug after treatment with 25 μM Tfp for 24 hours permitted resumption of growth. Tfp also inhibited Fc-mediated phagocytosis in a dose-dependent fashion (Figure 2). Half-maximal inhibition occurred at a concentration of 15 μM Tfp; the effect was apparent 1 hour after introducing the drug and was maximal after 4 hours of exposure. In contrast, taxol, a drug that promotes polymerization of tubulin *in vitro* and in cultured cells,[31] did not affect Fc-mediated phagocytosis.

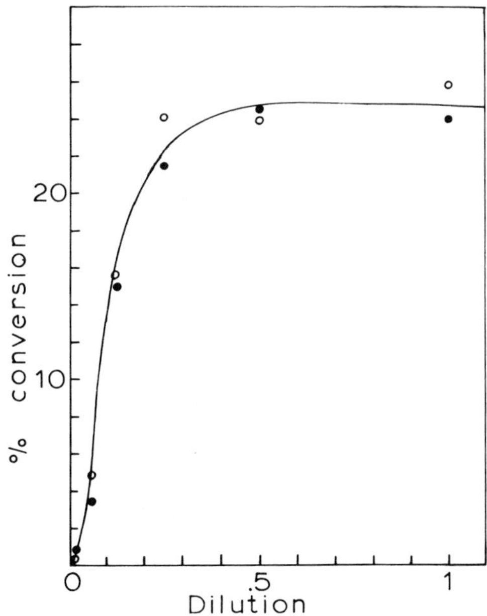


Figure 8. Activation of purified brain cyclic nucleotide phosphodiesterase by heated extracts derived from J774.2 and C2. Extracts of J774.2 and C2 (1.5 mg protein/ml) were heated for 1 minute at 90°C to eliminate endogenous phosphodiesterase activity. Following removal of the precipitate by centrifugation, aliquots of the supernatant fluid were added to the purified CaM-deficient brain phosphodiesterase (see Methods). Depicted is phosphodiesterase activity (ordinate) in the presence of equal dilutions of heated extracts of J774.2 (○) or C2 (●).

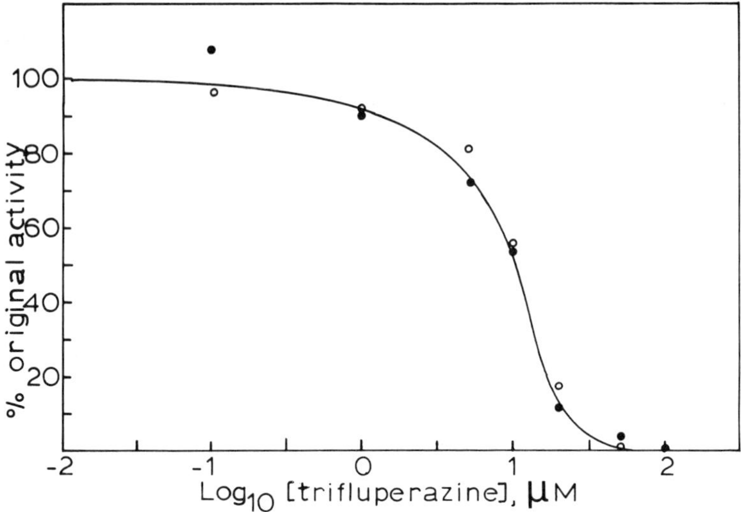


FIGURE 9. Sensitivity of the CaM activity in J774.2 and C2 to inhibition by trifluoperazine. The experiment was performed as described in FIGURE 7, except that the CaM in the heated supernatants was purified by batch adsorption to DEAE cellulose. At equal protein concentrations, the material eluted between 0.22 M and 0.40 M NaCl was assayed for ability to activate brain phosphodiesterase in the presence of increasing concentrations of Tfp. J774.2, (O); C2, (●).

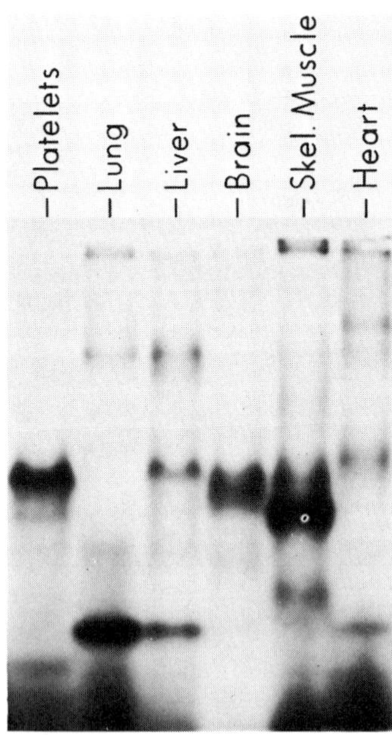


FIGURE 10. [$^{125}$I]calmodulin complexes in extracts of different tissues. Extracts were prepared, subjected to electrophoresis in the presence of $Ca^{2+}$ on a nondenaturing 7.5% polyacrylamide slab gel, and analyzed as described in METHODS. The radioautogram shows the complexes found in (from left to right) human platelets, and murine lung, liver, brain, skeletal muscle, and heart.

### *$Ca^{2+}$-CaM Sensitive Enzyme Activities in J774.2*

Tfp-sensitive, $Ca^{2+}$-CaM-dependent cyclic nucleotide phosphodiesterase, or adenylate cyclase activities, have not been detected in J774.2. This is consistent with the observation that Tfp had no effect on intracellular cyclic AMP content (FIGURE 3). Other $Ca^{2+}$-CaM-sensitive enzymic activities were, however, apparent. FIGURE 4 illustrates $Ca^{2+}$-sensitive protein phosphorylation in extracts of J774.2. This phosphorylation can be inhibited by EGTA or by 25 μM Tfp. A distinct subset of protein phosphorylation is stimulated by cyclic AMP. The cyclic AMP-induced phosphorylations but not the $Ca^{2+}$-dependent ones are diminished in the cyclic AMP-dependent protein kinase variants of J774.2. With careful cell breakage in the presence of 0.34 M sucrose and protease inhibitors (see METH-

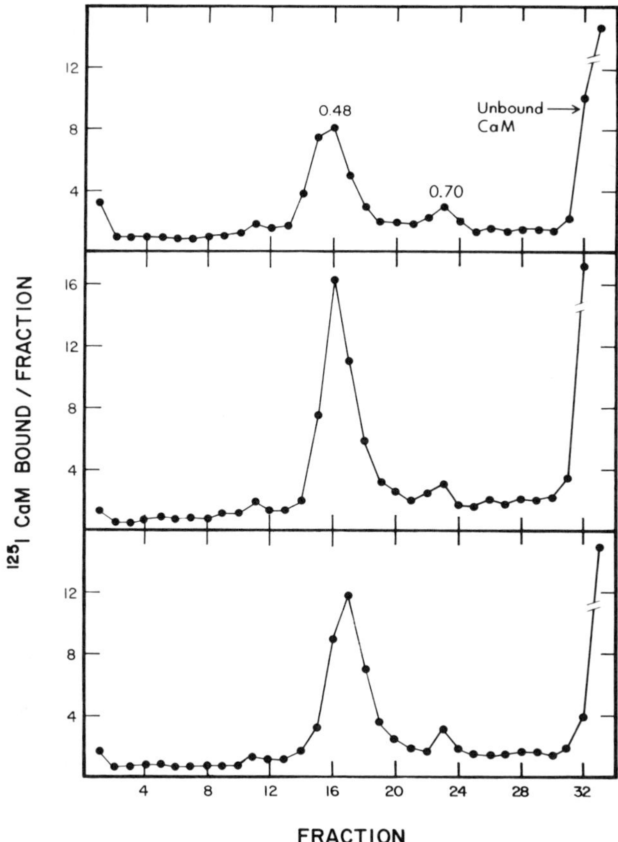


FIGURE 11. [$^{125}$I]CaM complexes in J774.2 and peritoneal macrophages. Extracts were prepared, subjected to electrophoresis on cylindrical nondenaturing 7.5% polyacrylamide gels, and analyzed by assaying $^{125}$I (cpm × $10^{-3}$) as described in METHOD (Top panel) J774.2; (middle panel) thioglycollate-stimulated peritoneal macrophages from BALB/c mice; (bottom panel) mixture of equal amounts of extracts of J774.2 and peritoneal macrophages. Unbound [$^{125}$I]CaM migrates with the dye front, as indicated. The two CaM complexes at $R_f$ 0.48–0.50 and $R_f$ 0.7 are indicated.

ODS), a $Ca^{2+}$-sensitive myosin light chain kinase activity can be detected in cell extracts of J774.2 (FIGURE 5). The activity is stimulated about 5-fold by $Ca^{2+}$ and this enhancement is completely inhibited by 25 $\mu$M Tfp.

*Selection of Variants Resistant to the Growth Inhibitory Effects of Tfp*

In an effort to obtain variants in either CaM or proteins that interact with CaM, cells were mutagenized with EMS (see METHODS) and cloned directly in soft agar containing 30 $\mu$M Tfp and 100 $\mu$g/ml Tween 80. Resistant clones were grown in nonselective medium and then tested for resistance to growth inhibition by 30 $\mu$M Tfp. One clone, C2, grew as well in 30 $\mu$M Tfp as it did in its absence (FIGURE 6). Since the C2 cells took up [$^3$H]Tfp as avidly as did J774.2, we

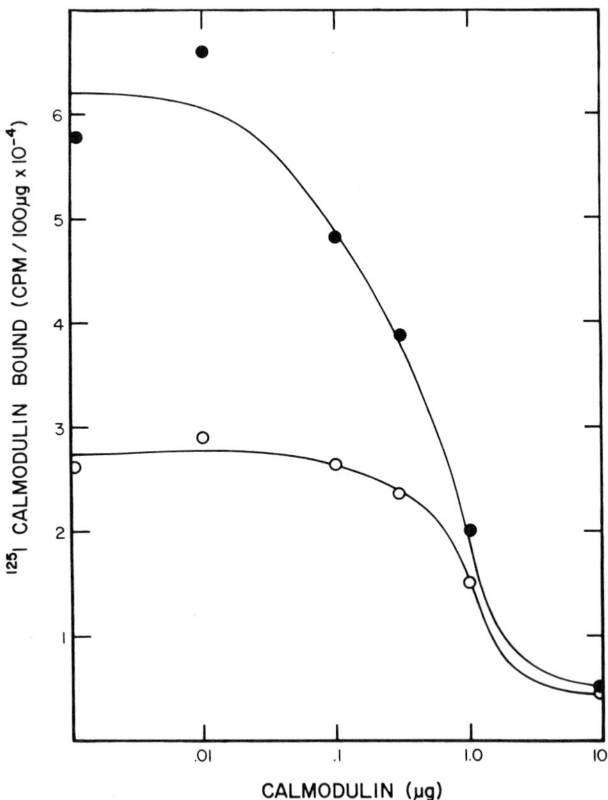


FIGURE 12. Displacement of [$^{125}$I]CaM-binding in extracts of J774.2 by nonradioactive CaM. Samples were prepared for electrophoresis on nondenaturing cylindrical gels as described in METHODS, except that the indicated amounts of purified brain CaM were added to the samples along with the [$^{125}$I]CaM. The open circles (-O-) show displacement of binding from a 50,000 × g supernatant fluid of J774.2, and the closed circles (-●-) show displacement from the same extract following chromatography on DEAE-cellulose which removed approximately 90% of the endogenous CaM.

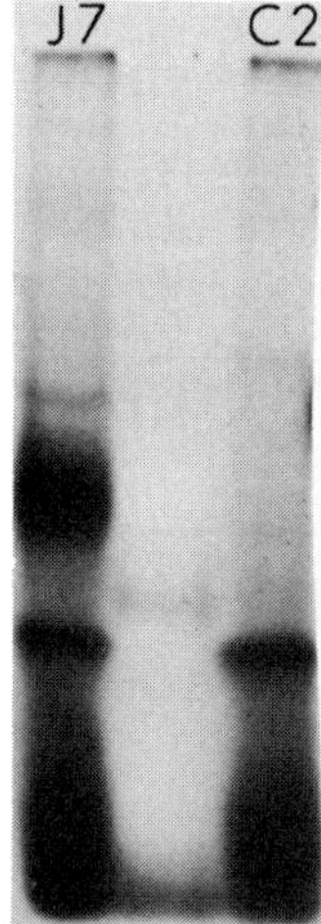


FIGURE 13. [$^{125}$I]CaM complexes in parental J774.2 and trifluoperazine-resistant variant, C2. Samples were prepared as in FIGURE 10. Shown is a radioautogram prepared from the dried gel: left lane, J774.2; right lane, C2.

concluded that the resistant phenotype might result either from an alteration in the content, activity, or sensitivity to Tfp of the CaM in C2 or from a deficiency of a CaM-binding protein whose interaction with CaM was not essential for growth. To address the first possibility, a radioimmunoassay for calmodulin was established using goat antibody to homogeneous bovine brain CaM. The standard

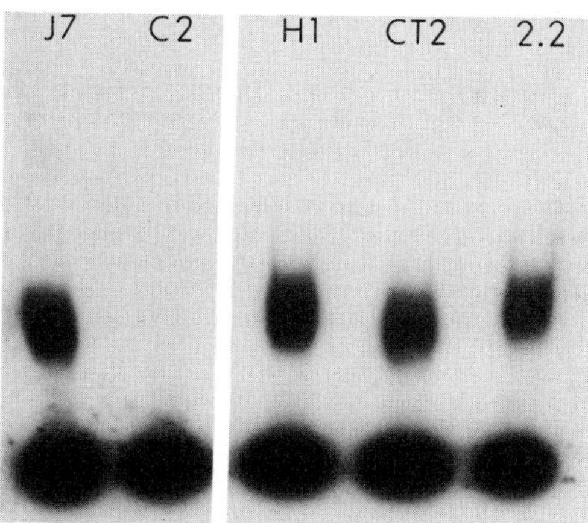


FIGURE 14. [$^{125}$I]CaM complexes in variants of J774.2. Samples were prepared as in FIGURE 10 but electrophoresis was performed on a nondenaturing 5–15% gradient slab gel, so that the $R_f$ 0.7 complex seen on a 7.5% gel is not resolved from the free [$^{125}$I]CaM. Lane 1, parental, J774.2, lane 2, Tfp-resistant variant, C2; lane 3, a cyclic AMP- dependent protein kinase deficient variant, $H_1$[5]; lane 4, an adenylate cyclase deficient variant, $CT_2$[5]; lane 5, a nonphagocytic variant, 2.2, selected with tubericidin.[4]

curve obtained with this assay is depicted in FIGURE 7. Using this method, it was determined that CaM constitutes 1.1% of the soluble protein in both J774.2 and C2. A calmodulin-dependent cyclic nucleotide phosphodiesterase purified from bovine brain was used as a test system for the activity of CaM. The abilities of heated extracts from J774.2 and C2 to activate cyclic nucleotide phosphodiesterase were also identical (FIGURE 8) as was the sensitivity to inhibition of this activation by Tfp (FIGURE 9).

The possibility that C2 contained an altered CaM-binding protein was then

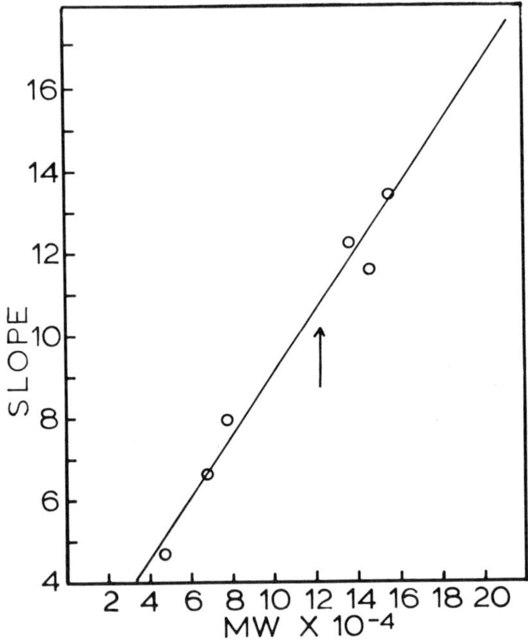


FIGURE 15. Estimation of the molecular weight of the major CaM complex ($R_f$ 0.5) in J774.2. Standard proteins of known molecular weight were subjected to electrophoresis on nondenaturing polyacrylamide gels of different porosities as described in METHODS. The change in mobility as a function of pore size is plotted as a function of molecular weight (MW). The arrow indicates the position of the principal [$^{125}$I]CaM complex in J774.2.

investigated. To scan CaM-binding proteins we modified the electrophoretic technique of La Porte and Storm.[32] Iodinated CaM was admixed with cell extracts in the presence of EGTA; $Ca^{2+}$ was then added and the extract subjected to polyacrylamide gel electrophoresis (see METHODS). To assess the ability of this procedure to detect discrete sets of CaM-binding proteins in different tissues, a variety of murine tissues were examined (FIGURE 10). It can be seen that each tissue contained a distinct group of [$^{125}$I]CaM complexes. The CaM-binding proteins visualized by this technique do not represent the entire repertoire of CaM-interacting proteins—others become apparent only upon further purification. Two principal complexes of [$^{125}$I]CaM were found in extracts of J774.2: a major complex with an $R_f$ of 0.5 and a less prominent complex with an $R_f$ of 0.7 (FIGURE 11). Extracts of thioglycollate-induced peritoneal macrophages contained the same two CaM-binding activities as J774.2 (FIGURE 11). The [$^{125}$I]CaM-binding activity was eliminated by either omission of $CaCl_2$, addition of 6.5 M urea, heating at 95°C for 1 minute, or incubation of extracts with trypsin (0.1 mg/ml, 60 minutes at 4°C). The [$^{125}$I]CaM in the complexes could be displaced by nonradioactive brain CaM added to either unpurified cell extracts or extracts chromatographed on DEAE cellulose to remove endogenous CaM (FIGURE 12). Thus, both [$^{125}$I]CaM-binding activities behave like proteins that interact with CaM in a $Ca^{2+}$-dependent fashion. Extracts of C2 were found to lack the major CaM complex at $R_f$ 0.5 (FIGURE 13). The loss of binding activity in C2 is specific for the $R_f$ 0.5 activity, since the binding at $R_f$ 0.7 is retained. It is unlikely that the absence of this activity is due to the presence of a soluble inhibitor or the activation of a protease since CaM-binding in mixtures of J774.2 and C2 extracts was additive. The difference between wild-type and variant is consistent and does not depend upon whether the variant was recently exposed to the drug. When extracts of other variants of J774.2 (phagocytosis defective, protein kinase defective, and adenylate cyclase defective) were examined, all contained complexes at $R_f$ 0.5 and $R_f$ 0.7 (FIGURE 14). The Stokes radius of the CaM-binding protein, missing in extracts of C2, was estimated by the method of Hedrick and Smith[20] to be approximately 51 Å, corresponding to a molecular weight of 125,000 for a globular protein (FIGURE 15). From the amount of CaM bound to the $R_f$ 0.5 CaM-binding protein in extracts of J774.2, we estimate that the protein may constitute 0.3% to 0.6% of the soluble cell protein, assuming a 1:1 stoichiometry of CaM to binding protein.

In conclusion, Tfp, a drug that interferes with the activity of $Ca^{2+}$-CaM has been shown to inhibit growth and Fc-mediated phagocytosis in the macrophage-like cell line J774. A variant resistant to the growth-inhibitory properties of Tfp lacks a CaM-binding activity present in parental cells. Efforts are currently underway to establish (1) the relationship between Tfp-resistance and the absence of the $R_f$ 0.5 CaM-binding protein; (2) the properties of Tfp-resistant cells; (3) the identity of the $R_f$ 0.5 protein and (4) the role of CaM in the differentiated functions of J774.

## REFERENCES


1. RALPH, P. & I. NAKOINZ. 1975. Nature (London) **257:** 393–394.
2. BLOOM, B. R., B. DIAMOND, R. MUSCHEL, N. ROSEN, J. SCHNECK, G. DAMIANI, O. ROSEN & M. SCHARFF. 1978. Fed. Proc. **37:** 2763–2771.
3. ROSEN, N., J. SCHNECK, B. R. BLOOM & O. M. ROSEN. 1978. J. Cyclic Nucl. Res. **5:** 345–358.
4. MUSCHEL, R. J., N. ROSEN & B. R. BLOOM. 1977. J. Exp. Med. **145:** 175–186.
5. ROSEN, N., J. PISCITELLO, J. SCHNECK, R. J. MUSCHEL, B. R. BLOOM & O. M. ROSEN. 1979. J. Cell. Physiol. **98:** 125–136.
6. WEISS, B., R. FERTEL, R. FIGLIN & P. UZUNOV. 1974. Mol. Pharmacol. **10:** 615–625.
7. HIDAKA, H., T. YAMAKI, M. ASANO & T. TOTSUKA. 1978. Blood Vessels **15:** 55–64.
8. BROSTROM, M. A., C. O. BROSTROM, B. BRECKENRIDGE & D. J. WOLFF. 1978. Adv. Cyclic Nucl. Res. **9:** 85–99.
9. LEVIN, R. M. & B. WEISS. 1980. Neuropharmacology **19:** 169–174.
10. KOBAYASHI, R., M. TAWATA & H. HIDAKA. 1979. Biochem. Biophys. Res. Commun. **88:** 1037–1045.
11. GIETZEN, K., A. MANSARD & H. BADER. 1980. Biochem. Biophys. Res. Commun. **94:** 674–681.

12. Hidaka, H., T. Yamaki, M. Naka, T. Tanaka, H. Hayashi & R. Kobayashi. 1980. Mol. Pharmacol. **17:** 66–72.
13. Wong, P. Y. K. & W. Y. Cheung. 1979. Biochem. Biophys. Res. Commun. **90:** 473–480.
14. Levin, R. M. & B. Weiss. 1977. Mol. Pharmacol. **13:** 690–697.
15. Levin, R. M. & B. Weiss. 1978. Biochim. Biophys. Acta **540:** 197–204.
16. Salomon, Y., C. Londos & M. Rodbell. 1974. Anal. Biochem. **58:** 541–548.
17. Dedman, J. R., J. D. Potter, R. L. Jackson, J. D. Johnson & A. R. Means. 1977. J. Biol. Chem. **252:** 8415–8422.
18. Wang, J. H. & R. Desai. 1977. J. Biol. Chem. **252:** 4175–4184.
19. Watterson, D. M. & T. C. Vanaman. 1976. Biochem. Biophys. Res. Commun. **73:** 40–46.
20. Chafouleas, J. G., J. R. Dedman, R. P. Munjaal & A. R. Means. 1979. J. Biol. Chem. **254:** 10262–10267.
21. Dedman, J. R., M. J. Welsh & A. R. Means. 1978. J. Biol. Chem. **253:** 7515–7521.
22. Rangel-Aldao, R., D. Schwartz & C. S. Rubin. 1978. Anal. Biochem. **87:** 367–375.
23. Laemmli, U. K. 1970. Nature **227:** 680–685.
24. Davis, B. J. 1964. Ann. N.Y. Acad. Sci. **121:** 405–427.
25. Coffino, P. R., R. Baumel, R. Laskow & M. D. Scharff. 1972. J. Cell. Physiol. **79:** 429–440.
26. Thilly, W. G., J. G. Deluca, H. Hoppe & B. W. Penman. 1979. Mutation Res. **50:** 137–144.
27. Ling, V., J. E. Aubin, A. Chase & F. Sarangi. 1979. Cell **18:** 423–430.
28. Bensadoun, A. & D. Weinstein. 1976. Anal. Biochem. **70:** 241–250.
29. Bradford, M. M. 1976. Anal. Biochem. **72:** 248–254.
30. Hedrick, J. L. & A. J. Smith. 1968. Arch. Biochem. Biophys. **126:** 155–164.
31. Schiff, P. B., J. Fant & S. B. Horwitz. 1979. Nature (London) **277:** 665–667.
32. La Porte, D. C. & D. R. Storm. 1978. J. Biol. Chem. **253:** 3374–3377.

## Discussion of the Paper

Dr. D. M. Watterson: Could you tell us how many phenothiazine-binding proteins there are in your macrophage cell line? Are there other ones besides calmodulin?

Dr. M. G. Speaker: We have found that a number of proteins besides CaM bind to a fluphenazine-Sepharose column, and further, that there are obvious differences between parental J7 and Tfp-resistant C2.

Dr. Nemeth (*Yale University, New Haven, CT*): Did you use trifluoperazine and did you try W7?

Dr. Speaker: We have not tested these drugs in our gel assay for CaM binding. We have found, however, that W7 inhibits phagocytosis.

Dr. Zendegui: Have you looked at "resident" macrophages at all?

Dr. Speaker: We have found that the pattern of CaM binding in thioglycollate-induced peritoneal macrophages is identical to that in J774.2. We have not yet looked at "resident" macrophages, but we are certainly going to determine whether CaM binding patterns are altered by immunological activation of the "resident" cells.

# THE ROLE OF CALMODULIN IN PROSTAGLANDIN METABOLISM*

Patrick Y-K Wong, Warren H. Lee, and Patricia H-W Chao

*Department of Pharmacology*
*New York Medical College*
*Valhalla, New York 10595*

Wai-Yiu Cheung

*Department of Biochemistry*
*St. Jude Children's Research Hospital and*
*University of Tennessee Center for Health Sciences*
*Memphis, Tennessee 38101*

Some forty years ago, Von Euler discovered prostaglandins in sheep seminal vesicles.[1] These compounds are oxygenated derivatives of arachidonic acid and other twenty-carbon unsaturated fatty acids that contain a five-member cyclopentane structure. Later, Hamberg and Samuelsson[2] found that the primary prostaglandins, designated the E, F, and D series, are derived from a common intermediate: endoperoxide (FIGURE 1).

In 1975, the same group isolated "prostaglandin endoperoxides" $PGG_2$ and $PGH_2$ from aggregated platelets.[3] They later found that the endoperoxides were rapidly converted by a platelet microsomal enzyme, thromboxane synthetase, into a highly unstable compound: thromboxane $A_2$. Thromboxane $A_2$ ($TxA_2$), which is short-lived, with a half-life in aqueous solution of about 35 sec, is rapidly converted to a stable, but biologically less active derivative: thromboxane $B_2$ ($TxB_2$)[3]. The same investigators also showed that in many cells including human platelets, the majority of the prostaglandin endoperoxide is also converted to thromboxanes.[4] Apparently, thromboxane $A_2$ is a major cellular regulator in many tissues. Since prostaglandin endoperoxides ($PGG_2$ and $PGH_2$) and thromboxane $A_2$ cause arteries to constrict and platelets to aggregate, they play key roles in diseased states such as hypertension, thrombosis, and stroke.

More recently, Vane and his colleagues discovered a new enzyme system in the microsomal fraction of pig aortas that converts prostaglandin endoperoxides into an unstable prostacyclin ($PGI_2$).[5] Prostacyclin has a short half-life, losing its biological activity in aqueous solution within 2 to 3 min upon transformation to a more stable product, 6-keto-$PGF_{1\alpha}$ ($6KF_{1\alpha}$). The effect of $PGI_2$ is not limited to the aortas, it also relaxes rabbit vascular muscle such as mesenteric and coeliac strips and coronary arteries. Although it does not contract aorta or pulmonary artery, $PGI_2$ is the most potent inhibitor of platelet aggregation so far identified, and is ten times more potent than $PGE_1$ as a stimulator of adenylate cyclase.[6] The generation of $PGI_2$ from the arterial wall may serve as a natural homeostatic agent by preventing an overdeposition of platelets on the undamaged vascular

*Supported in part by grants HL 18845, HL 22075, NS 08059, DE 04591 from the U.S. Public Health Service and by the American Heart Association and ALSAC. P. Y-K Wong is a recipient of the National Institute of Dental Research Special Dental Research Award. W. Y. Cheung is a faculty scholar, Josiah Macy, Jr. Foundation, at the Department of Pharmacology, University of Washington, Seattle, Washington.



0077-8923/80/0356-0179 $1.75 © 1980, NYAS

wall as a result of an increased cyclic AMP level. Recently, $PGI_2$ has been suggested to be a circulating hormone as it appears to escape degradation during the pulmonary circulation.[7] However, two recent studies demonstrated that the circulating level of $PGI_2$ is too low to produce any physiological effects.[8,9]

Recent studies on the metabolism of $PGI_2$ in blood vessels revealed that a substantial level of $PGI_2$ was metabolized by a 15-hydroxyprostaglandin dehydrogenase (15-OH-PGDH) present in the vascular wall to 6,15-diketo-$PGF_{1\alpha}$, a biologically inactive metabolite.[10] The oxidation of $PGI_2$ by 15-OH-PGDH may be a major determinant of its levels in blood vessels and, therefore, of crucial importance in regulating the extent and duration of the action of $PGI_2$. Thus, the blood vessel has its own enzymic system for regulation of $PGI_2$ levels *in situ*. The demonstration of the dehydrogenase in blood vessels provides evidence for the inactivation of $PGI_2$ at, or near, the site of its biosynthesis.

The biosynthesis of prostaglandin, thromboxane, and prostacyclin depends on the availability of arachidonic acid. The release of arachidonic acid from phospholipids is brought about by an acylhydrolase such as phospholipase $A_2$,

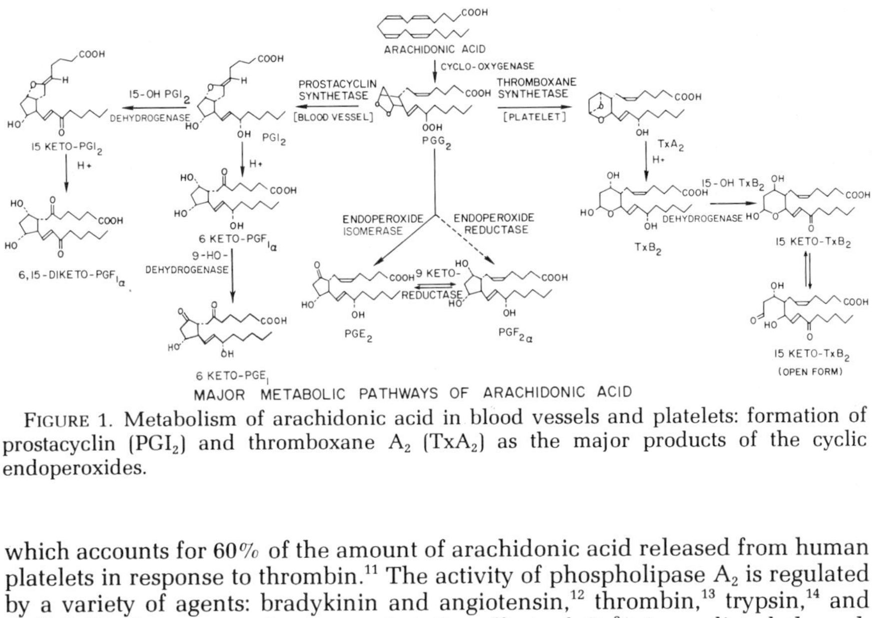

MAJOR METABOLIC PATHWAYS OF ARACHIDONIC ACID

FIGURE 1. Metabolism of arachidonic acid in blood vessels and platelets: formation of prostacyclin ($PGI_2$) and thromboxane $A_2$ ($TxA_2$) as the major products of the cyclic endoperoxides.

which accounts for 60% of the amount of arachidonic acid released from human platelets in response to thrombin.[11] The activity of phospholipase $A_2$ is regulated by a variety of agents: bradykinin and angiotensin,[12] thrombin,[13] trypsin,[14] and $Ca^{2+}$.[15] We have recently shown that the effect of $Ca^{2+}$ is mediated through calmodulin, an ubiquitous $Ca^{2+}$-binding protein originally discovered as an activator of phosphodiesterase.[15] It was later found to regulate numerous enzyme systems and cellular processes.[16] It now appears that calmodulin is a major mediator of $Ca^{2+}$ functions in the eukaryotes.

In this communication, we study the role of calmodulin in prostaglandin metabolism. Specifically, we investigate the effect of calmodulin on platelet aggregation, and on the metabolism of arachidonic acid in human platelets. In addition, we present preliminary evidence suggesting that calmodulin inhibits a 15-hydroxyprostaglandin dehydrogenase of bovine lung. The same enzyme metabolizes prostaglandins ($PGE_2$ and $PGF_{2\alpha}$)[17] as well as prostacyclin ($PGI_2$) in blood vessel wall.[10]

## Experimental Procedures

### *Platelet Preparation*

Blood was drawn from volunteers who abstained from acetylsalicylic acid or other drugs for the preceding 10 days. Nine parts of whole blood were mixed with 1 part of 3.8% sodium citrate; 5-ml fractions were transferred to plastic tubes, and were centrifuged at 150 g for 10 min. The platelet-rich plasma (PRP) was removed with a siliconized pipette. Platelet-poor plasma (PPP) was prepared by centrifuging the PRP at 1200 × g for 10 min. The final platelet count in PRP was adjusted to $2 \times 10^8$/ml with PPP.

### *Measurement of Thromboxane $B_2$ and Other Prostaglandin Released from Washed Human Platelets*

Platelet-rich plasma was passed through a Sepharose 2B column (1.5 × 20 cm) previously equilibrated with Tyrode's buffer, pH 7.4.[15] The platelets emerged in the void volume of the column and were well separated from plasma proteins. A fraction of the platelet suspension (0.5 ml) was incubated with (1-$^{14}$-C)-arachidonic acid (1.4 nmole, 0.36 $\mu$Ci, specific activity 56.5 mCi/mmole, New England Nuclear) for 10 min at 37°C with constant stirring in a Payton dual channel aggregometer in the presence or absence of 2.3 $\mu$M calmodulin. After aggregation of the platelets, as monitored by an increase in light transmission, the reaction was terminated by acidification to pH 3.0 with 0.1 M HCl, and the suspension was extracted 3 times, each with 2 ml of ethyl acetate. The organic phase was pooled and dried under $N_2$, and the reaction products were separated by thin-layer chromatography, using a solvent system containing ethyl acetate: acetic acid (99:1, vol/vol).[18] The reaction products were identified with authentic $TxB_2$, $PGE_2$, $PGF_{2\alpha}$, and $PGD_2$ on the same thin-layer plate (Brinkman Instruments, NJ). This solvent system clearly separates $TxB_2$ from $PGE_2$, $PGF_{2\alpha}$ and $PGD_2$.[18] The radioactive peaks were localized with a Packard 7230 radiochromatogram scanner. The zones corresponding to $TxB_2$, $PGE_2$, $PGF_{2\alpha}$, and $PGD_2$ were cut out, suspended in 10 ml of 0.4% Omnifluor and 20% Triton X-100 toluene liquid scintillation liquid, and counted in a Beckman LS-7500 liquid scintillation counter.

### *Preparation of Bovine Lung 15-Hydroxyprostaglandin Dehydrogenase*

Bovine lung, obtained from a local slaughterhouse, was cut up into small pieces, rinsed twice with 50 mM Tris-HCl buffer (pH 7.4) until free of blood and homogenized in four volumes of 50 mM Tris-HCl (pH 7.4) containing 0.1 mM dithiothreitol (DTT) with a Polytron homogenizer operated at top speed for 2 min. The homogenate was centrifuged at 7000 × g for 20 min to remove connected tissues and again at 105,000 × g for 60 min. 15-Hydroxyprostaglandin dehydrogenase was purified from the supernatant fluid by ammonium sulfate precipitation and DEAE-cellulose column chromatography according to Lee and Levine.[19] The partially purified enzyme was used in this study.

### *Assay for 15-Hydroxyprostaglandin Dehydrogenase*

The reaction mixture contained 4 mM $NAD^+$, [$^3$H]$PGE_2$, 50 mM Tris-HCl (pH 7.4), 0.1 mM DTT, 2.3 $\mu$M calmodulin, 0.1 mM $CaCl_2$, and enzyme (100 $\mu$g) in a final volume of 1 ml. After incubation at 37°C for 60 min, the reaction was stopped by acidification with 0.1 N HCl to pH 3.0, and extracted 3 times with 3 ml of ethyl acetate. The extract was dried under a stream of $N_2$, the residue was redissolved in 0.1 ml of $CHCl_3$:$CH_3OH$ (1:1 vol/vol), applied to a thin-layer chromatographic plate (0.25 mm thick, 20 × 20 cm, silica gel precoated plastic sheets, Brinkman, NJ) and developed by a solvent system consisting of iso-octane:ethyl acetate:acetic acid:water (25:55:10:50, vol/vol). Radioactive zones of $PGE_2$ and its metabolite 15-keto-$PGE_2$ were located by radiochromatogram scanning (Packard, model 7230) and identified by comparison with the mobilities of authentic $PGE_2$, 15-keto-$PGE_2$ and 15-keto-13,14-dihydro-$PGE_2$. The zone corresponding to the major metabolite 15-keto-$PGE_2$ ($R_f$ = 0.45) was cut out and suspended in 10 ml of 0.4% Omnifluor and 10% Triton X-100 for counting. The data were expressed as picomoles of 15-keto-$PGE_2$ formed. All determinations were duplicated, each value representing the mean.

## RESULTS

Human platelets contain cyclo-oxygenase and thromboxane synthetase which convert arachidonic acid to $PGH_2$ and $TxA_2$. $TxA_2$ is short-lived and is rapidly hydrolyzed to a more stable product—$TxB_2$.[3,4] When washed platelets were incubated with [$^{14}$C]arachidonic acid, $TxB_2$ was the only metabolite identified on thin-layer chromatography (FIGURE 2a). This result supports the notion that cyclo-oxygenase and $TxA_2$-synthetase are the major enzymes of arachidonic acid metabolism in human platelets.[20] In the presence of exogenous calmodulin (2.3 $\mu$M), the conversion of arachidonic acid to $TxB_2$ was nearly doubled (FIGURE 2b). We have previously shown that calmodulin stimulated phospholipase $A_2$ in human platelets.[15] The effect of calmodulin on the metabolism of arachidonic acid was investigated further by the use of various thromboxane synthetase inhibitors: 1-octyl-imidazole,[21] 9,11-iminoepoxy prosta-5, 13-dienoic acid (9,11-iminoepoxy),[22] and 9,11-azo prosta-5, 12,dienoic acid (9,11-diazo or azo-analog I).[18] As shown in FIGURE 2c, 1-octyl-imidazole (2.5 $\mu$M) decreased the formation of $TxB_2$ while it increased that of $PGF_{2\alpha}$, $PGE_2$, and $PGD_2$, presumably as a result of diverting $PGH_2$ to endoperoxide isomerase and endoperoxide reductase. In the presence of both the inhibitor and calmodulin the formation of $PGF_{2\alpha}$, $PGE_2$ and $PGD_2$ was further augmented (FIGURE 2d). Other inhibitors of thromboxane synthetase, 9,11-iminoepoxy (7.2 $\mu$M) and 9,11-diazo (4 $\mu$M) gave essentially similar results: An increase of $PGE_2$ and a concomitant decline of $TxB_2$. In the presence of calmodulin, the levels of $PGF_{2\alpha}$, $PGE_2$, and $PGD_2$ were increased correspondingly. In the presence of any of these three inhibitors, $PGE_2$ appeared to be the major metabolite of $PGH_2$ (FIGURES 2d, 2f, and 2h), a finding consistent with their known specificity.[18,21,22]

The calmodulin-induced formation of thromboxane in human platelets was further investigated by the use of TMB-8, an agent believed to be a $Ca^{2+}$-

antagonist.[23] TMB-8 blocked the release of ADP from platelet granules[24] without any significant effect on thromboxane synthetase activity.[25] As shown in FIGURE 3, TMB-8 reduced the formation of $TxB_2$, presumably by immobilization of intracellular $Ca^{2+}$[23,25] (FIGURE 3e). However, the addition of calmodulin restored the formation of $TxB_2$ to the control level with little or no effect on the formation of $PGF_{2\alpha}$, $PGE_2$, and $PGD_2$ (FIGURES 3a, 3b, 3c, and 2d). TABLE 1 summarizes the results of three different experiments using the various inhibitors. In all cases, the inhibitors decreased the formation of $TxB_2$ while it increased that of $PGF_{2\alpha}$, $PGE_2$, and $PGD_2$; $PGE_2$ being the major product. In the presence of calmodulin, the levels of these compounds were further increased.

FIGURE 4 shows the conversion of $PGE_2$ to 15-keto-$PGE_2$ catalyzed by 15-hydroxyprostaglandin dehydrogenase purified from bovine lung (upper panel). In the presence of calmodulin, the radioactivity associated with 15-keto-$PGE_2$

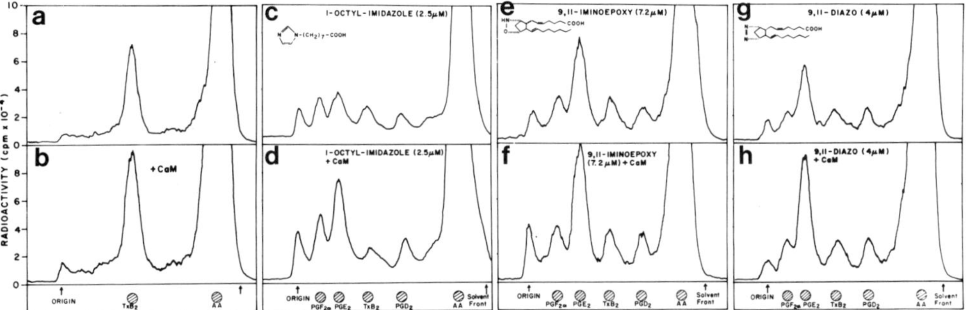


FIGURE 2. Radiochromatogram scans of reaction products from washed human platelets incubated with [$^{14}$C]-arachidonic acid in the presence of calmodulin (CaM) with or without thromboxane synthetase inhibitors.

[$^{14}$C]-arachidonic acid (0.36 $\mu$Ci/1.4 nmole) was added to washed platelets (0.5 ml) in an aggregometer cuvette stirred at 1,200 rpm and incubated at 37°C for 15 min, alone (a); or after a 5-min incubation with 2.3 $\mu$M of CaM (b).

In studies with thromboxane synthetase inhibitors, they were added to washed platelets 10 min before [$^{14}$C]-arachidonic acid (c, e & g). In studies using both thromboxane synthetase inhibitors and CaM, CaM was added 5 min after the inhibitors and incubated for an additional 5 min at 37°C before the addition of [$^{14}$C]-arachidonic acid (d, f & h). The reaction was further incubated for 10 min at 37°C with constant stirring. At the end of the incubation, the reaction products were acidified, extracted, and separated by thin-layer chromatography as described in EXPERIMENTAL PROCEDURES. The radioactivity on the thin-layer plates was located by scanning with a Packard 2730 radiochromatogram scanner. Radioactive zones were identified with authentic standards.

was markedly reduced (middle panel). The activity of the 15-PGDH was further decreased when the $Ca^{2+}$ in the incubation was increased from $10^{-6}$ to $10^{-5}$ M (lower panel).

The effect of the concentration of $Ca^{2+}$ on the activity of 15-PGDH was further investigated. FIGURE 4 shows that in the presence of calmodulin, $Ca^{2+}$ at $10^{-5}$ M decreased 15-PGDH activity some 70%. Increasing the concentration of $Ca^{2+}$ to $10^{-2}$ M did not inhibit the enzyme activity further. In the absence of calmodulin, $Ca^{2+}$ alone did inhibit the enzyme activity. In fact, $Ca^{2+}$ at a concentration higher than $10^{-4}$ M slightly enhanced the dehydrogenase activity. The reason for the apparent stimulation of the enzyme by $Ca^{2+}$ is not clear.

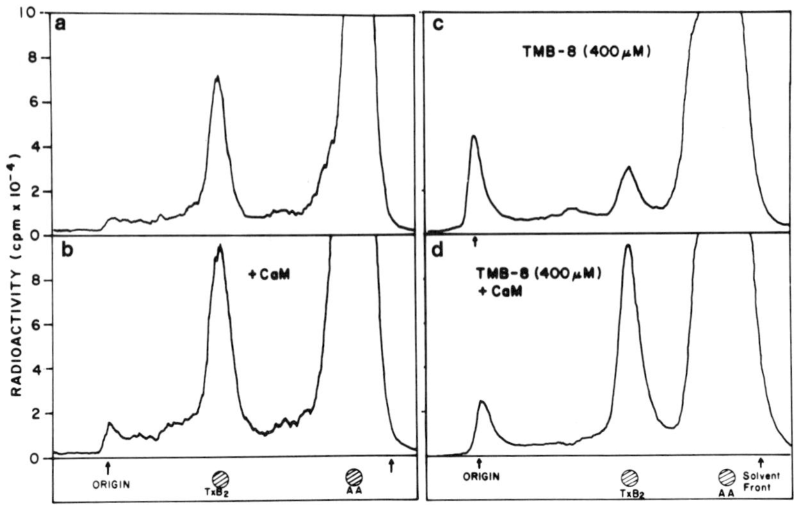


FIGURE 3. Radiochromatogram scan of reaction products from washed platelets in the absence of CaM (2.3 μM) (a), in the presence of CaM (2.3 μM) (b), in the presence of TMB-8 (400 μM) (c), and in the presence of TMB-8 and CaM (2.3 μM) (d). TMB-8, CaM, and [$^{14}$C]-arachidonic acid were added sequentially as described in the legend to FIGURE 2.

TABLE 1

THROMBOXANE AND PROSTAGLANDIN SYNTHESIS INDUCED BY CALMODULIN IN THE PRESENCE OR ABSENCE OF THROMBOXANE SYNTHETASE INHIBITORS AND TMB-8 IN HUMAN PLATELETS

| Other Additions | Prostaglandins Formed (pmoles)* | | | |
|---|---|---|---|---|
| | $PGF_{2\alpha}$ | $PGE_2$ | $TxB_2$ | $PGD_2$* |
| None | 4.1 ± 0.5 | 8.2 ± 0.6 | 68.6 ± 4.7 | 7.8 ± 0.4 |
| CaM (2.3 μM) | 5.3 ± 0.1‡ | 10.2 ± 1.3† | 100.0 ± 1.9‡ | 8.3 ± 0.2† |
| azo-analog-I (4 μM) | 18.7 ± 2.5 | —5.7 ± 6.7 | 26.3 ± 1.3 | 15.7 ± 1.7 |
| azo-analog-I (4 μM) + CaM (2.3 μM) | 18.7 ± 3.3 | 64.0 ± 3.0† | 31.4 ± 2.6† | 21.8 ± 1.3‡ |
| 9,11 IEPA (7.2 μM) | 31.9 ± 1.3 | 35.3 ± 3.3 | 28.5 ± 0.1 | 19.1 ± 1.2 |
| 9,11 IEPA (7.2 μM) + CaM (2.3 μM) | 38.6 ± 2.3 | 60.8 ± 2.8‡ | 32.0 ± 1.0† | 27.0 ± 1.8‡ |
| 1-Octyl-Imidazole (2.5 μ) | 28.1 ± 2.4 | 36.0 ± 3.4 | 22.4 ± 3.4 | 17.0 ± 2.4 |
| 1-Octyl-Imidazole (2.5 μM) + CaM (2.3 μM) | 44.8 ± 3.9† | 47.6 ± 5.9† | 28.5 ± 4.2† | 22.1 ± 5.6† |
| TMB-8 (400 μM) | 4.1 ± 0.8 | 9.8 ± 0.8 | 33.4 ± 1.0 | 8.3 ± 0.6 |
| TMB-8 (400 μM) + CaM (2.3 μM) | 5.1 ± 0.6 | 13.4 ± 1.7† | 58.4 ± 4.8† | 9.1 ± 2.8† |

*Mean Values ± SE of 3 experiments.

†$P < 0.005$, statistically different from control without calmodulin.

‡$P < 0.001$, statistically different from control without calmodulin.

Washed platelets were incubated alone or with calmodulin at 37° C with constant stirring at 1,200 rpm. The inhibitor was added 5 min before calmodulin; the mixture was further incubated for an additional 15 min. The reaction was stopped by acidification to pH 3.0, and extracted with ethyl acetate. The reaction products were separated on TLC as described in FIGURE 1. Radioactivity was determined as described under EXPERIMENTAL PROCEDURES.

TABLE 2 shows that the inhibition of 15-PGDH by calmodulin was completely reversed by trifluoroperazine, which is known to bind calmodulin specifically, rendering it biologically inactive.[26] Trifluoroperazine did not affect the basal activity of 15-PGDH. This is in line with the effect of trifluoroperazine on the activities of other calmodulin-regulated enzymes.[16]

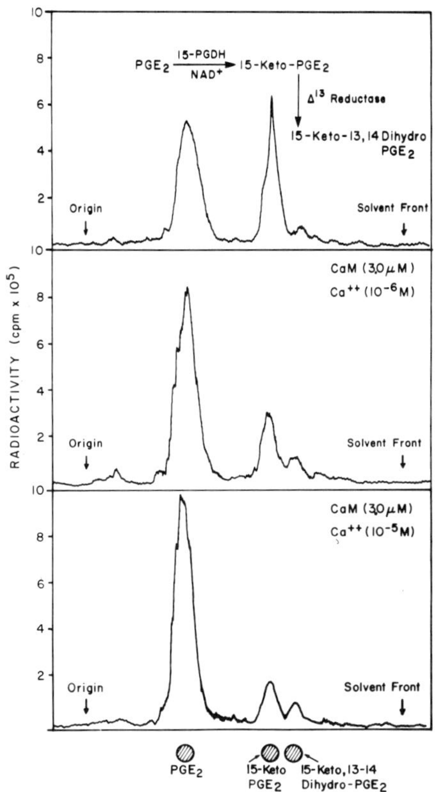


FIGURE 4. Radiochromatogram scan of radioactive products isolated after incubation of $[^3H]PGE_2$ with purified bovine lung 15-hydroxyprostaglandin dehydrogenase (a) in the presence of $Ca^{2+}$ ($10^{-6}$ M) (upper panel); b) in the presence of $Ca^{2+}$ ($10^{-6}$ M) and calmodulin (3 $\mu$M) (middle panel); and c) in the presence of $Ca^{2+}$ ($10^{-5}$ M) and calmodulin (3 $\mu$M) (lower panel).

## DISCUSSION

$Ca^{2+}$ regulates many cellular processes including platelet aggregation.[27] Rodan and Feinstien[28] showed that $Ca^{2+}$ inhibits the $PGE_1$-stimulated synthesis of cAMP probably by suppressing the activity of adenylate cyclase. More recently, Lyon and Shaw[29] reported that $Ca^{2+}$ flux preceded the onset of serotonin release in rabbit platelets. Local anesthetics which inhibit $Ca^{2+}$ release from the skeletal

TABLE 2

EFFECT OF CALMODULIN ON THE ACTIVITY OF 15-HYDROXYPROSTAGLANDIN DEHYDROGENASE FROM BOVINE LUNG

| Other Additions | 15-Keto-$PGE_2$ Formed (pmol) |
|---|---|
| None | 678 |
| $Ca^{2+}$ (1 mM) | 644 |
| Calmodulin (3 $\mu$M) | 673 |
| TFP (200 $\mu$M) | 587 |
| Calmodulin + $Ca^{2+}$ | 292 |
| Calmodulin + $Ca^{2+}$ + TFP | 609 |

Each reaction mixture, in a final volume of 1 ml, contained 0.5 $\mu$Ci of [$^3$H]$PGE_2$, 50 mM Tris-HCl (pH 7.4), approximately 250 $\mu$g purified enzyme and 1 mM $Ca^{2+}$, 3 $\mu$M calmodulin, or trifluroperazine (TFP). TFP was dissolved in 50 mM Tris-HCl buffer (pH 7.4). The reaction mixture was incubated for 60 min at 37° C with constant shaking. 15-Hydroxyprostaglandin dehydrogenase is expressed as pmoles of 15-keto $PGE_2$ formed.

muscle also inhibit platelet aggregation and secretion.[30] The mechanism of how $Ca^{2+}$ relates to platelet aggregation has not been clarified.

The effect of exogenous calmodulin on the augmentation of thromboxane formation in the intact platelet needs further investigation; calmodulin either exerts its effect on the exterior surface of plasma membrane, or it is transported into the platelet and acts intracellularly. The augmentation of thromboxane synthesis by calmodulin is consistent with our earlier findings that calmodulin activates phospholipase $A_2$, leading to an increase of $PGH_2$ and $TxA_2$. However, exogenous arachidonic acid may be utilized by the cyclo-oxygenase to form $PGH_2$; and our experiments with washed platelets do not exclude the possibility that calmodulin may enhance the activity of cyclo-oxygenase.

Gorman *et al.* recently reported that TMB-8 is not a $TxA_2$ synthetase inhibitor,[25] but rather a $Ca^{2+}$ antagonist which immobilizes intracellular $Ca^{2+}$.[23] White and Gerrard suggested that $TxA_2$ acts as a $Ca^{2+}$ ionophore, mobilizing $Ca^{2+}$ from the storage sites of human platelets, most likely the dense tubular system.[31] The finding that calmodulin effectively restored the synthesis of $TxA_2$ in the presence of TMB-8 suggests that calmodulin either counteracts the effect of the drug or it activates phospholipase $A_2$ independent of the $Ca^{2+}$ otherwise released from the storage sites (FIGURE 4).

The present finding that calmodulin inhibits 15-hydroxyprostaglandin dehydrogenase adds to the growing list of calmodulin-regulated enzymes. The dehydrogenase is of particular interest, not only because of its important role in the metabolism of prostacyclin and other prostaglandins, but also because it is the first enzyme whose activity is inhibited rather than activated by calmodulin. The concentration of $Ca^{2+}$ required to inhibit the activity of 15-PGDH in the presence of calmodulin is within the physiological range ($10^{-6}$ M).

The finding that calmodulin inhibits 15-PGDH has physiological implications. 15-PGDH is present abundantly in the vascular wall[10] and other tissues, and figures importantly in controlling the intracellular level of prostacyclin.[32] When the vascular wall is damaged as a result of injury, the exposure of the dehydrogenase in the endothelial cell to the high concentration of $Ca^{2+}$ in the plasma would inhibit 15-PGDH, causing an increase of $PGI_2$ *in situ*. $PGI_2$ released from the endothelial cells would activate adenylate cyclase of the platelets at the site of injury, increasing the cAMP level in these platelets. Minkes and co-workers showed that cAMP depressed platelet levels of arachidonic acid, presumably by inhibiting its release from membrane phospholipids.[33] The reduction of arachidonic acid would diminish the synthesis of thromboxane $A_2$, and thus prevent further platelet clumping on the vascular wall, an important step towards ameliorating thrombosis.

Although this notion appears attractive, it is speculative. It is necessary to demonstrate that calmodulin inhibits 15-PGDH of blood vessels in a manner similar to that of bovine lung, and that prostacyclin and thromboxane interact at the level of endothelium and platelets in the control of platelet aggregation. It has been pointed out that calmodulin not only serves as a versatile intracellular

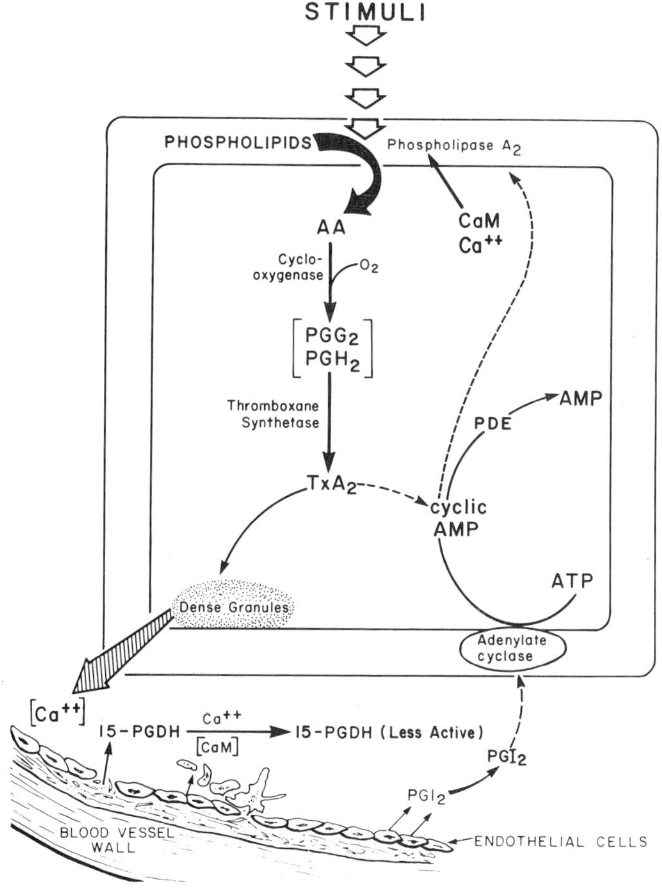


FIGURE 5. Scheme showing the metabolic pathways of arachidonic acid in human platelets. When thromboxane synthetase is blocked by its inhibitors, $PGH_2$ is diverted to the formation of $PGE_2$, $PGF_{2\alpha}$, and $PGD_2$. Calmodulin may exert its effect on phospholipase $A_2$, or it may also act on another enzyme involved in phospholipid or arachidonic acid metabolism.

calcium receptor, controlling diverse cellular processes, but also serves as a molecular link between the cAMP, $Ca^{2+}$, and prostaglandin-regulated systems.[15] The modulation of phospholipase $A_2$ and 15-hydroxyprostaglandin dehydrogenase by calmodulin further supports this notion.

## Acknowledgments

Prostaglandins and thromboxane synthetase inhibitors were gifts from Drs. U. Axen, J. Pike and J. Babcock of the Upjohn Co. 1-Octyl-imidazole was supplied by Dr. E. Ku of Ciba-Geigy Corp., NY and bovine brain calmodulin by Dr. R. Wallace, St. Jude Children's Research Hospital. The assistance of Ms. Pat Nicholas and Ms. Nancy Gentile in the preparation of the manuscript is gratefully acknowledged.

## References

1. Von Euler, U. S. 1934. Arch. Exp. Pathol. Pharmakol. **175:** 78–84.
2. Hamberg, M. & B. Samuelsson. 1973. Proc. Natl. Acad. Sci. USA **70:** 899–903.
3. Hamberg, M. & B. Samuelsson. 1974. Proc. Natl. Acad. Sci. USA **71:** 3400–3404.
4. Needleman, P., S. Moncada, S. Bunting, J. R. Vane, M. Hamberg & B. Samuelsson. 1976. Nature **261:** 558–560.
5. Gryglewski, R. J., S. Bunting, S. Moncada, R. J. Flower & J. R. Vane. 1976. Prostaglandins **12:** 685–713.
6. Tateson, J. E., S. Moncada & J. R. Vane. 1977. Prostaglandins **13:** 377–382.
7. Gryglewski, R. J., R. Korbut & A. Ocetkiewicz. 1978. Nature **273:** 765–767.
8. Smith, J. B., M. L. Ogletree, A. M. Lefer & K. C. Nicolaou. 1978. Nature **274:** 64–65.
9. Steer, M. L., D. E. MacIntyre, L. Levine & E. W. Salzman. 1980. Nature **283:** 194–195.
10. Wong, P. Y-K, F. F. Sun & J. C. McGiff. 1978. J. Biol. Chem. **253:** 5555–5557.
11. Mekean, M. L., J. B. Smith & M. J. Silver. 1980. *In* The Proceeding of Golden Jubilee International Congress on Essential Fatty Acid and Prostaglandins. **74:** 45.
12. Isakson, P. C., A. Raz, S. E. Denny, A. Wyche & P. Needleman. 1977. Prostaglandins **14:** 853–871.
13. Bills, T. K., J. B. Smith & M. J. Silver. 1976. Biochim Biophys. Acta **424:** 303–314.
14. Pickett, W. C., R. L. Jessee & P. Cohen. 1976. Biochem. J. **160:** 405–408.
15. Wong, P. Y-K & W. Y. Cheung. 1979. Biochem. Biophys. Res. Commun. **90:** 473–480.
16. Cheung, W. Y. 1980. Science **207:** 19–27.
17. Wong, P. Y-K & J. C. McGiff. 1977. Biochim. Biophys. Acta **500:** 436–439.
18. Gorman, R. R., G. L. Bundy, D. C. Peterson, F. F. Sun, O. V. Miller & F. A. Fitzpatrick. 1977. Proc. Natl. Acad. Sci. USA **74:** 4007–4011.
19. Lee, S. C. & L. Levine. 1974. J. Biol. Chem. **249:** 1369–1375.
20. Samuelsson, B., M. Goldyne, E. Granstrom, M. Hamberg, S. Hammarstrom & C. Molmsten. 1978. Ann. Rev. Biochem. **47:** 997–1029.
21. Yoshimoto, T., S. Yamamoto & O. Hayaishi. 1978. Prostaglandins **16:** 529–540.
22. Fitzpatrick, F., R. Gorman, G. Bundy, T. Honohan, J. McGuire & F. F. Sun. 1979. Biochim. Biophys. Acta **573:** 238–244.
23. Charo, I. F., R. D. Feinman & T. C. Detwiler. 1976. Biochem. Biophys. Res. Commun. **72:** 1462–1467.
24. Pickett, W. C., R. L. Jesse & P. Cohen. 1977. Biochem. Biophys. Acta **486:** 209–213.
25. Gorman, R., W. Wierenga & O. V. Miller. 1979. Biochim. Biophys. Acta **572:** 95–104.
26. Levin, R. M. & B. Weiss. 1976. Mol. Pharmacol. **12:** 581–589.
27. Rittenhouse-Simmons, S., F. A. Russell & D. Deykin. 1977. Biochem. Biophys. Acta **488:** 370–380.

28. RODAN, G. A. & M. B. FEINSTEIN. 1976. Proc. Natl. Acad. Sci. USA **73:** 1829–1833.
29. LYON, R. M. & J. O. SHAW. 1980. J. Clin. Invest. **65:** 242–255.
30. FEINSTEIN, M. B., J. FIEKERS & C. FRASER. 1976. J. Clin. Invest. **197:** 215–228.
31. WHITE, J. G. & J. M. GERRARD. 1978. *In* Platelets: A Multidisciplinary Approach. G. de Gaetano & S. Garattini, Ed.:17–34, Raven Press, New York, N.Y.
32. WONG, P. Y-K, F. F. SUN, K. U. MALIK, L. CAGEN & J. C. MCGIFF. 1979. *In* Prostacyclin. J. R. Vane & S. Bergstrom, Ed.:133–145, Raven Press, New York, N.Y.
33. MINKES, M., N. STANFORD, M. M-Y CHI, G. J. ROTH, S. RAZ, P. NEEDLEMAN & P. W. MAJERUS. 1977. J. Clin. Invest. **59:** 449–454.


## DISCUSSION OF THE PAPER

DR. NEMETH (*Yale University, New Haven, CT*): Investigators at Washington University have argued on kinetic grounds that phospholipase A2 is not sufficient for the release of arachidonic acid for prostaglandin metabolism and, in platelets, that phospholipase C, via diglyceride lipase, release arachidonic acid.

DR. P. Y. K. WONG: Phospholipase A2 will not deliver enough arachidonic acid for the metabolism to generate endoperoxide and prostaglandins. This is still a controversial situation. As far as I know, phospholipase A2 accounts for more than 60% of the release of arachidonic acid to make endoperoxide and the other 35–40% may come from phospholipase C. [See Reference 11.]

DR. JONDORF (*Albert Einstein College of Medicine, Bronx, NY*): I have a question about your interpretation of the effect of calmodulin on thromboxane since it's by intact platelets in the presence of radiolabeled exogenous AA. Wouldn't the interpretation be that if calmodulin indeed stimulates phospholipase, you would get dilution of radiolabeled arachidonic acid and less conversion of labeled arachidonic acid to thromboxane, rather than more?

DR. WONG: That's true. However, when you use a low dose of this very specific radioactive labeled arachidonic acid, dilution will be insignificant. We still have not ruled out the possibility that calmodulin has also activated another enzyme system, namely, the cyclo-oxygenase.

# ON THE MECHANISM OF ACTIVATION OF CYCLIC NUCLEOTIDE PHOSPHODIESTERASE BY CALMODULIN

Jerry H. Wang* and Rajendra K. Sharma

*Department of Biochemistry*
*University of Manitoba*
*Winnipeg, Manitoba, Canada R3E 0W3*

Charles Y. Huang, Vincent Chau, and P. Boon Chock

*Laboratory of Biochemistry*
*National Heart, Lung, and Blood Institute*
*National Institutes of Health*
*Bethesda, Maryland 20205*

## INTRODUCTION

Early studies on calmodulin have centered primarily on the regulation of mammalian cyclic nucleotide phosphodiesterase. Calmodulin was discovered as an activator protein of this enzyme by Cheung.[1,2] Shortly afterwards Kakiuchi and coworkers[3,4] demonstrated a $Ca^{2+}$-activatable cyclic nucleotide phosphodiesterase and a protein factor capable of enhancing the activation of the enzyme by $Ca^{2+}$. Subsequently, Teo *et al.*[5,6] purified the protein activator to homogeneity and showed that it is a $Ca^{2+}$-binding protein. In addition, the activation of phosphodiesterase was found to depend on the presence of $Ca^{2+}$.[6,7] These results led to the proposal of a general mechanism for the activation of phosphodiesterase by calmodulins as represented by Scheme 1.[8]

$$Ca^{2+} + CM \rightleftharpoons Ca^{2+}\cdot CM^*$$

$$Ca^{2+}\cdot CM^* + PDE \rightleftharpoons Ca^{2+}\cdot CM^*\cdot PDE \rightleftharpoons Ca^{2+}\cdot CM^*\cdot PDE^*$$

(Scheme 1)

where CM and PDE stand for calmodulin and phosphodiesterase, respectively and the superscript * denotes the activated states for these proteins.

However, based on our present understanding of this enzyme system, the proposed mechanism must be considered as tentative. For example, the multiple $Ca^{2+}$-binding to calmodulin and the stoichiometry of the interactions between calmodulin and phosphodiesterase should be incorporated into this general mechanism. The interaction between phosphodiesterase and calmodulin in the absence of $Ca^{2+}$, albeit very weak, should not have been ignored. Since $Ca^{2+}$ can enhance the association between calmodulin and phosphodiesterase, phosphodiesterase would be expected to enhance the binding of $Ca^{2+}$ to calmodulin. The relative affinity between phosphodiesterase and calmodulin in the presence and absence of $Ca^{2+}$ would determine the magnitude of the effect of phosphodiesterase on the binding of $Ca^{2+}$ to calmodulin.

Although recent studies have significantly extended our understanding of the physiological functions and physicochemical properties of calmodulin, progress

*On leave from University of Manitoba at the Laboratory of Biochemistry, National Heart, Lung and Blood Institute, 1979–1980.



0077-8923/80/0356-0190 $1.75/0 © 1980, NYAS

on the elucidation of the mechanisms of calmodulin action has not been substantial. This is at least due in part to the lack of reasonable quantity of homogeneous preparations of the calmodulin-dependent cyclic nucleotide phosphodiesterase. Only very recently has homogeneous phosphodiesterase been prepared.[9-11] Little physicochemical characterization of the enzyme has been carried out.

In this communication, a method for the purification of calmodulin-dependent phosphodiesterase from bovine brain is described. Some physicochemical properties of the enzyme including its interaction with calmodulin are presented. Kinetic procedures are developed to determine the stoichiometry and equilibrium constant of the interaction of calmodulin and phosphodiesterase in the enzyme reactions, as well as the rate constants for the association of phosphodiesterase and calmodulin and the dissociation of the protein complex. In addition, a more general regulatory scheme for the activation of the enzyme by calmodulin taking into consideration the multiple $Ca^{2+}$-binding to calmodulin and the interaction of calmodulin and phosphodiesterase in the absence of $Ca^{2+}$ has been proposed.

## *Purification of Cyclic Nucleotide Phosphodiesterase*

A new procedure has been developed for the purification of calmodulin-dependent cyclic nucleotide phosphodiesterase from bovine brain.[12] TABLE 1 summarizes the data for a typical preparation of the enzyme from about 3 kg of frozen whole brains. The purified enzyme, when fully activated by calmodulin, has specific activities of 300 to 400 $\mu$mol of cAMP hydrolyzed/min/mg protein, which are somewhat higher than those reported for the other preparations of the enzymes.[9-11]

Calmodulin-dependent cyclic nucleotide phosphodiesterase may lose or diminish its response to calmodulin during its purification. Although crude phosphodiesterase may be activated more than 10 fold, pure preparations of the enzyme have been variously reported to be activated 2 to 6 fold by calmodulin.[9-13] The enzyme preparation purified by the present procedure, however, can be activated by calmodulin over 10 fold (FIGURE 1).

The most efficient purification step in the procedure is the calmodulin-Sepharose 4B column chromatography which usually achieves about 100-fold purification (TABLE 1). This affinity chromatography is used to isolate proteins capable of undergoing $Ca^{2+}$-dependent association with calmodulin. However, the extract of bovine brain contains many proteins that can bind to calmodulin. Among these proteins, the most abundant is the calmodulin binding protein I, which was originally discovered as a heat-labile inhibitor protein of cyclic nucleotide phosphodiesterase.[14,15] This inhibitor protein may be separated from the enzyme by Affigel Blue chromatography (FIGURE 2A). The marked difference in concentration-dependence of the calmodulin activation of the enzyme before and after the Affigel Blue chromatography (FIGURE 2B) is also indicative of the removal of other calmodulin binding components from the enzyme sample. The fraction containing the inhibitory activity may be used to prepare homogeneous preparations of calmodulin binding protein I.[16]

The cyclic nucleotide phosphodiesterase prepared by the present procedure appeared homogeneous when it was analyzed by disc gel electrophoresis, SDS-gel electrophoresis, and gel isoelectrofocusing. FIGURE 3 shows the electrophoretic patterns of a typical enzyme preparation.

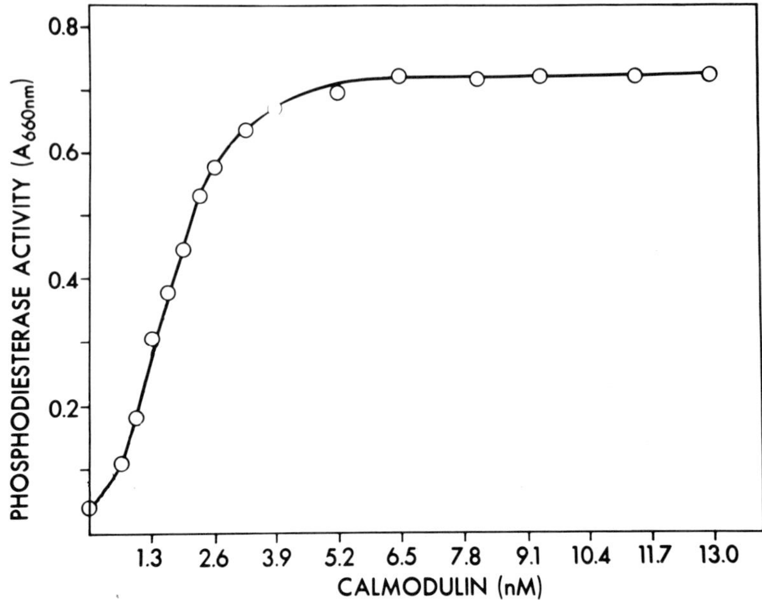


FIGURE 1. The activation of purified bovine brain cyclic nucleotide phosphodiesterase.

### *Physicochemical Properties of the Phosphodiesterase*

Some of the physicochemical properties of the purified calmodulin-dependent bovine brain phosphodiesterase, along with methods used for the analyses are summarized in TABLE 2. The molecular weight of the protein is about 120,000. The single protein band observed on the SDS-electrophoretic gel has a mobility corresponding to that of a 58,000 dalton polypeptide. These results are consistent with the suggestion[11] that the enzyme is composed of two apparently identical subunits. Bovine heart cyclic nucleotide phosphodiesterase has also been purified to homogeneity and suggested to be composed of two subunits of molecular weight about 60,000.[10]

TABLE 1

PURIFICATION OF BOVINE BRAIN PHOSPHODIESTERASE

| | Total Activity ($\mu$mol.$PO_4$/min) | Specific Activity ($\mu$mol.$PO_4$/min/mg) | Yield (%) | Fold of Purification |
|---|---|---|---|---|
| Crude Extract | 7,616 | 0.1 | 100 | 1 |
| DEAE-Cellulose Chromatography | 4,000 | 0.38 | 52 | 3.8 |
| Affi-Gel Blue Chromatography | 1,360 | 2.13 | 18 | 21.3 |
| Calmodulin Affinity Chromatography | 1,583 | 180 | 21 | 1,800 |
| G-200 Sephadex Gel Filtration | 746 | 360 | 10 | 3,640 |

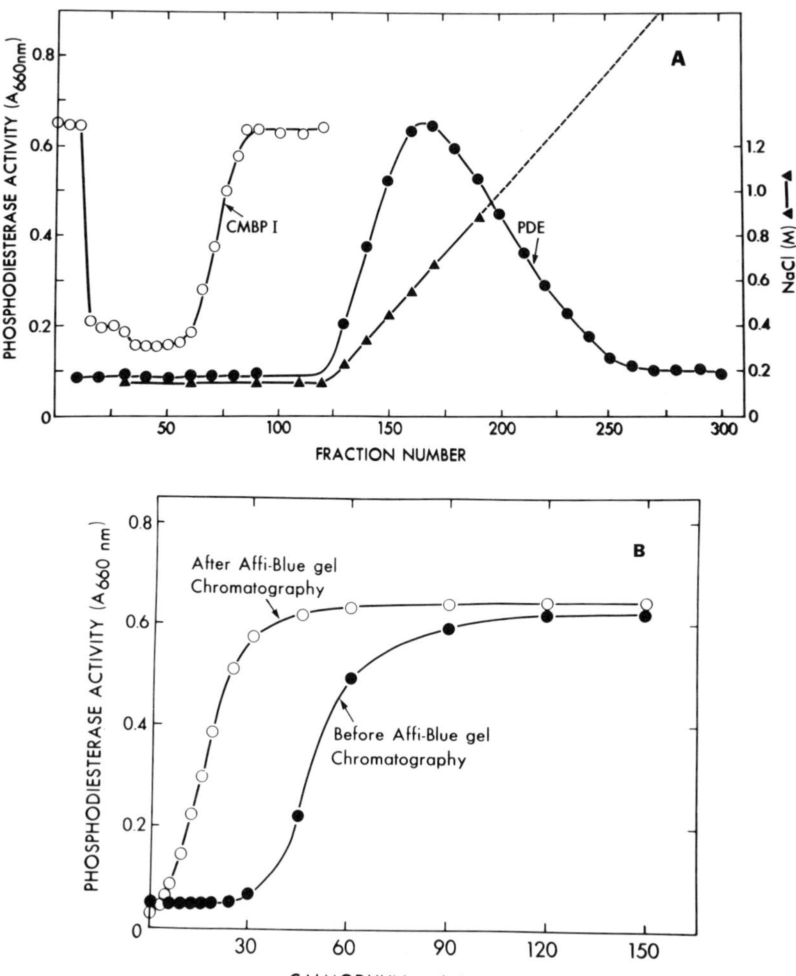


FIGURE 2. The separation of phosphodiesterase (●) and calmodulin-binding protein I (○) by Affigel Blue chromatography. (B) Comparison of the calmodulin activations of the enzyme before and after the chromatography.

### *The Stoichiometry of Interaction Between Calmodulin and Phosphodiesterase*

In media containing high concentrations of $Ca^{2+}$ ($>10^{-5}$ M), the interaction between calmodulin and cyclic nucleotide phosphodiesterase is enhanced to form a complex. The complex does not appear to dissociate significantly upon centrifugation on gel filtration so that its molecular weight may be determined by these methods. FIGURE 4 shows the elution profiles of the mixtures of calmodulin and purified phosphodiesterase from a Sephadex G-200 column with buffers containing either $10^{-4}$ M $Ca^{2+}$ (FIGURE 4B) or $10^{-4}$ M EGTA (FIGURE 4A). In the presence of EGTA, calmodulin and phosphodiesterase were separately eluted. In

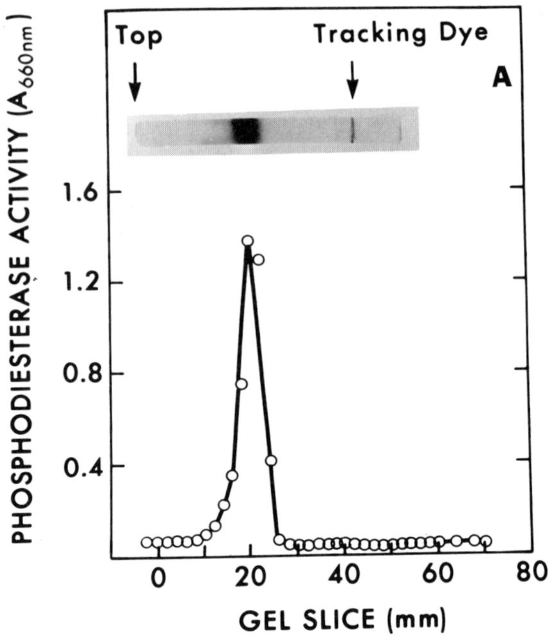


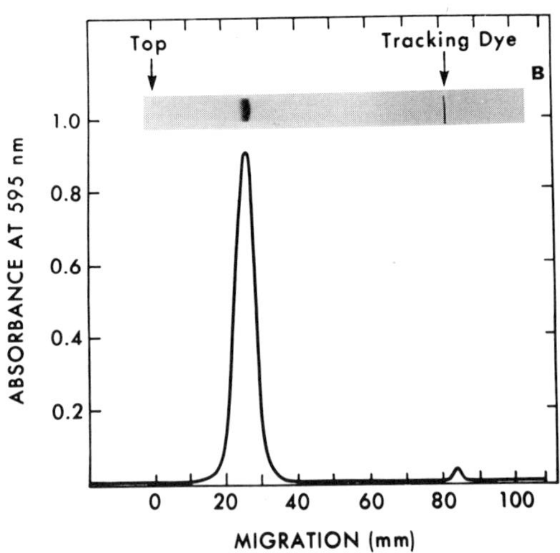


FIGURE 3. (A) Analysis of the purity of the phosphodiesterase preparation by disc gel electrophoresis, Upper: protein stain, lower: enzyme assay for a sliced gel, (B) SDS-gel electrophoresis of the enzyme preparation, upper: protein stain, lower: densitometric tracing.

the presence of $Ca^{2+}$, the two proteins cochromatographed, suggesting the formation of the protein complex. The elution volume of the complex corresponded to a protein of Stokes radius 48 Å. Similarly, the sedimentation constant of the phosphodiesterase-calmodulin complex was determined by sucrose-density centrifugation to be 8.0S.

The molecular weight of the phosphodiesterase-calmodulin complex has been calculated from the sedimentation constant and the Stokes radius to be 159,000.[17] Thus, it is suggested that the protein complex has a composition of $(CM)_2$ PDE.

To further substantiate the suggestion, the gel filtration eluates containing the protein complex (FIGURE 4) were pooled and analyzed for the contents of calmodulin and phosphodiesterase. The pooled sample was concentrated and subjected to SDS-gel electrophoresis. As was expected, the gel showed two protein bands corresponding to phosphodiesterase and calmodulin (FIGURE 5). The mass ratio of phosphodiesterase to calmodulin determined from the densitometric tracing is 3.4, consistent with a molar ratio of 1 to 2. From these results, it may be concluded that the calmodulin-dependent bovine brain phosphodiesterase is capable of binding two moles of calmodulin per mole of the enzyme. Presumably, each subunit of the enzyme can bind one molecule of calmodulin. Bovine heart phosphodiesterase has also been shown by LaPorte *et al.*[10] using the cross-reacting technique to be capable of binding one calmodulin per subunit of the enzyme.

TABLE 2

PHYSICAL PARAMETERS OF CALMODULIN-DEPENDENT PHOSPHODIESTERASE

| | | |
|---|---|---|
| Sedimentation Constant (s) | Sucrose density centrifugation | 6.9 |
| Stokes Radius (Å) | Gel filtration | 44.2 |
| Partial Specific Volume (ml/g) | From Amino Acid composition | 0.726 |
| Molecular Weight | From Sedimentation constant and Stokes radius | 124,000 |
| | Sedimentation equilibrium | 115,000 |
| Subunit weight | SDS-gel electrophoresis | 58,000 |
| Isoelectric point (pH) | Isoelectric focusing | 4.85 |
| Absorbance at 278 nm for 1% solution | | 9.6 |

In order to test the relationship between the binding of calmodulin and the enzyme activation, the stoichiometry of the interaction between the two proteins has been determined in the enzyme reaction. This protein interaction was analyzed according to the continuous variation procedure of Job's using the activation of phosphodiesterase by calmodulin to monitor the calmodulin binding.[18,19]

To apply the procedure of Job's, the binding of the ligand to the protein is monitored at various molar ratios of the ligand to the protein while the total molar concentrations of the ligand and the protein are maintained at a constant level. When the ligand binding is plotted against the molar fractions of the ligand or/and of the protein, the two limiting slopes, as the protein and ligand concentrations approaching zero, will meet at a point. The molar ratio of the ligand to the protein at this intersection point is used to determine the stoichiometry of the ligand binding to the protein.[18,19]

FIGURE 6 shows the Job plot for the activation of the phosphodiesterase by calmodulin at saturating concentrations of $Ca^{2+}$. As can be seen in this figure, the intersection point occurs at a molar ratio of calmodulin to phosphodiesterase of 2.

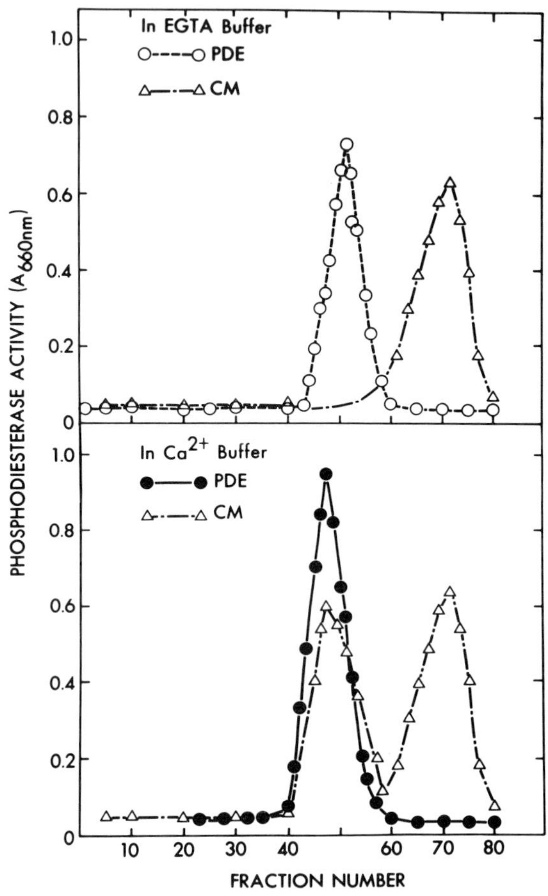


FIGURE 4. Sephadex G-200 gel filtration of aliquots (1 ml) of a mixture of phosphodiesterase (1.2 μM) and calmodulin (5.5 μM) with buffers containing either 0.1 mM EGTA (upper) or 0.01 mM $Ca^{2+}$ (lower).

If, however, molar concentration of the enzyme subunit instead of the dimeric enzyme was used in the experiment, the intersection point occurred at a molar ratio of calmodulin and phosphodiesterase of about 1. Thus, it may be concluded that two molecules of calmodulin may bind to one molecule of the phosphodiesterase, and that each binding of calmodulin results in enzyme activation.

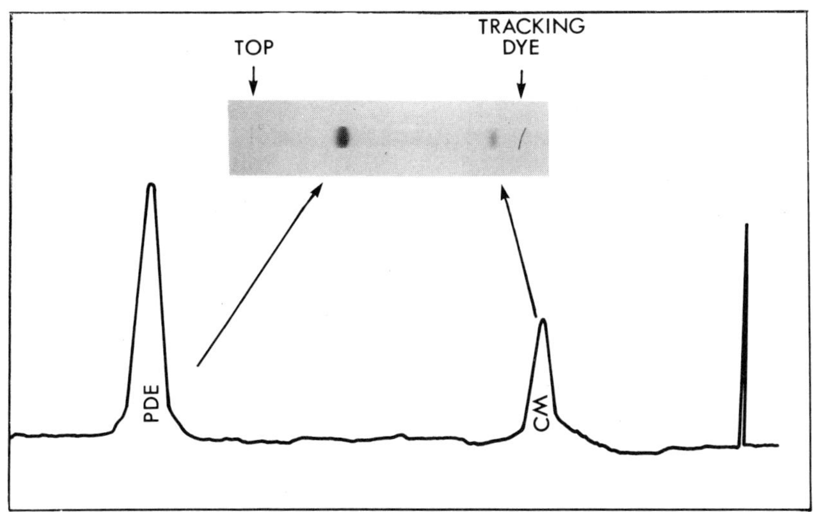


FIGURE 5. SDS-gel electrophoretic analysis of the calmodulin-phosphodiesterase complex, upper: protein stain, lower: densitometric tracing.

## *Equilibrium and Kinetic Constants of the Interaction Between Calmodulin and Phosphodiesterase*

The observations that each subunit of cyclic nucleotide phosphodiesterase may bind a calmodulin molecule and that the binding of calmodulin appears to

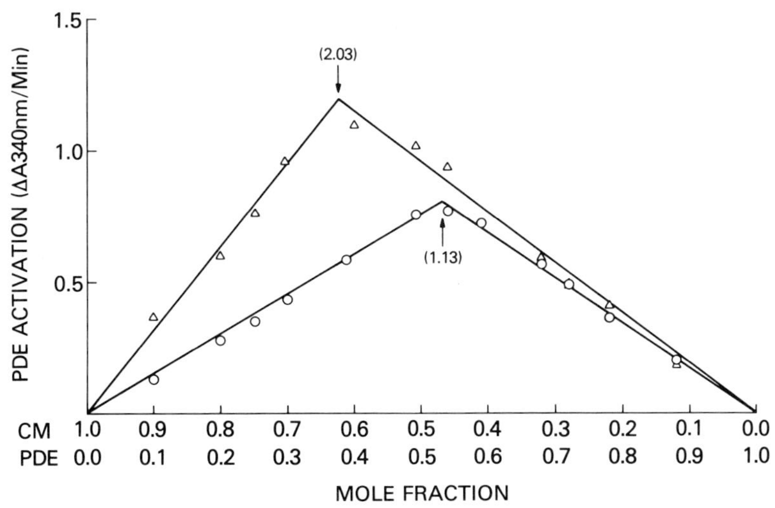


FIGURE 6. Job plot for the activation of phosphodiesterase by calmodulin at 15°C, pH 8.0, with the total concentration of calmodulin and phosphodiesterase maintained at 4.17 nM (Δ), or with the total concentration of calmodulin and the enzyme subunit at 4.17 nM (O).

be directly proportional to the activation of the enzyme suggest that the interaction of these two proteins in the phosphodiesterase reaction in the presence of saturating $Ca^{2+}$, may be analyzed by the following simplified Equation 1

$$Ca^{2+}\cdot CM + PDE \overset{K_d}{\rightleftharpoons} Ca^{2+}\cdot CM\cdot PDE, \tag{1}$$

where PDE represents subunit of the enzyme.

The association constant for this protein-protein interaction may be determined from the dose-dependent curve of the activation of phosphodiesterase by calmodulin. At 50% of the maximal enzyme activation, the concentration of free calmodulin equals the value of $K_d$. Since the amount of calmodulin bound to phosphodiesterase at the point is equivalent to 50% of the total enzyme concentration, the $K_d$ value may be readily calculated from the concentrations of total enzyme and of total calmodulin at 50% of maximal activation.

The values of the dissociation constant for the complex of phosphodiesterase and calmodulin determined by this method were found to vary depending on the batch and the age of the enzyme preparation. The dissociation constant determined at pH 8 and 30°C, in the presence of saturating levels of $Ca^{2+}$ was about 1 nM.

In addition to the dissociation constant, kinetic constants for the interaction between cyclic nucleotide phosphodiesterase and calmodulin have been determined by monitoring the time course of the phosphodiestarase reaction. A continuous assay coupling the phosphodiesterase reaction with myokinase, pyruvate kinase, and lactate dehydrogenase reactions were used to follow the progress of the phosphodiesterase reaction. To determine the rate of the activation of phosphodiesterase by calmodulin, the progress of the basal enzyme reaction was followed briefly, then calmodulin was introduced and the increase in the rate of phosphodiesterase reaction was determined. As is shown in FIGURE 7, it took up to 30 minutes after the addition of calmodulin for the phosphodiesterase reaction to reach the final activated rate when the concentrations of calmodulin and phosphodiesterase were low. The slow transition from the basal to the activated rate of phosphodiesterase reaction was used to determine the kinetic constant for the association of phosphodiesterase and calmodulin.

The activations of phosphodiesterase by calmodulin in the presence of saturating concentrations of $Ca^{2+}$ may be described by two consecutive reactions: an initial association of the two proteins followed by the conformational change of the protein complex (Equation 2):

$$Ca^{2+}\cdot CM + PDE \underset{k_{-1}}{\overset{k_1}{\rightleftharpoons}} Ca\cdot CM\cdot PDE \underset{k_{-2}}{\overset{k_2}{\rightleftharpoons}} Ca\cdot CM\cdot PDE \tag{2}$$

If the rate limiting step of the consecutive reactions is the conformational change of the protein complex, the activation of phosphodiesterase by calmodulin would conform to a first order reaction. On the other hand, if the rate limiting step is the association of the two proteins, the overall kinetics of the activation of phosphodiesterase by calmodulin would be expected to follow a second order reaction. The general rate equations depicting the progress curve of phosphodiesterase reaction activated by calmodulin according to either a first order or a second order reaction has been derived.

Three progress curves of the calmodulin-activated phosphodiesterase reaction are depicted in FIGURE 7. Using computer fitting, we found that the data conform to the rate equation describing the activation of the enzyme by calmodulin as a second order process. The second order rate constant for the interaction was determined to be on the order of $4 \times 10^6$ $M^{-1}$ $sec^{-1}$.

Note that, for the experiments presented in FIGURE 7, the concentrations of calmodulin and phosphodiesterase subunit were equal and that the concentrations of the two proteins were high enough to bring about essentially maximal activation of the enzyme. These conditions were imposed so that the rate equation for the activation of phosphodiesterase by a second order process was significantly simplified.

To determine the kinetic constant for the dissociation of calmodulin-phosphodiesterase complex, the decrease in the rate of calmodulin-activated phosphodiesterase reaction resulting from the addition of calmodulin binding

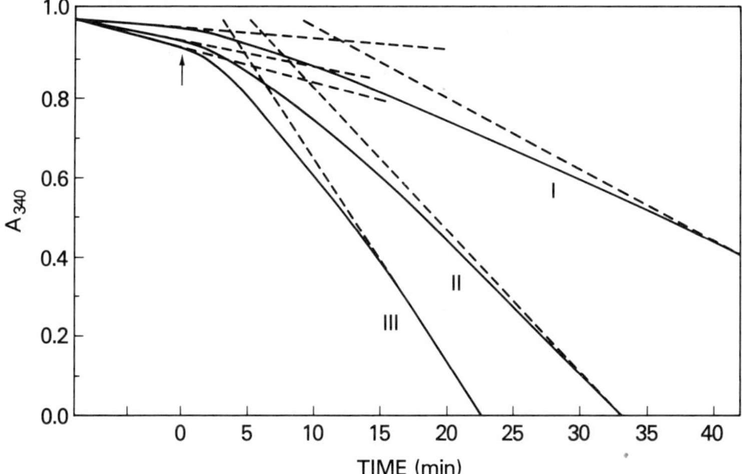


FIGURE 7. The rate of calmodulin activation of phosphodiesterase. Time courses of three phosphodiesterase reactions at different concentrations of the enzyme: 0.295 (I), 0.59 (II), and 0.785 nM (III) were monitored and then 0.59, 1.18, and 1.57 nM of calmodulin respectively were added to the reactions. The arrow indicates the time of calmodulin addition. Broken lines represent extrapolated time courses for the basal and fully activated reactions.

protein I (CMBP-I) was monitored and analyzed. CMBP-I was originally discovered as an inhibitor protein of calmodulin-dependent cyclic nucleotide phosphodiesterase.[16]

Subsequently, the CMBP-I was found capable of undergoing $Ca^{2+}$-dependent association with calmodulins.[14,20] The inhibition of the calmodulin-activated phosphodiesterase reaction by CMBP-I in the presence of saturating levels $Ca^{2+}$ may therefore be described schematically by the following reactions (Scheme 2):

$$Ca^{2+} \cdot CM \cdot PDE \rightleftharpoons Ca^{2+} \cdot CM + PDE$$

$$Ca^{2+} \cdot CM + CMBP\text{-}I \rightleftharpoons Ca^{2+} \cdot CM \cdot CMBP\text{-}I$$

(Scheme 2)

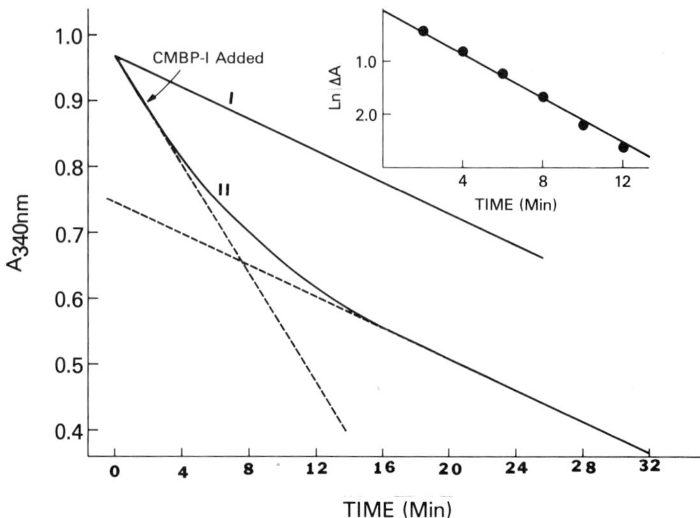


FIGURE 8. The rate of inactivation of calmodulin-activated reaction by calmodulin binding protein I: Basal (I) and calmodulin-activated reactions (II) were monitored as indicated and then calmodulin binding protein I (CMBP-I) was added to reaction (I) to initiate the inactivation of the enzyme. Arrow indicates the addition of CMBP-I. Broken lines represent the extrapolated progress curves of the calmodulin-activated reaction and the CMBP-I inactivated reaction.

This scheme suggests that bindings of calmodulin to phosphodiesterase and to CMBP-I are mutually exclusive and that there is no direct interaction between phosphodiesterase and CMBP-I.

As is shown in FIGURE 8, when large amount of CMBP-I was added to a calmodulin-activated phosphodiesterase reaction, the rate of the enzyme reaction decreased gradually to approach that of the basal phosphodiesterase reaction. Since the concentration of CMBP-I used was high enough to rapidly bind the free calmodulin in the reaction mixture, the rate of the inactivation of phosphodiesterase would, according to Scheme 2, reflect the rate of dissociation of the phosphodiesterase-calmodulin complex. As is expected, the enzyme inactivation following the addition of CMBP-I could be described kinetically by a first order process (FIGURE 8, inset). The first order rate constant of this reaction was calculated to be about $3.2 \times 10^{-3}$ sec$^{-1}$.

TABLE 3

EQUILIBRIUM AND KINETIC CONSTANTS FOR THE INTERACTION BETWEEN CALMODULIN AND PHOSPHODIESTERASE

| | |
|---|---|
| Association Rate Constant: $k_1$ | $\sim 4 \times 10^6\ M^{-1}\ sec^{-1}$ |
| Dissociation Rate Constant: $k - 1$ | $\sim 3.2 \times 10^{-3}\ sec^{-1}$ |
| Dissociation Constant: $K_D$ | |
| Calculated from Rate Constants: $(k - 1)/k_1$ | ~0.8 nM |
| Determined from Activation Curve | ~1 nM |

Using the kinetic constants for the forward and reverse reactions of the interaction between calmodulin and phosphodiesterase, we were able to calculate the dissociation constant of this reaction. TABLE 3 shows that the $K_d$ values calculated from the kinetic constants fell within the same range as that from the dose-dependent activation of phosphodiesterase by calmodulin. These results suggest that the mechanisms proposed for the activation of phosphodiesterase by calmodulin and for the inactivation of the enzyme by CMBP-I in the presence of saturating concentrations of $Ca^{2+}$ are basically correct.

## *General Scheme for the Enzyme Activation*

The previously proposed mechanism for the activation of phosphodiesterase by calmodulin as indicated by Scheme 1 is intended to suggest the possible sequence of events leading from an increase in intracellular $Ca^{2+}$ concentration to the stimulation of cyclic nucleotide hydrolysis.[8] The scheme is not suitable for the quantitative analysis of the relationship between the enzyme activation and the various interactions among $Ca^{2+}$, calmodulin, and phosphodiesterase. A more general scheme is therefore proposed as the basis for the analysis of the mechanisms of the enzyme activation:

$$\begin{array}{ccccccccc}
\mathrm{CM} & \xleftrightarrow{K_1 = 7.5\ \mu\mathrm{M}} & \mathrm{Ca^{2+}\cdot CM} & \xleftrightarrow{K_2 = 2.7\ \mu\mathrm{M}} & \mathrm{Ca_2^{2+}\cdot CM} & \xleftrightarrow{K_3 = 30\ \mu\mathrm{M}} & \mathrm{Ca_3^{2+}\cdot CM} & \xleftrightarrow{K_4 = 30\ \mu\mathrm{M}} & \mathrm{Ca_4^{2+}\cdot CM} \\
\updownarrow K_e = \sim 10^{-5}\mathrm{M} & & \updownarrow K_d & & \updownarrow K_d & & \updownarrow K_b & & \updownarrow K_a \\
\mathrm{CM\cdot PDE} & \xleftrightarrow[K_1']{} & \mathrm{Ca^{2+}\cdot CM\cdot PDE} & \xleftrightarrow[K_2']{} & \mathrm{Ca_2^{2+}\cdot CM\cdot PDE} & \xleftrightarrow[K_3']{} & \mathrm{Ca_3^{2+}\cdot CM\cdot PDE} & \xleftrightarrow[K_4']{} & \mathrm{Ca_4^{2+}\cdot CM\cdot PDE}
\end{array}$$

(Scheme 3)

where the symbols used are same as those indicated in Schemes 1 and 2, and $K$'s represent the dissociation constants governing the designated individual reversible reactions. The entrance of $Ca^{2+}$ and PDE into the reactions is not indicated to simplify the schematic representation.

Scheme 3 shows the various possible interactions in the phosphodiesterase activation system, many of these interactions are not included in the previously proposed schemes (Scheme 1). Several investigators[6,21-25] have demonstrated the binding of four $Ca^{2+}$ to calmodulin. In a recent study,[26] it has been shown that the binding of $Ca^{2+}$ to calmodulin exhibits positive and negative cooperativities. We have also studied the binding of $Ca^{2+}$ to calmodulin using the intrinsic tyrosine fluorescence change of calmodulin as a measure of the degree of $Ca^{2+}$ saturation and the dansylated troponin C as $Ca^{2+}$ indicator. The dissociation constants for the $Ca^{2+}$-binding determined by these methods are given in Scheme 3. It can be seen that the positive and negative cooperativities in the $Ca^{2+}$ binding are confirmed in our measurement.

Although the calmodulin-dependent phosphodiesterase, a dimeric enzyme, is capable of binding two molecules of calmodulin, the bindings of calmodulin are independent and appear to result in equal enzyme activation (see preceding sections). Therefore, the enzyme is treated as a monomeric unit in Scheme 3.

It is likely that little calmodulin-phosphodiesterase complex is formed in cells containing low concentrations of $Ca^{2+}$. In preliminary experiments using air-fuge technique,[27] we could not detect significant binding of calmodulin to phosphodiesterase in solutions containing 1 mM EGTA even at protein concentrations as high as 2 $\mu$M. Nevertheless, the interaction between calmodulin and phosphodiesterase in the absence of $Ca^{2+}$ has to be considered on the basis of chemical thermodynamics. A minimum of two pathways should be considered for free calmodulin to be converted into the fully-liganded calmodulin-phosphodiesterase complex $Ca^{2+}\cdot CM\cdot PDE$ (Scheme 3): one via the binding of $Ca^{2+}$ to calmodulin followed by the association of the fully-liganded calmodulin to phosphodiesterase, the other through the association of calmodulin and phosphodiesterase followed by the binding of $Ca^{2+}$ to the CM·PDE complex. The free energy change of this conversion, however, is independent of the chemical pathways. Consequently, the following relationship should hold:

$$K_e K_1' K_2' K_3' K_4' = K_a K_1 K_2 K_3 K_4 \quad \textbf{(3)}$$

Since the value of $K_a$ is in the order of $10^{-10}$ to $10^{-9}$ M and that of $K_e$ appears to have an upper limit of $\sim 10^{-5}$ M, the product of the four dissociation constants for the binding of $Ca^{2+}$ is expected to be about ten thousand-fold lower than that for the enzyme-calmodulin complex. From this analysis, it is clear that phosphodiesterase must have a marked effect on the binding of $Ca^{2+}$ to calmodulin. It may be suggested, therefore, that the $Ca^{2+}$ binding property of free calmodulin is not sufficient to describe the mechanisms of the enzyme activation by calmodulin and $Ca^{2+}$.

Many of the equilibrium constants in Scheme 3 are yet to be determined. However, this general scheme has already been used to elucidate several important features concerning the regulation of phosphodiesterase by calmodulin. We have examined the enzyme activation over wide concentration ranges of $Ca^{2+}$ and calmodulin, and analyzed the data by using a kinetic equation derived according to Scheme 3. The results of the study have been briefly discussed[29] and will be published later. It suffices to indicate here that one of the conclusions drawn from the kinetic study is that the dominant calmodulin-activated enzyme species is the fully-liganded calmodulin-phosphodiesterase complex $Ca^{2+}\cdot CM\cdot PDE$ (Scheme 3). Others have already shown that the activation of phosphodiesterase by $Ca^{2+}$ is highly cooperative.[25,26,28]

As has been indicated above, the affinities of calmodulin and phosphodiesterase in the presence and absence of $Ca^{2+}$ differ by about ten thousand-fold. However, due to the existence of four $Ca^{2+}$ binding sites, the dissociation constants of $Ca^{2+}$ for the individual sites in the calmodulin-phosphodiesterase complex do not have to be more than an order of magnitude higher than those in the free calmodulin (see Equation 3). Consequently, the enzyme activation may respond to physiological concentrations of $Ca^{2+}$, and the dissociation of $Ca^{2+}$ from the activated enzyme can be quite rapid when cellular $Ca^{2+}$ concentrations is lowered. Thus, the multiple $Ca^{2+}$ binding of calmodulin plays an important role in ensuring the rapid reversal of the enzyme activation as well as of the protein association.

The utilization of all four $Ca^{2+}$ and the strong positive cooperativity in the $Ca^{2+}$ concentration dependence in the enzyme activation provide the phosphodiesterase with a very effective on-and-off switch: i.e., the enzyme may be activated and deactivated over a narrow range of $Ca^{2+}$ concentration. As a consequence of the $Ca^{2+}$ dependent reversible association of calmodulin and phosphodiesterase, the $Ca^{2+}$ concentration range for the "activation switch" may be shifted by the change in free calmodulin concentration. In this respect, it is interesting to note that the transformation of chicken embryo fibroblasts by Rous sarcoma virus is accompanied with a marked increase in the cellular concentration of calmodulin.[30] It seems possible that calmodulin-dependent phosphodiesterase in the transformed cells, could respond to much lower concentrations of cellular $Ca^{2+}$ than that in normal chicken fibroblasts.

It should be noted that the general scheme (Scheme 3) described for cyclic nucleotide phosphodiesterase can be applied also to the study of the mechanism of activation of other calmodulin-dependent enzymes.

## References


1. CHEUNG, W. Y. 1970. Biochem. Biophys. Res. Commun. **29:** 478–482.
2. CHEUNG, W. Y. 1971. J. Biol. Chem. **246:** 2858–2869.
3. KAKIUCHI, S. & R. YAMAZAKI. 1970. Proc. Jpn. Acad. **46:** 387–392.
4. KAKIUCHI, S. & R. YAMAZAKI. 1970. Biochem. Biophys. Res. Commun. **41:** 1104–1110.
5. TEO, T. S., T. H. WANG & J. H. WANG. 1973. J. Biol. Chem. **248:** 588–595.
6. TEO, T. S. & J. H. WANG. 1973. J. Biol. Chem. **248:** 5950–5955.
7. KAKIUCHI, S., R. YAMAZAKI, Y. TESHIMA & M. UENISHI. 1973. Proc. Natl. Acad. Sci. USA **70:** 3526–3530.
8. WANG, J. H., T. S. TEO, H. C. HO & F. C. STEVENS. 1975. Adv. Cyclic Nucleotide Res. **5:** 179–194.
9. KLEE, C. B., T. H. CROUCH & M. H. KRINKS. 1979. Biochemistry **18:** 722–729.
10. LAPORTE, D. C., W. A. TOSCANO & D. R. STORM. 1979. Biochemistry **18:** 2820–2825.
11. MORRILL, M. E., S. T. THOMPSON & E. STELLWAGEN. 1979. J. Biol. Chem. **254:** 4371–4374.
12. SHARMA, R. K., T. H. WANG, E. WIRCH & J. H. WANG. 1980. J. Biol. Chem. **255:** 5916–5923.
13. HO, H. C., E. WIRCH, F. C. STEVENS & J. H. WANG. 1977. J. Biol. Chem. **252:** 43–50.
14. WANG, J. H. & R. DESAI. 1976. Biochem. Biophys. Res. Commun. **72:** 926–932.
15. KLEE, C. B. & M. H. KRINKS. 1978. Biochemistry **17:** 120–126.
16. SHARMA, R. K., R. DESAI, D. W. WAISMAN & J. H. WANG. 1979. J. Biol. Chem. **254:** 4276–4282.
17. SIEGEL, L. M. & K. J. MONTY. 1966. Biochim. Biophys. Acta **429:** 461–473.
18. JOB, P. 1928. Ann. Chim. (Paris) **9:** 113.
19. ASMUS, E. 1961. Z. Anal. Chem. **183:** 321.
20. WANG, J. H. & R. DESAI. 1977. J. Biol. Chem. **252:** 4175–4184.
21. LIN, Y. M., Y. P. LIU & W. Y. CHEUNG. 1974. J. Biol. Chem. **249:** 4943–4954.
22. WATTERSON, D. M., W. G. HARRELSON, P. M. KELLER, F. SHARIEF & T. C. VANAMAN. 1976. J. Biol. Chem. **251:** 4501–4513.
23. WOLFF, P. J., P. G. POIRIER, C. O. BROSTROM & M. A. BROSTROM. 1977. J. Biol. Chem. **252:** 4108–4117.
24. KLEE, C. B. 1977. Biochemistry **16:** 1017–1024.
25. DEDMAN, J. R., J. D. PORTTER, R. L. JACKSON, J. D. JOHNSON & A. R. MEANS. 1977. J. Biol. Chem. **252:** 8415–8429.
26. CROUCH, T. H. & C. B. KLEE. 1980. Biochemistry. (In press.)
27. CLARKE, R. G. & G. J. HOWLETT. 1970. Arch. Biochem. Biophys. **195:** 235–242.
28. BROSTROM, C. O. & D. J. WOLFF. 1976. Arch. Biochem. Biophys. **172:** 301–311.
29. HUANG, C. Y., V. CHAU, P. B. CHOCK, R. K. SHARMA & J. H. WANG. 1980. Fed. Proc. **39:** 1658.
30. WATTERSON, D. M., L. J. VAN ELDIK, R. E. SMITH & T. C. VANAMAN. 1976. Proc. Natl. Acad. Sci. USA **73:** 2711–2715.

## Discussion of the Paper

Dr. C. Klee: I think you mentioned that it is a $k_2$ which is a rate limiting step. Did I understand it correctly?

Dr. J. H. Wang: No. It is the interaction that's rate limiting. That's why it's second order.

Dr. Klee: It's a binding that would take about two to three minutes to get the enzyme activated in the presence of saturating calcium?

Dr. Wang: Because it is a second order reaction, it depends very much on the concentration of the component that we use.

Dr. P. Epstein (*University of Connecticut, Storrs, CT*): About a year ago, there was a report from Dr. Watterson's laboratory that this particular enzyme could be phosphorated by cyclic AMP-dependent protein kinase. When it gets phosphorylated there is an apparent loss in ability to be activated for cyclic AMP hydrolysis, but, not cyclic GMP hydrolysis.

Dr. Wang: We did try to phosphorylate PPE with the cyclic nucleotide dependent protein kinase. We could actually incorporate one phosphate per subunit of the enzyme. However, we could not see any change in activity and we examined several different conditions and couldn't see any difference in activity between the phosphorylated and nonphosphorylated form of PDE.

# $Ca^{2+}$ REGULATION OF CYCLIC NUCLEOTIDE METABOLISM*

Charles H. Keller, David C. LaPorte, W. A. Toscano, Jr.,†
Daniel R. Storm, and Keith R. Westcott

*Department of Pharmacology*
*University of Washington*
*Seattle, Washington 98195*

## INTRODUCTION

Calmodulin (CaM) mediates $Ca^{2+}$ stimulation of several enzymes, including the adenylate cyclase activities in brain,[1,2] adrenal medulla,[3] and pancreatic islets.[4] Previous studies concerning calmodulin regulation of brain adenylate cyclase have used chelator-washed particulate preparations or solubilized preparations.[1,2,5-7] There is evidence that these preparations may contain a mixture of calmodulin sensitive and insensitive adenylate cyclase activities.[5,8] Therefore, it has not been possible to characterize calmodulin stimulation of the enzyme without making assumptions concerning the relative contributions of each form to total enzyme activity. For example, Brostrom *et al.* operationally define CaM independent activity as that observed in the absence of CaM and the CaM dependent activity as the increment in activity obtained upon addition of CaM.[9] This convention assumes that the CaM sensitive enzyme is absolutely dependent upon CaM.

We recently described a procedure for isolation of calmodulin sensitive adenylate cyclase from bovine brain.[8] This material was selected on the basis of the ability of the calmodulin sensitive enzyme to form a complex with CaM-Sepharose in the presence of $Ca^{2+}$. Modifications of these chromatographic procedures have produced a calmodulin sensitive adenylate cyclase that is free of contaminating calmodulin. This material is suitable for studies concerning the interaction of calmodulin with the calmodulin sensitive adenylate cyclase without any uncertainties introduced by the presence of calmodulin insensitive adenylate cyclase or contaminating calmodulin. In this study, calmodulin stimulation of this preparation is characterized.

Since development of any quantitative model for CaM interactions with the various CaM binding proteins will require knowledge of the free energy coupling between binding of $Ca^{2+}$ and these CaM binding proteins to CaM, we have determined this thermodynamic parameter for binding of $Ca^{2+}$ and troponin I to CaM by measuring $Ca^{2+}$ binding to CaM in the presence and absence of troponin I. This system was chosen for initial investigation because troponin I can be obtained in high yield, does not bind $Ca^{2+}$, and forms a $Ca^{2+}$-dependent complex with CaM[10] with a 1:1 stoichiometry.[11,12] The results and implications of these experiments are presented.

*Supported by National Science Foundation Grant PCM-78 03188 and National Institutes of Health Grant HL 23606.

†Present address: Department of Toxicology, Harvard University School of Public Health, Boston, MA 02115



0077-8923/80/356-205/0 © 1980, NYAS

## Experimental Procedures

### *Materials*

All chemicals were of the finest grade available. Cyclic [$^3$H]AMP, [$^{45}$Ca]$Cl_2$, [acetic-2-$^{14}$C]EDTA and $^{32}P_i$ were purchased from ICN, [$^3$H]$_2$O was from NEN. Trypsin (hog pancreas grade IX), soybean trypsin inhibitor, Lubrol PX, alumina (neutral WN-3), egg phosphatidyl choline, and CNBr-activated sepharose were obtained from Sigma. Dowex AG-SOW-X4 Bio-Gel P-2, and Affi-Gel Blue were obtained from Bio Rad. Gpp(NH)p was obtained from PL Biochemicals, and DEAE-Sephadex A-25 was from Pharmacia. Dialysis membrane was a product of Spectrum Medical Industries.

### *Methods*

*Adenylate Cyclase Assay*

Adenylate cyclase was assayed by the method of Salomon *et al.*[13] using $\alpha$-[$^{32}$P]ATP prepared by the method of Symons.[14] All assays contained 20 mM Tris-HCl pH 7.5, 1 mM $\beta$-mercaptoethanol, and 0.1% bovine serum albumin. Creatine kinase (0.1%) and phosphocreatine (20 mM) were present as an ATP regenerating system. The ATP used in the assays was purified using DEAE-Sephadex A-25 and Dowex AG-50 chromatography prior to use. Assays performed to monitor activity in purification steps contained 1 mM ATP, 2 mM cAMP, 10 mM $MnCl_2$ and 1 mM EDTA. All other assays were performed using varying concentrations of $MnCl_2$ or $MgCl_2$, EDTA or EGTA, and $CaCl_2$ as indicated. In some assays, the ATP concentration was less than 1 mM. In these cases, the cAMP concentration was maintained at twice the ATP concentration. All assays were performed for 10 min at 30°C unless otherwise noted. Formation of cyclic AMP was linear with time during this incubation period. Each data point is the mean of triplicate determinations with a standard error of 5%. Protein was determined by the method of Peterson,[15] using bovine serum albumin as the standard.

*Preparation of Calmodulin and Troponin I*

Calmodulin was purified from bovine brain as previously described.[16] Calmodulin concentration was determined by absorbance at 276 nm ($E^{1\%}_{276nm}$ = 1.8).[17] Troponin I was purified by the method of Wilkinson.[18]

*Partial Purification of Adenylate Cyclase*

Calmodulin sensitive adenylate cyclase was partially purified as previously described with the following modifications.[8] Affi-Gel Blue columns were washed with three column volumes of 20 mM Tris-HCl, pH 7.35, 0.1% Lubrol PX, 5 mM NaF, 1 mM $MgCl_2$, 2 mM EGTA, 0.25 M sucrose, and 2.5 mM dithiothreitol prior to elution with ATP and KCl. CaM-Sepharose was prepared as previously described.[8] Four concentrated adenlylate cyclase preparations prepared by Affi-Gel Blue chromatography were adjusted to 2 mM $CaCl_2$ and 0.5 mM EGTA and loaded onto a 10 ml packed bed CaM-Sepharose column. The resin was washed with five volumes of 20 mM Tris-HCl, pH 7.35, 0.1% Lubrol PX, 1 mM $MgCl_2$, 2 mM $CaCl_2$, 0.5 mM EGTA, 0.25 M sucrose, and 2.5 mM dithiothreitol prior to elution with the same buffer containing 2 mM EGTA but no $CaCl_2$. The adenylate cyclase activity that did not adsorb to the resin in the presence of $CaCl_2$, designated Fraction I, was pooled. The activity that adsorbed to the resin and was specifically eluted with EGTA, designated Fraction II, was pooled and concentrated five fold by ultrafiltration with an Amicon XM-50 membrane. Calmodulin content of various partially purified adenylate cyclase fractions was determined by stimulation of cyclic nucleotide phosphodiesterase as previously described.[8,16]

### *Limited Proteolysis of CaM-Sepharose Fraction II*

CaM-Sepharose fraction II, desalted on Bio-Gel P-2 into 20 mM Tris-HCl, pH 7.5, 0.1% Lubrol PX, 1 mM $MgCl_2$, and 0.25 M sucrose (2 μg protein in 200 μl) was incubated with 0.05 μg hog pancreas trypsin for periods up to five minutes at 30°C. The proteolysis was stopped by addition of 0.25 μg soybean trypsin inhibitor. An identical sample was treated in an analogous manner in the absence of trypsin to serve as a control.

### *$Ca^{2+}$ Binding to CaM*

Binding of $[^{45}Ca]Cl_2$ to CaM was quantitated by equilibrium dialysis. All reagents were prepared and stored in polypropylene vessels previously washed in 0.25 N HCl and distilled water. The microdialysis cells were washed successively in 0.1% Lubrol PX, distilled water, 95% ethanol, distilled water, 0.25 N HCl, and distilled water before use. CaM was prepared for these experiments by dialysis at 6°C against 10 mM morpholino propane sulfonic acid (MOPS) pH 7.2, 50 mM KCl, and 50 μM EDTA, followed by dialysis against 10 mM MOPS, pH 7.2, and 50 mM KCl. To insure adequate removal of $Ca^{2+}$, the CaM was then passed through a 1.4 × 3 cm parvalbumin-Sepharose column. This column was regenerated before use as described by Lehky *et al.*[19] except that $[^{14}C]$EDTA was included in order to monitor removal of chelator from the column. $[^{45}Ca]Cl_2$ was added to the dialyzed CaM in order to monitor removal of $Ca^{2+}$ by the parvalbumin-Sepharose column. The calcium concentrations of protein solutions and reagents were determined by atomic absorption using a Perkin-Elmer model 305B atomic absorption spectrophotometer with a graphite furnace. Total calcium contamination from protein solutions and buffer components ranged from 1.2 to 2.6 μM in different experiments. Equilibrium dialysis was carried out at 25°C in microdialysis cells (100 μl solution/side) in 20 mM MOPS, pH 7.0, 150 mM KCl, 0.1 mM dithiothreitol, $[^{45}Ca]Cl_2$ (1.2 × $10^5$ cpm/cell) and varying concentrations of $CaCl_2$. $[^3H]_2O$ was included to monitor recovery.

## Results

### *Separation of CaM Sensitive and Insensitive Adenylate Cyclase*

CaM sensitive and insensitive forms of adenylate cyclase were resolved by CaM-Sepharose chromatography (Figure 1). Affi-Gel Blue chromatography was used to remove endogenous CaM prior to CaM-Sepharose chromatography. Although Affi-Gel Blue can be used for this purpose, there are several disadvantages with this resin. There is considerable batch to batch variation in the properties of Affi-Gel and the resin must be pretreated with bovine serum albumin and washed extensively with 8 M urea before use. More recently, we have determined that CaM can be removed from detergent-solubilized adenylate cylase preparations on DEAE-Sephacel columns run in the presence of EDTA (Toscano and Storm, unpublished observations). CaM depleted preparations obtained by ion exchange chromatography gave CaM-Sepharose elution profiles similar to that reported in FIGURE 1. Approximately 80% of the applied adenylate cyclase activity did not absorb to CaM-Sepharose in the presence of $Ca^{2+}$ (F-I) and the remaining 20% was eluted with excess EGTA or EDTA (F-II). When F-II was desalted and reapplied to CaM-Sepharose, all of the activity was absorbed in the presence $Ca^{2+}$ and eluted with EGTA. In seven independent experiments the specific activities of F-II varied from 100,000 to 250,000 pmoles cAMP/mg/10 min. The preparation can be stored up to six months at −60°C with no loss in activity.

The $Ca^{2+}$ and CaM sensitivities of F-I and F-II are reported in FIGURE 2. F-I was not stimulated by $Ca^{2+}$ in the absence of presence of added CaM. In contrast, F-II was not stimulated by $Ca^{2+}$ in the absence of CaM but was stimulated approximately four fold in the presence of CaM. It is notable that F-II expressed a basal activity in the absence of CaM and that CaM amplified this activity.

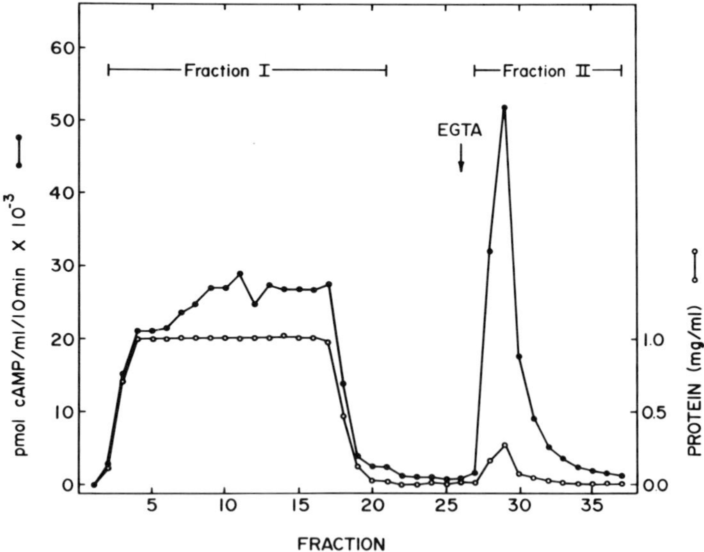


FIGURE 1. CaM-Sepharose chromatography. Partially purified adenylate cyclase (2,775,000 units) was submitted to CaM-Sepharose as described in METHODS. Fraction size was 4.8 ml. Recovery of adenylate cyclase applied to the column was 100%. The specific activities of Fractions I and II were 34,000 and 151,000 pmol/mg/min, respectively. Adenylate cyclase activities were determined in the presence of 1 mM ATP, 10 MM $MnCl_2$, and 1 $\mu$g CaM.

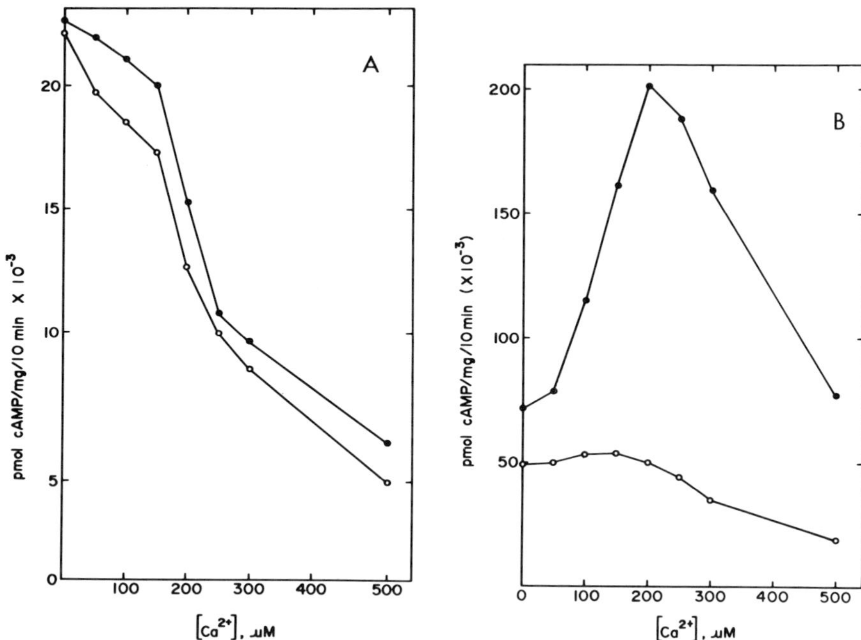


FIGURE 2. $Ca^{2+}$-sensitivities of Fraction I and II adenylate cyclase. A, Fraction I was adjusted to 2 mM EGTA and desalted into 20 mM Tris-HCl, pH 7.5, 0.1% Lubrol PX, and 0.25 M sucrose on Bio Gel P2 to remove excess $Ca^{2+}$. B, Fraction II was desalted into 20 mM Tris-HCl, pH 7.5, 0.1% Lubrol PX, and 0.25 M sucrose on Bio Gel P2 to remove excess EGTA. Adenylate cyclase was assayed in triplicate using 1 mM ATP, 5 mM $MgCl_2$, and 200 µM EGTA. $Ca^{2+}$, as $CaCl_2$, was included in the assays as indicated by the abscissa in the absence (○) and presence (●) of 1 µg CaM.

### *$Mn^{2+}$ Sensitivity of the CaM Sensitive Adenylate Cyclase.*

Adenylate cyclase requires a divalent cation for activity.[20] Initial studies indicated that CaM stimulation of adenylate cyclase occurred when $Mn^{2+}$ was substituted for $Ca^{2+}$. Since similar observations were made with the CaM sensitive phosphodiesterase and CaM binds $Mn^{2+}$,[21] the $Mn^{2+}$ sensitivity of F-II adenylate cyclase was examined (FIGURE 3). In the presence of 0.1 mM ATP and no calmodulin, maximal stimulation of adenylate cyclase occurred at about 1.0 mM $MnCl_2$ and half-maximal activation was at 0.2 mM $MnCl_2$. Addition of CaM stimulated the enzyme approximately 60% at saturating $Mn^{2+}$ concentrations. CaM also lowered the concentration of $Mn^{2+}$ required for half-maximal activation of adenylate cyclase. No inhibition of adenylate cyclase activity was observed at $Mn^{2+}$ levels as high as 10 mM.

### *CaM Concentration Dependence for Stimulation of Adenylate Cyclase*

The CaM dose response curves for stimulation of F-II adenylate cyclase were determined in the presence of optimal $Ca^{2+}$ or $Mn^{2+}$ and in the absence of either cation (FIGURE 4). In the presence of $Ca^{2+}$ or $Mn^{2+}$, half maximal stimulation of adenylate cyclase required 47 nM CaM. These concentrations of CaM are almost two orders of magnitude higher than those required for half-maximal stimulation of the cyclic nucleotide phosphodiesterase.[22] Furthermore, activation of adenylate cyclase by CaM occurred over a greater concentration range than would be expected for a simple saturating bimolecular process.[23] This CaM concentration dependence is most likely due to the presence of other CaM binding proteins in F-II and antagonism of CaM-adenylate cyclase interactions by Lubrol PX. Lubrol PX does antagonize CaM-phosphodiesterase interactions and there is evidence that the interface between CaM and these proteins is a hydrophobic domain.[11] It is interesting that very high levels of CaM stimulated adenylate cyclase in the absence of $Ca^{2+}$ or $Mn^{2+}$. This activation was not due to contaminating $Ca^{2+}$ since it still occurred when the EGTA concentration was increased to 1 mM. The apparent affinity of CaM for adenylate cyclase in the absence of $Ca^{2+}$ or $Mn^{2+}$ was at least two orders of magnitude lower than in their presence.

Davis and Daly[24] have reported that $K^+$ specifically inhibits the basal activity of the calcium-dependent phosphodiesterase. Therefore, the effects of NaCl and KCl on calmodulin stimulation of adenylate cyclase were examined (FIGURE 5). $Na^+$ and $K^+$ up to 250 mM did not inhibit basal activity of the CaM-sensitive adenylate cyclase. Both $Na^+$ and $K^+$ appeared to antagonize CaM-adenylate cyclase interactions. The concentration of CaM required to achieve half-maximal stimulation of adenylate cyclase increased 3.5 and 14 fold at salt concentrations of 0.125 and 0.250 M respectively. Since NaCl and KCl were equally effective, it can be assumed that increases in ionic strength antagonize

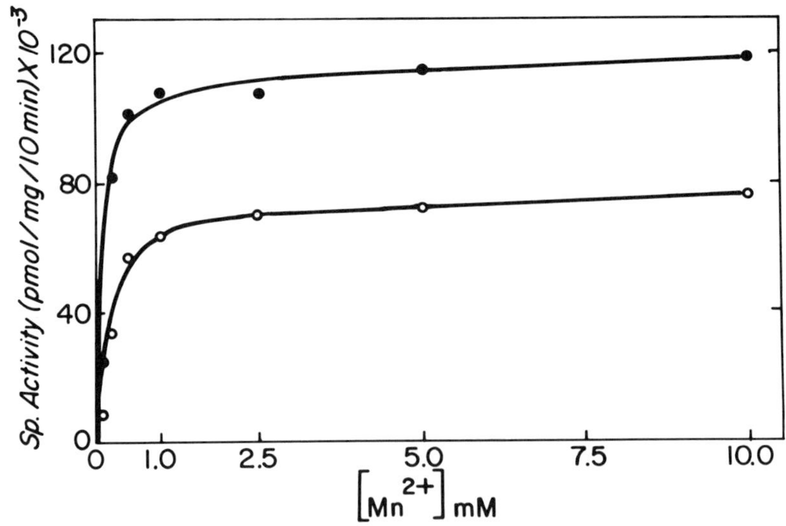


FIGURE 3. $Mn^{2+}$ concentration dependence of Fraction II adenylate cyclase. Fraction II was desalted into 20 mM Tris-HCl, pH 7.5, 0.1%, Lubrol PX, and 0.25 M sucrose on Bio Gel P-2 to remove excess EGTA. Adenylate cyclase was assayed in triplicate using 0.1 mM ATP, 0.2 mM cAMP, 0.1 mM EDTA, and increasing concentrations of $MnCl_2$ in the absence (O) and presence (●) of 3$\mu$g CaM.

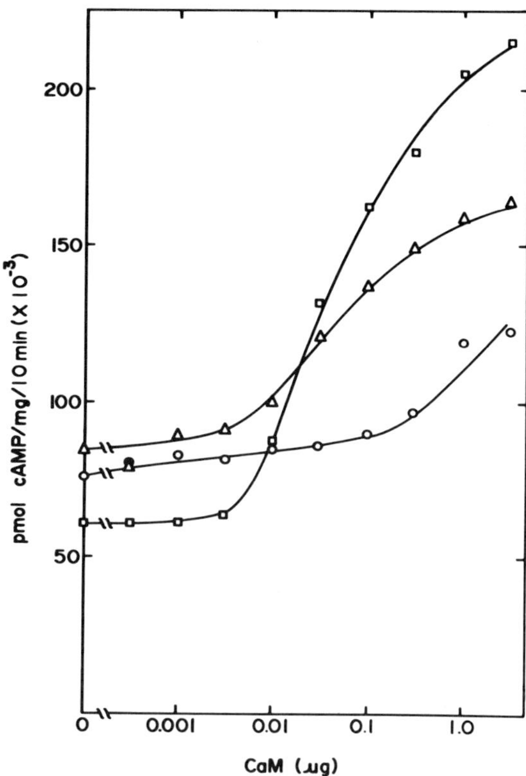


FIGURE 4. CaM sensitivity of Fraction II adenylate cyclase. Fraction II was desalted into 20 mM Tris-HCl, pH 7.5, 0.1% Lubrol PX, and 0.25 M sucrose on Bio Gel P2 to remove excess EGTA. Adenylate cyclase was assayed using 1 mM ATP and either 5MM $MgCl_2$, 0.225 mM $CaCl_2$, and 0.2 mM EGTA (□); 5 mM $MgCl_2$ and 0.2 mM EGTA (○); or 10 mM $MnCl_2$ and 1 mM EDTA (△). CaM was included in the assays as indicated by the abscissa.

CaM-adenylate cyclase interactions. Indeed, adenylate cyclase absorbed to CaM-Sepharose was eluted with 0.2 M KCl.

### *Influence of CaM on Kinetic Parameters of Adenylate Cyclase*

The effect of CaM, $Ca^{2+}$, and $Mn^{2+}$ on the $K_m$ and $V_m$ of the CaM-sensitive adenylate cyclase preparation were determined by measuring enzyme activity as a function of ATP concentration. Lineweaver-Burke plots were linear and the major effect of $Ca^{2+}$ and calmodulin was to increase the $V_m$ four fold with no effect on the $K_m$ for ATP (FIGURE 6). The apparent $K_m$ for ATP was significantly lower in $Mn^{2+}$ compared to $Mg^{2+}$, and a combination of $Mn^{2+}$ and CaM increased the $V_m$ 60% with no effect on the $K_m$ (TABLE 1). The $K_m$s determined for F-I and F-II were comparable in the presence of $Mg^{2+}$ or $Mn^{2+}$ but the $V_m$ of F-II was significantly higher. These data indicate that CaM, in the presence of

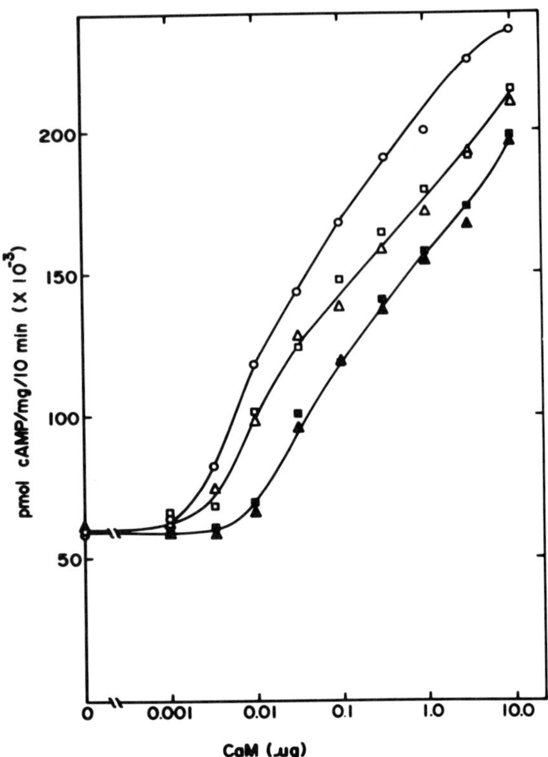


FIGURE 5. Effect of KCl and NaCl on CaM stimulation of adenylate cyclase. Fraction II was desalted into 20 mM Tris-HCl, pH 7.5, 0.1% Lubrol PX, and 0.25 M sucrose on Bio Gel to remove excess EGTA. Adenylate cyclase was assayed using 1 mM ATP, 5 mM $MgCl_2$, 0.225 mM $CaCl_2$, 0.2 mM EGTA. The assays contained CaM as indicated in the abscissa with no added salt (○), 125 mM KCl (□), 125 mM NaCl (□), 250 mM KCl (■), or 250 mM NaCl (▲).

optimal $Ca^{2+}$, increases the turnover number of the enzyme with no apparent effect on the affinity of the enzyme for ATP. In contrast, CaM increased the $V_m$ of the $Ca^{2+}$-sensitive phosphodiesterase and decreased its $K_m$ for cAMP.[25]

Since CaM did not affect the $K_m$ for ATP, the influence of ATP on CaM dose response curves was examined. The CaM dose response curves were equivalent over a wide range of ATP concentrations (FIGURE 7). The observations that CaM did not affect the $K_m$ for ATP, and that ATP had no effect on the CaM concentration dependence suggests that there is no energy coupling between binding of ATP and CaM to adenylate cyclase. It should be emphasized that this conclusion is inferred from activity data and direct binding of either ATP or CaM to adenylate cyclase was, of course, not measured.

*Proteolysis of CaM-Dependent Adenylate Cyclase*

Limited proteolysis of the $Ca^{2+}$-sensitive phosphodiesterase activates the enzyme with loss in CaM sensitivity.[26,27] In contrast, limited proteolysis of the

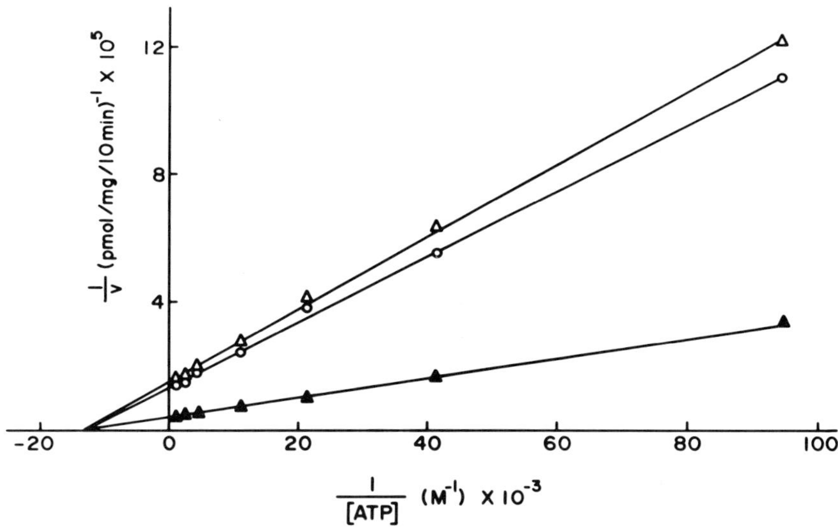


FIGURE 6. Lineweaver-Burke plots for Fraction II adenylate cyclase activity. Fraction II was desalted into 20 mM Tris-HCl, pH 7.5, 0.1% Lubrol PX, and 0.25 M sucrose on Bio Gel P2. ATP concentrations were varied from 0.01 to 1.0 mM with the ratio of cAMP/ATP constant at two. Adenylate cyclase was assayed using 5 mM $MgCl_2$, 0.225 mM $CaCl_2$, and 0.20 mM EGTA with (▲) or without (△) 10 μg CaM or 0.2 mM EGTA and 5 mM $MgCl_2$ (○). Lines are linear least squares best fit.

CaM-dependent adenylate cyclase had little or no effect on basal activity but destroyed CaM sensitivity (FIGURE 8). These data are consistent with the observation that CaM stimulation of brain adenylate cyclase requires the presence of a trypsin-sensitive regulatory subunit in addition to the catalytic subunit.[28] This regulatory subunit is most likely the guanyl nucleotide regulatory subunit.

TABLE 1

KINETIC PARAMETERS OF ADENYLATE CYCLASE

| Conditions | $K_m$(μM) | $V_m$(pmol/mg/10 min) |
|---|---|---|
| Fraction II | | |
| $Mg^{2+}$, EGTA | 78.3 | 76,160 |
| $Mg^{2+}$, $Ca^{2+}$, EGTA | 74.5 | 66,050 |
| $Mg^{2+}$, $Ca^{2+}$, EGTA, CaM | 77.9 | 251,380 |
| $Mn^{2+}$, EDTA | 30.1 | 100,200 |
| $Mn^{2+}$, EDTA, CaM | 32.0 | 159,240 |
| Fraction I | | |
| $Mg^{2+}$, EDTA | 72.5 | 23,100 |
| $Mn^{2+}$, EDTA | 42.1 | 30,030 |

Adenylate cyclase was assayed as described in METHODS. Fraction I and II adenylate cyclase were isolated by CaM-Sepharose chromatography as described in METHODS. ATP concentrations were varied from 0.01 to 1.0 mM. $K_m$ and $V_m$ were determined from least squares analysis of double-reciprocal plots of the data under each condition. When present: $[Mg^{2+}]$ = 5 mM; $[Mn^{2+}]$ = 10 mM; $[Ca^{2+}]$ = 0.225 mM; [EGTA] = 0.2 mM; [EDTA] = 1 mM.

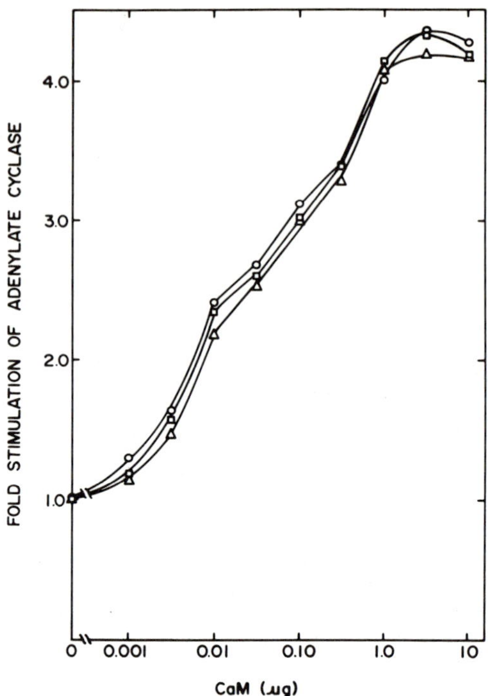


FIGURE 7. The effect of ATP on CaM stimulation of adenylate cyclase. Fraction II adenylate cyclase was desalted into 20 mM Tris-HCl, pH 7.5, 0.1% Lubrol PX, and 0.25 M sucrose. Adenylate cyclase was assayed at varying concentrations of CaM using 5 mM $MgCl_2$, 0.225 mM $CaCl_2$, 0.2 mM EGTA in the presence of 0.026 mM (○), 0.092 mM (□), or 0.89 mM (△) ATP.

### *Binding of $Ca^{2+}$ to CaM*

Binding of $Ca^{2+}$ to CaM was quantitated in the absence and presence of a stoichiometric concentration of troponin I (FIGURE 9). Scatchard plots of both sets of data were linear within experimental error except at low saturation ($\bar{\nu} < 0.5$), where errors in calcium concentration would be greatest. Free CaM exhibited 4 $Ca^{2+}$ binding sites with a $K_d$ of 12.0 μM. These results are qualitatively similar to those reported by Dedman *et al.* for rat testis CaM.[30] The CaM:troponin I complex displayed 4 equivalent $Ca^{2+}$ binding sites with a $K_d$ of 2.4 μM. No binding of $Ca^{2+}$ to free troponin I was detected. Doubling of the molar ratio of troponin I to CaM had no effect on the $Ca^{2+}$ binding parameters. These data indicate that the affinity of CaM for $Ca^{2+}$ is increased five fold upon formation of the CaM:troponin I complex.

### *Calculation of the Free Energy of Coupling*

These data can be analyzed by the method proposed by Weber.[23] The following free energy changes can be defined for formation of the $Ca_4^{2+}$:CaM:troponin I complex: ΔG°(C) is the free energy change upon binding of one mole of $Ca^{2+}$ to CaM, ΔG°(I) is the free energy change on binding of troponin I to CaM; ΔG°(C/I) is the free energy change on binding of one mole of $Ca^{2+}$ to the CaM:troponin I complex; $\Delta G°(I/C_4)$ is the free energy change on binding of troponin I to $Ca_4^{2+}$: CaM; and $\Delta G°\ (C_4I)$ is the free energy change upon formation of the $Ca_4^{2+}$:CaM:troponin I complex from the individual reactants. The $K_d$s determined for $Ca^{2+}$ binding to CaM and to the CaM:troponin I complex correspond to ΔG°(C) and ΔG°(C/I). Conservation of standard free energy requires that:

$$4(\Delta G°(C/I) - \Delta G°(C)) = \Delta G°(I/C_4) - \Delta G(I) = \Delta G°IC$$

ΔG°IC is the free energy coupling between binding of $Ca^{2+}$ and troponin I to CaM. From the data in FIGURE 9 the free energy coupling is −0.95 Kcal/mole $Ca^{2+}$. Thus, the affinity of troponin I for CaM should be increased 600 fold when CaM binds four moles of $Ca^{2+}$. The $K_d$ for the $Ca^{2+}$ dependent binding of

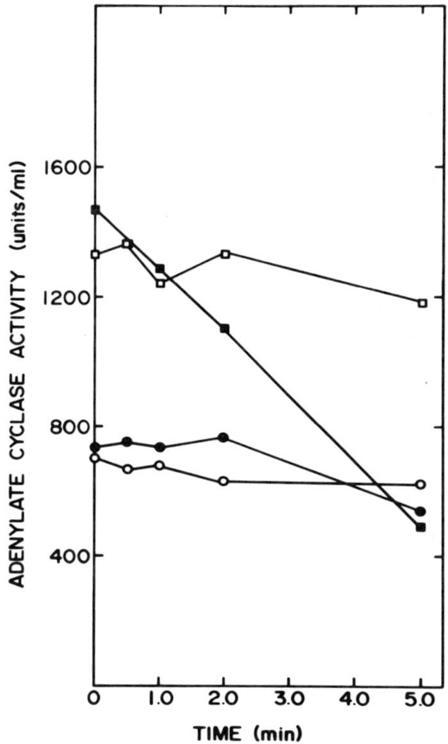


FIGURE 8. The effect of trypsin on CaM stimulation of adenylate cyclase. Desalted Fraction II adenylate cyclase was incubated at 30°C with (■,●) and without (□,○) 0.05 μg trypsin for various periods of times as described in METHODS. Trypsin treatment was terminated with 0.25 μg soybean trypsin inhibitor. The trypsin treated samples were then assayed in the presence of 1 mM ATP, 5 mM $MgCl_2$, and 0.225 mM $CaCl_2$ and 1 μg CaM (□,■) or 0.2 mM EGTA (○,●).

troponin I is approximately $1 \times 10^{-8}$ M.[11] Therefore, the $K_d$ for the complex in the absence of $Ca^{2+}$ should be approximately $6 \times 10^{-6}$ M.

## CONCLUSIONS

CaM-sensitive and insensitive forms of brain adenylate cyclase can be effectively separated using CaM-Sepharose. The calcium dependent binding of the enzyme to CaM-Sepharose is the first direct evidence indicating complex formation between CaM and adenylate cyclase. The CaM-sensitive adenylate cyclase apparently exhibits a basal activity in the absence of CaM and this activity is amplified by CaM·$Ca^{2+}$. CaM does not affect the $K_m$ for ATP but does increase the $V_m$. Limited proteolysis destroys CaM sensitivity without affecting basal adenylate cyclase activity. There are a number of similarities between $Ca^{2+}$ stimulation of adenylate cyclase and hormone regulation of the enzyme. In both cases, the influence of the activator is mediated by a specific regulatory subunit that binds the ligand. These regulatory subunits can dissociate from the adenylate cyclase system and it is clear that CaM has enhanced affinity for the enzyme in the presence of $Ca^{2+}$. Finally, the guanyl nucleotide regulatory subunit is required for both $Ca^{2+}$ and hormone stimulation of adenylate cyclase suggesting that these processes may have a similar molecular basis.

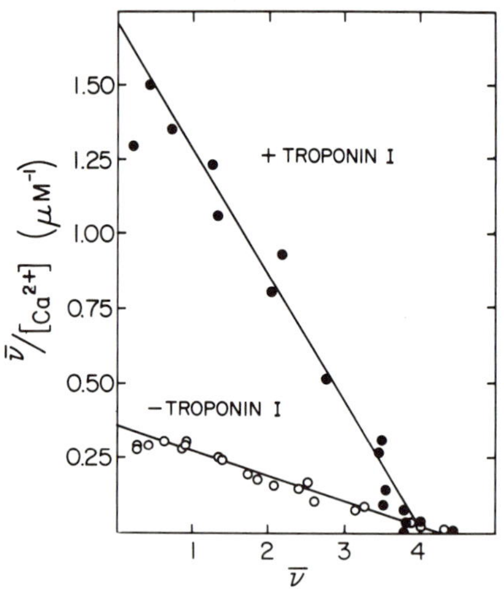


FIGURE 9. $Ca^{2+}$ binding to CaM in the presence (●) and absence (○) of troponin I. Binding of $Ca^{2+}$ was determined by equilibrium dialysis as described in METHODS and the data was plotted by the method of Scatchard.[29] $\bar{\nu}$ represents moles of $Ca^{2+}$ bound per mole CaM and $[Ca^{2+}]$ is the free $Ca^{2+}$ concentration. The experimental points are triplicate determinations. CaM was at 10 $\mu$M and troponin I, when present, was 10 $\mu$M. Linear regression analysis of data points in which $\bar{\nu}$ was greater than 0.5 yielded the lines shown; in both cases the coefficient of determination was 0.98.

The free energy coupling for binding of $Ca^{2+}$ and troponin I to CaM was determined and is $-0.95$ Kcal/mole $Ca^{2+}$. This is relatively small compared to the energies for binding of CaM to $Ca^{2+}$ or troponin I, but is typical of the free energy coupling found for other protein systems, which are generally 1 to 1.5 Kcal/mol.[23] The fact that the free energy coupling associated with each $Ca^{2+}$ binding site is small may explain why is was necessary for CaM to evolve with four $Ca^{2+}$ sites energetically coupled to protein-protein complex formation. Another implication of these results is that positive cooperativity for $Ca^{2+}$ binding would be predicted when $Ca^{2+}$ binding to a sub-stoichiometric number of sites is sufficient to promote CaM:protein complex formation. This was not observed in this study because the protein concentrations required for determination of $Ca^{2+}$ binding may have exceeded $K$(I), the $Ca^{2+}$-independent dissocation constant for the CaM:troponin I complex. Such positive cooperativity would only be observed if binding of $Ca^{2+}$ produced a substantial change in the aggregation state of the system. Similar thermodynamic characterizations of other CaM-protein complexes will be of importance in defining the intricate interactions between the components of each system. Although some quantitative differences may exist in these other systems, the fundamental properties are probably the same as those discussed above. All four $Ca^{2+}$ binding sites appear to promote protein-protein complex formation, each site contributing the same amount of energy. The free energy coupling may differ somewhat in other systems, but in view of the narrow range of such values found in other systems,[23] it probably does not vary much from $-1$ Kcal/mole for each $Ca^{2+}$ binding site.

## Acknowledgement

We wish to thank Diane Toscano and Berta Wierman for skillful technical assistance.

## References

1. Brostrom, C. O., Y. C. Huang, B. M. Breckenridge & D. J. Wolff. 1975. Identification of a calcium-binding protein as a calcium-dependent regulator of brain adenylate cyclase. Proc. Natl. Acad. Sci. USA **72:** 64–68.
2. Cheung, W., L. S. Bradham, T. J. Lynch, Y. M. Lin & E. A. Tallant. 1975. Protein activator of cyclic nucleotide phosphodiesterase of bovine rat brain also activates its adenylate cyclase. Biochem. Biophys. Res. Commun. **66:** 1055–1062.
3. LeDonne, N. C., Jr. & C. J. Coffee. 1979. Properties of the bovine adrenal medulla adenylate cyclase. Fed. Proc. **38:** 317.
4. Valverde, I., A. Vandermeers, R. Anjaneyulu & W. J. Malaisse. 1979. Calmodulin activation of adenylate cyclase in pancreatic islets. Science **206:** 225–227.
5. Brostrom, C. O., M. A. Brostrom & D. J. Wolff. 1977. Calcium-dependent adenylate cyclase from rat cerebral cortex, J. Biol. Chem. **252:** 5677–5686.
6. Lynch, T. J., E. A. Tallant & W. Y. Cheung. 1977. Rat brain adenylate cyclase. Arch. Biochem. Biophys. **182:** 124–133.
7. Wallace, R. W., T. J. Lynch, E. A. Tallant & W. Y. Cheung. 1978. An endogenous inhibitor protein of brain adenylate cyclase and cyclic nucleotide phosphodiesterase. Arch. Biochem. Biophys. **187:** 328–334.
8. Westcott, K. R., D. C. LaPorte & D. R. Storm. 1979. Resolution of adenylate cyclase sensitive and insensitive to $Ca^{2+}$ and CaM by CaM-Sepharose affinity chromatography. Proc. Natl. Acad. Sci. USA **76:** 204–208.
9. Brostrom, M. A. , C. O. Brostrom & D. J. Wolff. 1978. Calcium-dependent adenylate

cyclase from rat cerebral cortex: activation by guanine nucleotides. Arch. Biochem. Biophys. **191:** 341–350.

10. AMPHLETT, G. W., T. C. VANAMAN & S. V. PERRY. 1976. Effect of the troponin C-like protein from bovine brain (brain modulator protein) on the $Mg^{2+}$-stimulated ATPase of skeletal muscle actomyosin. FEBS Lett. **72:** 163–168.
11. LAPORTE, D. C., B. M. MASO & D. R. STORM. Biochemistry. (In press.)
12. KELLER. C. H., D. C. LAPORTE & D. R. STORM. Unpublished observations.
13. SALOMON, Y., C. LONDOS & M. RODBELL. 1974. A highly sensitive adenylate cyclase assay. Anal. Biochem. **58:** 541–548.
14. SYMONS, R. H. 1973. Improved synthesis of $^{32}P$ labeled 3′, 5′ cyclic AMP, 3′, 5′-cyclic GMP and other 3′, 5′ cyclic ribo and deoxyribo nucleotides of high specific activity. Biochem. Biophys. Acta **320:** 535–539.
15. PETERSON, G. L. 1977. A simplification of the protein assay method of Lowry *et al.* which is generally more applicable. Anal. Biochem. **83:** 346–356.
16. LAPORTE, D. C. & D. R. STORM. 1976. Detection of calcium dependent regulatory binding components using $^{125}I$ labeled CDR. J. Biol. Chem. **253:** 3374–3377.
17. WATTERSON, D. M., W. G. HARRELSON, JR., P. M. KELLER, F. SHARIEF & T. C. VANAMAN. 1976. Structural similarities between the $Ca^{2+}$-dependent regulatory proteins of cyclic nucleotide phosphodiesterase and actomyosin ATPase. J. Biol. Chem. **251:** 4501–4513.
18. WILKINSON, J. M. 1974. The preparation and properties of the components of troponin B. Biochim. Biophys. Acta **359:** 379–388.
19. LEHKY, P., M. COMTE, E. H. FISCHER & E. A. STEIN. 1977. A new solid-phase chelator with high affinity and selectivity for calcium:parvalbumin-polyacrylamide. Anal. Biochem. **82:** 158–169.
20. SUTHERLAND, E. W., T. RALL & T. MENON. 1962. Adenylate cyclase. I. Distribution, preparation, and properties. J. Biol. Chem. **237:** 1220–1227.
21. WOLFF, D. J., P. G. PORIRIER, C. O. BROSTROM & M. A. BROSTROM. 1977. Divalent cation binding properties of bovine brain $Ca^{2+}$-dependent regulator protein. J. Biol. Chem. **252:** 4108–4117.
22. TEO, T. S. & J. H. WANG. 1973. Mechanism of activation of a cyclic adenosine 3′:5′-monophosphate phosphodiesterase from bovine heart by calcium ions—Identification of the protein activator as a $Ca^{2+}$ binding protein. J. Biol. Chem. **248:** 5950–5955.
23. WEBER, G. 1975. Energetics of ligand binding to proteins. Adv. Prot. Chem. **209:** 1–83.
24. DAVIS, C. W. & J. W. DALY. 1978. Calcium-dependent cAMP phosphodiesterase: Inhibition of basal activity at physiological levels of potassium ions. J. Biol. Chem. **253:** 8683–8686.
25. KLEE, C. B., T. H. CROUCH & M. H. KRINKS. 1979. Catalytic properties of bovine brain $Ca^{2+}$-dependent cyclic nucleotide phosphodiesterase. Biochemistry **18:** 722–729.
26. CHEUNG, W. Y. 1970. Cyclic 3′5′ nucleotide phosphodiesterase: Evidence for and properties of a protein activator. Biochem. Biophys. Res. Commun. **33:** 533–538.
27. SAKAI, T., H. YAMANAKA, R. TANAKA, H. MAKIMO & H. KASAI. 1977. Stimulation of cyclic nucleotide phosphodiesterases from rat brain by activator protein, proteolytic enzymes and a Vitamin E derivative. Biochim. Biophys. Acta **483:** 121–134.
28. TOSCANO, W. A., JR., K. R. WESTCOTT, D. C. LAPORTE & D. R. STORM. 1979. Evidence for a dissociable protein subunit required for calmodulin stimulation of brain adenylate cyclase. Proc. Natl. Acad. Sci. USA **76:** 5582–5586.
29. SCATCHARD, G. 1949. The attractions of proteins for small molecules and ions. Ann. N.Y. Acad. Sci. **51:** 660–672.
30. DEDMAN, J. R. J. D. POTTER, R. L. JACKSON, J. D. JOHNSON & A. R. MEANS. 1977. Physiocochemical properties of rat testis $Ca^{+}$-dependent regulator protein of cyclic nucleotide phosphodiesterase. J. Biol. Chem. **252:** 8415–8422.

## DISCUSSION OF THE PAPER

DR. D. FLOCKHART: What is the affinity of calmodulin for adenylate cyclase?

DR. D. R. STORM: The apparent affinity of calmodulin for adenylate cyclase is about two orders of magnitude lower than the phosphodiesterase. I don't place too much emphasis on that. First, this is in the presence of detergent which antagonized calmodulin-protein interactions. Secondly, this is not a pure enzyme and there are probably other calmodulin binding proteins. So, I honestly don't feel that that concentration dependence has much physical meaning. The enzyme thus far has only been purified to about 5%.

DR. M. G. LUTHRA (*University of Arizona, Tucson, AZ*) I worry when you use calcium and fluorescent probes to monitor conformational changes in the protein. It is very difficult because the binding of calcium to most of these fluorescent probes will vary dependent upon the hydrophobic environment. You might be, actually, monitoring the localization of calcium in the calmodulin rather than the conformational changes in the calmodulin.

DR. STORM: Calcium does not interact with these fluorescent dyes.

DR. B. WEISS (*Medical College of Pennsylvania, Philadelphia, PA*): I would like to stress that whereas hydrophobic interactions may be one aspect of the binding, it is by no means the most important one. Bob Levin and I studied many compounds; many compounds that are equally nonpolar and have an equal oil and water solubility coefficient, but they don't bind equally to calmodulin.

DR. STORM: I agree. I think the only key that one can use if you are not looking at regulation of an enzyme by calmodulin is the strength of the binding constants. But, the crucial point, is that the binding is calcium-dependent.

DR. M. MARTIN (*University of Texas at Houston, Houston, TX*): Your results tend to indicate that calmodulin binds directly to the catalytic subunit of phosphodiesterase. However, in most tissues, there are also calmodulin independent forms of phosphodiesterase. What are your thoughts on these enzyme forms?

DR. STORM: I think that the calmodulin sensitive phosphodiesterase is a distinct isozyme. I don't know whether there is any relationship between the catalytic subunit of that enzyme and any of the other phosphodiesterases.

# CALCIUM DEPENDENT REGULATION OF BRAIN AND CARDIAC MUSCLE ADENYLATE CYCLASE*

James D. Potter, Michael T. Piascik, Patricia L. Wisler, Stephen P. Robertson and Carl L. Johnson

*Department of Pharmacology and Cell Biophysics*
*University of Cincinnati College of Medicine*
*Cincinnati, Ohio 45267*

It is now generally established that $Ca^{2+}$ plays a pivotal role in the regulation of cyclic nucleotide metabolism, although our specific knowledge of this process is still incomplete. Studies of adenylate cyclase from a variety of tissues and at various stages of purification have demonstrated that calcium modulates its enzymatic activity. In some tissues, notably brain, $Ca^{2+}$ appears to affect adenylate cyclase in a biphasic manner; low $Ca^{2+}$ concentrations stimulate, while higher concentrations inhibit enzymatic activity. Several studies have presented evidence that the stimulatory action of $Ca^{2+}$ is conferred on adenylate cyclase by calmodulin, a low molecular weight $Ca^{2+}$ binding protein homologous to skeletal muscle troponin C (see Cheung *et al.*[1,2] for recent reviews). Interestingly, a soluble form of phosphodiesterase, which can hydrolyze both cAMP and cGMP, is also activated in a $Ca^{2+}$ dependent manner by calmodulin.[2]

Previous studies on these two enzymes have not clarified the precise $Ca^{2+}$ dependence of the opposing reactions, due to the great difficulties in regulating the free $Ca^{2+}$ concentration, $[Ca^{2+}]$, below $10^{-6}$ M. Since the cytosolic $[Ca^{2+}]$ in all tissues probably varies between $<10^{-7}$ M to $>10^{-6}$ M, depending upon the state of activation of the cell, it is important to carefully determine the effects of $[Ca^{2+}]$ in this range on these two systems. From previous experience[3] on the regulation of muscle contraction by $Ca^{2+}$ in this concentration range, it was apparent that the only way to accurately study the role of $Ca^{2+}$ was to use metal chelators, such as EGTA, to buffer the $[Ca^{2+}]$. Using this system we report here our results on the $[Ca^{2+}]$ dependence of guinea pig brain and cardiac adenylate cyclase and calmodulin dependent phosphodiesterase.

## Activation of Brain Adenylate Cyclase by Calcium

The activation of guinea pig brain adenylate cyclase as a function of $[Ca^{2+}]$, regulated by EGTA, is illustrated in Figures 1–3. In Figure 1, two different particulate preparations were studied. In one case, in an attempt to bind all available calmodulin to all of its available $Ca^{2+}$ dependent binding sites in the preparation, 0.1 mM $CaCl_2$ was included in all solutions and this preparation is referred to as $Ca^{2+}$ membranes (CaM). Their calmodulin content was 1.57 $\mu$g/mg protein.[4] In the case of these membranes (Figure 1) as well as the similarly prepared untreated CaM in Figures 2 and 3, $[Ca^{2+}]$ has a biphasic effect on the adenylate cyclase activity, activation beginning at $<10^{-7}$ M $[Ca^{2+}]$ followed by

*This work was supported by grants from the National Institutes of Health (HL 22619-3A-3E, HL 22136, HL 07382, HL 00414) and from the American Heart Association (78-1167, 74–177).



0077-8923/80/0356-0220 $1.75/0 © 1980, NYAS

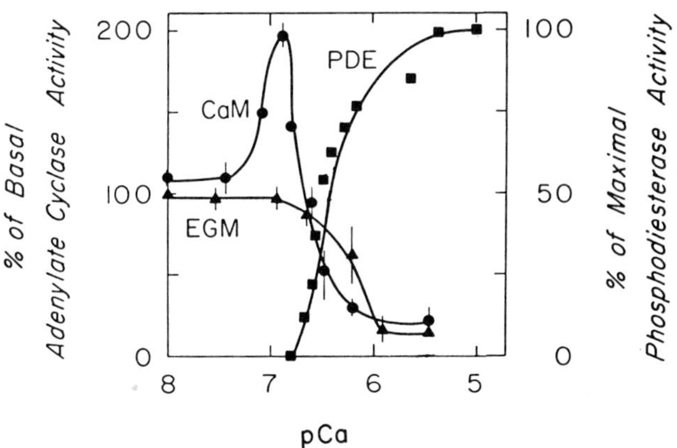


FIGURE 1. The $Ca^{2+}$ dependence of guinea pig brain adenylate cyclase (AC) and phosphodiesterase (PDE). For assay of AC activity, particulate fractions were prepared from frozen guinea pig brains. These were homogenized in 9 volumes of 10 mM imidazole (pH 7.5) containing 0.1 mM $CaCl_2$ and centrifuged at 27,000 × g for 20 min. The pellet was resuspended in the same volume and centrifuged twice. The final pellet was resuspended (15–25 mg/ml) in 5 mM Tris-HCl (pH 7.5) and 0.25 M sucrose, quick frozen, and stored at −20°C. These are referred to as $Ca^{2+}$-membranes (CaM). In certain experiments CaM were further washed 3 times with 1 mM EDTA followed by 3 washes with 1 mM EGTA, yielding a final pellet resuspended in 5 mM Tris-HCl (pH 7.4) and 0.25 M sucrose. These are referred to as EGTA-membranes (EGM). AC was assayed in reaction mixtures (250 μl) which contained 2mM EGTA, 150 mM MOPS (pH 7.20), 83 mM Tris base, 1 mM $Na_2ATP$, 2 mM $MgCl_2$, 10 μM GTP, 10 mM theophylline, 1 mM cAMP, 10 mM phosphocreatine, 12.5 μg creatine phosphokinase, [$^3$H]cAMP (10,000 cpm to monitor recovery), and brain particulate fractions (0.5–0.8 mg/ml). Various amounts of $CaCl_2$ were added to yield a calculated free metal ion concentration.[3,4] These mixtures were preincubated for 5 min prior to the addition of [α-$^{32}$P]ATP (1.2 × $10^6$ cpm). The reaction was allowed to proceed for 15 min at 30°C and was normally stopped by addition of 100 μl 1% SDS. EGM (▲); CaM (●). Reaction volumes were made up to 1 ml by addition of $H_2O$. [$^{32}$P]cAMP was purified by the double column procedure of Salomon *et al.*[5] The calcium-dependence of soluble[4] calmodulin dependent PDE (■) was determined under similar reaction conditions (ionic strength, pH, etc.) as AC. Reaction mixtures contained 150 mM MOPS (pH 7.20), 83 mM Tris, 2 mM $MgCl_2$ or 6 mM Mg acetate, 2 mM EGTA, 10 mM BME (2-mercaptoethanol), 0.3 units adenylate deaminase, 75 μM cAMP, 75–100 μg PDE, various amounts of $CaCl_2$ to yield the calculated free metal concentrations,[3,4] and 285 ng calmodulin in a total volume of 1 ml.

inhibition at $>10^{-7}$ M [$Ca^{2+}$]. The $Ca^{2+}$ dependent activation phase is clearly due to calmodulin since washing CaM with chelators to produce EGTA-membrane (EGM), presumably dissociating calmodulin bound in the presence of $Ca^{2+}$, abolished the activation (FIGURE 1). The activation could be restored by the addition of exogenous calmodulin to EGM (data not shown). In addition, when CaM were treated with either trifluoperazine (FIGURE 2) or troponin I (TnI) (FIGURE 3), agents known to bind to calmodulin and to block its biological effects,[2,6] the $Ca^{2+}$ dependent activation was also abolished. Thus, the $Ca^{2+}$ activation process of brain adenylate cyclase is clearly mediated by calmodulin.

When the $Ca^{2+}$ dependence of soluble calmodulin dependent phosphodiesterase was studied under nearly identical conditions (FIGURE 1) (see also Reference 4) to those employed in the cyclase assays, the [$Ca^{2+}$] at which half maximal

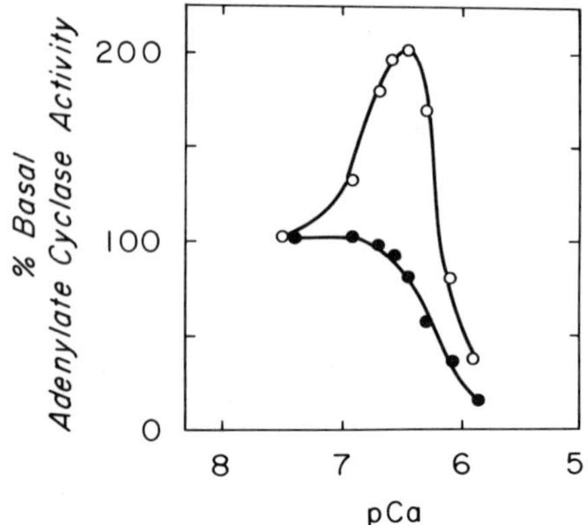


FIGURE 2. Effect of trifluoperazine (TFP) on the $Ca^{2+}$ dependence of brain AC. CaM were prepared as described in the legend to FIGURE 1 and were incubated in the presence (●) or absence (○) of TFP for 30′ at 23°C. AC was then assayed as in FIGURE 1, [TFP] = 250 μM.

stimulation occurred (0.3 μM) was higher than that required for half maximal activation of the cyclase (<0.1 μM). Thus, although both processes are regulated by calmodulin, their $Ca^{2+}$ requirements are quite different. We can only speculate as to the mechanism by which the affinity of $Ca^{2+}$ for calmodulin is increased but it clearly is. An analogous situation may be found with myofibrillar proteins where the formation of the troponin I-troponin C complex results in a 10-fold increase in $Ca^{2+}$ affinity of all four $Ca^{2+}$ sites on troponin C.[3] Similarly, association of calmodulin (with $Ca^{2+}$ bound for example to any one of its four equivalent sites) to its cyclase binding site might increase the affinity of this calmodulin for calcium.

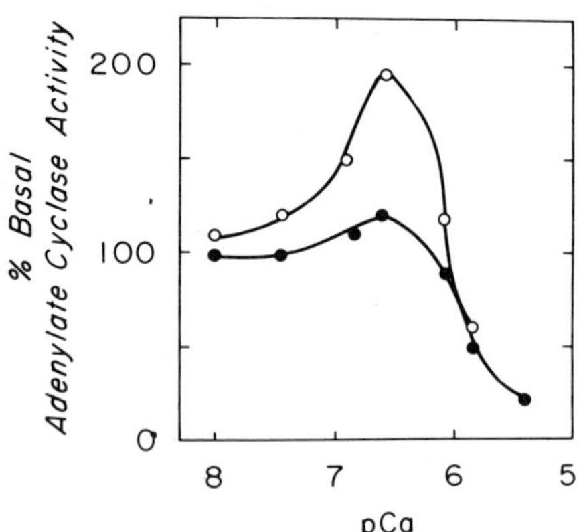


FIGURE 3. Effect of Troponin I (TnI) on the $Ca^{2+}$ dependence of guinea pig brain adenylate cyclase. CaM were prepared as described in the legend to FIGURE 1. These were preincubated in the presence (●) or absence (○) of cardiac TnI (0.25 mg TnI/mg membranes) for 20 min at 30°C and the AC assayed as described in FIGURE 1.

It should also be pointed out that it is possible that our *calculated* free $Ca^{2+}$ concentrations may be somewhat different from the *true* free $Ca^{2+}$ concentrations, but the use of the same concentration of EGTA and similar conditions in both the phosphodiesterase and adenylate cyclase assays shows that whatever the *actual* free $Ca^{2+}$ concentrations are, the assays will have the same relative relationship to each other with respect to $Ca^{2+}$ concentration.

## Inhibition of Brain Adenylate Cyclase by Calcium

Previous studies have suggested that $Ca^{2+}$ concentrations in excess of 10 $\mu$M were required to inhibit adenylate cyclase, although the actual free $Ca^{2+}$ concentration was not known. Since it was unlikely that intracellular [$Ca^{2+}$] concentrations ever exceed 10 $\mu$M, the $Ca^{2+}$ inhibition of cyclase was not thought to be physiologically important, although in one case[7] where EGTA was used to regulate the free $Ca^{2+}$ concentration, inhibition of cardiac adenylate cyclase was reported to occur at submicromolar [$Ca^{2+}$]. In Figures 1–3 the $Ca^{2+}$ and calmodulin dependent activation of brain adenylate cyclase was established. Figures 1–3 also illustrate the inhibition of adenylate cyclase by $Ca^{2+}$ to be in the submicromolar range. In CaM the activation phase is followed by an inhibitory phase that is half maximal at [$Ca^{2+}$] ~0.3 $\mu$M (Figures 1–3). When CaM are depleted of calmodulin (Figure 1, EGM) or when CaM are treated with either trifluoperazine (Figure 2) or TnI (Figure 3), the $Ca^{2+}$ dependent activation phase is lost; however, the $Ca^{2+}$ dependent inhibitory phase is maintained. Again, in all three cases, half maximal inhibition occurred at [$Ca^{2+}$] ~0.3 $\mu$M (Figures 1–3). Interestingly, the [$Ca^{2+}$] at half maximal inhibition of adenylate cyclase occurs at the same [$Ca^{2+}$] as does the activation of calmodulin dependent phosphodiesterase (Figure 1). Since EGM contained a significant quantity of calmodulin that was not released by chelators (0.24 $\mu$g/mg protein),[4] it occurred to us that calmodulin might also be mediating the inhibition of adenylate cyclase, although it would presumably be bound to an inhibitory site on some component of the cyclase (in contrast to the activating site) that is inaccessible to EGTA, trifluoperazine, and TnI, since none of these treatments abolished the inhibitory phase. This binding site for calmodulin might be similar to the binding site for calmodulin in phosphorylase kinase,[8] since treatment of this complex with EGTA alone is not sufficient to dissociate calmodulin. It is also possible that another high affinity $Ca^{2+}$ binding site(s), located on a protein distinct from calmodulin, is responsible for the observed inhibition. However, we have also shown[4] that the $Ca^{2+}$ binding sites on calmodulin and the protein involved in adenylate cyclase inhibition have the same affinity for $Sr^{2+}$, thus again suggesting their identity. Further studies will be required to clarify this point.

Several important conclusions emerge from this study. Firstly, brain adenylate cyclase is not only activated, but is also inhibited by physiological concentrations of $Ca^{2+}$. Secondly, the $Ca^{2+}$ dependent activation of adenylate cyclase occurs at a lower [$Ca^{2+}$] than does the activation of calmodulin dependent (cAMP, cGMP) phosphodiesterase.

In contrast to many published reports describing the effects of divalent cations on adenylate cyclase and phosphodiesterase, we report the calculated concentrations of *free cation* which influence these systems. The use of a cation-buffering concentration of EGTA (2 mM) in the enzyme assays and in the measurements of binding to calmodulin, mentioned later, enabled us to control the levels of free metal; the estimated values of free cation concentration were calculated using a computer program.[3] The primary factor that could alter the results obtained in these experimental systems would be the lack of consideration in the calculations of any reaction component(s) that affects the calculated free $Ca^{2+}$ concentration. However, the contribution of all known chelators have been taken into account. In any event, since the reaction components are essentially the same in the different assays (AC, PDE, etc.), and since the same calculations of $[Ca^{2+}]$ are made in each case, the results are therefore all directly comparable to each other, something that is not true of the other studies on these two enzymes.

The effective $[Ca^{2+}]$ concentrations reported here are much lower than found in previous studies and illustrate the importance of using metal buffers to regulate the free metal ion concentrations. The high EGTA concentration used in these experiments would also eliminate any competition between $Ca^{2+}$ and endogenous divalent cations (e.g., $Mn^{2+}$). Previous studies detailing $Ca^{2+}$-brain adenylate cyclase interactions have been done in the absence of chelators with contaminating *total* $Ca^{2+}$ concentrations in the 1–10 $\mu$M range. This $Ca^{2+}$ is probably buffered by the membranes and ATP in the reaction mixture. Thus, the actual free $Ca^{2+}$ concentrations in the solutions could have been lower than the total added $CaCl_2$ reported. It should be pointed out that many of the studies have added $Ca^{2+}$ to EGTA mixtures without actually calculating the free $Ca^{2+}$ concentration. Recently Westcott *et al.*[9] have detailed the response of brain adenylate cyclase to increasing amounts of total added $CaCl_2$ in the presence of a fixed concentration of EGTA. Using their reported values of total EGTA and $CaCl_2$ (and assuming the concentrations of EGTA and $CaCl_2$ were accurate, that contaminating endogenous $Ca^{2+}$ was insignificant, and that the pH remained constant at the reported value), we have estimated the free $Ca^{2+}$ concentration-dependence of their results with adenylate cyclase and found them very similar to those presented in FIGURES 1–3. In addition, there have been studies that have not used EGTA which report $Ca^{2+}$ having little, if any, ability to stimulate brain adenylate cyclase. That is exactly what would be expected if the actual free $Ca^{2+}$ concentration in those experiments fell between the activation and inhibitory phases reported here.

In light of our findings we propose the following model (FIGURE 7) for $Ca^{2+}$ modulation of cAMP levels in the brain. Subsequent to stimulation of the cell, intracellular $[Ca^{2+}]$ rises from resting state levels $<0.1$ $\mu$M. During the first portion of the time course of $Ca^{2+}$ influx and/or mobilization, at intracellular concentrations of 0.08 to 0.2 $\mu$M, $Ca^{2+}$ complexes with calmodulin and this complex, in turn, activates adenylate cyclase activity in the plasma membrane. As intracellular $Ca^{2+}$ levels continue to increase above 0.2 $\mu$M, the activity of the cyclase begins to be inhibited (perhaps mediated by calmodulin bound to a distinct inhibitory metal site on the cyclase). Concurrently with adenylate cyclase inhibition by $Ca^{2+}$, a $Ca^{2+}$-calmodulin complex activates soluble cyclic nucleotide (cAMP and cGMP) phosphodiesterase activity. Thus a coordinated regulation of adenylate cyclase and phosphodiesterase activities by $Ca^{2+}$ may modulate cellular cyclic nucleotide concentrations within a fairly narrow range.

## The Calcium Regulation of Cardiac Adenylate Cyclase

That $Ca^{2+}$ plays a decisive role in regulating cardiac cyclic nucleotide metabolism has become increasingly evident in recent years. Namm and his associates[10] found that augmented supplies of $Ca^{2+}$ to perfused rat hearts caused a decline in their content of cyclic AMP. Conversely, Endoh et al.[11] observed increased tissue concentrations of cyclic AMP in rabbit papillary muscle when $Ca^{2+}$ was removed from the perfusate. Harary et al.[12] found that in beating, cultured rat heart cells, the concentrations of cyclic AMP progressively diminished as the concentration of $Ca^{2+}$ in the medium was raised from $10^{-4}$ M to $10^{-3}$ M. At the enzymatic level, $Ca^{2+}$ inhibited adenylate cyclase activity ($K_i = 3 \times 10^{-4}$ M) in the particulate fraction of guinea pig ventricle.[13] Sulahke and Dhalla[14] working with dog heart sarcotubular membranes, observed that $Ca^{2+}$ concentrations of $2.5 \times 10^{-4}$ M and higher progressively inhibited adenylate cyclase activity. Unlike the previous two studies, which did not use any chelator to buffer the [$Ca^{2+}$], inhibition of cardiac adenylate cyclase by much lower [$Ca^{2+}$] concentrations was reported by Tada et al.[7] who, as mentioned earlier, used EGTA to regulate the [$Ca^{2+}$]. Basal, epinephrine-stimulated, and fluoride-stimulated cyclase activities of their guinea pig ventricle particulate fractions were inhibited at $Ca^{2+}$ concentrations between $10^{-7}$ M and $10^{-4}$ M.

The apparent discrepancies among these reported effective [$Ca^{2+}$] concentrations have led us to reexamine the $Ca^{2+}$ regulation of guinea pig cardiac adenylate cyclase, using the identical techniques described above for the studies on brain adenylate cyclase. The $Ca^{2+}$ dependence of guinea pig left ventricle adenylate cyclase is shown in FIGURE 4. These data were obtained on an STE preparation (see legend to FIGURE 4) and, as can be seen $Ca^{2+}$ inhibits the cyclase, with half maximal inhibition occurring at [$Ca^{2+}$] ~0.4 $\mu$M. This is precisely the [$Ca^{2+}$] at which brain adenylate cyclase is inhibited by $Ca^{2+}$. Again, since these measurements on the cardiac cyclase were made under identical conditions as those of the brain, the results are directly comparable. Note that there is no $Ca^{2+}$ dependent activation phase. This was true for STC preparations (analogous to CaM prepared in the presence of 0.1 mM $CaCl_2$), suggesting that there is no calmodulin activation binding site on the cardiac cyclase analogous to that found in brain. This is further supported by the fact that large excesses of calmodulin added to the STC preparation failed to produce an activation of the cyclase. Thus, it appears that only the inhibitory phase of $Ca^{2+}$ is seen in the cardiac adenylate cyclase. It is important also to point out that this occurs at submicromolar and therefore physiological levels of free $Ca^{2+}$ and would be regulated by the [$Ca^{2+}$] fluctuations that are known to occur during the diastolic-systolic cycle ($<10^{-7}$ M to $>10^{-6}$ M).

Since the cardiac adenylate cyclase had the same [$Ca^{2+}$] dependence of inhibition as the brain cyclase, and since this is exactly the same [$Ca^{2+}$] range where calmodulin dependent phosphodiesterase is activated, it is possible that calmodulin may also mediate the cardiac cyclase inhibition. There are two lines of evidence which support this possibility. The first is that STE preparations (well washed with EGTA, see legend to FIGURE 4) contain significant amounts of residual calmodulin (0.13 $\mu$g/mg protein) thus making it possible for some of this calmodulin to participate in the inhibition. Secondly, the concentration dependence of the inhibition of cardiac cyclase and of binding to calmodulin of a variety of metals ($Ca^{2+}$, $Sr^{2+}$, and $Ba^{2+}$) have been studied. As can be seen in FIGURES 4 and 5 (see also TABLE 1) the [$Me^{2+}$] concentration dependence of both the inhibition of cyclase and of the enhancement of calmodulin tyrosine fluorescence (an indirect measure of $Me^{2+}$ affinity) were remarkably similar, suggesting the identity of the metal binding sites responsible for inhibition of the cyclase and calmodulin. Not only were the $Me^{2+}$ affinities the same for the two processes, but so were the slopes of the two types of responses to the different cations (TABLE 1 and FIGURES 4 & 5). In addition, the ability of $Sr^{2+}$ or $Ba^{2+}$ to produce the same effect as $Ca^{2+}$, was reduced to the same extent in both the cyclase and fluorescence measurements. Thus, as in the case of the inhibition of brain cyclase, there is ample evidence to show that calmodulin mediates the inhibition of cardiac cyclase as well. As in the case of phosphorylase kinase[8] it is not possible to unequivocally prove that calmodulin is the $Ca^{2+}$ receptor for this inhibition, even though it is a very likely possibility. As is the case in brain

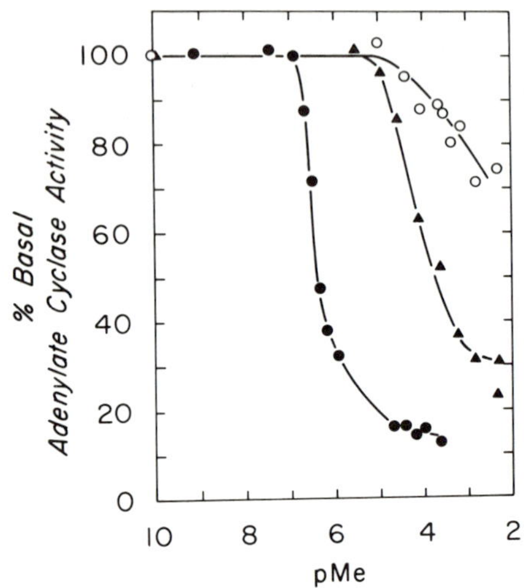


FIGURE 4. The $Ca^{2+}$, $Sr^{2+}$, and $Ba^{2+}$ dependence of guinea pig cardiac AC. For assay of AC activity particulate fractions were prepared from guinea pig left ventricle. The ventricles were placed in 9 volumes of 250 mM sucrose, 5 mM Tris, 1 mM EGTA, pH 7.2 (STE), homogenized by three 3-sec bursts of a Brinkman polytron set at 10, followed by 10 up and down strokes in a Potter-Elvehjem glass homogenizer fitted with a Teflon pestle. The homogenate, filtered through four layers of cheesecloth, was centrifuged at 1000 × g for 20 min, after which the supernatant was discarded and the pellet resuspended in 9 volumes of STE. Centrifugation and resuspension of pellets were repeated twice and the final pellet was suspended in 4 volumes original tissue weight of STE to give an approximate protein concentration of 10 mg/ml. These procedures were carried out at 4°C. At this point, the particulate fraction preparations were either kept chilled and used within 2 hr or were quick-frozen and stored at −20°C until used. In certain experiments, the STE homogenizing medium was replaced by 250 mM sucrose, 5 mM Tris, pH 7.2 (ST) or 250 mM sucrose, 5 mM Tris, 0.1 mM $CaCl_2$, pH 7.2 (STC). The AC activity of STE was measured as described in the legend to FIGURE 1. Various amounts of $CaCl_2$ (●), $SrCl_2$ (▲), or $BaCl_2$ (○) were added to yield the calculated[3,4] free metal concentration indicated on the abscissa. In the experiments with $Sr^{2+}$ and $Ba^{2+}$ 100 μl of Lubrol-WX and 20 mM EDTA was used to terminate the assay, instead of 1% SDS.

cyclase, it is possible that there is a $Ca^{2+}$ binding site(s) on a protein distinct from calmodulin that has identical $Me^{2+}$ binding properties.

The $Ca^{2+}$ dependence of the cardiac cyclase was studied under conditions that are known to stimulate the basal enzyme activity to see if the same $Ca^{2+}$ inhibition would be observed; also solubilized cyclase was compared to unstimu-

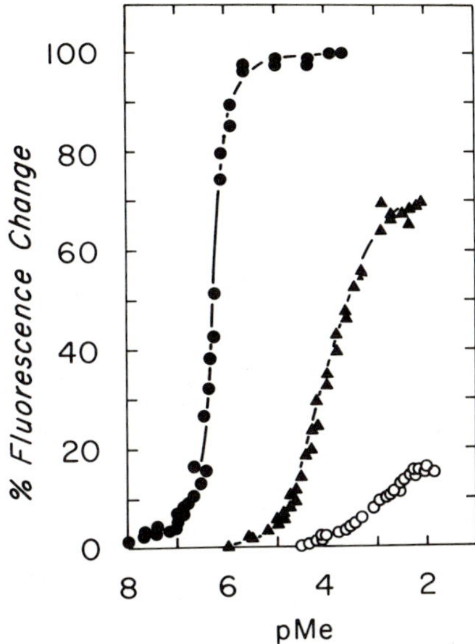


FIGURE 5. The $Ca^{2+}$, $Sr^{2+}$, and $Ba^{2+}$ dependence of calmodulin tyrosine fluorescence. Divalent cation binding to purified calmodulin was estimated indirectly by studying the $Me^{2+}$ dependence of its tyrosine fluorescence change.[4] The calmodulin sample, maintained at 30°C, was excited at 280 nm and emission was monitored at 304 nm for each addition of 2-4 $\mu$l $CaCl_2$ (●), $SrCl_2$ (▲) or $BaCl_2$ (○) to the 3 ml protein solution. The free metal concentration was calculated as in FIGURE 4. The solution contained 0.75 mg calmodulin, 150 mM MOPS, 83 mM Tris, pH 7.20, 2 mM EGTA, 1 mM $Na_2ATP$, and 2 mM $MgCl_2$. The pH was checked after each addition of cation and if necessary was adjusted to 7.20 with 45% KOH. The total volume of KOH added never exceeded 5 $\mu$l.

lated STE (FIGURE 6). As can be seen, the [$Ca^{2+}$] dependence of the solubilized and particulate cyclase were essentially identical. Also, the [$Ca^{2+}$] dependence of control STE, $MgCl_2$, histamine, and NaF (data not shown) stimulated STE were essentially identical. Thus, under a variety of stimuli, the $Ca^{2+}$ dependent inhibition of cardiac cyclase was not altered and this is also true in the case of the brain cyclase inhibition (data not shown).

Thus, several important conclusions emerge from this study. First, the cardiac

TABLE 1

| | $Me^{2+}$ Inhibition of Cardiac Adenylate Cyclase | | $Me^{2+}$ Enhancement of Calmodulin Tyrosine Fluorescence | |
|---|---|---|---|---|
| $Me^{2+}$ | *[$Ca^{2+}$] ½ max | Hill coefficient | *[$Ca^{2+}$] ½ max | Hill coefficient |
| $Ca^{2+}$ | 0.4 $\mu$M | 2.4 | 0.5 $\mu$M | 2.44 |
| $Sr^{2+}$ | 0.1 mM | 1.0 | 0.1 mM | 1.02 |
| $Ba^{2+}$ | 0.3 mM | 1.0 | 0.8 mM | 0.94 |

*Estimated from the transition midpoints in FIGURES 4 and 5.

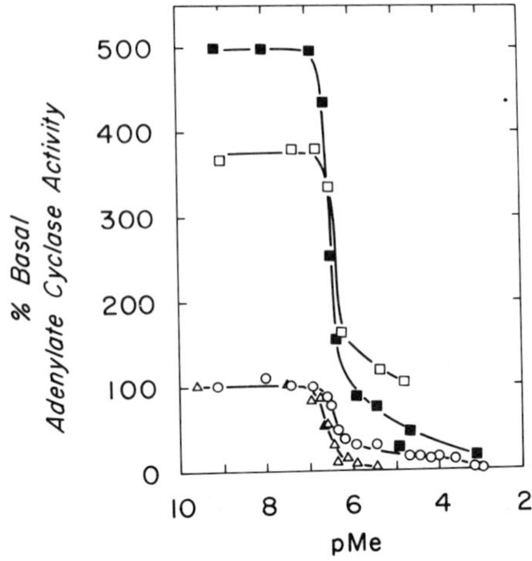


FIGURE 6. The $Ca^{2+}$ dependence of cardiac AC. STE particulate fraction was prepared as described in the legend to FIGURE 4. Solubilized AC was prepared from STE by making it 1% with Lubrol-WX, followed by occasional vortexing over a period of 30 min, then centrifuged for 60 min at 100,000 × g. The resulting supernatant was used as the solubilized AC preparation. The AC activity of particulate (STE) AC (○) and solubilized AC (A), STE in the presence of 6 mM $MgCl_2$ (□) and 1 mM histamine (■) were assayed as in FIGURE 1.

adenylate cyclase is inhibited over the same [$Ca^{2+}$] range, as is the brain cyclase, and calmodulin may mediate the inhibition in both cases. Secondly, the cardiac cyclase is not stimulated by $Ca^{2+}$ via calmodulin as is the case in brain.

## SUMMARY

The very close interdependence of $Ca^{2+}$ and hormones in the overall metabolism of cyclic nucleotides has recently been emphasized by Cheung.[1] Clearly the results presented here show that [$Ca^{2+}$] in the physiological range ($<10^{-7}$ M to $>10^{-6}$ M) has profound effects on the activity of adenylate cyclase from both brain and cardiac muscle. Whereas both brain and cardiac cyclase exhibit a $Ca^{2+}$ dependent inhibition (perhaps mediated by calmodulin), only the brain cyclase is activated by $Ca^{2+}$ via calmodulin. With both cyclases there is an inverse relationship between the inhibition of cyclase and the activation of calmodulin dependent (cAMP and cGMP) phosphodiesterase as a function of $Ca^{2+}$ concentration. Because the $IC_{50}$'s for $Ca^{2+}$ are the same in both heart and brain, the possibility exists that the $Ca^{2+}$ inhibitory site of both cyclases is similar and perhaps identical.

Considering the ability of $Ca^{2+}$ to both stimulate and inhibit cyclase, one could imagine that in different species, tissues, or regions of the same tissue, there could exist multiple populations of cyclase, that is a cyclase which would only show $Ca^{2+}$ dependent inhibition, $Ca^{2+}$ dependent stimulation, or the biphasic response to $Ca^{2+}$ (FIGURE 7).

The fact that $Ca^{2+}$ still regulates adenylate cyclase after various stimuli (histamine, NaF, etc.) suggests that $Ca^{2+}$ may function to regulate the cyclase over

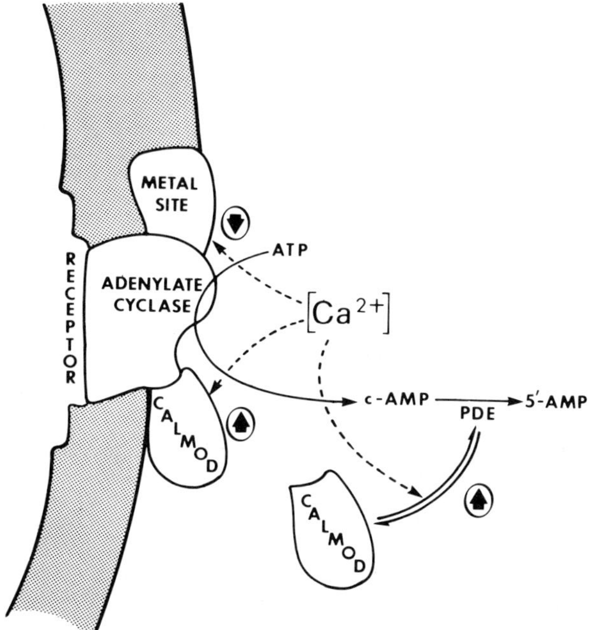


FIGURE 7. Proposed model for the regulation of adenylate cyclase by [$Ca^{2+}$]. $Ca^{2+}$ acts at three different sites in the model: 1) at low [$Ca^{2+}$] (0.08–0.2 $\mu$M) $Ca^{2+}$ binds to calmodulin and activates adenylate cyclase (in brain only); 2) in brain and heart slightly higher [$Ca^{2+}$] (0.2 $\mu$M) inhibits adenylate cyclase by combining with an inhibitory $Ca^{2+}$ binding site; 3) concomitant with this inhibition, $Ca^{2+}$ bound to cytoplasmic calmodulin activates a cyclic nucleotide phosphodiesterase. In two ways (inhibition of cyclase and activation of PDE) $Ca^{2+}$ can act to decrease cyclic nucleotide levels.

shorter time periods (regardless of its state of stimulation) and that other affectors of cyclase (e.g., hormones) would serve to regulate the cyclase over longer time periods.[2]

## REFERENCES

1. CHEUNG, W. Y., T. J. LYNCH & R. W. WALLACE 1978. Adv. Cyclic Nucleotide Res. **9:** 233–251.
2. CHEUNG, W. Y. 1980. Science **207:** 19–27.
3. POTTER, J. D. & J. GERGELY. 1975. J. Biol. Chem. **250:** 4628–4633.
4. PIASCIK, M. T., P. L. WISLER, C. L. JOHNSON & J. D. POTTER. 1980. J. Biol. Chem. **255:** 4176–4181.
5. SALOMON, Y., C. LONDOS & M. RODBELL. 1974. Anal. Biochem. **58:** 541–548.
6. KERRICK, W. G. L., L. BOLLES, P. CASSIDY, R. L. COBY, S. GRAZARIK, P. E. HOAR & D. A. MALENCIK. 1980. Fed. Proc. **39:** 2042.
7. TADA, M., M. A. KIRCHBERGER, J. M. SORIO & A. M. KATZ. 1975. Circ. Res. **36:** 8–17.
8. COHEN, P., A. BURCHELL, J. G. FOULKES, P. T. W. COHEN, T. C. VANAMAN & A. C. NAIRN. 1978. FEBS Lett. **92:** 287–293.

9. WESTCOTT, K. R., D. C. LAPORTE & D. R. STORM. 1979. Proc. Natl. Acad. Sci. USA **76:** 204–208.
10. NAMM, D. H., S. E. MAYER & M. MALTBIE. 1968. Mol. Pharmacol. **4:** 522–530.
11. ENDOH, M., O. E. BRODDE, D. REINHARDT & H. J. SCHUMANN. 1976. Nature (London) **261:** 716–717.
12. HARARY, I., J.-F. RENAUD, E. SATO & C. A. WALLACE. 1976. Nature (London) **261:** 60–61.
13. DRUMMOND, G. S. & L. DUNCAN. 1970. J. Biol. Chem. **245:** 976–983.
14. SULAKHE, P. V. & N. S. DHALLA. 1973. Biochem. Biophys. Acta **293:** 379–396.

## DISCUSSION OF THE PAPER

DR. G. FISKUM (*Johns Hopkins School of Medicine, Baltimore, MD*): Could someone summarize the different tissues where calmodulin regulation of both adenylate cyclase and phosphodiesterase has been observed and, in particular, has it been observed in liver?

DR. J. D. POTTER: I think mostly neurosecretory tissues show the activation phase, but, other tissues do not.

DR. P. SIEKEVITZ (*Rockefeller University, New York, NY*): We also have to look not only at the tissue, but, on the subcompartments of the tissue, that is, whether these two enzymes are in the same compartment or not. For example, in brain tissue the adenylate cyclase looks like it is a membrane enzyme, whereas the PDE looks like it's particulate bound. Of course, these are right next to each other, but, still, they are compartmentalized.

DR. W. Y. CHEUNG: I think it has been very clearly demonstrated that with the phosphodiesterase if you prepare the enzyme under conditions where the enzyme has been acted on by protease, one way or the other, then, the sensitivity to calmodulin would change, would decrease. It is quite conceivable in some tissues, for example, like platelets with a very active protease, that when you break up the platelets even though the phosphodiesterase is sensitive to calmodulin you would get a preparation that appears to be calmodulin insensitive.

DR. FLOCKHART: One can't help but draw analogies all the way along here with the interaction of protein kinase with substrates. What determines the site of action of calcium within a tissue which has multiple substrates for calmodulin is very important in this discussion. This is the reason I was asking Dr. Storm about the affinities of calmodulin for the various enzymes. At the moment it looks very confusing in that calmodulin has a very similar affinity for many of the enzymes. So, it would be very hard for calcium to determine a site.

DR. CHEUNG: An important aspect of the work is, given the multiplicity of calmodulin receptors, what are the relative affinities for these receptors for calmodulin?

Perhaps, there are additional factors that might change the relative affinities of these various receptors to calmodulin?

DR. MARSHAK (*Rockefeller University, New York, NY*): There have been reports of proteolytic activation of adenylate cyclase in a number of tissues. So, I would like to point out that some of these data could be accounted for by a calcium-dependent proteolytic action.

DR. POTTER: Are you saying that the inhibition that we are seeing could be due to the calcium-dependent protease?

DR. MARSHAK: Well in some tissues there are two phase systems just like you showed, that is, an initial phase of proteolytic activation followed by a later phase of proteolytic inhibition of adenylate cyclase.

DR. POTTER: All I can say is that the membranes are prepared over a long period of time in the presence of calcium and, then, they are washed with EGTA before being assayed. They are then only in the presence of calcium in the assay mixture for a very short period of time. So, if any proteolysis was to have occurred, it would have occurred before the assay—not during the assay. So, I doubt that the inhibition could be due to a $Ca^{2+}$-dependent protease.

# CALMODULIN AND THE PLASMA MEMBRANE CALCIUM PUMP*

Frank F. Vincenzi, Thomas R. Hinds, and Beat U. Raess

*Department of Pharmacology*
*University of Washington*
*Seattle, Washington 98195*

## Introduction

One of the many effects of calmodulin (CaM) was originally observed by Bond and Clough.[1] They found that a protein in human red blood cells (RBCs) activated the $(Ca^{2+}+Mg^{2+})$-ATPase of the RBC membrane while it had no effect on other ATPases. The $(Ca^{2+}+Mg^{2+})$-ATPase is generally regarded as the $Ca^{2+}$ pump ATPase[2,3] and Bond and Clough were prompted to speculate that the activator protein might be a regulator of the $Ca^{2+}$ pump. The existence in human RBCs of a protein that activates the $(Ca^{2+}+Mg^{2+})$-ATPase was confirmed. It was purified[4] and shown to be identical with CaM by a number of approaches.[5-7]

CaM was found to bind to RBC membranes in a reversible and $Ca^{2+}$ dependent fashion.[8,9] This led to speculation that $Ca^{2+}$ dependent binding of CaM to the inner membrane surface was an "on switch" for the $Ca^{2+}$ pump. It was further speculated that removal of $Ca^{2+}$ from the cell resulted in unbinding of CaM and a switching off the $Ca^{2+}$ pump ATPase.[10] While probably not correct in detail, these speculations led us to further examine the effects of CaM on the $Ca^{2+}$ pump ATPase. Various features of CaM activation of $(Ca^{2+}+Mg^{2+})$-ATPase are reviewed and illustrated in the present report.

## Methods

Methods employed in these experiments have been published in large part. The preparation of isolated human RBC membranes[8,9] and assay of membrane-bound ATPases[11] were carried out as described recently. Briefly, membranes were isolated using 20 mM imidazole buffer, pH 7.4. They were incubated for various times at 37°C in medium containing: NaCl, 80 mM; KCl, 15 mM; $MgCl_2$, 3 mM, ATP, 3 mM; histidine-imidazole, 18 mM, pH 7.1; EGTA, 0.1 mM; with or without added $Ca^{2+}$, ouabain (0.1 mM) and CaM. Unless otherwise noted $[Ca^{2+}]$s were determined with a $Ca^{2+}$ selective electrode.[12] Sodium dodecyl sulfate polyacrylamide gel electrophoresis (SDS-PAGE) was carried out according to Fairbanks *et al.*[13] using 4% gels. CaM was prepared by ion-exchange and gel filtration chromatography.[14,15]

## CaM Activation of $(Ca^{2+}+Mg^{2+})$-ATPase

The left-hand panel of Figure 1 demonstrates several features of two operationally defined ATPase activities of RBC membranes. ATPase activity was

*Supported in part by U.S. Public Health Service Grants AM16436 and AM07270.



0077-8923/80/0356-0232 $1.75/0 © 1980, NYAS

plotted as a function of [$Ca^{2+}$] in the absence and the presence of 3 concentrations of CaM. Ouabain was present to inhibit ($Na^{+}+K^{+}+Mg^{2+}$)-ATPase activity. In the absence of added [$Ca^{2+}$] the activity is the ($Mg^{2+}$)-ATPase, which was not influenced by CaM. Addition of $Ca^{2+}$ in the absence of CaM caused increased ATPase activity. This activity above the ($Mg^{2+}$)-ATPase is termed the basal ($Ca^{2+}+Mg^{2+}$)-ATPase activity. In a concentration-dependent fashion CaM activates ATPase activity in the presence of $Ca^{2+}$. It should be noted that the magnitude of CaM activation of the ($Ca^{2+}+Mg^{2+}$)-ATPase is highly dependent on [$Ca^{2+}$]. High [$Ca^{2+}$]s result in less than maximal activity in the presence of CaM. We shall return to this point. $Ca^{2+}$ dependence of both basal and CaM activated ($Ca^{2+}+Mg^{2+}$)-ATPase has been studied in other laboratories.[16] The question of whether there are two ATPases, one with high and one with low affinity, and/or whether CaM induces the high affinity has been the source of some controversy. This has been considered elsewhere.[17]

Comparison of the left- and right-hand panels of FIGURE 1 demonstrates another important aspect of CaM activation of ($Ca^{2+}+Mg^{2+}$)-ATPase. Data in the

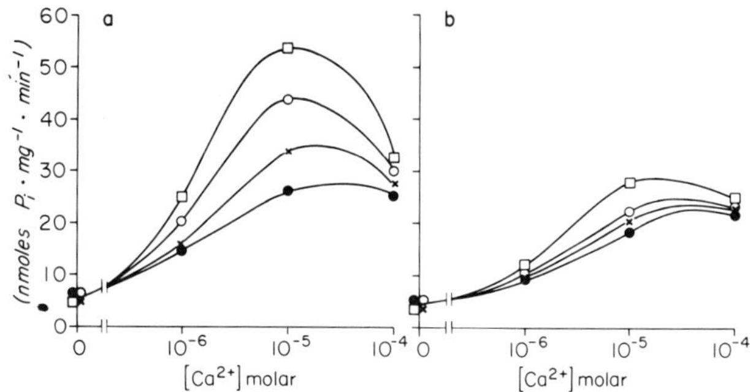


FIGURE 1. Human RBC membrane ATPase: Influence of $Ca^{2+}$, CaM and chlorpromazine (CPZ). Total ATPase activity was plotted as a function of free $Ca^{2+}$ in the absence (●——●) and the presence of three different concentrations of calmodulin: 0.093 (X——X); 0.33 (○——○); and 0.93 μg/ml (□——□). Panel a: in the absence of CPZ: Panel b: in the presence of $1 \times 10^{-4}$ M CPZ. Intercepts on the ordinates represent ($Mg^{2+}$)-ATPase activities no $Ca^{2+}$; 0.1 mM EGTA present. ($Mg^{2+}$)-ATPase activity may be subtracted from the total to obtain specific ($Ca^{2+}+Mg^{2+}$)-ATPase activity.

right hand panel were obtained in the presence of $10^{-4}$ M chlorpromazine. The data show that in this concentration chlorpromazine antagonizes CaM activation almost completely while producing minimal inhibition of ($Mg^{2+}$)-ATPase or basal ($Ca^{2+}+Mg^{2+}$)-ATPase. The result is consistent with the interpretation that chlorpromazine, like other phenothiazine tranquilizers, binds to CaM and reduces its binding to the ATPase. Weiss and co-workers first demonstrated that phenothiazines bind to CaM in the presence of $Ca^{2+}$.[19] Inhibition of $Ca^{2+}$ dependent cyclic nucleotide phosphodiesterase by phenothiazines was related to this effect and it was reported that a number of non-phenothiazine neuroleptic (tranquilizer) drugs inhibited phosphodiesterase.[20] It was implied that neuroleptic potency correlated with anti-CaM potency. However, Norman *et al.*, using phosphodiesterase,[21] and we using ($Ca^{2+}+Mg^{2+}$)-ATPase,[22] concluded that anti-CaM potency does not correlate with neuroleptic potency. Phenothiazines such as chlorpromazine (FIGURE 1) and trifluoperazine antagonized CaM, but non-phenothiazine neuroleptics did not.[22] Considering the hydrophobic nature of drug binding sites present on CaM in the presence of $Ca^{2+}$,[23] and the antagonism of CaM by several non-neuroleptic drugs[24] it seems reasonable to suppose that a variety of fairly nonspecific antagonists of CaM will eventually be identified. Whether any specific pharmacologic properties of some drugs may be traced to anti-CaM activity remains to be determined. A caveat is in order. Nonspecific inhibition of the $Ca^{2+}$ pump and its $(Ca^{2+}+Mg^{2+})$-ATPase was obtained with high concentrations of phenothiazines, even in the absence of added CaM.[22,25] In short, not all effects of phenothiazines in biological systems should be assumed to be related to CaM.

## CALCIUM-INDUCED MEMBRANE DAMAGE

Decreased responsiveness of the $Ca^{2+}$ pump ATPase to CaM may occur in certain diseases. We investigated the CaM activation of the $(Ca^{2+}+Mg^{2+})$-ATPase activity of RBC membranes from patients with sickle cell anemia. The study was prompted because of the reported abnormal $Ca^{2+}$ content of sickle cells.[26] Hemolysates from sickle RBCs contained normal CaM activity, showing that hemoglobin S does not interfere with the effect of CaM. On the other hand, membranes from sickle RBCs contained apparently normal basal $(Ca^{2+}+Mg^{2+})$-ATPase activity, but it was activated very little by CaM. This was true whether the source of CaM was the hemolysate of normal or sickle cells, or purified CaM.[27] We speculated that the decreased response of membrane ATPase might be secondary to increased $Ca^{2+}$ in sickle cells. A model of membrane protein damage by intracellular $Ca^{2+}$ has been developed by Lorand and co-workers.[28] They found in RBCs a transglutaminase, which is activated by relatively high levels of $Ca^{2+}$. The enzyme promotes covalent crosslinking between proteins with the formation of a very high molecular weight polymer.

We confirmed and extended the work of Lorand *et al.* Fresh RBCs were washed and incubated at 37°C for four hours in isotonic NaCl with $1 \times 10^{-5}$ M A23187 in the absence or presence of various amounts of added $CaCl_2$. By inference, $Ca^{2+}$ in the medium had access to the cytoplasm under these conditions. Scans of Coomassie blue-stained SDS-PAGE patterns of membrane proteins of cells incubated with 3 mM $Ca^{2+}$ revealed several changes compared to the control cells (FIGURE 2). Using the terminology of Fairbanks *et al.*,[13] these included: appearance a high molecular weight band at the very top of the gel (X); reduction of band 3; increase of an unnumbered band after band 4.2; appearance or increase in an unnumbered band (labeled Y) between band 7 and hemoglobin; and increased hemoglobin band. Other differences in the pattern from $Ca^{2+}$ exposed cells included a disappearance of band 4.1 and increases in bands 4.5, 5, 6, and 7.

The appearance or increase in the hemoglobin band was not surprising because membrane preparations from $Ca^{2+}$ pretreated cells have a distinctly pink appearance. From our view the most intriguing observation was the appearance of a new or considerably increased band between band 7 and hemoglobin. This band, which we called "Y," had an $R_f$ value of 0.81 which was identical to the $R_f$ value for CaM on the same type of gels. The appearance of a band between band 7 and hemoglobin on gels from $Ca^{2+}$ exposed cells was also observed by Allen and Cadman.[29] They were unable to identify any enzymatic function for band 8 (as it is called in their work) but were able to rule out that

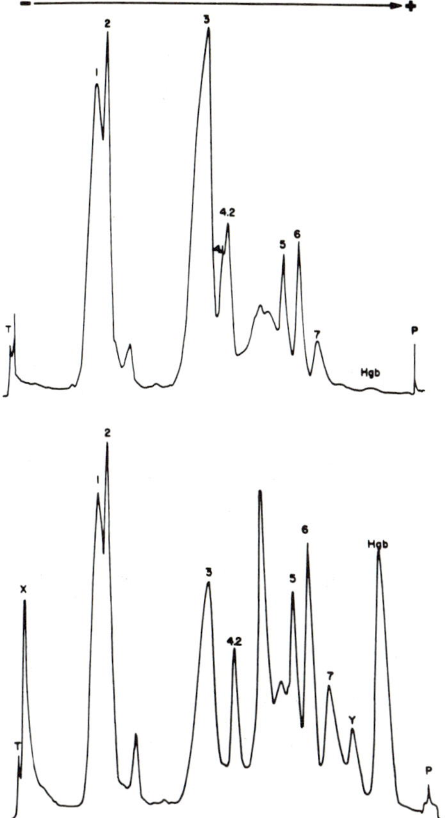


FIGURE 2. Effect of $Ca^{2+}$ loading on human RBC membrane protein electrophoretic patterns. SDS-PAGE patterns of human RBC membrane proteins. Fresh RBCs were washed and then incubated at 37° for 4 hours in isotonic NaCl (10% hematocrit) with $1 \times 10^{-5}$ M A23187 in the absence (top panel) and presence (lower panel) of 3mM $CaCl_2$. Membranes were then isolated. Solubilized material was electrophoresed, stained and numbered according to the method of Fairbanks *et al.*[13] using 4% gels.

band 8 was a proteolytic digestion product. They estimated the molecular weight of band 8 to be 24,000. Based on $R_f$ value in our system we would estimate 17,000. On the basis of the previous and present results it is suggested that band 8 and band Y are the same and could represent CaM. The appearance of a substantial peak of what appears to be CaM suggests that with 3 mM $Ca^{2+}$ a large amount of CaM binds to the membrane. It should be noted that this is much more than necessary to activate the $Ca^{2+}$ pump ATPase. A similar peak is not visible on gels of fully activated membranes isolated without preincubation.[30]

Experiments were designed to investigate the possible loss of ATPase activity in $Ca^{2+}$ loaded RBCs. Outdated, washed RBCs were incubated for 4 hours at 37°C in the presence of A23187 ($1 \times 10^{-5}$ M) and various amounts of added $CaCl_2$. Subsequently prepared membranes were tested for various ATPase

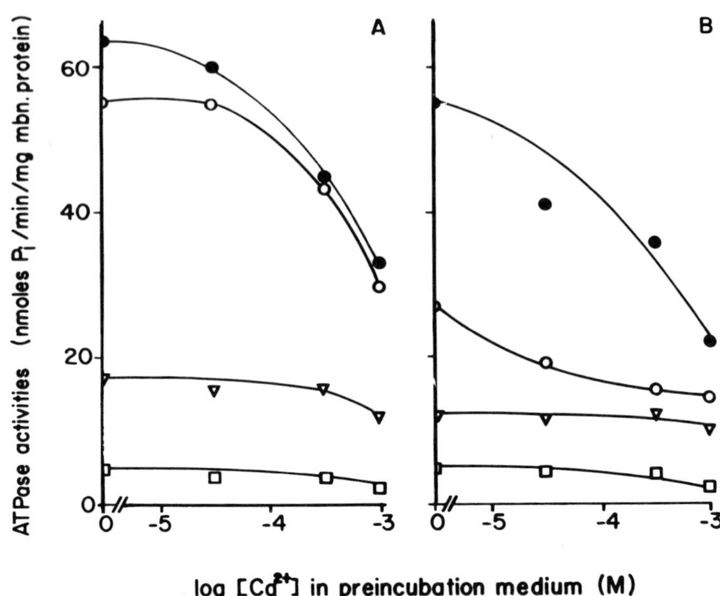


FIGURE 3. Influence of preincubation with A23187 and $CaCl_2$ on ATPase activities of human RBC membranes: Comparison of normal and EGTA washed membranes. Outdated RBCs were preincubated for 4 hours at 37° with $1 \times 10^{-5}$ M A23187 at various nominal $[Ca^{2+}]$s. Various ATPase activities were determined in membranes isolated using 20 mM imidazole (Panel A) or 20 mM imidazole with 1 mM EGTA in the first two hemolysates (Panel B). $(Ca^{2+}+Mg^{2+})$-ATPase (○), $(Na^{+}+K^{+}+Mg^{2+})$-ATPase (▽) and $(Mg^{2+})$-ATPase (□) activities were determined in the absence of added CaM and $(Ca^{2+}+Mg^{2+})$-ATPase activity was also determined in the presence of added CaM (●). Results show that CaM binds to RBC membranes during preincubation and/or hemolysis following preincubation and can be removed by exposure to EGTA. Results also show selective loss of $(Ca^{2+}+Mg^{2+})$-ATPase activity and its responsiveness to CaM as a function of preincubation $[Ca^{2+}]$.

activities. In one such experiment (not shown) $Ca^{2+}$ dependent loss of $(Ca^{2+}+Mg^{2+})$-ATPase activity was observed. None of the membranes appeared to respond significantly to CaM added during the ATPase assay. Presumably, CaM was already bound to these membranes during preincubation and had not been removed.[8,9] In a similar experiment RBCs were preincubated identically and were then divided into two aliquots. One aliquot was used for standard preparation of membranes (20 mM imidazole) whereas the second aliquot was used for preparation of membranes in which the first two hemolysis buffers also contained 1 mM EGTA. From earlier experience it was anticipated that exposure to EGTA would result in removal of endogenous CaM.[9] Results obtained from this experiment are depicted in FIGURE 3. Panel A shows various ATPase activities of membranes prepared by standard hemolysis. The results are qualitatively comparable to those described above. Closed symbols represent $(Ca^{2+}+Mg^{2+})$-ATPase activities that were assayed in the presence of maximally activating amounts of CaM. Panel B shows ATPase activities from membranes exposed to EGTA during hemolysis. These membranes showed considerably lower ($Ca^{2+}$ + $Mg^{2+}$)-ATPase activity in the absence of added CaM (○) but could be stimulated by CaM(●).

Comparison of A and B shows that exposure to ionophore, even in the absence of added $CaCl_2$ resulted in CaM binding. In addition, in Panel B it should be noted that membranes from RBCs which were preincubated with high $CaCl_2$ and ionophore showed decreased ($Ca^{2+}$ + $Mg^{2+}$)-ATPase and markedly decreased activation by added CaM. SDS-PAGE revealed a high molecular weight polymer[28] only in membranes from cells exposed to $3 \times 10^{-4}$ and $1 \times 10^{-3}$ M $Ca^{2+}$.

From FIGURE 2 it is clear that under conditions of high intracellular [$Ca^{2+}$] endogenous proteins, such as hemoglobin, can associate with the membrane. Since ATPase activities in FIGURE 3 are expressed on the basis of membrane protein, erroneous specific activities could result from this binding. To assess this, acetylcholinesterase activity and sialic acid content was determined. The results (not shown) demonstrated that, based on such indices, $Ca^{2+}$ loading of RBCs yielded membranes that possess impaired $Ca^{2+}$ pump ATPase.[15] These results show that exposure to relatively high [$Ca^{2+}$]s on the inside of the RBC have detrimental effects not only on the gross protein structure of the membrane but also on functional aspects of the membrane that are presumably responsible for maintenance of low [$Ca^{2+}$] inside. It may be that some of this loss of activity is due to disulfide bond formation.[31] This could produce functional changes that would not be apparent on gels such as in FIGURE 2, which were run under reducing conditions.

## CALMODULIN BINDING AND ACTIVATION OF THE CALCIUM PUMP ATPASE

An extreme case of CaM binding to RBC membranes was demonstrated in FIGURE 2. It is interesting to consider the extent of CaM binding to RBC membranes under more physiological conditions. We recently performed calculations based on solution mass action kinetics that were aimed at estimating CaM binding. The assumptions for these calculations are summarized in TABLES 1 and 2. Activation of the $Ca^{2+}$ pump ATPase occurs as a consequence of the binding of CaM to the ATPase[32] and we assumed binding only at pump sites. The apparent $K_d$ for CaM activation of the pump is approximately $4 \times 10^{-9}$,[33] far below the 3 $\mu$M[15] to 7$\mu$M[14] content of the cell. We assumed the $Ca^{2+}$ binding properties of CaM reported by Dedman *et al.*,[34] (4 equivalent and independent binding sites $K_d = 2.38 \times 10^{-6}$ M). An important question is whether one or more $Ca^{2+}$ ions is necessary and sufficient to promote CaM binding and/or activation of the $Ca^{2+}$ pump ATPase. This question was raised by Dedman *et al.* with regard to other

TABLE 1

ESTIMATION OF $CaM(Ca^{2+})_{n\geq2}$ IN THE HUMAN RBC

| Assumption | Reference |
|---|---|
| RBC[$Ca^{2+}$] = $2.5 \times 10^{-7}$ M | (35) |
| RBC[CaM] = $7 \times 10^{-6}$ M | (14) |
| $K_d$ $Ca^{2+}$ CaM = $2.38 \times 10^{-6}$ M | (34) |
| *therefore:* | |
| [$CaM(Ca^{2+})_{n\geq2}$] = $3.5 \times 10^{-7}$ M | |

TABLE 2

ESTIMATION OF CaM BINDING TO $Ca^{2+}$ PUMP SITES IN THE HUMAN RBC

| Assumption | Reference |
|---|---|
| $CaM(Ca^{2+})_{n\geq2}$ binds to pump ATPase | (18) |
| $[CaM(Ca^{2+})_{n\geq2}] = 3.5 \times 10^{-7}$ M | (18) |
| One CaM binds to one pump site | |
| Total pump sites = 400 | (36) |
| RBC volume = $1 \times 10^{-13}$ liters | |
| $K_d$ CaM = $4 \times 10^{-9}$ M | (34) |
| *therefore:* | |
| >98% of pump sites will contain CaM. | |

cell functions.[34] We calculated the percent of total CaM present in each form at various $[Ca^{2+}]$s: CaM; $CaM(Ca^{2+})_1$; $CaM(Ca^{2+})_2$; etc. The curve for percent activation of CaM-dependent $Ca^{2+}$ transport fit best with the sum of CaM $(Ca^{2+})_2$ + $CaM(Ca^{2+})_3$ + $CaM(Ca^{2+})_4$.[18] We assumed, therefore, that the binding of two or more $Ca^{2+}$ ions to CaM was necessary and sufficient for it to bind to and activate the $Ca^{2+}$ pump. We assumed that $Ca^{2+}$ induced *binding* of CaM and $Ca^{2+}$ induced *activation* of the plasma membrane $Ca^{2+}$ pump are equivalent. However, the "roll-over" phenomenon previously shown (FIGURE 1) is not predicted by such simple assumptions. It seemed appropriate to ask whether, for example, CaM $(Ca^{2+})_4$ may not bind or might not activate the ATPase, even if bound.

Another example of the "roll-over" phenomenon is shown in FIGURE 4. When CaM is present during an ATPase assay, high levels of $Ca^{2+}$ produce less than maximal activity. As will be shown, this is probably not due to decreased binding of CaM at high $[Ca^{2+}]$s. Results in FIGURE 5 were obtained with the same

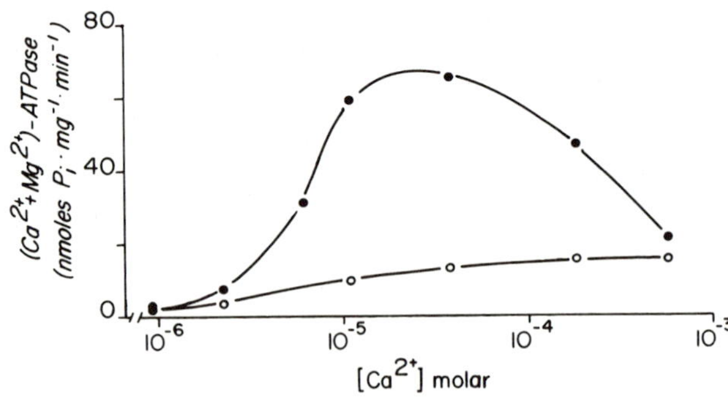


FIGURE 4. RBC $(Ca^{2+}+Mg^{2+})$-ATPase: Dependence on $Ca^{2+}$. $(Ca^{2+}+Mg^{2+})$-ATPase was determined in the absence (○——○) and presence (●——●) of added CaM (0.83 μg/ml) at various free $[Ca^{2+}]$s. In the presence of CaM the $K_{0.5}$ Ca was approximately $7 \times 10^{-6}$ M. At greater than $4 \times 10^{-5}$ M $Ca^{2+}$ less than maximal activation was seen in the presence of CaM. Results in this figure were obtained with the same membrane preparation used in Figure 5.

membrane preparation used in FIGURE 4. The membranes were preincubated with CaM at various [$Ca^{2+}$]s and were then separated from unbound CaM by centrifugation. The membranes were resuspended and ($Ca^{2+}+Mg^{2+}$)-ATPase activity was determined at optimal [$Ca^{2+}$] (preincubation, centrifugation, and resuspension had little effect on control membranes preincubated without CaM). The results show that the subsequently determined ATPase activity increased to a maximum as a function of the [$Ca^{2+}$] present during the preincubation with CaM. No "roll-over" of CaM binding was apparent. The results using purified CaM are thus in complete qualitative agreement with those of Scharff and Foder,[37] who determined the $Ca^{2+}$ dependence of the binding of endogenous CaM at the time of hemolysis of RBCs. In the results shown in FIGURE 5 the [$Ca^{2+}$] for half-maximal binding was approximately the same ($7 \times 10^{-6}$ M) as that for half-maximal activation of ($Ca^{2+}+Mg^{2+}$)-ATPase of membranes prepared from

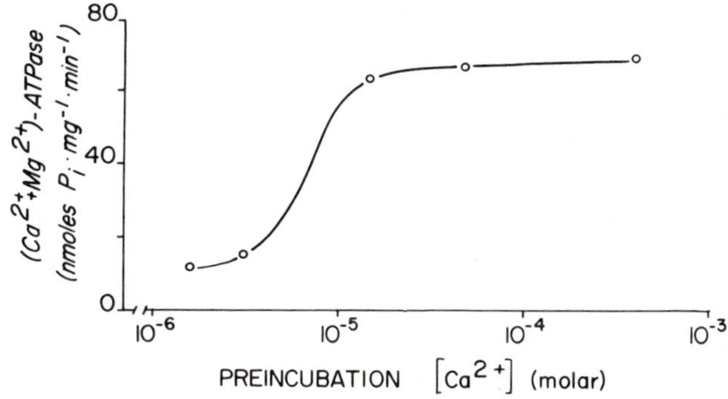


FIGURE 5. $Ca^{2+}$ dependent binding of CaM to isolated RBC membranes. Isolated RBC membranes were incubated for 30 minutes at 37° with 0.42 μg/ml CaM at various free [$Ca^{2+}$]s. Membranes were then separated by centrifugation, the supernate was removed and the pelleted membranes were resuspended and assayed for ($Ca^{2+}+Mg^{2+}$)-ATPase activity at $4 \times 10^{-5}$ M [$Ca^{2+}$] as in Figure 4. Preincubation media were identical to those for ATPase assay except for variable [$Ca^{2+}$] and the omission of ATP. Results show that the binding of purified CaM to isolated membranes remained high at high [$Ca^{2+}$].

the same cells, and determined in the presence of added CaM (FIGURE 4). The internal consistency suggests that one is looking at two manifestations of the same phenomenon in this experiment. The results probably can not be compared quantitatively with results from other workers who have determined free [$Ca^{2+}$] in other ways. We may note here the relative simplicity of being able to detect the binding of CaM to a membrane bound enzyme. Of course, more elegant binding assays are possible with radiolabeled CaM. The procedure shown here is simple and employs unmodified CaM. Obviously, comparison of isotopic and functional binding assays would be useful in determining the number of specific and nonspecific binding sites for CaM.

By inference, the results in FIGURES 4 and 5 show that the extent of CaM binding during preincubation increased as a function of [$Ca^{2+}$] and did not decrease at high levels of [$Ca^{2+}$]. Consistent with this interpretation, it was found that addition of CaM to the ATPase assay system increased the ($Ca^{2+}+Mg^{2+}$)-ATPase activity of membranes preincubated at 2.1 or $3.1 \times 10^{-6}$ M [$Ca^{2+}$] but had

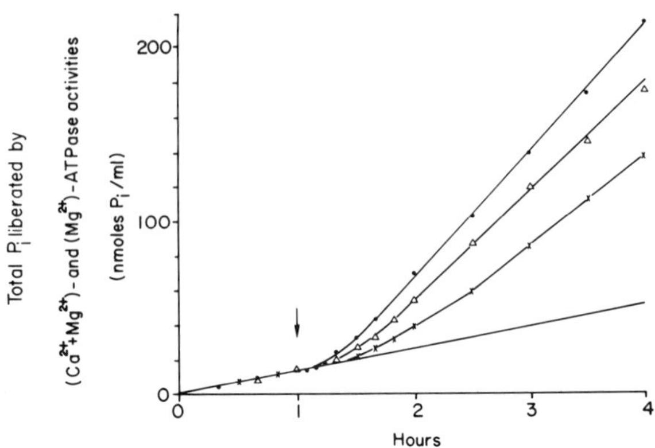


FIGURE 6. Time course of CaM activation of the $(Ca^{2+}+Mg^{2+})$-ATPase. Combined $(Ca^{2+}+Mg^{2+})$- and $(Mg^{2+})$-ATPase activities were monitored as indicated, over a 60 min period. The release of $P_i$ was linear with respect to time. At 61 minutes CaM was added to separate incubation flasks at 0.017 (X), 0.056 (Δ) and 0.17 (●) μg/ml, respectively. Results show a slow onset of the effect of CaM, especially at low [CaM]s.

little effect on the already elevated activity of membranes preincubated at $[Ca^{2+}]$ greater than $1.5 \times 10^{-5}$ M (data not shown). The results do not exclude the possibility that CaM is bound but not active at very high $[Ca^{2+}]$. We favor the interpretation that in high concentrations $Ca^{2+}$ acts at another site (probably not located on CaM) to inhibit the ATPase reaction. In view of the more pronounced "roll-over" in the presence of CaM it may be that this site is made available as a consequence of CaM binding to the ATPase.

Because the human RBC contains a great excess of CaM over the apparent $K_d$ for its binding to the $Ca^{2+}$ pump ATPase, short term regulation of the pump is probably obtained by variations in $[Ca^{2+}]$ rather than [CaM]. Presumably, if intracellular $[Ca^{2+}]$ increases, the system pumps $Ca^{2+}$ until the intracellular level falls to a point where the efflux rate matches influx. This is presumably quite slow in the human RBC compared to the maximal capacity of the pump.[17,18] Based on assumptions outlined in TABLES 1 and 2 we have concluded that, for all practical purposes, the $Ca^{2+}$ pump contains CaM as a subunit *in vivo*.[18] This conclusion is quite different from that we have expressed earlier[10] or that of Roufogalis[17] who recently reviewed the field. It should be obvious that the assumptions and calculations summarized in these tables may be incorrect in some details. While recognizing the tentativeness of the calculations it may be pointed out that the finite, albeit low $[Ca^{2+}]$ coupled with the very high [CaM] in the RBC (much higher than used in FIGURE 5, for example) is what apparently keeps CaM bound to the pump sites. If this conclusion is correct then basal $(Ca^{2+}+Mg^{2+})$-ATPase (that seen in CaM deficient membranes) may represent an artifact of membrane isolation.

## TIME COURSE OF CALMODULIN ACTIVATION OF $(Ca^{2+}+Mg^{2+})$-ATPASE

We had originally imagined that CaM binding to the ATPase is $Ca^{2+}$ dependent and rapidly reversible.[10] Models for CaM regulation of phosphodiesterase have been similar. However, as shown in FIGURE 6, the addition of a modest [CaM] to $(Ca^{2+}+Mg^{2+})$-ATPase assay system does not result in an immediate effect. FIGURE 7 shows $(Mg^{2+})$-ATPase, basal $(Ca^{2+}+Mg^{2+})$-ATPase, and CaM activated $(Ca^{2+}+Mg^{2+})$-ATPase activities as a percent of the total $P_i$ liberated during a 240 min period. The abscissa represents time after a 60 min preincubation with ATP but no CaM (panel A) or CaM but no ATP (panel B). In panel A, at time zero, CaM was added to give a final concentration of 0.056 $\mu$g/ml or 0.17 $\mu$g/ml. Appropriate control additions of $H_2O$ were made to the flasks with basal $(Ca^{2+}+Mg^{2+})$-ATPase and $(Mg^{2+})$-ATPase. In panel B membranes were preincubated for 60 min in the presence of 0.056 $\mu$g/ml or 0.17 $\mu$g/ml CaM and the reaction was started with the addition of ATP. The thin line slopes in panel B are the same data as in panel A, transposed for comparison, and represent $(Mg^{2+})$- and $(Ca^{2+}+Mg^{2+})$-ATPase activities respectively.

The obvious result from these experiments is that addition of CaM to the active basal $(Ca^{2+}+Mg^{2+})$-ATPase resulted in nonlinear ATP hydrolysis with respect to time (FIGURE 6 & FIGURE 7A). A similar result (not shown) was obtained if CaM was preincubated with membranes in the presence of $Mg^{2+}$ and ATP, but no $Ca^{2+}$. When $Ca^{2+}$ was added a slow onset of effect was seen. On the other hand, (FIGURE 7B) when membranes preparation were preincubated for 60 min with various amounts of CaM in the presence of $Ca^{2+}$ and then initiated with ATP, linear rates of $P_i$ liberation were observed. Thus, at [CaM]s below those needed for maximal activation of the $(Ca^{2+}+Mg^{2+})$-ATPase, CaM activation of the enzyme is not instantaneous.

As shown in FIGURE 8, it could be determined, although indirectly, that the nonlinearity seen in FIGURES 6 and 7A was probably due to a slow association

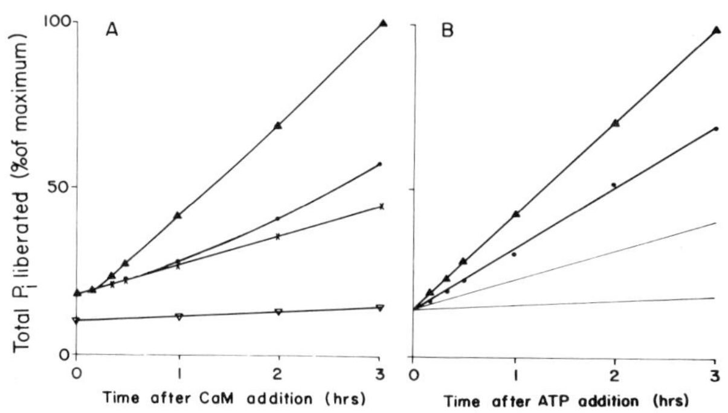


FIGURE 7. Time course of $(Ca^{2+}+Mg^{2+})$-ATPase activation: Preincubation with ATP or CaM. Data shown in Panel A are similar to those presented in Figure 6. $(Mg^{2+})$-ATPase and three different $(Ca^{2+}+Mg^{2+})$-ATPase reaction mixtures were preincubated for one hour in the presence of ATP (3mM). At 61 min two $(Ca^{2+}+Mg^{2+})$-ATPase reaction mixtures received CaM at final concentrations of 0.056 (●) and 0.17 (▲) $\mu$g/ml, respectively. $(Mg^{2+})$-ATPase (▽) and $(Ca^{2+}+Mg^{2+})$-ATPase in the absence of CaM (X) were also monitored for an additional 3 hours as indicated. Data in Panel B were obtained following preincubation with CaM at final concentrations of 0.056 (●) and 0.17 (▲) $\mu$g/ml for one hour. At 61 min the reactions were started with the addition of ATP (3mM) and monitored for 3 hours. Thin lines are the same slopes as in Panel A for $(Mg^{2+})$- and $(Ca^{2+}+Mg^{2+})$-ATPase respectively. Data are expressed as % of maximum $P_i$ liberated at 3 hours by the maximally activated ATPase. Results show that the influence of CaM on $(Ca^{2+}+Mg^{2+})$-ATPase activity is dependent on time and [CaM].

between the two macromolecules (CaM and ATPase). In FIGURE 8 preincubation times with an approximately half-maximal concentration of CaM were 5 min, 10 min, and 20 min, respectively. A time lag to a linear $(Ca^{2+} + Mg^{2+})$-ATPase rate was observed with the 5 and 10 min preincubations. When membranes were preincubated for 20 min, linearity was obtained almost immediately upon ATP addition.

From these experiments it becomes clear that sufficient preincubation is critical when CaM activation of $(Ca^{2+} + Mg^{2+})$-ATPase is to be measured. Apparently the binding reaction is fairly slow under the conditions of the ATPase assay, even at an optimal $[Ca^{2+}]$. The rate of dissociation of CaM from membranes is presumably very slow at high $[Ca^{2+}]$s, considering the apparent affinity of CaM for the site to which it binds to activate the ATPase, Dissociation is probably more rapid at low $[Ca^{2+}]$s. Nevertheless, the relatively slow binding of CaM to RBC

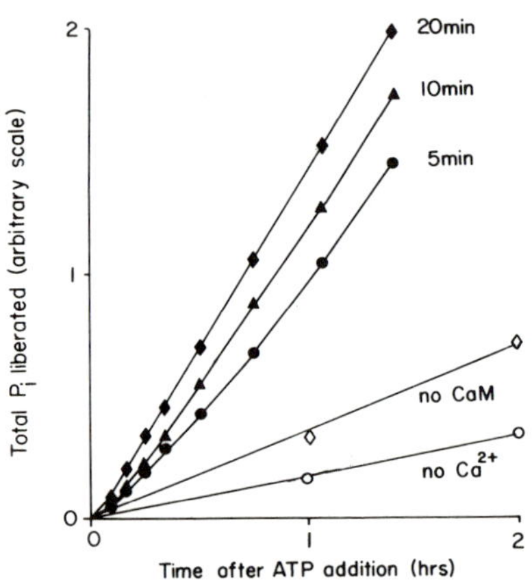


FIGURE 8. Calmodulin activation of RBC membrane $(Ca^{2+} + Mg^{2+})$-ATPase: Dependence on preincubation time. $(Mg^{2+})$-ATPase (○) and $(Ca^{2+} + Mg^{2+})$-ATPase (◇) in the absence of added CaM were monitored for two hours. $(Ca^{2+} + Mg^{2+})$-ATPase was also determined in membranes preincubated with 0.056 $\mu$g/ml CaM for 5 (●), 10 (▲) and 20 (◆) min, respectively. ATP was added to start the reaction at time zero.

membranes under our experimental conditions prompts us to think in terms of reversibly bound CaM as a subunit of the $Ca^{2+}$ pump rather than as on-off switch for each cycle of the pump *in vivo*. Whether basal ATPase and pump activities of isolated membranes are dependent on irreversibly bound CaM is less certain in our minds. More experiments on these questions are needed. In particular, high [CaM] and low $[Ca^{2+}]$ would better exemplify the conditions thought to be present in the RBC *in vivo*.

## SUMMARY

The data summarized and presented in this paper are consistent with the interpretation that CaM participates in the regulation of the plasma membrane calcium pump. Certain drugs, such as phenothiazines can antagonize CaM. $Ca^{2+}$

loading of RBCs promotes CaM binding to RBC membranes and results in decreased responsiveness of the $(Ca^{2+} + Mg^{2+})$-ATPase to CaM. The latter effect may be mediated by a $Ca^{2+}$ activated transglutaminase. Activation of $(Ca^{2+} + Mg^{2+})$-ATPase by CaM *in vitro* was shown not to be instantaneous, probably because of slow binding. CaM binding to isolated RBC membranes exhibits a $Ca^{2+}$ dependence that is similar to that for activation of the $(Ca^{2+} + Mg^{2+})$-ATPase, and CaM binding does not decrease at high $[Ca^{2+}]$s. Calculations based on assumed values for RBC $[Ca^{2+}]$, [CaM], and binding affinities of $Ca^{2+}$ for CaM and $CaM(Ca^{2+})_n$ for the $Ca^{2+}$ pump ATPase resulted in the tentative conclusion that most pump sites are occupied by CaM in the RBC *in vivo*. This conclusion, and the relatively slow time course of the CaM effect on $(Ca^{2+} + Mg^{2+})$-ATPase prompt us to suggest that, for all practical purposes, CaM is a subunit of the $Ca^{2+}$ pump ATPase *in vivo*.

## REFERENCES


1. BOND, G. H. & D. L. CLOUGH. 1973. Biochim. Biophys. Acta **323:** 592–599.
2. SCHATZMANN, H. J. & F. F. VINCENZI. 1969. J. Physiol. (London) **201:** 369–395.
3. LEE, K. S. & B. C. SHIN. 1969. J. Gen. Physiol. **54:** 713–729.
4. LUTHRA, M. G., G. R. HILDENBRANDT & D. J. HANAHAN. 1976. Biochim. Biophys. Acta **419:** 164–179.
5. JARRETT, H. W. & J. T. PENNISTON. 1978. J. Biol. Chem. **253:** 4676–4682.
6. GOPINATH, R. M. & F. F. VINCENZI. 1977. Biochem. Biophys. Res. Commun. **17:** 1203–1209.
7. LARSEN, F. L., B. U. RAESS, T. R. HINDS & F. F. VINCENZI. 1978. J. Supramolec. Struct. **9:** 269–274.
8. FARRANCE, M. L. & F. F. VINCENZI. 1977. Biochim. Biophys. Acta **471:** 49–58.
9. FARRANCE, M. L. & F. F. VINCENZI. 1977. Biochim. Biophys. Acta **471:** 59–66.
10. VINCENZI, F. F. 1978. Ann. N.Y. Acad. Sci. **307:** 229–231.
11. RAESS, B. U. & F. F. VINCENZI. 1980. J. Pharmacol. Methods. (In press.)
12. SIMON, W., D. AMMANN, M. OEHME & W. E. WORF. 1978. Ann. N.Y. Acad. Sci. **307:** 52–70.
13. FAIRBANKS, G., T. L. STECK & D. F. H. WALLACH. 1971. Biochemistry **10:** 2606–2617.
14. JUNG, N. S. G. T. 1978. Master's Thesis. University of Washington, Seattle.
15. RAESS, B. U. 1980. Doctoral Dissertation. University of Washington, Seattle.
16. QUIST, E. E. & B. D. ROUFOGALIS. 1975. Arch. Biochem. Biophys. **168:** 240–251.
17. ROUFOGALIS, B. D. 1979. Canad. J. Physiol. Pharmacol. **57:** 1331–1349.
18. VINCENZI, F. F. & T. R. HINDS. 1980. Calmodulin and plasma membrane calcium transport. *In* Calcium and Cell Function, W. Y. Cheung, Ed. **1:** 127–165. Academic Press, Inc., New York, N.Y.
19. LEVIN, R. M. & B. WEISS. 1977. Molec. Pharmacol. **13:** 690–697.
20. WEISS, B. & R. M. LEVIN. 1978. Mechanisms for selectively inhibiting the activation of cyclic nucleotide phosphodiesterase and adenylate cyclase by antipsychotic agents. *In* Advances in Cyclic Nucleotide Research. W. J. GEORGE & L. J. IGNARRO, Eds. Vol. **9:** 285–303. Raven Press. New York, N.Y.
21. NORMAN, J. A., A. H. DRUMMOND & P. MOSER. 1979. Molec. Pharmacol. **16:** 1089–1094.
22. RAESS, B. U. & F. F. VINCENZI. 1980. Molec. Pharmacol. (In press.)
23. LAPORTE, D. C. & D. R. STORM. 1980. Biochemistry (In press.)
24. WOLPI, M., R. I. SHA'AFI & M. B. FEINSTEIN. 1980. Fed. Proc. **39:** 958.
25. HINDS, T. R., B. U. RAESS & F. F. VINCENZI. 1980. J. Membrane Biol. (In press.)
26. EATON, J. W., T. D. SKELTON, H. S. SWOFFORD, C. E. KOLPIN & H. S. JACOB. 1973. Nature (London) **246:** 105–106.
27. GOPINATH, R. M. & F. F. VINCENZI. 1980. Amer. J. Hematol. **7:** 303–312.
28. LORAND, L., G. E. SIEFRING & L. LOWE-KRENTZ. 1978. J. Supramolec. Struct. **9:** 427–440.

29. ALLEN, D. W. & S. CADMAN. 1979. Biochim. Biophys. Acta **55:** 1–9.
30. FARRANCE, M. L. 1977. Doctoral Dissertation. University of Washington, Seattle.
31. PALEK, J., P. A. LIU & S. C. LIU. 1978. Nature (London) **274:** 505–507.
32. NIGGLI, V., J. T. PENNISTON & E. CARAFOLI. 1979. J. Biol. Chem. **254:** 9955–9958.
33. LARSEN, F. L. & F. F. VINCENZI. 1979. Science **204:** 306–309.
34. DEDMAN, J. R., J. D. POTTER, R. L. JACKSON, J. D. JOHNSON & A. R. MEANS. 1977. J. Biol. Chem. **252:** 8415–8422.
35. SIMONS, T. J. B. 1976. J. Physiol. (London) **256:** 227–244.
36. DRICKAMER, L. K. 1975. J. Biol. Chem. **250:** 1925–1954.
37. SCHARFF, O. & B. FODER. 1977. Biochim. Biophys. Acta **483:** 416–424.

## DISCUSSION OF THE PAPER

DR. J. PENNISTON (*Mayo Foundation, Rochester, MN*): Where you show a slow interaction between calmodulin and the vesicles, what kind of vesicles were you using? Had they been frozen or otherwise treated to reduce permeability barriers that might interfere with the interaction of calmodulin and the pump site? Is it possible that the slowness of the interaction might not be entirely due to the on rate at the pump site; that there might be barriers due to the membrane?

DR. F. F. VINCENZI: I think I would worry about the point you are raising unless I consider Wang's data with the isolated enzyme or unless I recall that calmodulin stimulates the ATPase and transport on the outside of these vesicles.

DR. M. G. LUTHRA: What kind of concentration of calmodulin are you talking about when you see this delayed effect?

DR. VINCENZI: Approximately .05 microgram/ml to .1 microgram/ml of calmodulin will give a slow onset. I think if you go back and look at the old phosphodiesterase papers which showed "instantaneous" activation you will see that rather high concentrations of calmodulin were employed.

DR. C. KLEE: I wanted to ask a question about the inhibition at high calcium concentration. Have you looked if it is possibly due to the presence of protease? Is it reversible if you lower the calcium concentration?

DR. VINCENZI: I think that calmodulin, when it binds to the ATPase, changes the conformation of the ATPase. When calmodulin has bound, there becomes available a site to which calcium binds with fairly low affinity. When it binds to that site it inhibits the enzyme. However, the experiment you suggest is obvious and easy and would give a different interpretation.

DR. KLEE: Yes, because that conformational change would be reversible.

DR. VINCENZI: Whereas the effects of a protease would not. Good point.

DR. D. R. STORM: I wanted to make a comment, actually directed towards the question Dr. Penniston raised. Both the data showing a time lag with the ATPase and the phosphodiesterase are completely consistent with diffusion as a regulating step. This is a novel situation. Here, the regulator is not a small molecule. This is a protein diffusing in solution and that's of physiological interest. So, I think it is quite consistent with what we would expect.

# CALMODULIN REGULATION OF THE $Ca^{2+}$ PUMP OF ERYTHROCYTE MEMBRANES*

John T. Penniston, Ernst Graf, and Toshifumi Itano

*Section of Biochemistry*
*Department of Cell Biology*
*Mayo Foundation*
*Rochester, Minnesota 55901*

## INTRODUCTION

The $Ca^{2+}$ pumping ATPase of plasma membranes is an important mechanism by which the intracellular concentration of $Ca^{2+}$ is controlled. In mature mammalian erythrocytes,[1] and in squid giant axon[2] this ATP requiring $Ca^{2+}$ pump is the main mechanism by which $Ca^{2+}$ is removed from the cell. Only the ($Ca^{2+}$-$Mg^{2+}$)-ATPase from erythrocytes has been studied extensively, but $Ca^{2+}$ requiring ATPases with similar properties have been found in plasma membranes from brain,[3] luteal cells,[4] adipocytes,[5] neutrophils,[6] heart,[29] and pancreatic islet membranes.[7] Calmodulin, in one of its many roles, interacts with this $Ca^{2+}$ pump;[8] we will discuss whether this interaction is likely to affect the properties of the pump *in vivo,* and how calmodulin may regulate the pump.

Because of their possible importance in this regulation, it is necessary briefly to discuss the evidence for states of the pump with high and low affinity for $Ca^{2+}$. High and low $Ca^{2+}$ affinity forms of the $Ca^{2+}$ ATPase from erythrocyte membranes were reported at an early point in the study of this activity,[9,10] but physiological significance for the low $Ca^{2+}$ affinity form was dismissed as probably being due to denaturation of the ATPase.[11,12] The question was reopened by Quist and Roufogalis who showed that the high affinity form could be eliminated by extraction of a factor from the membranes and regenerated by restoration of the extract.[13] They interpreted these results on the basis of two different ($Ca^{2+}$-$Mg^{2+}$)-ATPases, while acknowledging the possibility of two types of $Ca^{2+}$ sites as an alternative explanation. Scharff succeeded in preparing erythrocyte membranes in either the low or the high affinity state, depending on the method of preparation, and interpreted this as corresponding to two different states of the $Ca^{2+}$ pump.[14] In his subsequent study on a reversible shift between two states of this ATPase he reiterated the position that the two affinities represent two states of the same enzyme.[15] The factor that causes this change in affinity is the same factor that was discussed above as stimulating the ATPase, namely calmodulin.[16] All of these studies of $Ca^{2+}$ affinity were done on whole erythrocyte membranes; they were subject to uncertainties about the number of $Ca^{2+}$ and $Mg^{2+}$ ATPases present and to interference due to the presence of other enzymes in the membrane. The present study on the purified $Ca^{2+}$ pumping ATPase avoids these uncertainties.

The work leading up to the purification of the erythrocyte membrane $Ca^{2+}$ pump to homogeneity has been presented elsewhere.[17,18] We do not propose to review that material again here, but rather to present data concerning the

*Supported by National Institutes of Health Grants AM21820 and AM19785 and by the Mayo Foundation.



0077-8923/80/0356-0245 $01.75/0 © 1980 NYAS

interaction of calmodulin with ghosts and with the purified enzyme, and to discuss models for calmodulin regulation of the calcium pump.

## EXPERIMENTAL METHODS

### *Calmodulin and Calmodulin-Binding Proteins in Hemolysate*

For calmodulin analysis, hemolysates were obtained from the Mayo Clinic red blood cell laboratory, and were heat-treated by immersion of a 0.5-ml sample contained in a 12 × 75 mm disposable borosilicate culture tube in a boiling water bath for 2 min. The denatured protein was removed by centrifugation in a clinical centrifuge and the supernatant solution analyzed for calmodulin as previously described.[19]

For detection of calmodulin-binding proteins, a calmodulin-Sepharose column was prepared as previously described[17] and hemolysate was prepared by lysis of erythrocytes in a medium that was 40 mM in Tris-HCl (pH 7.5), 40 mM in imidazole, 3 mM in magnesium acetate and 0.1 mM in $CaCl_2$. Ghosts were removed from the hemolysate by centrifugation, and the supernatant was passed through the column; the vast majority of the protein did not bind to the column, which was washed with the same buffer until the protein eluted reached background levels. The bound proteins were then eluted from the column with solutions that were 1 mM in EGTA, 1 M in KCl plus 1 mM EGTA, and 6 M in urea. The first two solutions also contained the substances present in the hemolysis buffer, while the urea solution contained only urea and water. Protein was determined as previously described.[17]

### *Binding of [$^{125}$I]Calmodulin to Membranes*

Preparation of [$^{125}$I]calmodulin and performance of the binding experiment were as described.[20] The medium for the binding experiment contained 8 mg/ml bovine serum albumin, 6 mM $MgCl_2$, 0.8 mM $CaCl_2$, 100 mM NaCl, 20 mM KCl, 0.5 mM EGTA, 0.1 mM ouabain, 25 mM in a buffer composed of tris(hydroxymethyl)methyl-2-aminoethanesulfonic acid neutralized with triethanolamine, pH 7.4 (TES-TEA), and 0.3–0.5 mg/ml frozen and thawed ghosts, prepared as described previously.[21]

For the study of the dependence of binding on pCa, the above mixture was made 5.0 mM in EGTA, and the $CaCl_2$ concentration was varied to achieve the desired pCa. Free $Ca^{2+}$ was calculated using a computer program[22] which took account of all species involved in the equilibria between $H^+$, $Mg^{2+}$, $Ca^{2+}$, ATP, and EGTA; the affinity constants used were those for 25°C and 0.1 ionic strength taken from the compendia of Martell and Smith,[23,24] except that a value of $10^{4.85}$ was used for MgATP,[25] and $10^{10.36}$ for Ca-EGTA.[26]

### *ATPase Assays*

Free $Ca^{2+}$ was calculated and ghosts prepared as described above. Purified ATPase was prepared by a modification of the previously described method.[17] The principal modification was the introduction of a wash of the calmodulin-

Sepharose column with a solution containing 1 μM free $Ca^{2+}$ before elution of the ATPase with a $Ca^{2+}$-free solution. Assay of the ATPase of the ghosts was done in a medium that was 25 mM in TES-TEA (pH 7.4), 0.1 mM EGTA, 6 mM $MgCl_2$, and 6 mM ATP; incubation was 30 min at 37°C and release of $^{32}P_i$ from [γ-$^{32}$P]ATP was measured.[27] Calmodulin, when used, was 0.6 μg/ml. Because of the low EGTA concentration used, the calculated free $Ca^{2+}$ was checked by electrode (Radiometer F2112) at all free $Ca^{2+}$ concentrations above 1 μM. The measured values are plotted in FIGURE 1 (top).

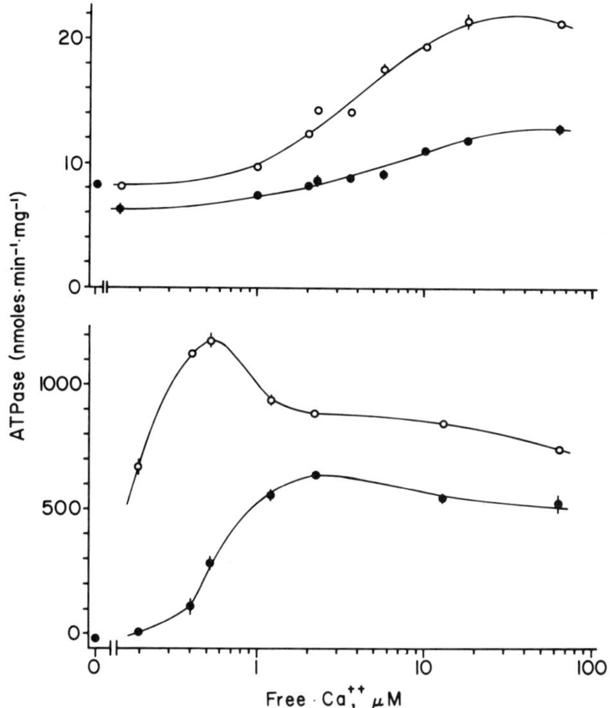


FIGURE 1. Effect of calmodulin on affinity for $Ca^{2+}$. Effects on the $Ca^{2+}$ ATPase of erythrocytes (top) and pure ATPase (bottom) are shown. Calmodulin increased the affinity for $Ca^{2+}$ of the Triton-solubilized pure ATPase. Open circles, with calmodulin; filled circles, without calmodulin. Where significant, standard errors of the mean are shown. (Three samples were averaged to obtain each point). Blanks containing no protein were subtracted from all samples.

The assays of the purified ATPase for FIGURE 1 were performed under the same conditions as for the ghosts, except that the EGTA concentration was raised to 6.2 mM, the free $Ca^{2+}$ was calculated, and the medium contained small amounts of Triton X-100, lipid, and KCl which were present in the ATPase preparation; the concentration of ATPase was 15 μg/ml and of calmodulin 2.4 μg/ml.

Assays of the temperature dependence of the purified ATPase were done in a medium that was 25 mM in TES-TEA, pH 7.4; 5 mM in EDTA, 1.0 mM in ATP, 12 $\mu$g/ml in enzyme, and 6 $\mu$g/ml in calmodulin (when called for). The sample tubes were 0.36 mM in free $Mg^{2+}$ and 0.44 $\mu$M in free $Ca^{2+}$, and blanks containing $Mg^{2+}$ but no $Ca^{2+}$ were subtracted (these blanks were not very different from the no ion or no protein blanks). The total concentrations required to give these results were 5.7 mM $MgCl_2$ and 0.6 mM $CaCl_2$ for the samples. Incubation times ranged from 15 min at 38°C to 130 min at 0°C.

## Results

The purified ATPase consists of a single subunit of molecular weight about 138,000; this ATPase has sites of its own capable of responding to $Ca^{2+}$ as is demonstrated by its possession of a substantial $Ca^{2+}$ ATPase activity. This ATP splitting activity is absolutely dependent upon the presence of $Ca^{2+}$ (Table 1) and, when reconstituted into liposomes, the purified ATPase accumulates $Ca^{2+}$ effectively.[18]

### *Effect of Calmodulin on Kinetic Properties of the Pump*

Since calmodulin is not required for the splitting of ATP and transport of $Ca^{2+}$ by the pump, its specific interaction with the pump probably serves a regulatory purpose. In support of this, the data on the purified enzyme show significant changes in the kinetic properties of the pump caused by addition of calmodulin.

Figure 1 compares the effect of calmodulin on the $Ca^{2+}$ pumping ATPase of ghosts and of pure enzyme. The results with ghosts are typical of those generally observed; they illustrate the difficulty of demonstrating a clear-cut change in the $K_m$ for $Ca^{2+}$ by means of studies on ghosts. When analyzed by Eadie-Hofstee plots (not shown) the ghosts in the absence of calmodulin show a mixture of high and low affinity characteristics, with a low maximal activity. However, the size of the $Mg^{2+}$ ATPase background compared to the small stimulation due to $Ca^{2+}$ makes a detailed interpretation of this plot difficult. In the presence of calmodulin the ($Ca^{2+}$-$Mg^{2+}$)-ATPase of ghosts shows a higher $Ca^{2+}$ stimulation and has a single affinity for $Ca^{2+}$.

Using the purified enzyme, it is much easier to show distinct changes caused by calmodulin. The bottom half of Figure 1 shows data obtained using the

Table 1

$Ca^{2+}$ Requirement of Ghosts and Purified ATPase

| | ATPase nmoles/(mg, min) | |
|---|---|---|
| | Ghosts | Pure ATPase |
| $Mg^{2+}$ | 5.5 | 4.6 |
| $Ca^{2+}$, $Mg^{2+}$ | 8.3 | 2227.0 |

Ghosts and pure enzyme were assayed as described in Experimental Methods. For ghosts, the concentration of free $Ca^{2+}$ was 10 $\mu$M; for the pure ATPase free $Ca^{2+}$ was 1.19 $\mu$M. A no protein blank was subtracted from all values obtained.

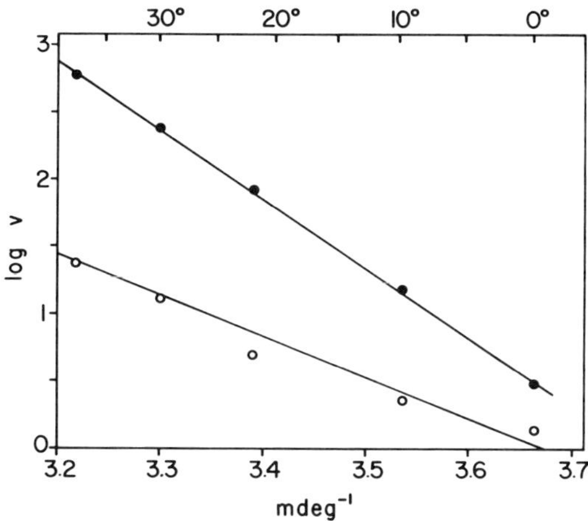


FIGURE 2. Effect of calmodulin on temperature dependence of $Ca^{2+}$ ATPase. Purified ATPase was assayed with calmodulin (filled circles) and without calmodulin (open circles). Three samples were averaged to obtain each point.

Triton-solubilized form of the pure ATPase. In this situation, the change in the apparent $K_m$ for $Ca^{2+}$ caused by calmodulin is about four-fold and the shift in the curve is better defined than is the case for ghosts. For example, at a free $Ca^{2+}$ concentration of 0.2 $\mu$M, the ATPase activity requires calmodulin absolutely. This result clearly indicates that the enzyme is capable of existing in two states, a lower $Ca^{2+}$ affinity state in the absence of calmodulin and a high $Ca^{2+}$ affinity one in its presence. Note that the $V_{max}$ of the reaction is increased substantially by calmodulin, but excess $Ca^{2+}$ inhibits the reaction so that the effect of calmodulin at very high $Ca^{2+}$ is slight. Similar differences in affinity with and without calmodulin have been obtained in Dr. Carafoli's laboratory for the ATPase reconstituted into liposomes.

In addition to causing an increase in $V_{max}$ and in $V_{max}/K_m$ for $Ca^{2+}$, calmodulin also increases the activation energy for ATP splitting by the enzyme. FIGURE 2 shows Arrhenius plots for ($Ca^{2+}$-$Mg^{2+}$)-ATPase with and without calmodulin. The energy of activation, which was 12.8 kcal/mol without calmodulin, nearly doubled, to 23.4, in the presence of calmodulin. The significance of this change is not clear, but it is further evidence that calmodulin causes substantial changes in the catalytic properties of the $Ca^{2+}$ pump.*

*In order to understand how a reaction can be catalyzed more efficiently and, at the same time, have a higher energy of activation, it is necessary to consider the form of the Arrhenius equation:

$$K = A\, e^{-E_a/RT}$$

where $K$ is the rate constant, A is a temperature independent factor, and $E_a$ is the activation energy. According to activated-complex theory, $E_a$ is nearly the same as the enthalpy of activation, while the factor A contains the entropy of activation, $\Delta S\ddagger$. In the case considered here, $E_a$ changes in a direction that would decrease the rate, while $\Delta S\ddagger$ changes in a way that would more than compensate for the changes in $E_a$.

*Calmodulin-Binding Sites in the Erythrocyte*

Since calmodulin significantly affected the properties of the calcium pump, it was desirable to know whether calmodulin was bound to the calcium pump in a normal erythrocyte. In order to approach this problem, it was necessary to consider the factors that influence the concentration of free calmodulin, $Ca^{2+}$, and binding sites. It is not possible to obtain a definitive resolution of this problem, but we will present here some relevant data.

When measured under controlled conditions, erythrocyte membranes prepared in EDTA contained about $4.1 \times 10^3$ high affinity calmodulin-binding sites per ghost, and the association constant of calmodulin to these high affinity sites (in the presence of excess $Ca^{2+}$) was about 5 nM.[20] When converted to a volume basis, (using the total volume of the erythrocyte), this number of high affinity binding sites represents a concentration of about .07 $\mu$M. The concentration of these high $Ca^{2+}$ affinity binding sites is very much less than the total concentration of calmodulin in the cell. This latter number, estimated for erythrocytes from 20 normal human subjects, was $0.174 \pm 0.055$ (standard deviation) $\mu$g calmodulin/mg of hemoglobin. Assuming 350 mg hemoglobin per ml of red cells, this converts to a calmodulin concentration of 3.6 $\mu$M, about a 50-fold excess of calmodulin over high affinity binding sites. If this much calmodulin was free in the erythrocyte, the calcium pump would be loaded with calmodulin even at the extremely low $Ca^{2+}$ levels that exist inside a normal erythrocyte.

In order to evaluate the actual amount of free calmodulin in the erythrocyte, let us consider the evidence for the presence of other binding sites for calmodulin. FIGURE 3 shows a Scatchard plot of the binding of iodocalmodulin to

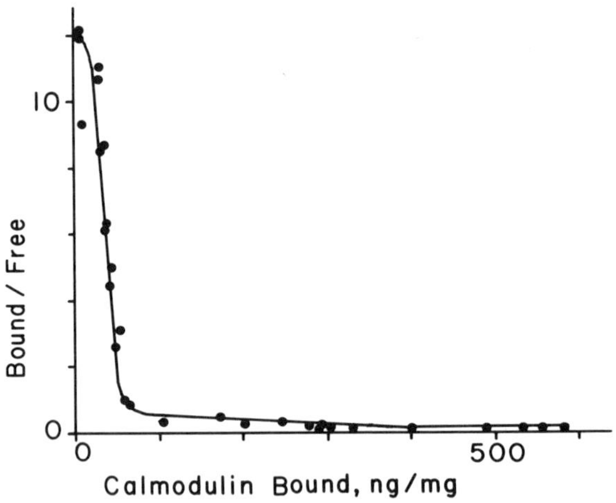


FIGURE 3. Binding of [$^{125}$I]calmodulin to erythrocyte ghosts. Two components of binding were seen, while the binding above 450 ng/mg showed no evidence of saturability. The maximum total calmodulin concentration utilized for this study was 3.7 $\mu$M, which corresponded to our estimate of the total calmodulin concentration in erythrocytes (see Text). For the vertical axis, the unit used for free calmodulin was nM while the unit used for bound was ng/mg.

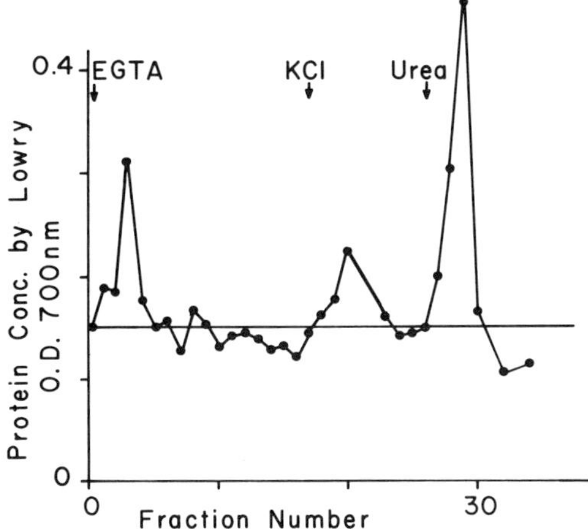


FIGURE 4. Elution of proteins which bound to a calmodulin-Sepharose column. Hemolysate was passed through the column in a $Ca^{2+}$-containing medium. The vast bulk of the protein was not bound; the figure shows subsequent elution from the column of the bound protein. The EGTA peak contained 83 ng protein per mg hemoglobin in hemolysate; the subunit molecular weights of the eluted proteins were 20,000, 50,000, and 100,000. The corresponding numbers for the KCl and urea eluates were 52 ng/mg of 70,000 MW and 161 ng/mg of 70,000–400,000 MW.

erythrocyte membranes. In this experiment the total calmodulin was increased to a concentration equal to that found in intact erythrocytes. In addition to the high affinity sites, there was a class of binding sites with a lower affinity for calmodulin ($K_d$ about 0.8 $\mu$M). In this batch of erythrocyte membranes the total binding to high and low affinity sites was 436 ng calmodulin/mg membrane protein. The calmodulin bound in excess of this amount did not show saturability and probably represented trapped, rather than bound, calmodulin. Using the same assumptions as were utilized for the high affinity sites,[20] this would correspond to about 0.2 $\mu$M binding sites (both high and low affinity) on the membrane, still a small concentration compared with the total concentration of calmodulin.

A larger number of binding sites was disclosed by analysis of hemolysate using a calmodulin affinity column. Only those proteins that interact with calmodulin should be retained by this column, and they may be eluted by solutions containing $Ca^{2+}$ chelators and salts. Specific calmodulin-binding proteins (such as the calmodulin-requiring phosphodiesterase) are eluted by EGTA, while strongly basic (nonspecific) calmodulin-binding proteins such as histone H2B are eluted primarily by 1 M KCl, with some of the histone remaining bound to the column until elution by urea. FIGURE 4 shows the elution of a significant amount of calmodulin-binding proteins from this column. Both specific and nonspecific calmodulin-binding proteins were present in the hemolysate in a substantial amount. Based on the amounts and molecular weights observed in each peak, and the minimal assumption that each mole of protein could bind 1 mole of calmodulin, there would be present in the hemolysate 1.5 $\mu$M calmodulin binding sites. If any of the proteins bind more than 1 mole of

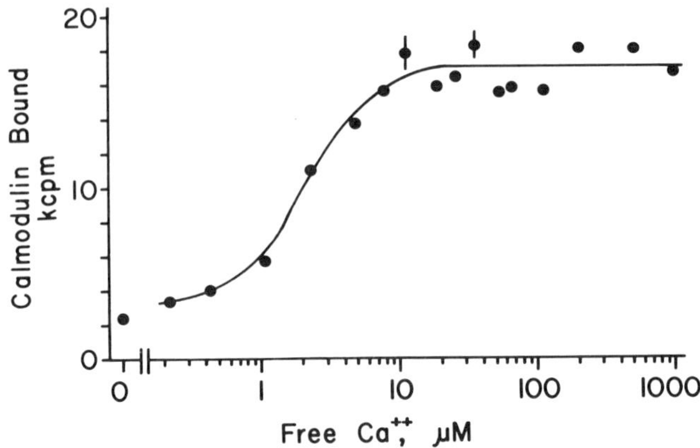


FIGURE 5. Binding of [$^{125}$I]calmodulin to erythrocyte ghosts. Total calmodulin was 0.50 μg per 1 ml assay, which also contained about 0.5 mg of ghost protein. Note that a higher calmodulin concentration would shift the curve to the left.

calmodulin per mole, the concentration of binding sites would be greater. The data presented are certainly not adequate for an accurate estimate of the number of calmodulin binding sites present in an erythrocyte, but it is clear that the total number of binding sites is comparable to the number of calmodulin molecules and that a considerable portion of these proteins does not require $Ca^{2+}$ for its binding of calmodulin. Thus the amount of free calmodulin in the cytosol of the erythrocyte is probably much less than the total calmodulin concentration.

FIGURE 5 shows the calcium dependence of the binding of calmodulin to erythrocyte membranes in the presence of an amount of calmodulin just sufficient to saturate the high affinity binding sites. These sites, which behave as if they are the calcium pumping ATPase, are half saturated at about 2 μM free calcium. Since the level of free intracellular $Ca^{2+}$ is less than 0.25 μM,[28] it is reasonable to suppose that under normal intracellular conditions most of the calcium pumping sites are not loaded with calmodulin, as was suggested by Roufogalis.[1]

## DISCUSSION

Our knowledge concerning the plasma membrane $Ca^{2+}$ pump has rapidly expanded in recent years. The availability of a well defined regulatory protein, calmodulin, has made possible purification of this enzyme and has opened the door to molecular studies of the ATPase.

It is certainly clear that both low $Ca^{2+}$ affinity and high $Ca^{2+}$ affinity states of the ATPase exist, and that the enzyme can be switched from one to the other by calmodulin. This property of the pure enzyme may account for the high and low calcium affinity states observed in ghosts by previous workers,[12-15] and discussed in the introduction.

The affinity of calmodulin for its binding sites on erythrocyte membranes is very high; sufficiently high that it will bind to the ATPase at low free $Ca^{2+}$

concentrations. The experiment shown in FIGURE 5 used 0.03 $\mu$M calmodulin, just enough to saturate the high affinity binding sites in the presence of excess $Ca^{2+}$. Under these circumstances, the calmodulin was fully bound at concentrations of $Ca^{2+}$ that would occur after a trauma such as sickling of the erythrocyte. In the intact erythrocyte, at least 0.03 $\mu$M calmodulin will be available to regulate the $Ca^{2+}$ pump; thus, we can be sure that calmodulin will bind to the $Ca^{2+}$ pump at least as well as shown in FIGURE 5. This assumes that the properties of the binding sites in isolated membranes have not been altered from those of the intact cell. Calmodulin bound to the pump *in vivo* will make the pump more effective in removal of $Ca^{2+}$ from the cell.

Since we have already seen that the $Ca^{2+}$ pump can operate in the absence of calmodulin, it is interesting to speculate whether this ability of the enzyme is put to any physiological use. The model discussed by Roufogalis[1] is attractive because it proposes a physiological role for this capacity of the pump.

His model suggests that in the resting state of the red blood cell, calmodulin does not bind to the $Ca^{2+}$ pump and the pump, therefore, has a low affinity for $Ca^{2+}$. In this state only a small proportion of the $Ca^{2+}$ binding sites on the pump would be loaded at any given time. FIGURE 6 shows a conception of the $Ca^{2+}$ pump in this resting state; under these circumstances, the pump would rely on its unassisted ability to remove $Ca^{2+}$. This situation would result in a very low rate of $Ca^{2+}$ efflux from the cell, but the excess capacity of the pump appears to be sufficiently great that it could easily keep up with normal levels of $Ca^{2+}$ influx.[1] If, because of some traumatic event, such as a sickle cell crisis, the $Ca^{2+}$ concentration became very much higher, calmodulin would become fully loaded with $Ca^{2+}$ and would bind to the $Ca^{2+}$ pump sites as shown in FIGURES 7 and 8. In this state, the pump would remove the excess $Ca^{2+}$ with maximal efficiency.

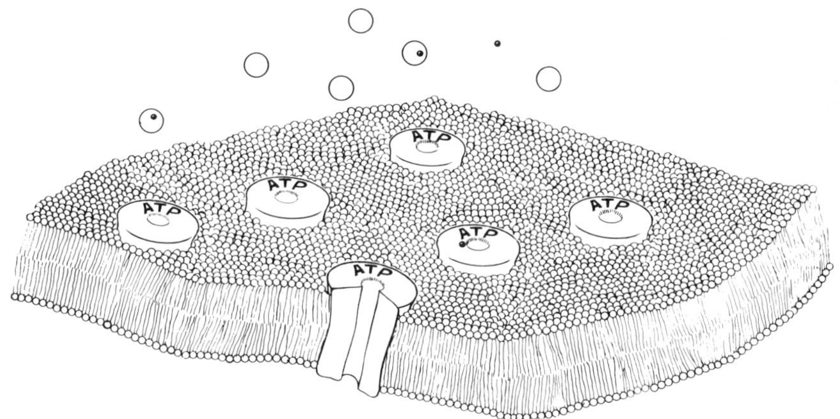


FIGURE 6. Drawing of hypothesized situations of the $Ca^{2+}$ pump under normal conditions of low intracellular $Ca^{2+}$. Relative sizes are not drawn to scale; the torus shape of the ATPase molecules was chosen arbitrarily, and the ATPase molecules are crowded much more closely than in a real membrane. The small black spheres represent Ca ions and the larger spheres calmodulin molecules. As illustrated, at low intracellular $Ca^{2+}$, only an occasional $Ca^{2+}$ site of the ATPase molecules would be occupied.

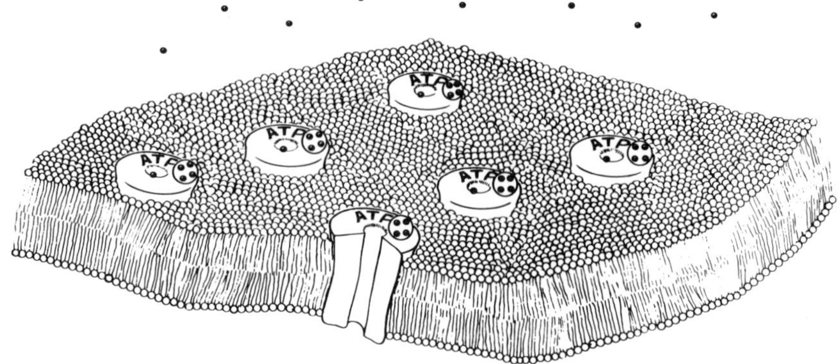

FIGURE 7. Situation with high intracellular $Ca^{2+}$ if the affinity of the ATPase for $Ca^{2+}$ was increased by calmodulin binding. According to this model the $Ca^{2+}$ translocated would not be one of those bound to calmodulin.

It is difficult to tell whether calmodulin would actually dissociate from the pump under normal intracellular conditions. The amount of binding of calmodulin at low intracellular $Ca^{2+}$ depends on the concentration of free calmodulin, which in turn depends on the concentration of calmodulin binding sites in the erythrocyte and the affinity of these sites for calmodulin. We have shown that there is a substantial concentration of such binding sites, but a final resolution of this question awaits more definitive evidence.

Vincenzi *et al.*[30] suggest that calmodulin is always bound to the $Ca^{2+}$ pump. They arrived at this conclusion because of their higher estimate of free calmodulin and because of measurements of ($Mg^{2+}$-$Ca^{2+}$)-ATPase that suggest a slow interaction of calmodulin and the $Ca^{2+}$ pump. We, on the other hand, suggest that a lower free calmodulin would allow dissociation of calmodulin from the $Ca^{2+}$ pump at low free $Ca^{2+}$. Our direct measurement of the kinetics of binding of

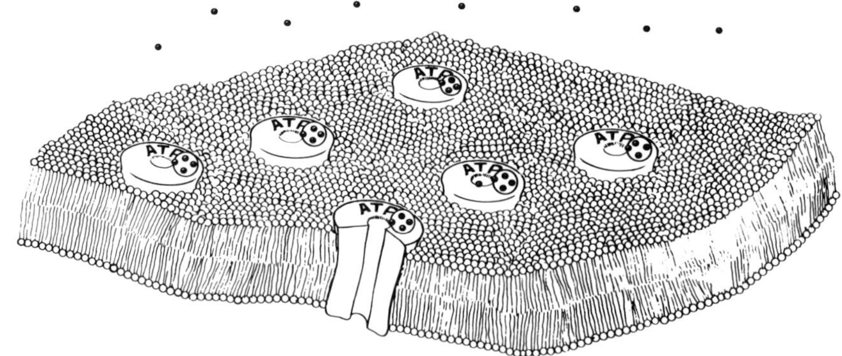

FIGURE 8. Situation with high intracellular $Ca^{2+}$ if the affinity of the ATPase for $Ca^{2+}$ was not increased by calmodulin binding. According to this model, the $Ca^{2+}$ bound to calmodulin would be translocated.

[$^{125}$I]calmodulin to ghosts (not shown) demonstrated that about two-thirds of the calmodulin was bound at the shortest time measured (about 1 minute). The difference between our conclusions and those of Vincenzi *et al.*,[30] demonstrates the need for more definitive evidence on this point.

Regardless of the state of the pump at low $Ca^{2+}$, FIGURES 7 and 8 represent models of the pump in its high affinity state. Since both calmodulin and the pump have $Ca^{2+}$ sites, either one might contribute the $Ca^{2+}$ that is transported. The binding of calmodulin may increase the affinity of the enzyme's own $Ca^{2+}$ binding site (FIGURE 7) or the calcium transported may be one of those bound to calmodulin rather than the calcium bound to the sites on the enzyme itself (FIGURE 8). A third alternative, not shown in the figure, is that calmodulin might pass one of its calcium ions to the calcium binding site of the enzyme and this calcium might then be translocated. The model shown in FIGURE 7 seems the most probable, since it would require the least alteration in the enzyme's mechanism, but essentially no evidence is available on this point.

## ACKNOWLEDGMENTS

We wish to thank Adelaida Filoteo, Karen Streit, and Carol McDonough for technical assistance.

## REFERENCES

1. ROUFOGALIS, B. 1979. Regulation of calcium translocation across the red blood cell membrane. Canad. J. Physiol. Pharmacol. **57:** 1331–1349.
2. DIPOLO, R. & L. BEAUGE. 1979. Physiological role of ATP-driven calcium pump in squid axon. Nature **278:** 271–273.
3. SOBUE, K., S. ICHIDA, H. YOSHIDA, R. YAMAZAKI & S. KAKIUCHI. 1979. Occurrence of a $Ca^{2+}$ and activator protein-activatable ATPase in the synaptic plasma membranes of brain. FEBS Lett. **99:** 199–202.
4. VERMA, A. K. & J. T. PENNISTON. (Manuscript in preparation.)
5. PERSHADSINGH, H. A. & J. M. MCDONALD. 1980. A high affinity calcium-stimulated magnesium-dependent adenosine triphosphatase in rat adipocyte plasma membranes. J. Biol. Chem. (In press.)
6. SCHNEIDER, C., C. MOTTOLA & D. ROMEO. 1979. Calcium ion-dependent ATPase activity and plasma-membrane phosphorylation in the human neutrophil. Biochem. J. **182:** 655–660.
7. MCDONALD, J. M., H. A. PERSHADSINGH, C. G. BRY, M. L. MCDANIEL & P. E. LACY. 1980. Identification of a high affinity $Ca^{2+}$ ATPase in plasma membranes from pancreatic islets. Fed. Proc. **39:** 958.
8. JARRETT, H. W. & J. T. PENNISTON. 1978. Purification of the $Ca^{2+}$-stimulated ATPase activator from human erythrocytes. Its membership in the class of $Ca^{2+}$-binding modulator proteins. J. Biol. Chem. **253:** 4676–4682.
9. WINS, P. 1969. The interaction of red cell membrane ATPase with calcium. Arch. Int. Physiol. Biochem. **77:** 245–250.
10. SCHATZMANN, H. J. & G. L. ROSSI. 1971. ($Ca^{2+}+Mg^{2+}$)-activated membrane ATPases in human red cells and their possible relations to cation transport. Biochim. Biophys. Acta **241:** 379–392.
11. SCHATZMANN, H. J. 1973. Dependence on calcium concentration and stoichiometry of the calcium pump in human red cells. J. Physiol. **235:** 551–569.
12. SCHARFF, O. 1972. The influence of calcium ions on the preparation of the ($Ca^{2+}+Mg^{2+}$)-activated membrane ATPase in human red cells. Scand. J. Clin. Lab. Invest. **30:** 313–320.

13. QUIST, E. E. & B. D. ROUFOGALIS. 1975. Calcium transport in human erythrocytes. Separation and reconstitution of high and low Ca affinity (Mg+Ca)-ATPase activities in membranes prepared at low ionic strength. Arch. Biochem. Biophys. **168:** 240–251.
14. SCHARFF, O. 1976. $Ca^{2+}$ activation of membrane-bound ($Ca^{2+}+Mg^{2+}$)-dependent ATPase from human erythrocytes prepared in the presence or absence of $Ca^{2+}$. Biochem. Biophys. Acta **443:** 206–218.
15. SCHARFF, O. & B. FODER. 1978. Reversible shift between two states of $Ca^{2+}$-ATPase in human erythrocytes and a membrane-bound activator. Biochim. Biophys. Acta **509:** 67–77.
16. ROUFOGALIS, B. D. & D. MAULDIN. 1980. Regulation by calmodulin of the calcium affinity of the calcium-transport ATPase in human erythrocytes. Canad. J. Biochem. (Submitted for publication.)
17. NIGGLI, V., J. T. PENNISTON & E. CARAFOLI. 1979. Purification of the ($Ca^{2+}$-$Mg^{2+}$)-ATPase from human erythrocyte membranes using a calmodulin affinity column. J. Biol. Chem. **254:** 9955–9958.
18. CARAFOLI, E., V. NIGGLI & J. T. PENNISTON. 1980. Reconstitution of the purified ($Ca^{2+}$-$Mg^{2+}$)-ATPase of the erythrocyte membrane. Ann. N. Y. Acad. Sci. (In press.)
19. ITANO, T., R. ITANO & J. T. PENNISTON. 1980. Interaction of basic polypeptides and proteins with calmodulin. Biochem. J. **189:** 455–459.
20. GRAF, E., A. G. FILOTEO & J. T. PENNISTON. 1980. Preparation of $^{125}$I-calmodulin with retention of full biological activity; its binding to human erythrocyte ghosts. Arch. Biochem. Biophys. **203:** 719–726.
21. JARRETT, H. W. & J. T. PENNISTON. 1976. A new assay for endocytosis in erythrocyte ghosts based on loss of acetylcholinesterase activity. Biochim. Biophys. Acta **448:** 314–324.
22. PERRIN, D. D. & I. G. SAYCE. 1967. Computer calculations in mixtures of metal ions and complexing species. Talanta **14:** 833–842.
23. MARTELL, A. E. & R. M. SMITH. 1974. Critical stability constants. Vol. 1: Amino Acids, Plenum Press. New York, New York.
24. SMITH, R. M. & A. E. MARTELL. 1975. Critical stability constants. Vol. 2: Amines. Plenum Press. New York, New York.
25. ADOLFSEN, R. & E. N. MOUDRIANAKIS. 1978. Control of complex metal ion equilibria in biochemical reaction systems: Intrinsic and apparent stability constants of metal-adenine nucleotide complexes. J. Biol. Chem. **253:** 4378–4379.
26. OGAWA, Y. 1968. The apparent stability constant of glycoletherdiaminetetraacetic acid for calcium at neutral pH. J. Biochem. Tokyo **64:** 255–257.
27. JARRETT, H. W. & J. T. PENNISTON. 1977. Partial purification of the ($Ca^{2+}$-$Mg^{2+}$) ATPase activator from human erythrocytes: its similarity to the activator of 3′:5′-cyclic nucleotide phosphodiesterase. Biochem. Biophys. Res. Commun. **77:** 1210–1216.
28. SIMONS, T. J. B. 1976. Calcium-dependent potassium exchange in human red cell ghosts. J. Physiol. **256:** 227–244.
29. CARONI, P. & E. CARAFOLI. 1980. An ATP-dependent $Ca^{2+}$-pumping system in dog heart sarcolemma. Nature **283:** 765–767.
30. VINCENZI, F. F., T. R. HINDS & B. U. RAESS. 1980. Calmodulin and plasma membrane $Ca^{2+}$ transport. Ann. N.Y. Acad. Sci. (This volume.)

## DISCUSSION OF THE PAPER

DR. D. R. STORM: What was the yield and fold purification?

DR. T. PENNISTON: The yield would be about 50%, I think, and the fold purification, I don't know. In this situation, the purification factor could be misleading because of activation or inactivation of the enzyme. It was probably about a thousand fold from the membranes.

DR. STORM: So, does this imply, since you only use a calmodulin column, that you don't have other calmodulin binding proteins in that preparation?

DR. PENNISTON: Yes, that's the conclusion that we've seen so far.

DR. STORM: You said that calmodulin increased the energy of activation by a factor of two. That implies that the calmodulin stimulation is an entropy effect.

DR. PENNISTON: That's right.

DR. T. MURTAUGH (*University of Wisconsin, Madison, WI*): I am just wondering if you have any indication in your purified enzyme that the calcium effect in the absence of calmodulin may be due to calmodulin as a subunit as we saw with the phosphorylase kinase?

DR. PENNISTON: We don't think so. We have no indications that such a thing is happening.

# CALMODULIN IN NATURAL AND RECONSTITUTED CALCIUM TRANSPORTING SYSTEMS*

E. Carafoli, V. Niggli, K. Malmström, and P. Caroni

*Laboratory of Biochemistry*
*Swiss Federal Institute of Technology (ETH)*
*Zurich, Switzerland CH-8092*

## INTRODUCTION

Efficient transport across biological membranes is a requirement for the messenger function of $Ca^{2+}$ (see Carafoli & Crompton[1] for review). The unique importance of this messenger function is underscored by the existence of a multiplicity of $Ca^{2+}$ transporting systems, located in different membrane systems, and different in mechanism. The regulation of these systems is one obscure aspect of the field, but it is clear that the efficiency of the messenger function would be heightened by the existence of regulatory mechanisms. In this respect, the original observation by Gopinath and Vincenzi[2] and Jarrett and Penniston[3] of a stimulation by calmodulin of the ($Ca^{2+}+Mg^{2+}$)-ATPase of the erythrocyte membrane, and the more recent observation by Katz and Remtulla[4] of similar effects on the $Ca^{2+}$ pumping system of heart sarcoplasmic reticulum, are of great interest.

At the present date, $Ca^{2+}$ transporting systems have been described in the following membranes: in mitochondria, where separate, and mechanistically different, routes for $Ca^{2+}$ uptake and $Ca^{2+}$ release exist; in the plasma membrane of different cells, where a "slow $Ca^{2+}$ channel" for $Ca^{2+}$ influx, a $Ca^{2+}$-ATPase for $Ca^{2+}$ efflux, and a $Na^+/Ca^{2+}$ exchange system (that is probably used in either direction) have been described; in sarcoplasmic reticulum, which possesses a specific $Ca^{2+}$-pumping ATPase, which has now been studied in great molecular detail; in endoplasmic reticulum, in which a specific ATP-dependent $Ca^{2+}$ pumping system has been documented; and in several bacteria, in which a $Ca^{2+}$-specific ATPase, and a $Ca^{2+}/H^+$ exchange, have been demonstrated.

The $Ca^{2+}$-pumping ATPase of the erythrocyte membrane has been purified by Niggli *et al.*,[5] and reconstituted in phospholipid bilayer vesicles (Carafoli *et al.*[6]). Its interaction with calmodulin can now be studied in the isolated and reconstituted system: some aspects of the work will be described here. The sarcoplasmic reticulum $Ca^{2+}$-ATPase has also been available in purified form for a long time, but no work on the interaction between calmodulin and the purified (and reconstituted) ATPase has been reported so far. However, the original observation on the stimulation of the $Ca^{2+}$ transport in heart sarcoplasmic reticulum by calmodulin[4] has recently been extended by LePeuch *et al.*,[7] and convincing evidence has been provided that the effect of calmodulin is *not* direct, but is mediated through a phosphorylation-dephosphorylation system. The present work also includes data on calmodulin in sarcoplasmic reticulum from voluntary muscle.

No indications have so far been given for the role, or even the presence, of calmodulin in any of the other $Ca^{2+}$ transporting systems mentioned above. The

*Part of the original research was supported by Swiss Nationalfonds Foundation Grant 3.282-0.78 and the Geigy Jubiläumsstiftung.



0077-8923/80/0356-0258 $01.75/0 © 1980, NYAS

results of preliminary investigations on mitochondria, on sarcolemma, and on endoplasmic reticulum, will be mentioned here.

## CALMODULIN IN THE RECONSTITUTED $(Ca^{2+}+Mg^{2+})$-ATPASE OF THE ERYTHROCYTE MEMBRANE

FIGURE 1 shows a reconstitution experiment on the purified $(Ca^{2+}+Mg^{2+})$-ATPase of the erythrocyte membrane, in which phosphatidyl serine liposomes, and phosphatidyl choline liposomes, have been used and compared. The figure shows that in the absence of calmodulin the specific activity of the purified enzyme is very different in the two phospholipid environments. When incorporated in phosphatidyl serine, the enzyme splits ATP at a rate of 3 μmoles per mg of protein per min (at 10 μM $Ca^{2+}$), when incorporated in phosphatidyl choline the enzyme works at a much slower rate (0.4 μmoles per mg protein per min). Evidently, the phosphatidyl choline environment is not favorable to the enzyme, in agreement with earlier findings of Ronner et al.[8] on the solubilized enzyme. Phosphatidyl serine, on the other hand, provides a better environment to the enzyme, most likely due to its acidic character.[8] The surprising finding, however, is that calmodulin added to the reconstituted system *does* activate the enzyme in the case of phosphatidyl choline liposomes, but does *not* in the case of the

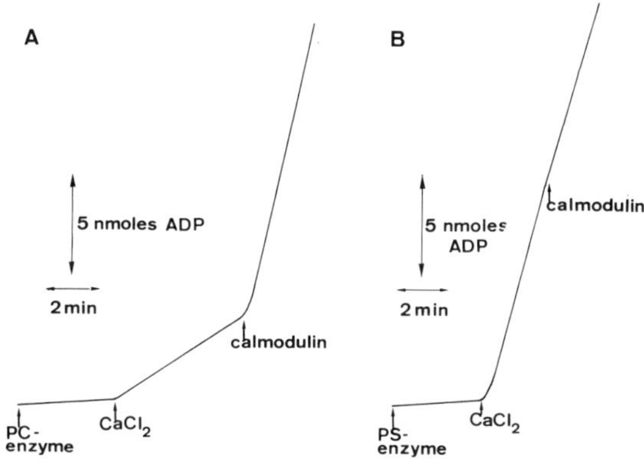


FIGURE 1. Stimulation of the $(Ca^{2+}+Mg^{2+})$-ATPase of erythrocytes by calmodulin. The isolated enzyme is reconstituted in phosphatidyl-serine or phosphatidyl choline liposomes as indicated in Reference 6. The $(Ca^{2+}+Mg^{2+})$-ATPase activity is determined with a coupled enzyme assay as indicated in Reference 5. The absorbance changes are monitored at 366 to 550 nm in an Aminco dual wavelength spectrophotometer. The system is calibrated by adding 5 nmoles ADP per 1 ml medium. The medium contains 500 μM EGTA, 500 μM HEEDTA (N'(-2 hydroxy ethyl) ethylene diamine N, N'N'-triacetic acid), 2 mM $MgCl_2$, 500 μM ATP, 130 mM KCl, 60 mM HEPES, pH 7, 0.2 mM NADH, 0.5 mM PEP (phosphoenolpyruvate), 1 I.U. pyruvate kinase/ml, 1 I.U. lactate dehydrogenase/ml, 37°C. The mixture is preincubated for 4 min at 37°C. Then, 1.5 μg reconstituted enzyme are added, together with 0.1 μM A23187 (Final volume, 1 ml). After about 3 min, 600 μM $CaCl_2$ is added, (corresponding to a $Ca^{2+}$ activity, measured with a $Ca^{+2}$ specific electrode, of 10 μM). After 2–5 additional minutes, 2 μg of calmodulin (bovine brain) are added.

phosphatidyl serine liposomes. In the presence of calmodulin, the activity of the phosphatidyl choline enzyme is stimulated about 7-fold, and is increased to 100% of that seen in the phosphatidyl serine system. Evidently, the phosphatidyl serine enzyme is already maximally activated in the absence of calmodulin, or, to say it differently, phosphatidyl serine and calmodulin are interchangeable in their ability to activate the enzyme. When the enzyme is reconstituted in asolectin liposomes, its behavior is similar to that of the enzyme incorporated in phosphatidyl serine liposomes: the enzyme appears to be already maximally stimulated in the absence of calmodulin. Most likely, this is due to the presence of about 18% acidic phospholipids (mainly phosphatidyl inositol and cardiolipin[9]) in asolectin.

TABLE 1 shows that the effect of calmodulin on the reconstituted phosphatidyl choline system is to shift the enzyme from a low affinity form ($K_m$ ($Ca^{2+}$) 10–14 $\mu$M) to a high affinity ($K_m$ ($Ca^{2+}$) 0.8–0.9 $\mu$M) form. The phosphatidyl serine enzyme, on the other hand, is already in the high affinity form in the absence of calmodulin, as shown by the $K_m$ of 0.4–0.6 $\mu$M.

In all the experiments reported so far, the enzyme has been solubilized and purified in the presence of Triton X-100. Experiments currently under way in this laboratory attempt to clarify the possible role of the detergent in the activation of the enzyme. In preliminary experiments it has been observed that Triton X-100 may indeed contribute to the activation of the enzyme. The problem now arises of understanding how acidic phospholipids can substitute for calmodulin in activating the purified and reconstituted ($Ca^{2+}$+$Mg^{2+}$)-ATPase. To this aim, it would be very convenient if the mechanism by which calmodulin itself activates the enzyme were known. This not being the case, yet, a great deal of what can be said is however bound to remain speculative.

As shown in TABLE 1, calmodulin and acidic phospholipids shift the ATPase enzyme to a high affinity form. In FIGURE 2, the data from an experiment similar to that of TABLE 1 have been plotted as activation ratios at different $Ca^{2+}$ activities, to emphasize the concept that calmodulin and acidic phospholipids activate the enzyme with the most marked effect in a comparatively narrow range of $Ca^{2+}$ activities, between 0.1 and 1 $\mu$M. What has to be stressed here, is that optimal activation by calmodulin takes place in an ambience of $Ca^{2+}$ activity where the $Ca^{2+}$ binding sites of calmodulin are probably not saturated by $Ca^{2+}$. The information now available[10] indeed indicates that the overall affinity of calmodulin for $Ca^{2+}$ is expressed by a $K_d$ of 4–18 $\mu$M, but one of the four sites has a much higher affinity for $Ca^{2+}$ than the others. On the other hand, at the $Ca^{2+}$ activity that stimulates maximally, phosphatidyl serine should bind $Ca^{2+}$ only marginally, since its $K_d$ is of the order of 10 $\mu$M. These considerations indicate

TABLE 1

EFFECTS OF CALMODULIN AND PHOSPHATIDYL SERINE ON THE AFFINITY OF THE ($Ca^{2+}$+ $Mg^{2+}$)-ATPASE OF ERYTHROCYTES FOR $Ca^{2+}$

| Enzyme In | Addition | $K_m$ ($Ca^{2+}$) |
|---|---|---|
| PC liposomes | none | 10–14 $\mu$M |
| PC liposomes | calmodulin | 0.8–0.9 $\mu$M |
| PS liposomes | none | 0.4–0.6 $\mu$M |
| PS liposomes | calmodulin | 0.4–0.6 $\mu$M |

Reconstitution of the enzyme, experimental conditions, and activity measurements as in the experiments of FIGURES 1 and 2. PC = phosphatidyl choline. PS = phosphatidyl serine.

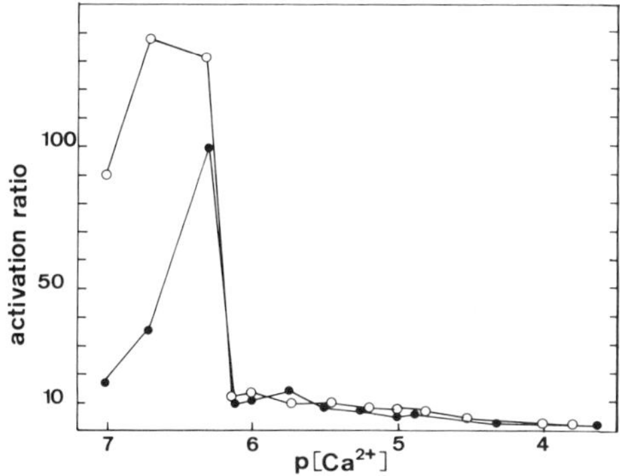


FIGURE 2. Activation of the isolated $(Ca^{2+}+Mg^{2+})$-ATPase of calmodulin and by phosphatidyl serine, at different $Ca^{2+}$ activities. $(Ca^{2+}+Mg^{2+})$-ATPase activities are determined with the same method and under the same conditions as described in the legend for FIGURE 1. Different amounts of $CaCl_2$ are added to the medium, to obtain the $Ca^{2+}$-activities given in FIGURE 2 (values determined with a $Ca^{2+}$-specific electrode). Activity is given in $\mu$moles ATP hydrolyzed per mg of protein per min. ○——○ , activity of enzyme reconstituted in phosphatidyl sevine (PS), minus calmodulin/activity of enzyme reconstituted in phosphatidyl choline (PC) minus calmodulin. ●——● , activity of enzyme reconstituted in PC, plus calmodulin/activity of enzyme reconstituted in PC, minus calmodulin.

that the common link between the activation by calmodulin and that by acidic phospholipids is probably *not* to be found in their ability to bind $Ca^{2+}$, which would have been a very convenient way of explaining the phenomena observed. Possibly both calmodulin and acidic phospholipids have the ability to induce a conformational change of the ATPase enzyme favorable to the expression of optimal activity.

## CALMODULIN IN HEART AND VOLUNTARY MUSCLE SARCOPLASMIC RETICULUM

As mentioned, Katz and Remtulla[4] found that brain calmodulin stimulates both the active binding and the accumulation (i.e., in the presence of oxalate) of $Ca^{2+}$ by heart sarcoplasmic reticulum and found that the stimulation is additive with respect to that given by the cyclic AMP-dependent protein kinase system. They also found that the activation, which is less than two-fold, is visible at all ranges of free $Ca^{2+}$ tested, (0.1 to 5.0 $\mu$M). More recently, LePeuch *et al.*,[5] found that the rate of $Ca^{2+}$ uptake by dog heart sarcoplasmic reticulum is indeed stimulated by both a cyclic AMP-dependent phosphorylation system, and by exogenous calmodulin plus $Ca^{2+}$. The calmodulin-$Ca^{2+}$ system operates through a membrane bound kinase, which phosphorylates phospholamban, a 22,000 MW membrane protein, at a site different from that phosphorylated by the cyclic AMP-dependent kinase. As a result, the rate of $Ca^{2+}$ uptake is increased, although the $Ca^{2+}$-ATPase is *not* influenced.

The possible presence of endogenous calmodulin in voluntary muscle (lobster) sarcoplasmic reticulum has been investigated in the experiment shown in

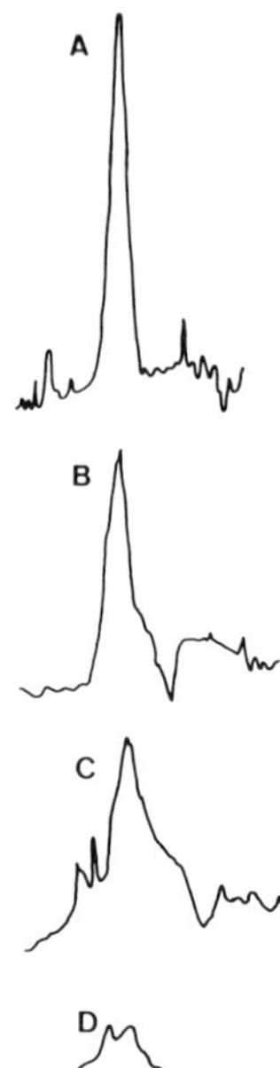


FIGURE 3. Lobster muscle sarcoplasmic reticulum is prepared according to the method of Peterson and Deamer.[11] 2 mg of sarcoplasmic reticulum are extracted at room temperature for 10 min with 20 mM Tris-HCl, pH 8.5 (C) or with 20 mM Tris-HCl, 1 mM EDTA pH 8.5 (D). The reticulum is collected by centrifugation, heated at 100°C for 4 min and centrifuged again. The supernatants are analyzed by polyacrylamide gel electrophoresis according to Laemmli.[12] Trace B refers to nonextracted sarcoplasmic reticulum, trace A to a sample of purified bovine brain calmodulin (1.2 $\mu$g).

FIGURE 3. A protein band corresponding to the MW of calmodulin is indeed present in extracts of boiled lobster muscle sarcoplasmic reticulum (Trace B). This band decreases sharply if the sarcoplasmic reticulum is treated for 10 min at room temperature with 20 mM Tris-HCl, pH 8.5, or with 20 mM Tris-HCl, pH 8.5 plus 1 mM K-EDTA. (Traces C & D). Separate experiments have shown that both treatments result in a marked decrease, or even a total elimination, of the $Ca^{2+}$-ATPase activity of sarcoplasmic reticulum, a finding that may be viewed against the claim[4] that *exogenous* calmodulin does not modify the $Ca^{2+}$-ATPase of heart sarcoplasmic reticulum. It has been found, however, that addition of the Tris-HCl, or Tris-HCl plus EDTA extract, does not restore the $Ca^{2+}$-ATPase of sarcoplasmic reticulum to normal levels.

That the protein extracted from lobster muscle and heart sarcoplasmic reticulum by the treatment mentioned above is indeed calmodulin is shown by experiments in which extracts of either lobster muscle, or heart sarcoplasmic reticulum boiled for 4 min, are shown to contain calmodulinlike activity, since they can activate the ($Ca^{2+}+Mg^{2+}$)-ATPase of erythrocyte ghosts, (TABLE 2) and phosphodiesterase (not shown).

Following the demonstration by Niggli *et al.*[5] that the ($Ca^{2+}+Mg^{2+}$)-ATPase activity of erythrocytes can be isolated on calmodulin-affinity chromatography columns, heart or lobster muscle sarcoplasmic reticulum, solubilized in detergents, have been passed through a calmodulin-affinity chromatography column. No $Ca^{2+}$ ATPase remained bound to the column, a finding that is hardly surprising in view of the probably indirect mechanism of calmodulin stimulation of the $Ca^{2+}$ pumping in sarcoplasmic reticulum. Work currently in progress in this laboratory attempts to identify the protein fraction(s) from the sarcoplasmic reticulum membrane that remain bound to the calmodulin-affinity chromatography column.

## Calmodulin and the Mitochondrial $Ca^{2+}$ Transporting Systems

Liver mitochondria take up $Ca^{2+}$ by an electrophoretic system which is sensitive to the polycation ruthenium red, and release it by a separate pathway, which is insensitive to ruthenium red, and is made evident by its addition. In heart mitochondria, the release pathway is activated by the addition of $Na^+$, which penetrates into mitochondria in exchange for $Ca^{2+}$. The regulation of the uptake and the release pathways is at the moment completely obscure, although effects of hormonal treatments *in vivo* have been described. In this laboratory, calmodulin has been tested on the uptake and on the release of $Ca^{2+}$, but in either case, no effects have been observed in the presence of calmodulin concentrations of up to 3 μg per mg of mitochondrial protein. One obvious objection here, is the impermeability of the outer mitochondrial membrane to exogenous calmodulin. Other experiments, however, indicate that calmodulin is either not present in mitochondria, or present only in negligible amounts. Extracts of boiled liver mitochondria fail to activate the $(Ca^{2+}+Mg^{2+})$-ATPase of erythrocytes, and do not show any protein band in the 17,000 MW region in SDS polyacrylamide gel electrophoresis.

## Calmodulin in Sarcolemma

Recently, purified vesicular preparations of heart sarcolemma have become available.[13] They have permitted detailed investigations on the mechanism of the $Na^+/Ca^{2+}$ exchange process,[14,15] and have been instrumental in the demonstration of the existence of a specific $Ca^{2+}$-ATPase,[16] which resembles that of erythrocytes in $Mg^{2+}$ requirement, $Ca^{2+}$ affinity, and vanadate sensitivity. It was, therefore, thought interesting to investigate whether heart sarcolemma contains calmodulin, and whether calmodulin activates the $Ca^{2+}$ pumping ATPase. A highly purified preparation of heart sarcolemma (specific activity of the specific

Table 2

Activation of the $(Ca^{2+}+Mg^{2+})$-ATPase of Erythrocytes by Extracts from Heat-Treated Sarcoplasmic Reticulum

| Addition | Amount | $(Ca^{2+}+Mg^{2+})$-ATPase Activity μmoles ATP Hydrolyzed/mg/min |
|---|---|---|
| none | — | 0.006 |
| calmodulin | 200 ng | 0.029 |
| extract from 1 mg heart sarcoplasmic reticulum protein | 174 μg | 0.017 |
| extract from 0.5 mg lobster muscle sarcoplasmic reticulum protein | 73 μg | 0.023 |

The preparation of sarcoplasmic reticulum is given in the legend for Figure 3. The $(Ca^{2+}+Mg^{2+})$-ATPase of erythrocyte ghosts is measured as detailed in the legend for Figure 1.

marker $Na^+/K^+$-ATPase 120 μmol phosphate liberated per mg of protein per hour), contains a heat-stable component that activates the $(Ca^{2+}+Mg^{2+})$-ATPase of erythrocyte ghosts. The activation is lost if the sarcolemmal preparation, prior to boiling, is extracted with a hypertonic solution containing EGTA. The observation indicates that sarcolemma indeed contains calmodulin, but offers no indication as to whether the activator is associated with the $Ca^{2+}$-pumping ATPase or has any role in it. This, however, is demonstrated by the experiment shown in FIGURE 4, in which the ATPase of "calmodulin-free sarcolemma" is clearly activated by calmodulin. Very interestingly, the affinity of the enzyme for $Ca^{2+}$ is influenced by calmodulin: the $K_m$, which is normally about 0.3 μM[14] is increased almost 30 times (~10 μM) in calmodulin-free membranes. This is reminiscent of the effect of calmodulin on the $(Ca^{2+}+Mg^{2+})$-ATPase of erythrocytes, where the enzyme is shifted from a low to a high affinity form in the presence of the activator (FIGURE 2). It can be concluded, then, that the $Ca^{2+}$-pumping ATPase of heart sarcolemma is activated by calmodulin and shifted by it to a form having higher $Ca^{2+}$ affinity.

## CALMODULIN IN ENDOPLASMIC RETICULUM

The $Ca^{2+}$ transport system of endoplasmic reticulum has not been characterized to the same extent as some of the $Ca^{2+}$ transporting systems in other membranes. It is, however, dependent on ATP, suggesting the involvement of a specific ATPase. Early experiments in this laboratory[17] have shown that the treatment of rat liver endoplasmic reticulum with a hypotonic solution results in the extraction of several proteins. Inclusion of EDTA in the extraction medium results in the appearance of 3 additional protein bands in the extract, 2 of them

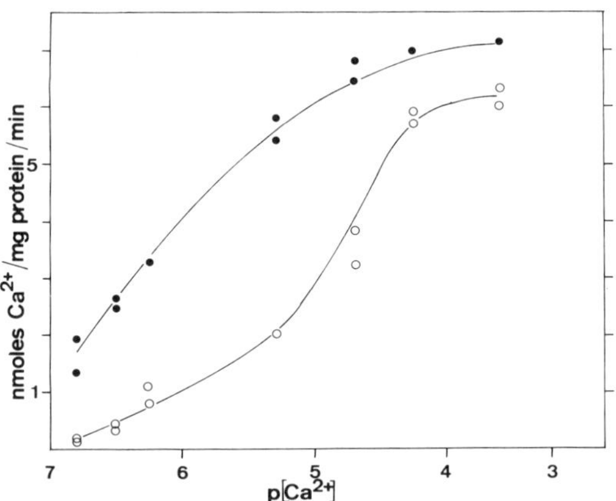


FIGURE 4. Stimulation of the ATP-dependent $Ca^{2+}$ uptake in heart sarcolemma by calmodulin. Sarcolemmal vesicles, isolated by a modification of the method of Jones *et al.*,[13] are depleted of calmodulin by a hypertonic (0.6 M KCl) treatment in the presence of 2 mM K-EGTA. $Ca^{2+}$ uptake is measured as described in Reference 16. The vesicles are preincubated in the reaction medium in the presence of 25 μM $CaCl_2$, with or without 1 μg bovine brain calmodulin, then EGTA is added to yield the desired final $Ca^{2+}$ concentration (determined separately with a $Ca^{2+}$ selective electrode). After 30 additional seconds the uptake is started by the addition of 1 mM K-ATP.

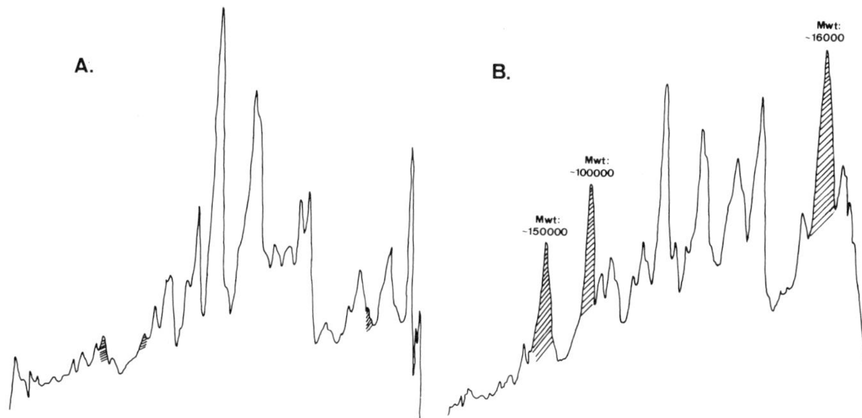


FIGURE 5. Rat liver endoplasmic reticulum is prepared by a conventional differential centrifugation procedure. 27 mg of protein/ml are extracted with 10 mM Tris-HCl, pH 7.8 (A) or with 5 mM Tris-HCl and 5 mM Na-EDTA pH 7.8,[8] overnight at 0°C, with gentle stirring. The extracts are lyophilized, and dissolved in distilled water, 140 $\mu$g of extracted protein are used for polyacrylamide gel electrophoresis according to Laemmli.[12] The shaded areas represent the protein bands that are enriched in the EDTA extract.

having high MW, and one in the range between 15,000 and 17,000 (FIGURE 5). The possibility that the latter fraction corresponds to calmodulin is supported by experiments in which rat liver endoplasmic reticulum is extracted with a slightly alkaline hypotonic solution containing EDTA. Analysis of the extract by high resolution two-dimensional electrophoresis has shown a protein band that clearly corresponds to calmodulin. It has also been found that the boiled extract, suitably concentrated, and dialyzed to remove excess EDTA, activates the ($Ca^{2+}+Mg^{2+}$)-ATPase of erythrocytes.

## CONCLUSIONS

Calmodulin has been demonstrated in the membranes of erythrocytes, of heart and voluntary muscle sarcoplasmic reticulum, in liver endoplasmic reticulum, and in heart sarcolemma. It is probably absent, or present in negligible amounts, in mitochondria. Whereas conclusive evidence demonstrates that calmodulin interacts directly with the membrane-bound and the isolated ($Ca^{2+}+Mg^{2+}$)-ATPase of erythrocytes and activates it, the $Ca^{2+}$ pumping ATPase of heart sarcoplasmic reticulum is probably activated by calmodulin in an indirect way. In heart sarcolemma, calmodulin activates the $Ca^{2+}$-pumping ATPase, and shifts it to a form having high $Ca^{2+}$ affinity. The other $Ca^{2+}$ transporting systems of sarcolemma ($Na^+/Ca^{2+}$ exchange, slow $Ca^{2+}$ channel), and the $Ca^{2+}$ transporting systems of bacteria, have not yet been tested for calmodulin sensitivity.

## REFERENCES

1. CARAFOLI, E. & M. CROMPTON. 1978. Current Topics in Membrane and Transport **10:** 151–216.

2. GOPINATH, R. M. & F. F. VINCENZI. 1977. Biochem. Biophys. Res. Commun. **77:** 1203–1209.
3. JARRETT, H. W. & J. T. PENNISTON. 1977. Biochem. Biophys. Res. Commun. **77:** 1210–1216.
4. KATZ, S. & M. A. REMTULLA. 1978. Biochem. Biophys. Res. Commun. **83:** 1373–1379.
5. NIGGLI, V., J. T. PENNISTON & E. CARAFOLI. 1979. J. Biol. Chem. **254:** 9955–9958.
6. CARAFOLI, E., V. NIGGLI & J. T. PENNISTON. 1980. Ann. N.Y. Acad. Sci. (In press.)
7. LE PEUCH, C., J. HAIECH & J. G. DEMAILLE. 1979. Biochemistry **18:** 5150–5157.
8. RONNER, P., P. GAZZOTTI & E. CARAFOLI. 1977. Arch. Biochem. Biophys. **179:** 578–583.
9. EYTAN, E. & E. RACKER. 1976. J. Membr. Biol. **26:** 319–333.
10. LIN, Y. M., Y. P. LIU & W. Y. CHEUNG. 1974. J. Biol. Chem. **249:** 4945–4954.
11. PETERSON, S. W. & D. W. DEAMER. 1977. Arch. Biochem. Biophys. **179:** 218–228.
12. LAEMLLI, U. K. 1970. Nature **227:** 680–685.
13. JONES, L. R., H. R. BESCH, J. W. FLEMING, M. M. MCCONNAUGHEY & A. M. WATANABE. 1979. J. Biol. Chem. **254:** 530–539.
14. REEVES, J. P. & J. L. SUTKO. 1979. Proc. Natl. Acad. Sci. USA **76:** 590–594.
15. PITTS, B. J. R. 1979. J. Biol. Chem. **254:** 6232–6235.
16. CARONI, P. & E. CARAFOLI. 1980. Nature **283:** 765–767.
17. BILLETER, R. 1975. Master's Thesis. ETH, Zurich.

## DISCUSSION OF THE PAPER

DR. F. F. VINCENZI: What happens if you treat phosphatidyl serine liposomes with phenothazines? Can you prevent the phosphatidyl serine effect?

DR. E. CARAFOLI: This is one of the next experiments we would like to carry out.

DR. VINCENZI: If calmodulin is involved in the regulation of intracellular calcium pump then an obvious question is: Does calmodulin also modulate or modify sodium/calcium exchange?

DR. CARAFOLI: My lab has run a test on calmodulin influence on the sodium/calcium exchange and we found none. I would hate to be specific on this point now because that was only one experiment.

DR. C. KLEE: I would like to ask Dr. Carafoli to comment on the different mechanisms of activation in the two types of ATPase, the plasma membrane and the sarcoplasmic reticulum ATPase.

DR. CARAFOLI: To the best of our knowledge there is a direct interaction of calmodulin with the calcium ATPase in erythrocytes whereas there is no direct interaction in sarcoplasmic reticulum. I may also add that the two ATPases have different molecular weights, different phospholipid requirements, and different vanadate sensitivity.

# MECHANISM OF STIMULATION OF CALCIUM TRANSPORT IN CARDIAC SARCOPLASMIC RETICULUM PREPARATIONS BY CALMODULIN*

Sidney Katz

*Division of Pharmacology*
*University of British Columbia*
*Vancouver, British Columbia V6T 1W5*
*Canada*

## INTRODUCTION

Calmodulin has been shown to stimulate the activity of a number of membrane-bound enzyme systems in many tissues.[1] Recently, calmodulin was found to modulate the activity of $(Mg^{2+}+Ca^{2+})$-ATPase in red blood cell membrane preparations and $Ca^{2+}$ transport in inside-out red blood cell vesicle preparations.[2-6] Some ATP-dependent calcium transport systems are plasma-membrane bound and function, as in the red blood cell, to maintain low intracellular levels of free calcium. In other cases, $Ca^{2+}$-transport systems have been found to be associated with intracellular organelles. One such system is the active $Ca^{2+}$ transport associated with the sarcoplasmic reticulum of muscle cells. Considerable work has been done to elucidate the molecular mechanism of coupling between the intermediate steps of the $(Mg^{2+}+Ca^{2+})$-ATPase reaction and the transport of calcium across the membrane of the sarcoplasmic reticulum.[7-15] In this reaction, for every mole of ATP hydrolyzed, 2 moles of $Ca^{2+}$ are removed from the external solution into the sarcoplasmic reticulum vesicles against a large concentration gradient.[7,3] During the active transport of calcium a high energy phosphoprotein intermediate, [EP], is formed.[14] This [EP] formation is activated by the binding of $Ca^{2+}$ to a site of high affinity ($K_m$ = 1–3 $\mu$M) located at the outer surface of the membrane. Addition of hydroxylamine leads to the decomposition of the phosphoprotein formed, indicating that it is an acyl phosphate intermediate.[15]

The accumulation of calcium by the cardiac sarcoplasmic reticulum via the $(Mg^{2+}+Ca^{2+})$-ATPase $Ca^{2+}$-transport system is thought to play an important role in terminating the contractile event and producing relaxation. The $Ca^{2+}$-transport system effectively reduces the intracellular free calcium concentration in the region of the contractile proteins to less than $10^{-7}$ M.[10]

It has recently been shown that $Ca^{2+}$-transport activity at the level of the sarcoplasmic reticulum can be augmented by at least two mechanisms. Kirchberger *et al.*[16-18] and Tada *et al.*[19-21] have observed that cyclic AMP-dependent protein kinase will significantly increase the maximal activation of the calcium transport system of the sarcoplasmic reticulum. In addition, it has been known for many years that $K^+$ concentrations greater than 30 mM will significantly augment $Ca^{2+}$ transport in both skeletal[22,25] and cardiac sarcoplasmic reticulum.[26,27] The significance of this enhancement of $Ca^{2+}$ transport by these intracellular modulators remains to be determined, although it is thought that the greater the enhancement of the calcium uptake the faster will be the relaxation rate. It is

*Supported by a Grant from the British Columbia (Canada) Heart Foundation.



0077-8923/80/0356-0267 $01.75/0 © 1980, NYAS

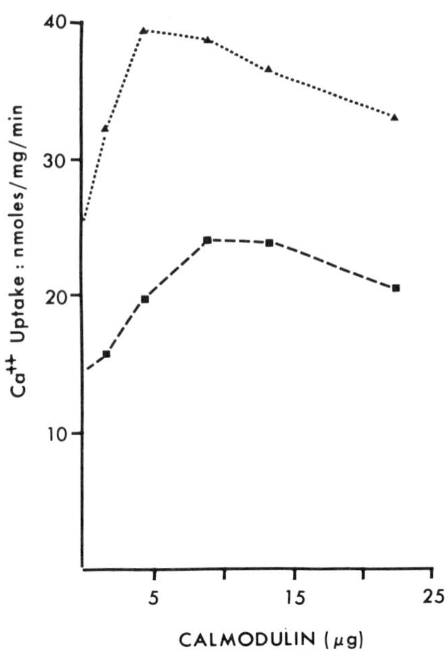


FIGURE 1. Effect of calmodulin on $Ca^{2+}$ uptake in microsomal preparations enriched in sarcoplasmic reticulum. Oxalate-facilitated calcium uptake was determined in the presence of bovine brain calmodulin in an incubation medium containing 40–50 $\mu$g protein of the microsomal preparation, 40 mM histidine-HCl, pH 6.8, 5 mM $MgCl_2$, 110 mM KCl, 5 mM Tris-ATP, 2.5 mM Tris-oxalate and 1 $\mu$M $Ca^{2+}$ free containing [$^{45}$Ca]$Cl_2$ (10 $C_i$/mmole). The desired free calcium concentration was maintained by the addition of ethylene-bis-($\beta$-aminoethyl ether)N,N'-tetraacetate (EGTA). Following a preincubation of 7 min at 30°C, the reaction was started by the addition of [$^{45}$Ca]$Cl_2$. After 5 min the reaction was terminated by filtering an aliquot of the reaction mixture through a Millipore filter (HA 45, Millipore Co.). The filter was then washed twice with 15 ml of 40 mM Tris-HCl, pH 7.2 and dried and counted for radioactivity in Aquasol (New England Nuclear Co.) using standard liquid scintillation counting techniques. $Ca^{2+}$ uptake was determined in the presence (▲---▲) and absence (■---■) of cyclic AMP-dependent protein kinase (type 1, 25 $\mu$g/0.5 ml) and cyclic AMP (1 $\mu$M). Result shown is a typical experiment.

also suggested that more complete uptake of calcium into the sarcoplasmic reticulum will lead to an increase in contractility in subsequent beats.[21]

Calmodulin is present in large quantities in cardiac tissue.[28,29] In view of the stimulatory effect of calmodulin on $Ca^{2+}$ transport in red cell membranes, the possible role of calmodulin in the regulation of calcium transport in cardiac sarcoplasmic reticulum was investigated. In addition, the relationship of calmodulin to other possible endogenous modulators of calcium transport in the cardiac cell was studied.

## METHODS

Dog cardiac microsomal preparations enriched in cardiac sarcoplasmic reticulum were prepared by the method of Harigaya and Schwartz[30] with some modifications.[31] It was determined that these preparations were stable for up to 2 weeks when stored in 40% sucrose at $-80°C$ at protein concentrations greater than 1.4 mg/ml. $Ca^{2+}$ uptake into sarcoplasmic reticulum vesicles was determined by a Millipore filtration technique (0.45 $\mu$m millipore filters) using oxalate, in most cases, as the permeant anion.[19] $^{45}CaCl_2$ was utilized as the substrate and the free calcium concentrations maintained by Ca-EGTA buffers.[31] $(Mg^{2+}+Ca^{2+})$-ATPase activity and $Ca^{2+}$ dependent phosphoprotein formation were determined by conventional methods using $[\gamma\text{-}^{32}P]ATP$ as substrate. All experimental parameters were tested in duplicate and each study repeated at least three times. Calmodulin was obtained either from commercial sources (Sigma) or as a gift (Dr. J. Wang, University of Manitoba and Dr. John Penniston, Mayo Clinic), or was prepared from red blood cell hemolysates following the technique of Vincenzi and co-workers.[6] In all cases, the calmodulin used was found to have no indigenous $(Mg^{2+}+Ca^{2+})$-ATPase activity, not to stimulate $Mg^{2+}$-ATPase activity, and to significantly stimulate $Ca^{2+}$ transport and $(Mg^{2+}+Ca^{2+})$-ATPase activity in red blood cell membrane preparations.

## Results

### *Effect of Calmodulin on Calcium Transport in Cardiac Microsomes Enriched in Sarcoplasmic Reticulum*

Figure 1 indicates the effect of calmodulin on $Ca^{2+}$ transport in microsomes enriched in sarcoplasmic reticulum. Calmodulin prepared from either brain or

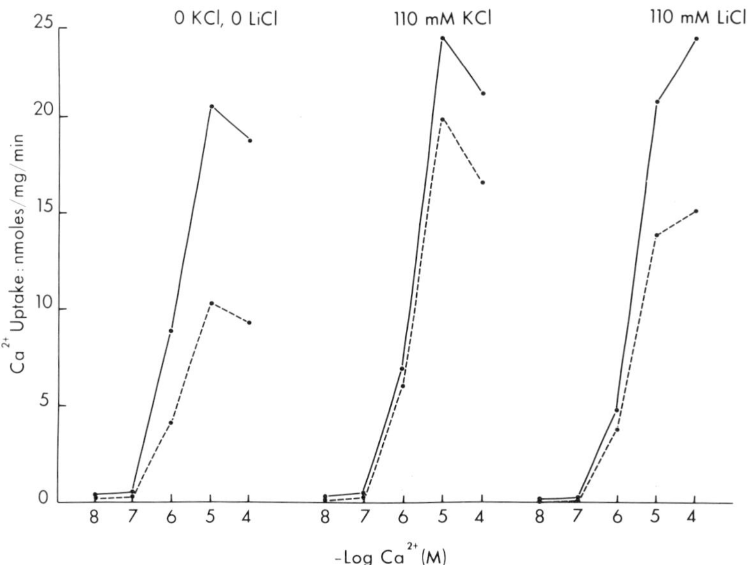


Figure 2. Effect of monovalent cations on $Ca^{2+}$ uptake in the presence, and absence of calmodulin. $Ca^{2+}$ uptake was determined as described in Figure 1 in the presence and absence of 110 mM KCl, 110 mM LiCl or 200 mM sucrose (0 KCl, 0 LiCl) with (●——●) and without (●---●) red cell calmodulin (2.0 $\mu$g/0.5 ml final incubation volume). Result shown is a typical experiment. Reprinted with permission from *Biochemistry*, © 1980, American Chemical Society.

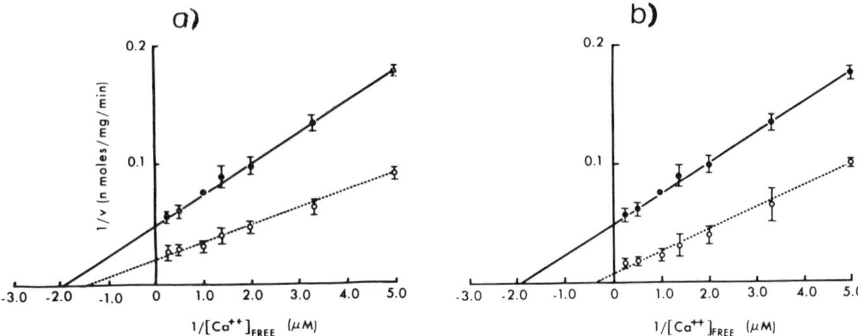


FIGURE 3. Effect of calmodulin on the $K_{diss}$ for $Ca^{2+}$ and the $V_{Ca^{2+}}$ uptake in microsomal preparations enriched in sarcoplasmic reticulum. $Ca^{2+}$ uptake was determined as described in the METHODS at various free $Ca^{2+}$ concentrations (0.2 μM–30.0 μM $Ca^{2+}$ free) and the data plotted as reciprocals of results obtained in the absence (●——●) and presence (O---O) of calmodulin (FIGURE 3a) and in the absence (●——●) and presence (O---O) of 110 mM KCl (FIGURE 3b). Data obtained in assays performed on separate days were normalized to that activity found in the absence of calmodulin at 1 μM $Ca^{2+}$ free. Result shown is a typical experiment.

red blood cells stimulated calcium uptake into sarcoplasmic reticulum in a dose-dependent manner. It was observed that following the peak of calcium uptake, further concentrations of calmodulin were slightly inhibitory. In our hands, calmodulin prepared from red blood cells was more potent than either commercial or noncommercial sources prepared from brain. Maximal levels of $Ca^{2+}$-transport stimulation were achieved with 2–2.5 μg of red blood cell calmodulin preparation compared to 9 μg of the bovine brain preparation. The significance and reason for this result is not known.

Cyclic AMP-dependent protein kinase has been shown to stimulate calcium uptake into sarcoplasmic reticulum.[16-21,31] This was observed as well in this present study. Calmodulin, though, enhanced calcium uptake in a manner that was additive to the effect of cyclic AMP-dependent protein kinase (FIGURE 1). The possibility thus exists that calmodulin and cyclic AMP-dependent protein kinase stimulate sarcoplasmic reticulum calcium transport by different mechanisms.[31]

### *Characterization of Calmodulin Stimulation of Calcium Transport in Cardiac Microsomes Enriched in Sarcoplasmic Reticulum*

FIGURE 2 indicates that calmodulin stimulates calcium transport activity at all free calcium concentrations tested. It was also noted that calmodulin stimulation was greatly enhanced in the absence of added $K^+$ (substituted by 200 mM sucrose or 110 mM LiCl). Subsequent studies indicated that in the presence of greater than 30 mM $K^+$, calmodulin stimulation of calcium transport was greatly inhibited. This would indicate that $K^+$ and calmodulin may be stimulating calcium transport through the same mechanism. Similar results were obtained when $Na^+$ was substituted for $K^+$. The fact that this was not an ionic strength effect was indicated by the lack of antagonism of calmodulin stimulation by 110 mM LiCl.

The relationship of $K^+$ to calmodulin stimulation of calcium transport into sarcoplasmic reticulum was further investigated. FIGURE 3 indicates that calmodulin stimulated the maximal activation of the transport system ($V_{CA^{2+}}$) with only a slight effect on the $K_{diss}$ for calcium. $K^+$, on the other hand, stimulated the $V_{Ca^{2+}}$ but decreased the $K_{diss}$ for calcium of the $Ca^{2+}$-transport system.

Calmodulin increased the initial rate of calcium uptake at all free calcium concentrations (FIGURE 4). This stimulation was greater at the lower free calcium concentrations tested. This result was very similar to that produced by cyclic AMP-dependent protein kinase.[32]

In these present studies we consistently observed a marked effect of calmodulin on the $V_{Ca^{2+}}$ and only a slight effect on the $K_{diss}$ for calcium. In other tissues it has been observed that in the absence of endogenous calmodulin an effect on the $K_{diss}$ for calcium was readily observed.[33] This does not appear to be the case in sarcoplasmic reticulum, as other studies conducted in our laboratory have shown that no calmodulin was bound to sarcoplasmic reticulum under the preparatory procedures utilized.[34,35]

## *Effect of Calmodulin on the Calcium-Dependent Phosphoprotein Intermediate*

It appeared from the foregoing experiments and those from other laboratories that calmodulin stimulated calcium transport into cardiac sarcoplasmic reticulum

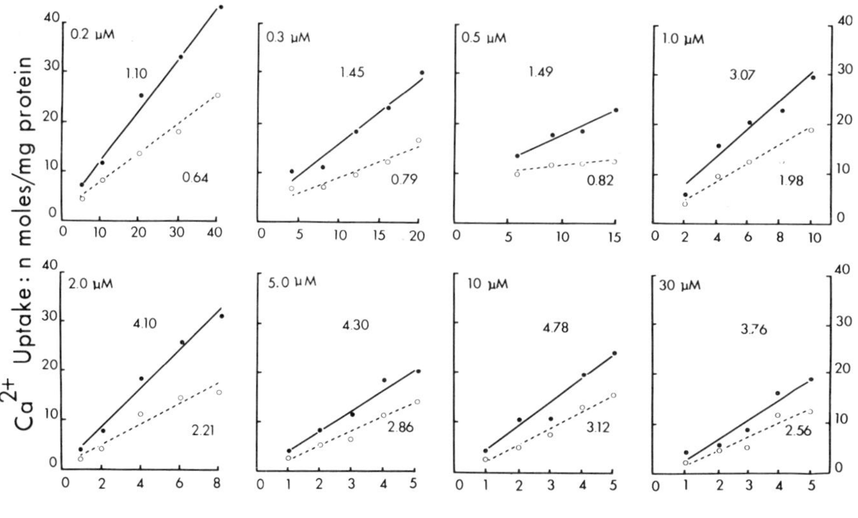


FIGURE 4. Effect of calmodulin on the initial rate of calcium uptake in microsomal preparations enriched in sarcoplasmic reticulum. $Ca^{2+}$ uptake (nmoles/mg protein) was measured at various times such that no greater than 5% of the total calcium present was accumulated by the microsomes during the incubation period. $Ca^{2+}$ uptake rates are indicated by their respective plots in the presence (●——●) and absence (O---O) of calmodulin (2 μg/0.5 ml) for each free $Ca^{2+}$ concentration used. In this experiment 50 mM phosphate was utilized as the $Ca^{2+}$ precipitating anion to allow detailed kinetic analysis.

by a direct effect on the calcium pump. The exact mechanism of this stimulation, though, remained to be elucidated. In this part of the study we therefore investigated the effect of calmodulin on the formation and decomposition of the $Ca^{2+}$-dependent phosphoprotein, $[E_{Ca}P]$, the membrane associated intermediate event of the $Ca^{2+}$-transport pathway.

The effect of calmodulin on the steady-state level of phosphoprotein present in cardiac microsomal preparations enriched in cardiac sarcoplasmic reticulum is shown in TABLE 1. In the presence of calcium (1.8 $\mu$M), calmodulin had little effect on the steady-state level of phosphorylation noted. Under the assay conditions used (2 $\mu$M ATP, 14°C and incubation times of 15 seconds), cyclic AMP and cyclic AMP-dependent protein kinase did not stimulate the level of phosphorylation above that noted in the presence of 0.1 mM EGTA. When calmodulin was present along with $Ca^{2+}$, cyclic AMP-dependent protein kinase, and cyclic AMP, the steady-state level of phosphoprotein observed was similar to that found in the presence of calcium alone. These results therefore indicate that calmodulin does not increase the steady-state level of the calcium-dependent phosphoprotein intermediate.

Calmodulin concentrations that stimulate $(Mg^{2+}+Ca^{2+})$-ATPase activity in this preparation were shown to have little effect on steady-state level of phosphoprotein present (FIGURE 5). The insert in FIGURE 5 illustrates the effect of calmodulin on the turnover rate ($v$/[EP]) of the calcium pump in this preparation. The $v$/[EP] ratio, the turnover rate of this reaction, increases from 5.6 to 10.5 at concentrations of calmodulin previously found to maximally stimulate $Ca^{2+}$ uptake in this preparation.

In view of these observations on the turnover rate of the $Ca^{2+}$ pump, we then investigated whether calmodulin increased the decomposition of the phosphoprotein intermediate when further formation of the $[^{32}P]$ phosphoprotein was prevented by the addition of high concentrations of cold ATP or EGTA (FIGURE 6). Where indicated, calmodulin was present throughout the reaction procedure. Dephosphorylation of the $Ca^{2+}$-dependent phosphoprotein in the presence of EGTA (2 mM) was slightly but not significantly enhanced in the presence of

TABLE 1

THE EFFECT OF CALMODULIN ON THE STEADY-STATE LEVEL OF $E_{Ca}P$ IN MICROSOMAL PREPARATIONS ENRICHED IN CARDIAC SARCOPLASMIC RETICULUM

| | Phosphorylation (pmoles/mg protein) |
|---|---|
| 0.1 mM EGTA (Control) | 132 |
| + $Ca^{2+}$ | 291 |
| + $Ca^{2+}$ + calmodulin | 258 |
| + cyclic AMP + protein kinase | 145 |
| + cyclic AMP + protein kinase + calmodulin + $Ca^{2+}$ | 295 |

Cardiac microsomes enriched in sarcoplasmic reticulum (0.09–0.13 mg/0.2 ml final incubation volume) were phosphorylated for 15 sec at 14°C in the presence of 2 $\mu$M [8-$[^3H]$, $\gamma$-$^{32}P$]ATP (1,000–3,000 cpm/pmole $^{32}P$ and 500–1,000 cpm/pmole $[^3H]$), 40 mM histidine-HCl, pH 6.8, 0.10 mM EGTA, and 10 $\mu$M $MgCl_2$. When calcium was added to the incubation medium it was present in a final concentration of 1.8 $\mu$M $Ca^{2+}$ free. When cyclic AMP and cyclic AMP-dependent protein kinase were present they were added in concentrations of 1 $\mu$M and 20 $\mu$g, respectively. The calmodulin concentration used was 1 $\mu$g. The result shown is a typical experiment of 3 similar experiments performed on different microsomal preparations.

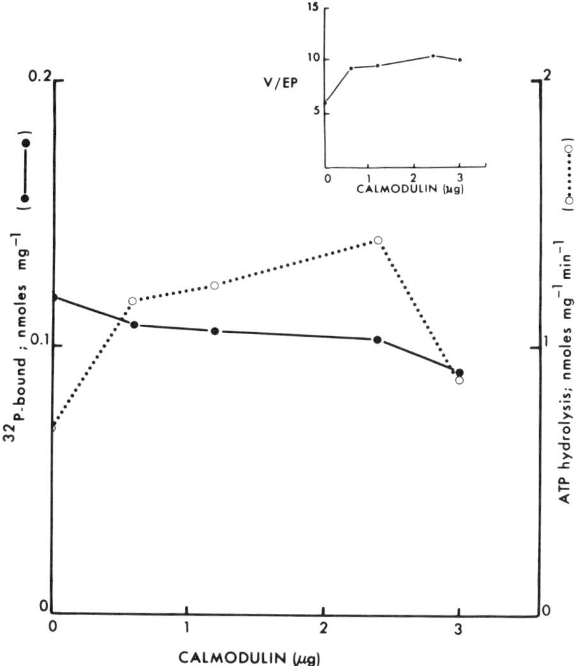


FIGURE 5. The effect of calmodulin concentration on the steady-state level of $Ca^{2+}$-dependent phosphoprotein and ($Mg^{2+} + Ca^{2+}$)-ATPase activity in microsomal preparations enriched in sarcoplasmic reticulum. Dog cardiac microsomal preparations were phosphorylated at 10°C in the presence of $Ca^{2+}$ (1.8 $\mu$M $Ca^{2+}$ free) and varying concentrations of calmodulin (0.5–2.5 $\mu$g/0.2 ml final incubation volume). Following termination of the reaction and centrifugation, an aliquot of the supernatant was assayed to determine the ($Mg^{2+} + Ca^{2+}$)-ATPase activity. The insert indicates the v/EP ratio at each calmodulin concentration. The result shown is a typical experiment of 3 similar experiments done on different preparations.

Phosphorylation refers to the $Ca^{2+}$-stimulated component following subtraction of values measured in the absence of added $CaCl_2$ (approximately 70 pmoles/mg in the presence of 0.1 mM EGTA). Reprinted with permission from *Biochemistry*, © 1980, American Chemical Society.

calmodulin. In the presence of 0.5 mM Tris-ATP, calmodulin significantly increased $Ca^{2+}$-dependent phosphoprotein decomposition compared to levels noted in the absence of added calmodulin ($P < 0.02$).

It was observed that the calmodulin effect on the decomposition of the $Ca^{2+}$-dependent phosphoprotein was more marked when Tris-ATP (0.5 mM) was the dephosphorylation agent rather than EGTA (2 mM). This could be due to the fact that calmodulin itself requires calcium for activation,[1,36,37] and not enough calcium is present in the incubation medium following the addition of the high EGTA concentrations to satisfy this requirement. These effects of calmodulin on the steady-state level of the calcium-dependent phosphoprotein and the decomposition of the intermediate of the $Ca^{2+}$-transport system have also been observed in red cell membranes (manuscript in preparation). In the latter system,

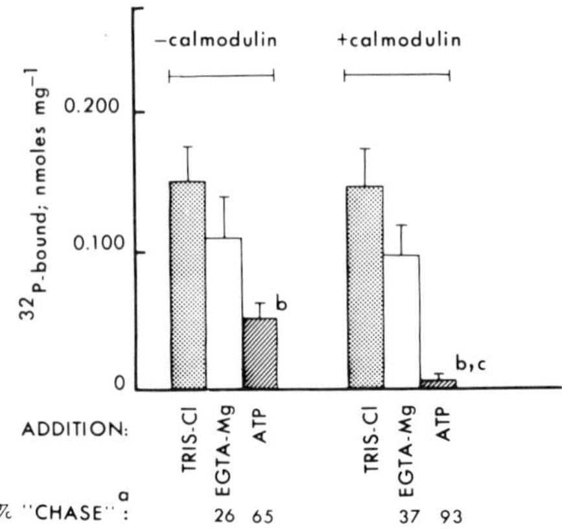


FIGURE 6. Effect of calmodulin on $Ca^{2+}$-dependent phosphoprotein decomposition in microsomal preparations enriched in sarcoplasmic reticulum. Dog cardiac microsomal preparations were phosphorylated in the presence of $Ca^{2+}$ (1.8 $\mu$M $Ca^{2+}$ free) and in the presence and absence of calmodulin (1.2 $\mu$g). The reaction proceeded at 10°C for 15 seconds when either 2 mM Tris-HCl, 2 mM EGTA in the presence of 2 mM $MgCl_2$ or 0.5 mM Tris-ATP were added to the reaction medium. The reaction was continued for an additional 5 seconds then terminated as described in the METHODS. The results shown are a mean ± S.E.M. of 4 similar experiments done on different microsomal preparations. Phosphorylation refers to the $Ca^{2+}$-stimulated component following subtraction of values measured in the absence of added $CaCl_2$ (approximately 70 pmoles/mg in the presence of 0.1 mM EGTA). [a]% dephosphorylation compared to that noted in the presence of 2 mM Tris-HCl. [b]Significant ($P < 0.01$) when compared to the control (2 mM Tris-HCl). [c]Significant ($P < 0.02$) when compared to the level noted with 0.5 mM Tris-ATP in the absence of calmodulin. Reprinted with permission from *Biochemistry*, © 1980, American Chemical Society.

calmodulin either present in the reaction medium throughout the incubation or added along with the "chase" media stimulated $Ca^{2+}$-dependent phosphoprotein dephosphorylation. This effect of calmodulin was found to be rapid, occurring within 5 seconds following addition of the "chase" media.

## DISCUSSION

Our results to date have shown that calmodulin stimulates calcium transport in cardiac microsomal preparations enriched in sarcoplasmic reticulum in a dose-dependent manner. This effect appears to be due to a direct action on the intermediate events of the calcium pump mechanism; the steady-state level of the $Ca^{2+}$-dependent phosphoprotein intermediate is unchanged in the presence of added calmodulin whereas the $(Mg^{2+}+Ca^{2+})$-ATPase activity is significantly increased. Decomposition of the phosphoprotein intermediate in the presence of high concentrations of Tris-ATP was found to be significantly increased by calmodulin. These studies indicate that calmodulin stimulates $Ca^{2+}$ transport into cardiac sarcoplasmic reticulum by increasing the rate of dephosphorylation of the $Ca^{2+}$-dependent phosphoprotein intermediate, thereby increasing the turnover rate of the transport process.

It therefore appears that calmodulin, present in large concentrations in the cardiac cell, may modulate calcium transport at the site of the removal of calcium from the contractile unit, the sarcoplasmic reticulum, thereby enhancing the relaxation rate of the muscle unit. Two other possible modulators of calcium transport at the level of the sarcoplasmic reticulum have recently been elucidated; they are, cyclic AMP-dependent protein kinase and $K^+$ (110 mM). The possible relationship of these 3 systems in the regulation of calcium transport into sarcoplasmic reticulum is illustrated in FIGURE 7. This diagram represents a simple scheme of the $(Mg^{2+}+Ca^{2+})$-ATPase and the possible interaction of the regulatory systems. In the presence of ATP, $Ca^{2+}$, and $Mg^{2+}$ the terminal phosphate of the ATP is cleaved, leading to the formation of the $Ca^{2+}$-dependent phosphoprotein intermediate, $[E_{Ca}P]$, on the outside of the sarcoplasmic reticulum membrane. The subsequent removal of $Ca^{2+}$ into the sarcoplasmic reticulum vesicles leads to the decomposition of the phosphoprotein intermediate and the production of $P_i$. This scheme is extremely simplified, as it is now known that at least two forms of phosphoprotein intermediate exist with differing $Mg^{2+}$ and ADP sensitivities.[38]

Cyclic AMP-dependent protein kinase has been shown to mediate the phosphorylation of sarcoplasmic reticulum protein.[21] The extent of this phosphorylation correlates closely with the increment of stimulation of the rate of calcium uptake.[18,21] The phosphoprotein formed was almost completely stable in trichloroacetic acid or hydroxylamine. These and other chemical characteristics distinguish this phosphoprotein from the acyl phosphate $(Mg^{2+}+Ca^{2+})$-ATPase intermediate, [EP]. In subsequent studies it was found that this increase in phosphoprotein formation was due to the phosphorylation of a single microsomal component, a 22,000 dalton protein named phospholamban.[21] Tada *et al.*[21] have proposed a mechanism for modulation of the cardiac sarcoplasmic reticulum

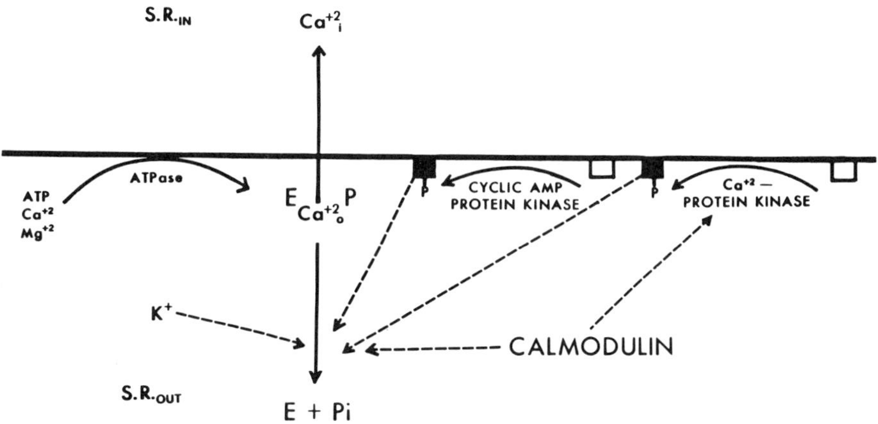


FIGURE 7. Possible mechanism of stimulation of regulators of calcium transport in cardiac sarcoplasmic reticulum. S.R. refers to sarcoplasmic reticulum. The ATPase pathway described is simplified. $Ca_O$ refers to calcium outside sarcoplasmic reticulum. $Ca_i$ refers to calcium inside the sarcoplasmic reticulum. CAMP-protein kinase refers to cyclic AMP-dependent protein kinase, $Ca^{2+}$-protein kinase refers to $Ca^{2+}$–calmodulin-dependent protein kinase. □ refers to phospholamban sites.

calcium pump by cyclic AMP-dependent protein kinase: according to this hypothesis, the 22,000 dalton component of sarcoplasmic reticulum phospholamban, serves as a modulator of the $Ca^{2+}$ pump by altering the ATPase activity tightly coupled to the $Ca^{2+}$-transport system. Under the influence of hormones and cyclic AMP, cytoplasmic and/or membrane-bound protein kinase phosphorylates a component of sarcoplasmic reticulum that can stimulate $(Mg^{2+} + Ca^{2+})$-ATPase activity and $Ca^{2+}$ uptake by increasing the turnover rate of this process. Phosphoprotein phosphatase then catalyzes the dephosphorylation of phospholamban. $Ca^{2+}$ uptake will proceed in the absence of phospholamban phosphorylation, which suggests that this protein is a modulator rather than an indigenous part of the $Ca^{2+}$-transport system.[20] The relationship of phospholamban phosphorylation to $Ca^{2+}$-uptake stimulation has been postulated by Tada *et al.*[39] to be due to an enhancement of the dephosphorylation of the $Ca^{2}$-dependent phosphoprotein intermediate.This appears to be the same mechanism by which calmodulin stimulates calcium transport in this preparation. We have observed, though, that the augmentation of calcium transport by calmodulin is additive to that of cyclic AMP-dependent protein kinase. It is thus conceivable that there is another mechanism by which cyclic AMP-dependent protein kinase or calmodulin stimulate calcium transport into sarcoplasmic reticulum. It is also possible that these systems stimulate different $(Mg^{2+} + Ca^{2+})$-ATPase sites on the membrane.

$K^+$, in concentrations close to physiological levels (110 mM), has been shown to stimulate $Ca^{2+}$ uptake and $(Mg^{2+} + Ca^{2+})$-ATPase activity in cardiac microsomal preparations enriched in sarcoplasmic reticulum.[26,27] No specific phosphoprotein formation has been attributed to this ion. It has recently been shown, though, that $K^+$ increases the rate of dephosphorylation of the $Ca^{2+}$-dependent phosphoprotein intermediate.[40] In these present studies, contrary to the results observed in the presence of cyclic AMP-dependent protein kinase, calmodulin and $K^+$ do not appear to produce additive effects on calcium transport activity; in the absence of $K^+$ the stimulation of calcium transport by calmodulin is markedly increased. It is possible that calmodulin stimulates calcium uptake in a similar manner to $K^+$ and thus does not function maximally in the presence of high $K^+$ concentrations. The physiological relevance of the role of calmodulin in this system therefore remains to be elucidated.

Thus all three potential regulators of $Ca^{2+}$ transport appear, at least in part, to have a similar mechanism: They increase the turnover rate of the calcium pump mechanism by enhancing the rate of dephosphorylation of the calcium-dependent phosphoprotein intermediate of the transport process. This may then be a unifying mechanism by which these agents affect calcium transport.

Le Peuch *et al.*[41] have recently suggested that calmodulin affects calcium transport at the level of the sarcoplasmic reticulum by stimulation of phospholamban phosphorylation via a $Ca^{2+}$–calmodulin-dependent membrane-bound protein kinase. This possibility is also illustrated in the diagram (FIGURE 7). In our studies, to date, under conditions where cyclic AMP-dependent protein kinase did not stimulate phosphoprotein phosphorylation the stimulatory effects of calmodulin on $(Mg^{2+} + Ca^{2+})$-ATPase activity were still observed. In fact, in the presence of calmodulin total phosphoprotein levels were unchanged. It is conceivable that the $Ca^{2+}$–calmodulin-dependent protein kinase was not active under the incubation conditions utilized.

The postulation of a direct effect of calmodulin on the $(Mg^{2+} + Ca^{2+})$-ATPase protein is corroborated by the findings of Lynch and Cheung,[37] who showed that in the presence of $Ca^{2+}$ an ATPase-calmodulin complex is formed; lowering the $Ca^{2+}$ level dissociates the two proteins, returning the enzyme activity to its basal level. The effects of $Ca^{2+}$ on the enzyme–calmodulin complex were found to be immediate and reversible. In other studies recently completed we have observed that calmodulin is not bound to sarcoplasmic reticulum under normal preparation procedures.[34,35] The binding of exogenous [$^{125}$I]calmodulin to these preparations was found to be $Ca^{2+}$ dependent and increased with increasing calcium concentration.[35] This binding was decreased in the presence of 110 mM $K^+$ at all free calcium concentrations studied. These observations again indicate that in order for calmodulin to exert its stimulatory effects, $Ca^{2+}$ is required and $Ca^{2+}$-calmodulin must be bound to the sarcoplasmic reticulum membrane. Presumably, this binding is specific for the $(Mg^{2+}+Ca^{2+})$-ATPase sites and reduction of extracellular free calcium levels will markedly reduce the calmodulin bound to the sarcoplasmic reticulum membrane. Further studies are required to indicate the selectivity of the calmodulin binding to sarcoplasmic reticulum membrane sites and to elucidate the influence of $Ca^{2+}$ and $K^+$ on this activity.

## Acknowledgments

The author gratefully acknowledges the significant contributions of Gary Lopaschuk, Betty Richter, and Mohamed Remtulla in the studies described.

## References


1. Cheung, W. Y. 1980. Science **207:** 19.
2. Gopinath, R. M. & F. F. Vincenzi. 1977. Biochem. Biophys. Res. Commun. **77:** 1204.
3. Luthra, M. G., G. R. Hildenbrandt & D. J. Hanahan. 1976. Biochim. Biophys. Acta **419:** 164.
4. Katz S., B. D. Roufogalis, A. D. Landman & L. Ho. 1979. J. Supramol. Struct. **10:** 215.
5. Jarrett, H. W. & J. T. Penniston. 1977. Biochem. Biophys. Res. Commun. **77:** 1210.
6. Hinds, T. R., F. L. Larsen & F. F. Vincenzi. 1978. Biochem. Biophys. Res. Commun. **81:** 455.
7. Hasselbach, W. & M. Makinose. 1963. Biochem. Z. **339:** 94.
8. Ebashi, S. & F. Lipmann. 1966. J. Cell. Biol. **14:** 389.
9. Weber, A., R. Herz & I. Reiss. 1966. Biochem. Z. **345:** 329.
10. Ebashi, S., M. Endo & I. Ohtsuki. 1969. Quart. Rev. Biophys. **2:** 251.
11. Inesi, G. 1972. Annu. Rev. Biophys. Bioeng. **1:** 191.
12. Yamada, S., T. Yamamoto & Y. J. Tonomura 1970. Biochem. **67:** 789.
13. Hasselbach, W. & M. Makinose. 1961. Biochem. Z. **333:** 518.
14. Yamamoto, T. & Y. Tonomura. 1978. J. Biochem. **64:** 137.
15. MacLennan, D. H. & P. C. Holland. 1975. Annu. Rev. Biophys. Bioeng. **4:** 377.
16. Kirchberger, M. A., M. Tada, D. I. Repke & A. M. Katz. 1972. J. Mol. Cell. Cardiol. **4:** 673.
17. Kirchberger, M. A., M. Tada & A. M. Katz. 1974. J. Biol. Chem. **249:** 6166.
18. Kirchberger, M. A. & G. Chu. 1976. Biochim. Biophys. Acta **419:** 559.
19. Tada, M., M. A. Kirchberger, D. I. Repke & A. M. Katz. 1974. J. Biol. Chem. **249:** 6174.
20. Tada, M., M. A. Kirchberger & A. M. Katz. 1975. J. Biol. Chem. **250:** 2640.
21. Tada, M., T. Yamamoto & Y. Tonomura. 1978. Physiol. Rev. **58:** 1.
22. Shigekawa, M. & L. J. Pearl. 1976. J. Biol. Chem. **251:** 6947.
23. Duggan, P. F. 1977. J. Biol. Chem. **252:** 1620.
24. Tomlinson, C. W., A. M. Bernatsky & N. S. Dhalla. 1976. Biochem. Pharmacol. **25:** 2271.
25. Green, R. A., J. J. A. Heffron & G. Mitchell. 1976. Gen. Pharmacol. **7:** 361.
26. Jones, L. R., H. R. Besch, Jr. & A. M. Watanabe. 1977. J. Biol. Chem. **252:** 3315.

27. JONES, L. R., H. R. BESCH & A. M. WATANABE. 1978. J. Biol. Chem. **253:** 1643.
28. STEVENS, F. C., T. M. WALSH, H. C. HO, T. C. TEO & J. H. WANG. 1976. J. Biol. Chem. **251:** 4495.
29. WATERSON, D. M., W. G. HARRELSON, JR., P. M. KELLER, F. SHARIEF & T. C. VANAMAN, 1976. J. Biol. Chem. **251:** 4501.
30. HARIGAYA, S. & A. SCHWARTZ. 1969. Circ. Res. **25:** 781.
31. KATZ, S. & M. A. REMTULLA. 1978. Biochem. Biophys. Res. Commun. **83:** 1373.
32. HICKS, M. J., M. SHIGEKAWA & A. M. KATZ. 1979. Circ. Res. **44:** 384.
33. LARSEN, F. L., S. KATZ & B. D. ROUFOGALIS. 1980. Ann. N.Y. Acad. Sci. (This volume).
34. LOPASCHUK, G. & S. KATZ. 1980. Proc. West. Pharmacol. Soc. **23:** 25.
35. LOPASCHUK, G. & S. KATZ. 1980. Ann. N.Y. Acad. Sci. (This volume).
36. TEO, T. S. & T. H. WANG. 1973. J. Biol. Chem. **248:** 5950.
37. LYNCH, T. J. & W. Y. CHEUNG. 1979. Arch. Biochem. Biophys. **194:** 165.
38. TONOMURA, Y. 1973. *In* Muscle Proteins, Muscle Contraction and Cation Transport. University Park Press. Baltimore, London & Tokyo. p. 305.
39. TADA, M., F. OMORI, M. YAMADA & H. ABE. 1979. J. Biol. Chem. **254:** 319.
40. SHIGEKAWA, M. & A. A. AKOWITZ. 1979. J. Biol. Chem. **254:** 4726.
41. LE PEUCH, C. J., J. HIECH & J. G. DEMAILLE. 1979. Biochemistry **18:** 5150.

## DISCUSSION OF THE PAPER

DR. D. FLOCKHART: I wondered what your evidence was for there not being a calcium-dependent protein kinase present in your preparation?

DR. S. KATZ: Under the incubation conditions that we used (two micromolar ATP, ten seconds at ten degrees) we have found in other studies that there is only one peak phosphorylated.

DR. E. CARAFOLI: It seems probable, at least, that there is calmodulin in voluntary muscle SR. Would you care to comment on the possibility that phospholamban-dependent mechanisms are operating also in nonheart sarcoplasmic tissue?

DR. KATZ: Yes. We are presently looking at a series of fast, slow, and mixed skeletal muscle systems to, perhaps, answer that particular question. We know that there is a controversy regarding the role of cyclic AMP-dependent protein kinase and phospholamban phosphorylation in other muscle systems. What we have noted to date, is that calmodulin does not appear to stimulate slow skeletal muscle sarcoplasmic reticulum, but it does affect fast skeletal muscle sarcoplasmic reticulum calcium transport.

# THE ROLE OF CALMODULIN IN REGULATION OF CARDIAC SARCOPLASMIC RETICULUM PHOSPHORYLATION*

Evangelia G. Kranias, Louise M. Bilezikjian, James D. Potter
Michael T. Piascik, and Arnold Schwartz

*Department of Pharmacology and Cell Biophysics*
*University of Cincinnati College of Medicine*
*Cincinnati, Ohio 45267*

## INTRODUCTION

Our past studies have concentrated on the role of sarcoplasmic reticulum (SR) in control of relaxation and contraction in normal and diseased cardiac muscle.[1,2] It is now established that the SR in cardiac muscle has the enzymatic machinery and the kinetic characteristics consistent with a major role in intracellular calcium control and relaxation, very similar to that long accepted for fast skeletal muscle.[3-5] It is of further importance, that the SR, particularly the membranes derived from cardiac muscle, may possess certain site or sites that, when phosphorylated by cyclic AMP-dependent protein kinase, may change in conformation so as to enhance the rate of calcium transport into the lumen of the vesicle.[4,5] One mechanism of regulation, therefore, may involve phosphorylation of membranes.[4,5]

Calmodulin appears to be an intracellular calcium receptor with multiple functions, one of which might involve the sarcoplasmic reticulum. Calmodulin has been shown to regulate the activities of cyclic nucleotide phosphodiesterase,[6,7] brain adenylate cyclase,[8-10] erythrocyte ($Ca^{2+}$-$Mg^{2+}$)-ATPase,[11,12] myosin light chain kinase,[13,14] nicotinamide adenine dinucleotide (NAD) kinase,[15] phospholipase $A_2$,[16] synaptosomal $Ca^{2+}$ ATPase[17] and phosphorylase kinase.[18] Calmodulin has also been shown to stimulate calcium transport in erythrocytes[19] and sarcoplasmic reticulum (SR) from cardiac muscle.[20,21] It is not surprising that the calmodulin effects on cardiac SR may involve phosphorylation of a site of $M_r$ 22,000 that has been implicated in regulation.[21] Phosphorylation of this membrane component, named phospholamban, by cyclic AMP-dependent protein kinases has been studied by several investigators, who have shown an enhancement of both calcium transport and $Ca^{2+}$-dependent ATPase activity, while maintaining the stoichiometric coupling ratio of Ca:ATP.[22-25] In a recent paper[21] it was suggested that calmodulin-stimulated $Ca^{2+}$-dependent phosphorylation of cardiac SR may occur at a different site on phospholamban than the cAMP-dependent phosphorylation and that phosphorylation of this site results in stimulation of $Ca^{2+}$ transport, but has no effect on ATPase activity. Thus, $Ca^{2+}$ transport by the cardiac SR may be under regulation by both cyclic AMP and

*Supported in part by U.S. Public Health Service Grants P01 HL 22619-02 (III A, III E, IV A), by a Grant-In-Aid from American Heart Association (78-1167) and by a Grant-In-Aid from the Southwestern Ohio Chapter of American Heart Association. Some of the studies reported in this communication form part of the Ph.D. Dissertation to be submitted by L. M. B.



0077-8923/80/0356-0279 $01.75/0 © 1980, NYAS

calcium and hence may involve calmodulin. The interrelationship of these regulatory processes is not clear. Katz and Remtulla[20] showed that maximal concentrations of calmodulin and cyclic AMP-dependent protein kinase were additive in their ability to stimulate calcium transport. On the other hand, Le Peuch *et al.*[21] claimed that cAMP-dependent phosphorylation did not activate $Ca^{2+}$ transport in the absence of calmodulin but only amplified the activation brought about by the calmodulin-dependent phosphorylation.

In this report we present our studies on the calmodulin-stimulated $Ca^{2+}$-dependent phosphorylation of cardiac SR. Our aim is to establish that phosphorylation of cardiac SR may be regulated by calmodulin. The criteria set by Cheung[26] for the calmodulin regulated reactions are: 1) tissue or cell must possess sufficient calmodulin in the appropriate locale, 2) depletion of endogenous calmodulin should render the experimental system susceptible to exogenous calmodulin, 3) sequestering calcium in the reaction system should return the calmodulin-induced activity to basal level, 4) addition of trifluoperazine should return the calmodulin-dependent activity to the steady-state level, and 5) the effect of calmodulin should be reversed by its antibody.

Our results presented here fulfill criteria 1) through 4) for the calmodulin regulated reactions.

## Methods

### *Preparation of Sarcoplasmic Reticulum Vesicles*

Sarcoplasmic reticulum (SR) from dog cardiac muscle was prepared as previously described[1,27] with the following modifications. The isolation was carried out in the presence of 0.3 M sucrose in 20 mM Tris-maleate buffer, pH 7.0. The microsomes enriched in SR were homogenized and washed in 0.6 M KCl twice, and the final pellet was then suspended in 30 mM Tris-maleate pH 7.0, containing 100 mM KCl and 0.3 M sucrose. The suspension was quickly frozen in liquid $N_2$ and stored at −75°C. This preparation was found to be stable for several weeks and no significant decrease in kinase activity was observed during storage. The final yield of SR protein ranged from 0.6 to 0.8 mg per gram of ventricular tissue.

### *Isolation of Calmodulin-Dependent Cyclic Nucleotide Phosphodiesterase*

Cyclic nucleotide phosphodiesterase (PDE) was isolated from guinea pig brain by a slight modification of the procedure of Dedman *et al.*[28] Animals were decapitated and brains placed in cold buffer containing 10 mM Tris-HCl (pH 7.5) and 10 mM $\beta$-mercaptoethanol. Brains were homogenized by two 5-second bursts using a Polytron (Brinkman) at a setting of 7. The homogenate was sonicated (Branson Sonic Power Co.) in three 1-minute pulses at a setting of 55. The sonicated homogenate was centrifuged for 15 min at 500 × g and the resulting supernatant centrifuged at 20,000 × g for 20 min. The supernatant was again centrifuged at 100,000 × g for 1 h and the supernatant was applied to a DE-52 column (30 × 2.5 cm) preequilibrated with 10 mM Tris-HCl containing 10 mM $\beta$-mercaptoethanol (BME) (pH 7.5). Following a wash with 200 ml of the same buffer, the PDE was eluted with a linear 0–0.4 M $(NH_4)_2SO_4$ gradient. Column fractions (3 ml) were tested for PDE activity. Calmodulin-dependent PDE was eluted between fractions 145–160. The fractions were pooled, quick-frozen, and stored at $-20°C$ until used.

### *Preparation of Calmodulin*

Calmodulin was prepared from bovine brain by the method of Lin *et al.*[29]

### *Quantitation of Calmodulin*

The calmodulin content of cardiac muscle and cardiac SR fractions was estimated by the ability to activate cyclic nucleotide PDE. SR membranes (1 mg/ml) were homogenized in 1% Lubrol and then left on ice for 30 min. The homogenate was subsequently boiled (>85°C) for 5 min and centrifuged at 12,000 × g for 15 min. The boiled supernatant was freeze-dried and then examined for the ability to activate calmodulin-dependent PDE. The activation of PDE was dependent on $Ca^{2+}$, as the addition of 2 mM EGTA to the reaction prior to the boiled supernatant fractions prevented activation. The amount of calmodulin present was estimated from a standard PDE activation curve using purified bovine brain calmodulin as a standard.

### *Assay of Cyclic Nucleotide Phosphodiesterase*

Phosphodiesterase activity was monitored by the spectrophotometric assay of Dedman and Means.[30] The reaction mixture contained 25 mM MOPS (morpholinopropane sulfonic acid), pH 7.0, 3 mM $MgCl_2$, 10 mM $\beta$-mercaptoethanol, 0.5 mM $CaCl_2$, 0.3 units adenylate deaminase, 75 $\mu$M cAMP, and 75–100 $\mu$g purified calmodulin-dependent PDE, in a total volume of 1.0 ml.

The standard PDE activation curve was constructed by adding increasing amounts (10–300 ng) of calmodulin to this assay mixture. Subsequently, fixed amounts (15–50 $\mu$l) of extracted SR membranes were tested for their ability to activate PDE. The hydrolysis rate at 25°C was determined by measuring the decrease in optical density at 265 nm.

### *Phosphorylation of Cardiac Sarcoplasmic Reticulum Membranes*

Phosphorylation was carried out at 30°C in the presence of 50 mM phosphate buffer (pH 7.0), 10 mM $MgCl_2$, 10 mM NaF containing 0.5 mg/ml sarcoplasmic reticulum vesicles, 1 $\mu$M cyclic AMP and 500 $\mu$M [$\gamma$-$^{32}$P]ATP. Cyclic AMP-dependent phosphorylation by either endogenous or exogenous (100 $\mu$g/ml) protein kinases was determined in the presence of 500 $\mu$M EGTA. Calmodulin-stimulated phosphorylation was determined in the presence of purified brain calmodulin and added $CaCl_2$ to achieve the desired free [$Ca^{2+}$] in the presence of 0.5 mM EGTA. Free calcium concentrations at pH 7.0 were calculated using the association constants of Sillen and Martell.[31] The reaction was terminated by the addition of 2.5 ml of 7% ice-cold perchloric acid containing 7% polyphosphate and 0.5 mg of carrier skeletal sarcoplasmic reticulum protein was added. The samples were centrifuged and the pellet was washed three times with 7% perchloric acid containing 7% polyphosphoric acid. The final pellet was dissolved in 0.5 ml of 10 mM NaOH containing 0.1 mM $Na_2PO_4 \cdot H_2O$; 10 ml of Aquasol II (New England Nuclear) was added and radioactivity was determined in a model 3320 Packard Tri-Carb liquid scintillation spectrometer.

In experiments in which cAMP-dependent and/or calmodulin stimulated phosphorylation of SR was correlated with calcium uptake, sarcoplasmic reticulum vesicles were preincubated in the same reaction medium described above for measuring sarcoplasmic reticulum phosphorylation, except that unlabeled ATP was used and sodium fluoride was omitted. Control vesicles were also incubated under identical conditions without ATP. After 2 minutes of incubation at 30°C, the mixture was centrifuged at 105,000 × g for 30 minutes and the pellet was homogenized gently in ice-cold 100 mM KCl, 20 mM Tris-maleate (pH 7.0). Sarcoplasmic reticulum protein recovery after this procedure was 90% and did not vary with the nature of the pretreatment. Assays for calcium uptake were subsequently carried out as described below.

### *Assay for Calcium Uptake*

Oxalate-facilitated calcium uptake was determined at 30°C in 40 mM Histidine-HCl (pH 7.0), 100 mM KCl, 6 mM $MgCl_2$, 5 mM ATP, 5 mM sodium azide, 2.5 mM Tris-oxalate, and a [$^{45}$Ca]-EGTA buffer system. A computer program was used to calculate the total calcium added to 200 $\mu$M ethyleneglycol-bis-($\beta$-aminoethylether)N,N′-tetraacetic acid (EGTA) in order to obtain various free [$Ca^{2+}$] at pH 7.0.[31] Final concentrations of sarcoplasmic reticulum membranes were 50 $\mu$g/ml. The reaction was initiated by adding ATP and was terminated by passing aliquots of the reaction mixture through a type HA Millipore filter (0.45 $\mu$m average pore diameter). The filters were washed with 20 ml containing 20 mM Tris-maleate (pH 7.0) and 3 mM $CaCl_2$, dried and counted for radioactivity in Aquasol (New England Nuclear).

### *Polyacrylamide Gel Electrophoresis of Phosphorylated Membranes*

Sarcoplasmic reticulum (1 mg) was incubated with [$\gamma$-$^{32}$P]ATP (200–1,000 cpm/pmol) in a total volume of 1 ml under the conditions of the phosphorylation assay as described above. After 2 minutes incubation at 30°C, the reaction mixture was cooled on ice and 0.5 ml of 20 mM Tris-maleate buffer (pH 7.4) containing 1% SDS and 1% BME were added. The sample was subsequently dialyzed overnight against the same buffer containing SDS and BME.

Polyacrylamide gel electrophoresis of $^{32}$P-labeled sarcoplasmic reticulum under denaturing conditions was carried out at pH 7.1 in the presence of 0.1% sodium dodecyl sulfate according to the method of Laemmli[32] and Weber and Osborn.[33] The molecular weight of the radioactive peak was determined by utilizing appropriate markers and constructing a relative mobility curve. Standards used were human serum albumin ($M_r$ 69,000), DNase 1 ($M_r$ 31,000), trypsin ($M_r$ 24,000), and egg white lysozyme ($M_r$ 14,300).

## RESULTS AND DISCUSSION

### *Preparation of Cardiac Sarcoplasmic Reticulum Vesicles*

Cardiac SR was prepared as previously described[1,27] except that sucrose (0.3 M) was included in the isolation buffers and the pH was 7.0. Vesicles isolated in this manner could be stored up to four weeks in 20 mM Tris-maleate/100 mM KCl/0.3 M sucrose (pH 7.0) at −70°C and showed no loss of autophosphorylation as compared to the fresh state.

The $Ca^{2+}$-dependent ATPase activity [0.15 μmol/(min mg)] was almost insensitive to 5 mM azide and 1 mM ouabain. The initial rate of $Ca^{2+}$ uptake was 0.16 to 0.2 μmol/(min mg) in the presence of 2.5 mM oxalate.

To determine whether isolation of SR vesicles in the presence of $Ca^{2+}$ or EGTA would affect the levels of calmodulin associated with SR, SR vesicles were isolated in the presence of: a) $Ca^{2+}$ (30 μM), b) EGTA (100 μM), and c) $Ca^{2+}$ (30 μM) and subsequently washed in $Ca^{2+}$ (30 μM) or EGTA (100 μM). Regardless of the isolation procedure used, SR was not totally depleted of calmodulin although when prepared in the presence of $Ca^{2+}$ it had higher (2×) levels of calmodulin than when prepared or washed in the presence of EGTA. However, the degree and concentration dependence of the calmodulin-stimulated phosphorylation was the same regardless of the isolation procedure. These results indicate that the $Ca^{2+}$-dependent protein kinase was firmly bound to the SR membrane and the levels of calmodulin present in the SR preparation were not sufficient to stimulate phosphorylation in the absence of added calmodulin. Therefore, SR vesicles were prepared as previously described[1,27] without including $Ca^{2+}$ or EGTA in the isolation buffers.

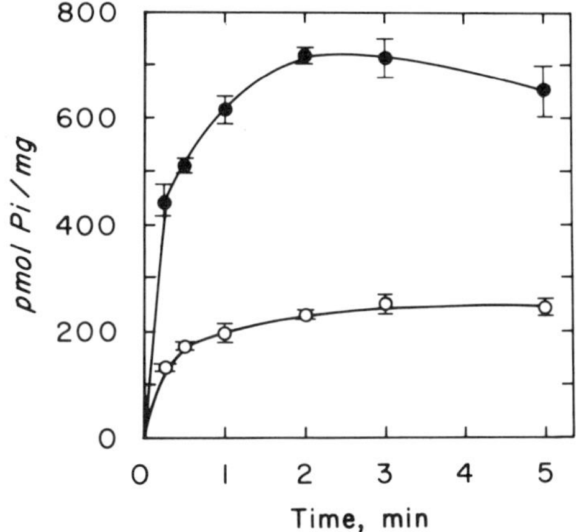


FIGURE 1. Phosphorylation of cardiac sarcoplasmic reticulum. SR vesicles were incubated, as described in METHODS, in the presence of 0.5 mM EGTA and 1 μM cAMP (○) or 1 μM free $Ca^{2+}$, 1 μM cAMP and $10^{-7}$ M calmodulin (●). Values represent the mean ± SEM for three determinations.

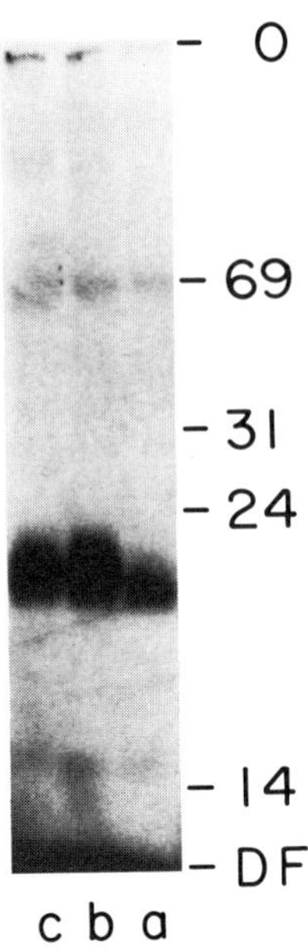


FIGURE 2. Autoradiogram of SDS-polyacrylamide gels of phosphorylated cardiac sarcoplasmic reticulum. SR vesicles were phosphorylated in the presence of 1 $\mu$M cAMP (a) or $10^{-7}$ M calmodulin in the presence (b) or absence (c) of 1 $\mu$M $Ca^{2+}$. Dye front (DF). Origin (O). The figures represent the molecular weights ($\times 10^3$) of the following markers: Human serum albumin (69,000), DNase 1 (31,000), Trypsin (24,000), and lysozyme (14,300).

### *Cyclic AMP-Dependent and Calmodulin Stimulated Phosphorylation of Cardiac SR Vesicles*

Cardiac SR was incubated in the presence of cAMP (1 $\mu$M) as well as in the presence of calmodulin ($10^{-7}$ M) and $Ca^{2+}$ (1 $\mu$M free $Ca^{2+}$). Cyclic AMP-dependent and calmodulin-stimulated phosphorylations of cardiac SR occurred so rapidly at 30°C (FIGURE 1) that it was difficult to measure initial rates of phosphate incorporation. About 60 percent of the maximum phosphate incorporation occurred within 15 sec, the first time point obtained. The speed by which these two reactions occur is very important if phosphorylation of SR is to be considered a regulatory mechanism in the cardiac muscle. Maximum phosphorylation occurred by 2 min and all data presented in this study represent levels of SR phosphorylation at an incubation period of 2 min.

Phosphorylation of SR vesicles reached 0.25 nmol/mg in the presence of cAMP and 0.7 nmol/mg in the presence of cAMP, calmodulin and $Ca^{2+}$. The extent of calmodulin-stimulated phosphorylation was the same whether cAMP-dependent phosphorylation was complete (by saturating concentrations of exogenous protein kinase) or not (by the endogenous protein kinase). Thus, calmodulin-stimulated phosphorylation probably occurs at a site different from cAMP-dependent phosphorylation.

### *Nature of the Phosphorylated Substrate in Cardiac SR*

Phosphorylated cardiac SR by either cAMP-dependent protein kinase and/or calmodulin stimulated protein kinase was resistant to extraction by hot acid or hydroxylamine indicating the presence of phosphoester bonds (data not shown). Phosphorylated SR was subjected to gel electrophoresis under denaturing conditions and a 22,000 $M_r$ protein, phospholamban, was identified as the only $^{32}$P-acceptor protein (FIGURE 2). The 22,000 $M_r$ protein was phosphorylated by the endogenous protein kinase in the presence of cAMP (FIGURE 2a). Addition of calmodulin ($10^{-7}$ M) in the presence of 500 $\mu$M EGTA had no profound effect on the level of cAMP-dependent phosphorylation (FIGURE 2c). However, addition of calmodulin in the presence of 1 $\mu$M free $Ca^{2+}$ caused a significant increase of $^{32}$P-incorporated in this protein (FIGURE 2b) indicating that calmodulin stimulated phosphorylation was $Ca^{2+}$ dependent. These results are in agreement with those by Le Peuch *et al.*[21] who also observed calmodulin stimulated phosphorylation of phospholamban. The authors reported that "phospholamban" is a dimer that can be dissociated by the mixture of ionic and nonionic detergents.

### *Properties of the Calmodulin Stimulated $Ca^{2+}$-Dependent Protein Kinase*

Sarcoplasmic reticulum vesicles were phosphorylated in the presence of various amounts of calmodulin. Calmodulin-stimulated phosphorylation of SR increased as the levels of added calmodulin increased over $5 \times 10^{-9}$ M and reached maximum at $5 \times 10^{-8}$ M (FIGURE 3). The contaminating calmodulin levels by the SR preparation were less than 10 nM, a concentration which was not sufficiently high to stimulate the $Ca^{2+}$-dependent protein kinase (FIGURE 3). Half-maximal stimulation occurred at $1.4 \times 10^{-8}$ M added calmodulin when assayed in the presence of 1 $\mu$M free $Ca^{2+}$. These data represent phosphorylation stimulated by calmodulin above that by the endogenous cAMP-dependent

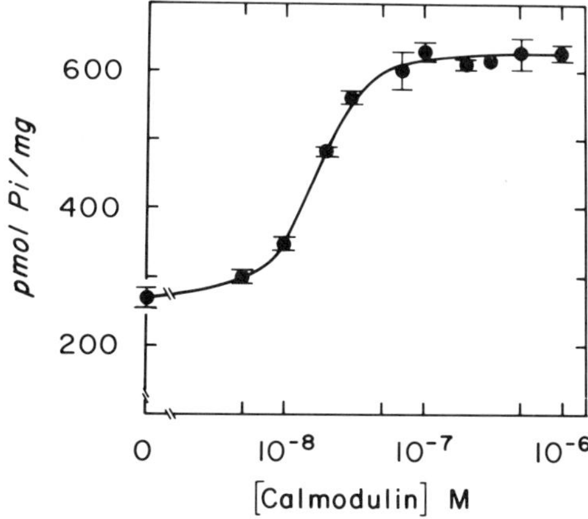


FIGURE 3. Effect of calmodulin concentration on phosphorylation of cardiac sarcoplasmic reticulum. Values represent the mean ± SEM for three determinations.

protein kinase (275 pmol $P_i$/mg). However, the results were similar when the study was conducted in the presence of exogenous cAMP-dependent protein kinase (data not shown). Membrane-bound $Ca^{2+}$-dependent protein kinases have been previously reported in SR from rabbit and mouse skeletal muscle.[34,35] However, those are different than the enzyme reported here because they do not require calmodulin.

### *$Ca^{2+}$-Dependence of Cyclic AMP-dependent and Calmodulin Stimulated Phosphorylations of Cardiac SR*

Cardiac SR vesicles, prepared as described under METHODS, were phosphorylated by the endogenous cAMP-dependent protein kinase in the presence of 1 μM cAMP. Addition of exogenous cAMP-dependent protein kinase stimulated SR phosphorylation about two-fold. The cAMP-dependent phosphorylation by either endogenous or exogenous protein kinase appeared to be $Ca^{2+}$-independent (FIGURE 4). Inclusion of calmodulin ($10^{-7}$M) in the phosphorylation assay had no effect on cAMP-dependent phosphorylation of cardiac SR by either endogenous or exogenous protein kinase when assayed in the absence of added $Ca^{2+}$. However, when the $[Ca^{2+}]$ was raised, a concentration-dependent increase in SR phosphorylation occurred in the presence of calmodulin (FIGURE 4). This concentration-dependent increase was not observed in the absence of added calmodulin indicating that the endogenous calmodulin levels (less than 10 nM), introduced by the SR preparation in the phosphorylation assay, were not sufficient to stimulate the $Ca^{2+}$-dependent protein kinase. These results are in agreement with the data in FIGURE 3 where it is shown that 10 nM of added calmodulin does not significantly stimulate SR phosphorylation above the levels obtained by the cAMP-dependent protein kinase.

Calmodulin-stimulated phosphorylation of SR increased as the free $[Ca^{2+}]$ was raised above $10^{-7}$ M and it reached maximum at $10^{-6}$ M free $[Ca^{2+}]$ (FIGURE

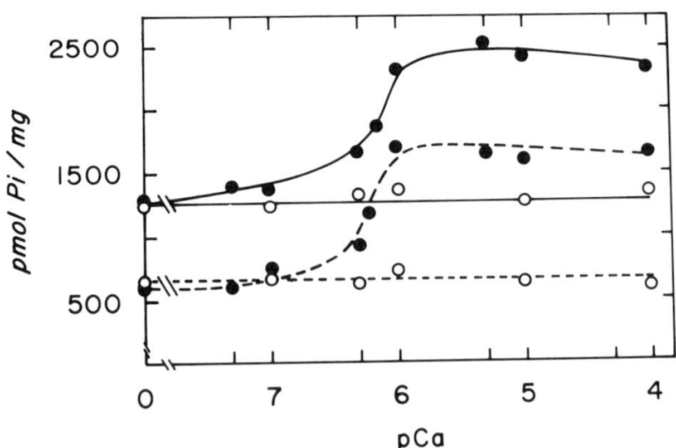


FIGURE 4. Effect of free $Ca^{2+}$ concentration on phosphorylation of cardiac sarcoplasmic reticulum. SR vesicles were phosphorylated in the presence of 1 μM cAMP (○--○) and with the following additions: 100 μg of cAMP-dependent protein kinase/ml (○—○); $10^{-7}$ M calmodulin (●--●); $10^{-7}$ M calmodulin and 100 μg of cAMP-dependent protein kinase/ml (●—●).

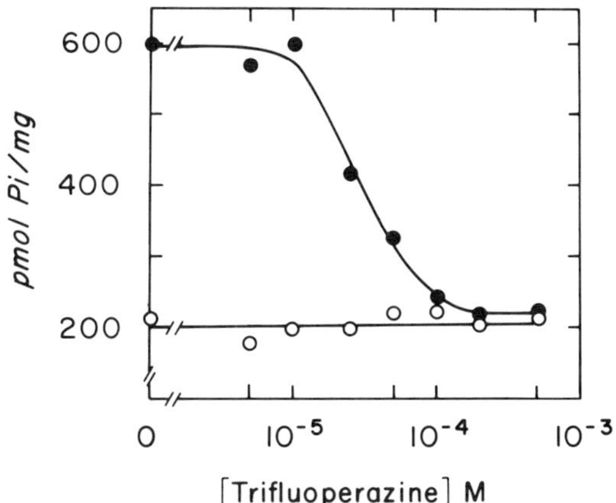


FIGURE 5. Effect of trifluoperazine concentration on phosphorylation of cardiac sarcoplasmic reticulum. SR vesicles were phosphorylated in the presence of 1 $\mu$M $Ca^{2+}$ and either 1 $\mu$M cAMP (○—○) or 1 $\mu$M cAMP and $10^{-7}$ calmodulin (●—●).

4). The extent of stimulation of SR phosphorylation was the same whether endogenous or exogenous cAMP-dependent protein kinase was used. The $K_D$ for $Ca^{2+}$ of the calmodulin-stimulated phosphorylation was about 0.6 $\mu$M in the presence of either endogenous or exogenous protein kinases. Since the calmodulin content of heart[36] is not likely to be a limiting factor, the calmodulin stimulated phosphorylation of SR is dependent only upon the concentration of free $Ca^{2+}$ ions. The concentration of free $Ca^{2+}$ that regulates calmodulin stimulated activity (FIGURE 4) is within the range of free cytosolic $Ca^{2+}$ during relaxation ($10^{-7}$ M) and contraction ($10^{-5}$ M) in the cardiac muscle. Therefore, calmodulin stimulated phosphorylation of phospholamban may be regulated by the concentration of free $Ca^{2+}$ during the cardiac cycle.

The $Ca^{2+}$-dependent protein kinase in cardiac SR may require calmodulin in a fashion similar to that proposed for the phosphodiesterase:[26]

$$(\text{calmodulin})_{\text{inactive}} + Ca^{2+} \rightleftharpoons (\text{calmodulin}^* \cdot Ca^{2+})_{\text{active}}$$

$$(\text{protein kinase})_{\text{inactive}} + (\text{calmodulin}^* \cdot Ca^{2+})_{\text{active}} \rightleftharpoons (\text{protein kinase}^* \cdot \text{calmodulin}^* \cdot Ca^{2+})_{\text{active}}$$

where the asterisk (*) stands for a new conformation, the active species.

### *Inhibition of Calmodulin Stimulated Phosphorylation*

It is known that all calmodulins stimulate the $Ca^{2+}$-dependent phosphodiesterase. This stimulation is blocked by trifluoperazine, a phenothiazine derivative used clinically as an antipsychotic agent. Calmodulin binds two moles of trifluoperazine with a dissociation constant of $10^{-6}$ M in the presence of $Ca^{2+}$; upon binding of trifluoperazine, calmodulin becomes biologically inactive.[37] Trifluoperazine inhibited the calmodulin stimulated phosphorylation of SR in a dose-dependent manner (FIGURE 5). The $IC_{50}$ for this inhibition was $3 \times 10^{-5}$ M trifluoperazine in agreement with the $IC_{50}$ for inhibition of PDE and red blood cell ATPase. Concentrations of trifluoperazine higher than $10^{-4}$ M prevented calmodulin stimulated phosphorylation of SR completely. Trifluoperazine had no effect on the cAMP-dependent phosphorylation of SR vesicles indicating the specificity of the drug for calmodulin. Similarly Troponin I, which has been shown to bind calmodulin specifically and inhibit its effects,[38] also caused inhibition of the calmodulin stimulated phosphorylation and had no effect on the cAMP-dependent activity (data not shown).

### *Effect of Calmodulin Stimulated Phosphorylation on Energy-Dependent Calcium Uptake by Cardiac Sarcoplasmic Reticulum*

Phosphorylation of dog cardiac sarcoplasmic reticulum by the endogenous protein kinase in the presence of cAMP resulted in a significant increase of

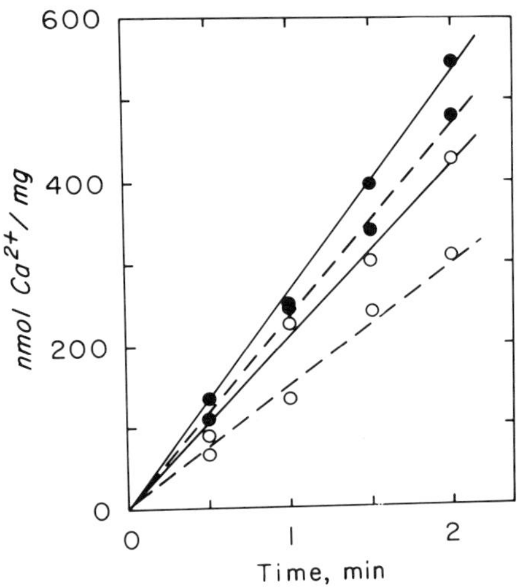


FIGURE 6. Effect of cardiac SR phosphorylation on initial rates of $Ca^{2+}$ transport by SR vesicles. SR vesicles were phosphorylated in the presence of 1 $\mu$M $Ca^{2+}$ and with the following additions: 1 $\mu$M cAMP (○—○); $10^{-7}$ M calmodulin (●--●); 1 $\mu$M cAMP and $10^{-7}$ M calmodulin (●—●). Control vesicles were incubated under identical conditions without ATP (○--○). The rate of $Ca^{2+}$ transport was measured as described in METHODS.

approximately 40% in the initial rate of calcium transport (FIGURE 6). Preincubation with calmodulin in the presence of calcium also resulted in an increase of 55% in the initial rate of calcium uptake (FIGURE 6). Phosphorylation by both cAMP-dependent and calmodulin stimulated protein kinases resulted in an increase of 80% in the initial rate of calcium uptake. These data indicate that calmodulin-dependent phosphorylation of cardiac SR vesicles stimulates the initial rate of calcium transport and its effect is additive to that of cAMP. Katz and Remtulla[20] have previously reported that calmodulin stimulated $Ca^{2+}$ transport in cardiac SR and its effect was additive to that by cAMP. However, the stimulatory effect was only obtained by high calmodulin concentrations (1 $\mu$M).

Recently, Le Peuch *et al.*[21] reported that calmodulin-dependent phosphorylation of cardiac SR slightly stimulated (16%) the initial rate of $Ca^{2+}$ transport in the absence of cAMP. The authors claimed that stimulation of $Ca^{2+}$ transport by cAMP could only be observed in the presence of calmodulin. However, we have shown here (FIGURE 6) that cAMP-dependent phosphorylation of cardiac SR, under conditions in which no calmodulin-dependent phosphorylation occurs, stimulates the initial rate of $Ca^{2+}$ transport. It remains to be seen whether calmodulin stimulates the $Ca^{2+}$ ATPase activity in SR in a manner analogous to calcium transport, thereby, maintaining the stoichiometric coupling ratio of Ca:ATP for the calcium pump. Le Peuch *et al.*[21] reported that the $Ca^{2+}$ ATPase was not stimulated by either calmodulin or calmodulin and cAMP, although $Ca^{2+}$ transport was enhanced under similar conditions. These data are difficult to interpret especially since several investigators have reported stimulation of the $Ca^{2+}$ ATPase in SR by cAMP.[22,25,33] The cAMP-dependent phosphorylation of "phospholamban" is thought to regulate cardiac SR function by regulating the rate of turnover of the $Ca^{2+}$ ATPase.[25,39] The molecular mechanism by which cAMP regulation of cardiac SR occurs is now under investigation.[39,40] However, the involvement of phosphorylation of the same protein by a calmodulin stimulated $Ca^{2+}$-dependent protein kinase in cardiac SR function is yet to be determined.

## REFERENCES


1. HARIGAYA, S. & A. SCHWARTZ. 1969. Circ. Res. **25:** *781–794*.
2. SCHWARTZ, A. 1971. *In* Calcium and the Heart. P. Harris & L. H. Opie, Eds.: 66–92. Academic Press, Inc. London.
3. HARIGAYA, S. & A. SCHWARTZ. 1973. *In* Organization of Energy-Transducing Membranes. M. Nakao & L. Pacher, Eds.: 117–126. University Park Press. Tokyo.
4. SCHWARTZ, A. 1976. Fed. Proc. **35:** 1279–1282.
5. VAN WINKLE, W. B. & A. SCHWARTZ. 1976. Ann. Rev. Physiol. **38:** 247–272.
6. CHEUNG, W. Y. 1970. Biochem. Biophys. Res. Commun. **38:** 533–538.
7. KAKIUCHI, S. & R. YAMAZAKI. 1970. Biochem. Biophys. Res. Commun. **41:** 1104–1110.
8. CHEUNG, W. Y., L. S. BRADHAM, T. J. LYNCH, Y. M. LIN & E. A. TALLANT. 1975. Biochem. Biophys. Res. Commun. **66:** 1055–1062.
9. BROSTROM, C. O., M. A. BROSTROM & D. J. WOLF. 1977. J. Biol. Chem. **252:** 5677–5685.
10. WESTCOTT, K. R., D. C. LE PORTE & D. R. STORM. 1979. Proc. Natl. Acad. Sci. USA **76:** 204–208.
11. GOPINATH, R. M. & F. F. VINCENZI. 1977. Biochem. Biophys. Res. Commun. **77:** 1203–1209.
12. LYNCH T. J. & W. Y. CHEUNG. 1979. Arch. Biochem. Biophys. **194:** *165–170*.
13. DABROWSKA, R., J. M. F. SHERRY, D. K. AROMATORIO & D. J. HARTSHORNE. 1978. Biochemistry **17:** 253–259.
14. HATHAWAY, D. R. & R. S. ADELSTEIN. 1979. Proc. Natl. Acad. Sci. USA **76:** 1653–1657.
15. ANDERSON, J. M. & M. J. CORMIER. 1978. Biochem. Biophys. Res. Commun. **84:** 595–602.
16. WONG, P. Y. K. & W. Y. CHEUNG. 1979. Biochem. Biophys. Res. Commun. **90:** 473–480.
17. SOBNE, K., S. ICHIDA, H. YOSHIDA, R. YAMAZAKI & S. KAKIUCHI. 1979. FEBS Lett. **99:** 199–202.
18. COHEN R., A. BURCHELL, J. G. FOULKES, P. T. W. COHEN, T. C. VANAMAN & A. C. NAIRN. 1978. FEBS Lett. **92:** 287–293.
19. HINDS, T. R., F. L. LARSEN & F. F. VINCENZI. 1978. Biochem. Biophys. Res. Commun. **81:** 455-461.
20. KATZ, S. & M. A. REMTULLA. 1978. Biochem. Biophys. Res. Commun. **83:** 1373–1379.
21. LE PEUCH, C. J., J. HAIECH & J. G. DEMAILLE. 1979. Biochemistry **18:** 5150–5157.
22. TADA, M., M. A. KIRCHBERGER, D. I. REPKE & A. M. KATZ. 1974. J. Biol. Chem. **249:** 6174–6180.

23. KIRCHBERGER, M. A. & M. TADA. 1976. J. Biol. Chem. **251:** 725–729.
24. SCHWARTZ, A., M. L. ENTMAN, K. KANIIKE, L. K. LANE, W. B. VAN WINKLE & E. P. BORNET. 1976. Biochim. Biophys. Acta. **426:** 57–72.
25. WRAY, H. L. & R. R. GRAY. 1977. Biochim. Biophys. Acta **461:** 441–459.
26. CHEUNG, W. Y. 1980. Science **207:** 19–27.
27. SUMIDA, M., T. WANG, F. MANDEL, J. P. FROEHLICH & A. SCHWARTZ. 1978. J. Biol. Chem. **253:** 8772–8777.
28. DEDMAN, J. R., J. L. FAKUNDING & A. R. MEANS. 1977. *In* Hormone Action and Molecular Endocrinology Workshop Syllabus. B. W. O'Malley & W. T. Schrader, Eds.: 901–946. The Endocrine Society. Bethesda, Md.
29. LIN, Y. M., Y. P. LIU & W. Y. CHEUNG. 1974. J. Biol. Chem. **249:** 4943–4954.
30. DEDMAN, J. R. & A. R. MEANS. 1977. J. Cyclic Nucleotide Res. **3:** 139–152.
31. SILLEN, L. G. & A. E. MARTELL. 1964. Stability Constants of Metal Ion Complexes. 2nd Ed. Serial Publication, No. 17. The Chemical Society. Burlington House. London.
32. LAEMLI, U. K. 1970. Nature **227:** 680–685.
33. WEBER, K. & M. OSBORN. 1969. J. Biol. Chem. **244:** 4406–4412.
34. HÖRL, W. H., H. P. JENNISSEN & L. M. G. HEILMEYER, JR. 1978. Biochemistry **17:** 759–766.
35. VARSÁNYI, M., U. GRÖSCHEL-STEWARD, L. M. G. HEILMEYER, JR. 1978. Eur. J. Biochem. **87:** 331–340.
36. GRAND, R. J. A., S. V. PERRY & R. A. WEEKS. 1979. Biochem. J. **177:** 521–529.
37. LEVIN, R. M. & B. WEISS. 1976. Mol. Pharmacol. **12:** 581–589.
38. KERRICK, W. G. L., L. BOLLES, P. CASSIDY, R. L. COBY, S. GAZAREK, P. E. HOAR & D. A. MALENCIK. 1980. Fed. Proc. **39:** 2042.
39. KRANIAS, E. G., F. MANDEL, L. I. TSAI & A. SCHWARTZ. 1980. Fed. Proc. **39:** 2173.
40. TADA, M., M. YAMADA, F. OHMORI, T. KUZUYA, M. INUI & H. ABE. 1980. J. Biol. Chem. **255:** 1985–1992.

## DISCUSSION OF THE PAPER

DR. P. COHEN: Have you looked at the reversal of these reactions by dephosphorylation? In particular, would you have any information as to whether the calmodulin-dependent phosphorylation or the cyclic AMP-dependent phosphorylation would be reversed more rapidly by, perhaps, endogenous phosphatase as present in the sacroplasmic reticulum?

DR. E. G. KRANIAS: We have not looked for the endogenous phosphatase for the calcium-dependent process.

DR. BARTOW (*Rockefeller University, New York, New York*): On your phosphorylation gel there was not only an increased phosphorylation of phospholamban, but a higher molecular weight band also.

DR. KRANIAS: Yes, we see that band high in the gels and there is also a faint band in the presence of cyclic AMP. In the presence of calmodulin that appeared to increase. It is in the region of about a hundred thousand. We are investigating that.

DR. F. F. VINCENZI: The calcium pump in the plasma membrane and the calcium pump in the sarcoplasmic reticulum may appear very similar and yet, depending upon whom you believe, the stoichiometry is two to one in the sarcoplasmic reticulum and one to one in the plasma membrane. I wonder if anyone would care to comment on the apparently much lower affinity of calmodulin for SR as compared to plasma membrane.

DR. S. KATZ (*University of British Columbia, Vancouver, B.C.*): Under the conditions that most people measure the calcium pump in sarcoplasmic reticulum, two moles of calcium are pumped for each mole of ATP. But when people are doing rapid kinetics work in the millisecond range a lot of the things that we have seen in the steady state might not necessarily be true at the fast rates. Perhaps in red cell membranes the stoichiometry is not always as you indicated. Your second point is a very good one and it is a worrisome one, that calmodulin is required in such high concentrations in sarcoplasmic reticulum in comparison to red cell membranes. I don't have any answer for that at the moment.

DR. VINCENZI: What is the estimate of the approximate amount of calmodulin in let's say cardiac muscle?

DR. COHEN: I think it is about ten micromolar in cardiac muscle.

DR. E. CARAFOLI: I think we know the following: We know that the SR ATPase does not stick to a calmodulin column. We have tested that and it doesn't stick. Secondly, we know the stoichiometry that is accepted to be two in the case of the SR ATPase and, at least in our hands, seems to be one for the erythrocyte ATPase. Thirdly, we know that vanadate, which inhibits at submicromolar concentrations in the case of erythrocyte ATPase, requires about fifty micromolar in the case of the SR ATPase. If you put it all together, I think, it seems quite clear that the mechanism must be different.

# ROLE OF CALMODULIN IN DOPAMINERGIC TRANSMISSION

I. Hanbauer, S. Pradhan, and H.-Y. T. Yang*

*Section on Biochemical Pharmacology*
*National Heart, Lung, and Blood Institute*
*National Institutes of Health*
*Bethesda, Maryland 20205*
*and*
**Laboratory of Preclinical Pharmacology*
*National Institute of Mental Health*
*St. Elizabeth's Hospital*
*Washington, D.C. 20035*

## Introduction

Abundant evidence indicates that in brain a great number of dopamine receptors are coupled with adenylate cyclase. Therefore, one of the most prominent physiological responses elicited by dopamine is an increased synthesis of cAMP due to the stimulation of a dopamine receptor-coupled adenylate cyclase (for review see Reference 1). Hence, adenylate cyclase can be viewed as a subunit of the supramolecular structure of the dopamine receptor. In general, the recognition site for a neurotransmitter with adenylate cyclase is coupled by guanine nucleotides[2,3] via a specific GTP-binding protein.[4] In the case of striatal dopamine receptors it was speculated that this protein probably also modulates the affinity of the recognition site to dopamine.[5] However, the GTP-binding protein is not the only membrane protein that modulates dopamine receptor function. Increasing evidence indicates that calmodulin, a thermostable $Ca^{2+}$ binding protein, also functions as a modulator of the $Ca^{2+}$-dependent adenylate cyclase,[6] protein kinase,[7] and phosphodiesterase[8,9] and, therefore, at least in nervous tissue participates in receptor function.

In brain two forms of calmodulin appear to be present: a soluble form and membrane-bound form. Studies on subcellular distribution of calmodulin show that this protein is preferentially located in the synaptosome-rich pellet with lesser amounts in the supernatant fraction (cytosol).[10]

With regard to the relationship of calmodulin to dopamine function, it is important to note that the membrane-bound form of calmodulin appears to function with adenylate cyclase, whereas the soluble form is required for phosphodiesterase activation. While there is a more immediate relationship between dopamine, adenylate cyclase, and calmodulin, the relationship between dopamine, calmodulin, and phosphodiesterase is remote. Substantial evidence indicates that calmodulin is involved in the dopamine receptor function by regulating the coupling of the membrane-bound adenylate cyclase to the transmitter recognition site.[11] In response to persistent dopamine receptor stimulation it appears to also regulate the link between increased cAMP content and phosphodiesterase activation.[12,13] Hence, calmodulin can be viewed as an important component in the transduction of short-term and long-term stimulation of specific post-synaptic dopamine receptors.



0077-8923/80/0356-0292 $01.75/0 © 1980, NYAS

## Production of Immunoglobulins Directed Toward Calmodulin

The conventional procedure to measure the calmodulin content in tissues takes advantage of its thermostability and is based on the activation of calmodulin-deficient phosphodiesterase.[14] However, in the bioassay the interference by the inhibitor protein for calmodulin and the deviation from optimal ion concentrations cannot always be eliminated. Therefore, it was of interest to establish an immunochemical method that allows measurement of calmodulin in tissues under various experimental conditions.

Calmodulin from porcine brain was purified to homogeneity according to the procedure by Klee.[15] It has poor antigenic properties due to being a low molecular weight ($M_r$ 17,000) and acidic protein pI = 3.9) and the absence or scarcity of tryptophan and tyrosine, respectively, in its amino acid composition. In order to obtain antibodies specific for calmodulin it was necessary to conjugate calmodulin covalently as a hapten to an antigenic protein, which when injected into an animal, will give rise to the formation of antibodies specific for the chemically coupled small protein. For this purpose homogeneous calmodulin was coupled with hemocyanin by I-ethyl-3-(3-dimethylaminopropyl) carbodiimide. The resulting conjugate was dissolved in saline, emulsified in complete Freund adjuvant (equal volumes), and was then injected intradermally into rabbits. Maximal antibody production occurred after 7 to 8 injections given at 2 week intervals. The immunoglobulins were partially purified by precipitation with 50% $(NH_4)_2SO_4$. This precipitate was dissolved in 0.9% NaCl solution (25% of original volume) and dialyzed against 0.9% NaCl solution with 3 changes.

The activity and specificity of anti-calmodulin was tested on polystyrene microplates which has been coated with calmodulin (100 ng/200 $\mu$l) for at least 48 hr. The details of this technique are described (*vida infra*). The results in FIGURE 1 show that the immunoglobulins cross-react with similar affinity with calmodulin and erythrocyte $Ca^{2+}$ binding activator protein. In fact, both proteins have been shown to have an identical amino acid sequence.[16] In contrast, the antibody directed toward calmodulin fails to cross-react under similar conditions with troponin C—a $Ca^{2+}$ binding protein with 70% amino acid sequence homology.

## Measurement of Calmodulin Content in Brain Tissues by an Enzyme-Linked Immunosorbent Assay (ELISA)

In order to study the role of calmodulin in receptor regulation it was important to have a reliable assay technique for this protein. The conventional radioimmunoassay involves the incorporation of radioiodine into the antigen, but the specific activity obtained can be relatively low. Other disadvantages include the relatively short half-life of 60 days, possible chemical damage to the protein by the radioiodine, and higher costs. In view of such problems we adopted an enzyme-linked immunosorbent assay (ELISA) for the immunological assessment of calmodulin.

The basic ELISA-test depends on the possibilities that antigen or antibodies can be attached to a solid-phase support and yet retain immunological activity and that either antigen or antibody can be linked to an enzyme and the resulting complex retains immunological and enzymatic activity. In fact, immunoglobulins can be linked to a variety of enzymes such as peroxidase, glucose oxidase, and alkaline phosphatase yielding stable and highly reactive reagents. It has to be

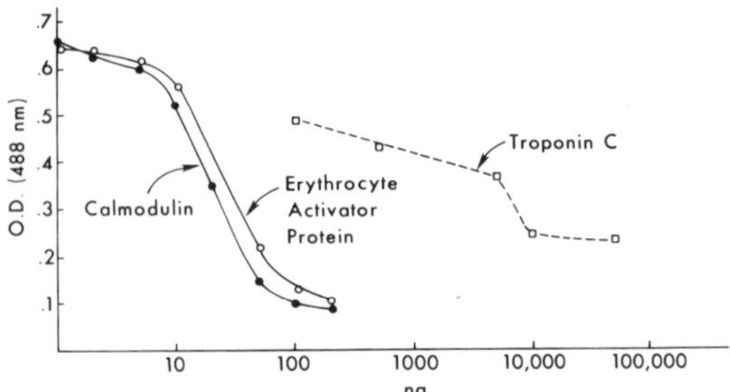


FIGURE 1. Specificity of calmodulin antiserum.

mentioned that the use of enzyme-labeled reactants overcomes some of the disadvantages of the radioimmunoassay but yields results with the same sensitivity as the radioimmunoassay.

For the measurement of calmodulin, the ELISA procedure described by Voller, Bidwell, and Bartlett[17] has been modified and is summarized in TABLE 1. In detail: 1) Polystyrene microplates are sensitized by passive adsorption with calmodulin (100 ng/200 $\mu$l carbonate-bicarbonate buffer pH 9.6 containing 0.02% $NaN_3$/well) for at least 48 hr at 4°C. 2) Aliquots of brain extracts are incubated with anti-calmodulin immunoglobulins in PBS-Tween 20 (137 mM NaCl, 1.5 mM $KH_2PO_4$, 8.1 mM $Na_2HPO_4 \cdot 12\, H_2O$, 2.7 mM KCl, and 0.05% Tween-20) at 4°C for about 18 hr. 3) Aliquots of each incubation mixture are transferred to the precoated microwells, where antibodies which are not conjugated with calmodulin react with the immobilized calmodulin on the well surface. 4) Horseradish peroxidase labeled anti-rabbit Ig conjugate is added to each well to react with the captured antibodies (step 3). 5) o-Phenylendiamine (0.001%) and $H_2O_2$ (0.003%) dissolved in $H_2O$ is added to each well. The rate of degradation of o-phenylendiamine is indicated by a color change, which is proportional to the antibody concentration (step 3). 6) The reaction is stopped by adding 4 M $H_2SO_4$ and the color change is assessed spectrophotometrically at 488 nm with an ELISA microreader (Biomedical Engineering and Instrumentation Branch, National Institutes of Health, Bethesda, Md).

In general, brain tissue is homogenized in 0.32 M sucrose (containing either 1 mM $Ca^{2+}$ or 1 mM EGTA) and synaptosome-rich membrane pellet were prepared by high-speed centrifugation. The resulting supernatant fraction is heat-inactivated (1 min at 90°C) and assayed for its calmodulin content—referred to as soluble calmodulin. The pellet fraction is resuspended in 0.05 M Tris buffer, pH 7.4, containing 0.1% Lubrol PX, heat-inactivated (1 min at 90°C), sonicated for 0.5 min, and centrifuged at 100,000 × g for 30 min. The resulting supernatant fraction containing solubilized calmodulin is referred to as membrane-bound.

The presence of Triton X-100 or Lubrol PX up to 2% and EGTA up to 10 mM did not interfere in the ELISA. Thus, this procedure is particularly suitable for calmodulin measurements in modified ionic conditions.

## CHANGES IN STRIATAL CALMODULIN CONTENT ELICITED BY STIMULATION OF DOPAMINE RECEPTORS

In caudate nucleus, persistent stimulation of dopamine receptors activates adenylate cyclase and causes cAMP to accumulate in the cytosol of the postsynaptic cell.[18] This accumulation of cAMP was shown to cause the activation of a cAMP-dependent protein kinase.[19] In fact, there exists indirect evidence that among the natural substrates of cAMP-dependent protein kinases there is a membrane protein that participates in an as yet undetermined way in the binding of calmodulin to membranes. It was shown that phosphorylation of membrane protein(s) catalyzed by a cAMP-dependent protein kinase triggers the release of calmodulin from the membrane.[10] In line with this mechanism are experiments showing that incubation of striatal membranes with dopamine in the presence of phosphorylating conditions results in the release of calmodulin from its membrane binding sites.[20] These findings initiated further studies to elucidate the physiological mechanisms that are involved in the translocation of calmodulin from its membrane binding sites to the cytosol. When striatal slices are incubated with dopamine ($10^{-7}$ M) or apomorphine ($10^{-7}$ M) for at least 30 min the content of soluble calmodulin increases[12] and the kinetic properties of phosphodiesterase change. Brain tissue contains multiple molecular forms of phosphodiesterases which differ with regard to their affinity for cAMP ($K_m$) and their ability to be regulated by calmodulin in the presence of $Ca^{2+}$.[21,22] The kinetic analysis of phosphodiesterases present in striatal slices yields a biphasic double reciprocal plot of phosphodiesterase activity versus cAMP concentrations, indicating the presence of at least two kinetic forms with a low $K_m$ (~10 $\mu$M cAMP) and a high $K_m$ (~80 $\mu$M cAMP) for cAMP.[12] In these slices, persistent stimulation of dopamine receptors elicits a monophasic double reciprocal plot of phosphodiesterase activity versus cAMP concentrations with a low $K_m$ for cAMP (~25 $\mu$M cAMP).[12] Thus, it appears that persistent stimulation of dopamine receptors triggers a cascade of biochemical events: membrane-bound calmodulin is released from the membranes into the cytosol, where in the presence of $Ca^{2+}$ it interacts with a specific form of phosphodiesterase, which in turn changes the catalytic properties of the enzyme, i.e., an increase in $V_{max}$ and in its affinity for cAMP.

To assess the physiological significance of these findings, we studied whether *in vivo* calmodulin translocates from membranes to the cytosol following the injection of drugs that modify dopaminergic transmission. A number of central acting stimulants—such as d-amphetamine and cocaine—were tested. These drugs in small doses augment motor activity, but in large doses cause stereotype movements.[23,24] Neither d-amphetamine nor cocaine stimulates dopamine receptors directly, but they act by triggering a release of dopamine. An index for the involvement of dopamine in the action of d-amphetamine and cocaine, is the increase of dopamine turnover-rate in caudate nucleus elicited by both drugs.[24]

TABLE 1

SCIENCE FOR CALMODULIN MEASUREMENT WITH ELISA-MICROMETHOD

|  | |
|---|---|
| I. | Adsorption of calmodulin onto microwells (48 hr at 4°C) |
| | ↓ Wash with PBS-Tween 20 |
| II. | Incubation of test samples in microwells (30 min at 24°C) |
| | ↓ Wash with PBS-Tween 20 |
| III. | Incubation with horseradish peroxidase labeled Anti-Rabbit Immunoglobulins (2 hr at 24°C) |
| | ↓ Wash with PBS-Tween 20 |
| IV. | Incubation with o-phenylenediamine + $H_2O_2$ (1 hr at 24°C) |
| | ↓ Stop reaction with 4 M $H_2SO_4$ |
| V. | Read OD at 488 nm |

d-Amphetamine has been studied more thoroughly than cocaine, and in these studies it was reported that while the drug causes a long lasting increase of dopamine turnover, only a transient increase in striatal cAMP could be measured.[25] This observation leads to the assumption that persistent, indirect stimulation of dopamine receptors elicited by d-amphetamine may trigger the activation of phosphodiesterase, which in turn is responsible for the rapid degradation of cAMP.

The results in TABLE 2 show that the injection of d-amphetamine increases the calmodulin content in the supernatant fraction, whereas the injection of cocaine fails to elicit a change in the soluble calmodulin pool. In contrast, cocaine significantly increases the calmodulin content of crude synaptic membrane fraction whereas d-amphetamine tends to lower the calmodulin content in the membrane fraction. As discussed above an increment of calmodulin in cytosol is associated with a change in the kinetic properties of phosphodiesterase, while calmodulin accumulated in synaptic membranes interacts with membrane-linked adenylate cyclase. In fact, studies carried out with d-amphetamine verified the first hypothesis. FIGURE 2 shows that in rats injected with d-amphetamine the supernatant fraction of striatal homogenates contains phosphodiesterase characterized by a monophasic double reciprocal plot of enzyme activity versus cAMP concentration. The finding of a monophasic double reciprocal plot can be used as an index for calmodulin-mediated activation of phosphodiesterase[12,13] and is in agreement with the increase in soluble calmodulin content elicited by d-amphetamine. In contrast, the treatment of rats with cocaine failed to change the apparent $K_m$ values of phosphodiesterase (FIGURE 2). The molecular mechanism whereby calmodulin accumulates in striatal membranes after injection of cocaine or whereby it is released into the cytosol after treatment with d-amphetamine is not completely understood. A possible explanation for the increase in soluble calmodulin content may be that d-amphetamine, by eliciting a transient increase in cAMP content, triggers the activation of cAMP-dependent protein kinase, which through a phosphorylating process may release calmodulin from its membrane binding sites. The accumulation of membrane-bound calmodulin caused by cocaine is difficult to interpret at the present time. Preliminary studies on the effect of cocaine on [$^3$H]spiroperidol binding activity in membrane fractions of striatal homogenates are shown in TABLE 3. The results show that the apparent $b_{max}$ of [$^3$H]spiroperidol binding is increased 30 min after injection of cocaine, whereas the affinity of the labeled ligand to the recognition site remained unchanged. This increase in [$^3$H]spiroper-

TABLE 2

CHANGES IN SOLUBLE AND MEMBRANE-BOUND CALMODULIN CONTENT BY DRUGS WHICH MODIFY STRIATAL DOPAMINE METABOLISM

| Drug (mg/kg) | Calmodulin (μg/mg protein ± SEM | |
|---|---|---|
| | Soluble | Membrane-Bound |
| Saline | 1.14 ± 0.17 | 5.5 ± 0.87 |
| Haloperidol (1) | 1.15 ± 0.16 | 4.8 ± 0.22 |
| d-Amphetamine (1.5) | 2.71 ± 0.58* | 4.5 ± 0.25 |
| Cocaine (30) | 0.94 ± 0.29 | 8.5 ± 0.48* |

*$P < 0.05$; $n = 6$

Rats were killed 30 min after cocaine injection (i.p.) and 40 min after haloperidol or d-amphetamine injection (i.p.)

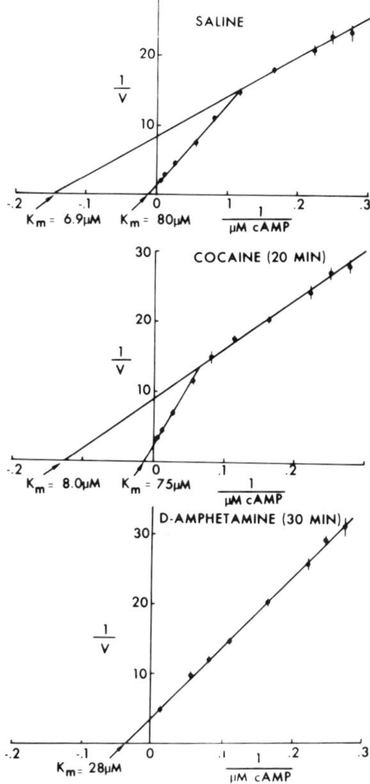


FIGURE 2. Double reciprocal plot of the initial velocity of phosphodiesterase in the supernatant fraction versus cAMP concentrations. Top: Effect of saline injection. Center: Effect of cocaine injection. Bottom: Effect of d-amphetamine injection. Rats were killed at times indicated in brackets. Caudate nucleus was dissected and homogenized in 0.32 M sucrose and centrifuged at 100,000 g for 30 min.

idol binding sites is in line with the increment in membrane-bound calmodulin. It is currently being studied in our laboratory whether cocaine increases the affinity of calmodulin for its binding site or whether it increases the synthesis rate of calmodulin. It will be important to establish whether the increment in membrane-bound calmodulin elicited by cocaine can be associated to supersensitivity of striatal adenylate cyclase to dopamine. Supersensitivity caused by long-term treatment with neuroleptics or lesions of the nigra-striatal pathway have been shown to be associated with an increase in [$^3$H]haloperidol binding sites[26] and an elevated calmodulin content as discussed below.

## CHANGES IN STRIATAL CALMODULIN CONTENT CAUSED BY LONG-TERM MODULATION OF DOPAMINE RECEPTORS

Several reports in the literature indicate that an increase in membrane-bound calmodulin content is related to a modified state of dopaminergic function which

TABLE 3

[^3H]SPIROPERIDOL BINDING IN RAT STRIATAL MEMBRANES

| Drug (mg/kg) | Apparent $B_{max}$ of [^3H]Spiroperidol (pmol/mg protein) | Equilibrium dissociation Constant ($K_D$) of [^3H]Spiroperidol |
|---|---|---|
| Saline | 414.5 ± 12 | 0.0017 ± 0.00012 |
| Cocaine (20) | 564.7 ± 31* | 0.0014 ± 0.00014 |

*$P < 0.005$; $n = 4$

Rats were killed 30 min after injection of cocaine. Striatal membranes were prepared as described by Creese *et al.*[5] The $B_{max}$ and $K_D$ values were derived from Scatchard analysis of data from 4 experiments.

is biochemically expressed as supersensitivity of the dopamine-sensitive adenylate cyclase in caudate nucleus. Long-term treatment with haloperidol[27] and transection of the nigro-striatal fibers[28] have been shown to increase the content of membrane-bound calmodulin. This increase coincides with a supersensitivity of striatal adenylate cyclase to stimulation by dopamine. TABLE 4 shows that the content of calmodulin increases in membranes prepared from caudate nucleus but fails to increase in membranes of hippocampus or cerebellum. In contrast, after long-term treatment with haloperidol, the calmodulin content of the supernatant fraction failed to change in the various brain areas (TABLE 5).

In order to understand the molecular mechanisms whereby calmodulin accumulates in striatal membranes after long-term treatment of rats with haloperidol, we measured the turnover of calmodulin. Rats which have been injected daily with haloperidol for 24 days were withdrawn from treatment with this drug for 4 days and then were injected intraventricularly with cycloheximide to block protein synthesis. Impairment of protein synthesis elicits a decline of the content of various proteins proportional to their rate of synthesis. The results in TABLE 6 show that the rate of decline of calmodulin content is faster in saline-treated than in haloperidol-treated rats. Thus, the fractional rate constant ($K.hr^{-1}$) of calmodulin is smaller in haloperidol-treated rats than in controls. Although, the fractional rate constant was lower after haloperidol treatment, the turnover rate for calmodulin was similar in both groups of rats because the steady-state concentration of this protein was higher in haloperidol-treated rats. From these results it can be inferred that long-term treatment with haloperidol does not increase the

TABLE 4

EFFECT OF LONG-TERM TREATMENT WITH HALOPERIDOL ON MEMBRANE-BOUND CALMODULIN CONTENT IN VARIOUS BRAIN AREAS

| Brain Area | Calmodulin ($\mu$g/mg protein ± SEM) | |
|---|---|---|
| | Saline | Haloperidol |
| Caudate Nucleus | 4.0 ± 0.78 | 6.9 ± 0.97* |
| Hippocampus | 3.2 ± 1.1 | 2.8 ± 0.26 |
| Cerebellum | 3.2 ± 0.56 | 3.7 ± 0.53 |

*$P < 0.05$; $n = 6$

Rats were injected daily with 1 mg/kg haloperidol for 24 days and were killed 4 days after withdrawal.

TABLE 5

EFFECT OF LONG-TERM TREATMENT WITH HALOPERIDOL ON THE SOLUBLE CALMODULIN CONTENT IN VARIOUS BRAIN AREAS

| | Calmodulin ($\mu$g/mg protein ± SEM) 100,000 × g supernatant | |
|---|---|---|
| Brain Area | Saline | Haloperidol |
| Caudate Nucleus | 1.56 ± 0.32 | 1.15 ± 0.16 |
| Hippocampus | 1.38 ± 0.21 | 1.21 ± 0.31 |
| Cerebellum | 2.07 ± 0.29 | 1.35 ± 0.21 |

$n = 5$

Rats were injected daily with 1 mg/kg haloperidol for 24 days and were killed 4 days after haloperidol withdrawal.

synthesis of calmodulin in striatal membranes, but may rather modify the binding affinity of calmodulin to its binding site at the membrane.

## IMPLICATIONS OF CALMODULIN IN DOPAMINE RECEPTOR FUNCTION

Calmodulin is currently being considered as an important modulator protein in synaptic functions because it mediates enzyme activation by $Ca^{2+}$. The role of calmodulin in the regulation of dopamine receptor function appears to depend on the compartmentation of this protein. Hence, the regulation of calmodulin has to be viewed in terms of the biochemical processes which modulate the distribution of this protein.

In this report we reviewed experimental conditions in which calmodulin translocates from membranes to cytosol and others in which calmodulin accumulates in membranes. If one assumes the existence of one pool of calmodulin, then its content in membranes or cytosol may depend on the binding capacity of membranes for the protein. The relevance of this implication is suggested by a decreased binding of calmodulin to membranes when they are in a phosphorylated state.

TABLE 6

CALMODULIN TURNOVER IN STRIATAL MEMBRANES: EFFECT OF LONG-TERM TREATMENT WITH HALOPERIDOL

| Hours after Cycloheximide (0.5 mg i.v.) | Calmodulin ($\mu$g/mg protein ± SEM) Saline | Haloperidol |
|---|---|---|
| 0 | 4.5 ± 0.40 | 7.7 ± 0.75* |
| 3.5 | 3.6 ± 0.25 | 6.9 ± 0.50* |
| 7.0 | 2.9 ± 0.30 | 5.9 ± 0.40* |
| t 1/2 (hr) | 12 | 19 |
| K $hr^{-1}$ | 0.058 | 0.036 |
| $T_r$ ($\mu$g/mg protein/hr) | 0.26 | 0.28 |

$n = 6$; *$P < 0.01$

Rats were injected daily with haloperidol (1 mg/kg) for 24 days. Four days after haloperidol withdrawal cycloheximide was injected intraventricularly and the rats were killed various times thereafter.

The scheme in FIGURE 3 depicts a working model that enables visualization of some of the molecular interactions of calmodulin in the membrane and in the cytosol that are operative in dopamine receptor function. The model shows that calmodulin is presumably bound to proteins at the inner leaflet of the membrane, where it plays a role in the activation of adenylate cyclase elicited by dopamine receptor stimulation. While the coupling of dopamine recognition sites to adenylate cyclase is regulated by guanine nucleotides and a specific guanine nucleotide binding protein, the activation of this enzyme appears to depend on

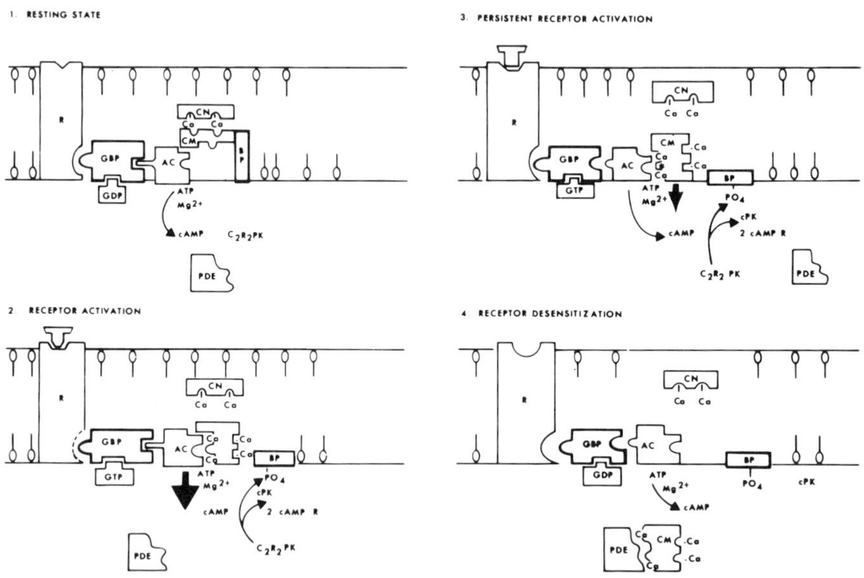


FIGURE 3. Working model of the role of calmodulin in dopamine receptor function R: receptor; AC: adenylate cyclase; GBP: guanine binding protein; CN: calcineurin; GDP: guanosine diphosphate; CM: calmodulin; GTP: guanosine triphosphate; BP: binding protein; PDE: phosphodiesterase; $C_2R_2$ PK: cAMP-dependent protein kinase.

the availability of calmodulin and $Ca^{2+}$. In fact, the release of calmodulin from membranes triggered by a phosphorylating process greatly decreases the activation of adenylate cyclase by dopamine or apomorphine. The model also shows, that persistent occupation of dopamine recognition sites leads to an increase in cAMP content in the cytosol which in turn activates cAMP-dependent protein kinase. Thus it can be assumed that drugs that lead to an increment in cAMP content and subsequently to an activation of cAMP-dependent protein kinase, may release a substantial amount of calmodulin from membranes into the cytosol. This translocation of calmodulin may be rate-limiting in regulating the activities of adenylate cyclase, phosphodiesterase, and indirectly also cAMP-dependent protein kinase. Therefore, this mechanism can be viewed as an important link involved in the "down-regulation" of the dopamine response.

## REFERENCES


1. KEBABIAN, J. W. 1977. Biochemical regulation and physiological significance of cyclic nucleotides in the nervous system. Adv. Cyclic Nucleotide Research **8:** 421–508.
2. RODBELL, M., M. C. LIN, Y. SOLOMON, C. LOUDOS, J. HARWOOD, P. MARTIN, B. R. RENDELL & M. BERMAN. 1975. Role in adenine and guanine nucleotide in the activity and response of adenylate cyclase systems to hormones: Evidence for multisite transition states. Adv. Cyclic Nucleotide Research **5:** 3–27.
3. RODBELL, M. 1980. The role of hormone receptors and GTP-regulatory proteins in membrane transduction. Nature **284:** 17–22.
4. PFEUFFER, T. 1979. FEBS Lett. **101:** 85–89.
5. CREESE, I., T. PROSSER & S. H. SNYDER. 1975. Dopamine receptor binding: Specificity, Localization and regulation by ions and guanyl nucleotides. Life Sciences **23:** 495–500.
6. BROSTROM, C. D., Y. C. HUANG, MCL. B. BRECKENRIDGE & D. J. WOLFF. 1975. Identification of a $Ca^{2+}$-binding protein as a $Ca^{2+}$-dependent regulator of brain adenylate cyclase. Proc. Natl. Acad. Sci. USA **72:** 64–68.
7. WAISMAN, D. M., T. J. SINGH & J. H. WANG. 1978. The modulator-dependent protein kinase. J. Biol. Chem. **253:** 3387–3390.
8. CHEUNG, W. Y. 1970. Cyclic 3′,5′-nucleotide phosphodiesterase: demonstration of an activator. Biochem. Biophys. Res. Commun. **38:** 533–538.
9. KAKIUCHI, S., R. YAMAZAKI, Y. TESHIMA & K. UENISHI. 1973. Regulation of nucleotide cyclic 3′,5′-monophosphate phosphodiesterase activity from rat brain by a modulatory and $Ca^{2+}$. Proc. Natl. Acad. Sci. USA **70:** 3526–3530.
10. GNEGY, M. E., J. A. NATHANSON & P. UZUNOV. 1977. Release of the phosphodiesterase activator by cAMP-dependent ATP: protein phosphotransferase from subcellular fractions of rat brain. Biochem. Biophys. Acta **497:** 75–85.
11. GNEGY, M., P. UZUNOV & E. COSTA. 1976. Regulation of dopamine stimulation of striatal adenylate cyclase by an endogenous $Ca^{2+}$-binding protein. Proc. Natl. Acad. Sci. USA **73:** 3887–3890.
12. HANBAUER, I., J. GIMBLE & W. LOVENBERG. 1979. Changes in soluble calcium-dependent regulator following activation of dopamine receptors in rat striatal slices. Neuropharmacology **18:** 851–857.
13. HANBAUER, I. & E. COSTA. 1980. Role of calmodulin in dopaminergic transmission. *In* Calcium and Cell Function (W. Y. Cheung, Ed.) Vol. **1.** Academic Press, Inc., New York, N.Y. (In press.)
14. UZUNOV, P., A. REVUELTA & E. COSTA. 1975. A role for the endogenous activator of 3′,5′-nucleotide phosphodiesterase in rat adrenal medulla. Mol. Pharmacol. **11:** 506–510.
15. KLEE, C. B. 1977. Conformational transition accompanying the binding of $Ca^{2+}$ to the protein activator of 3′,5′-cyclic adenosine monophosphate phosphodiesterase. Biochemistry **16:** 1017–1024.
16. VANAMAN, T. C., F. SHARIEF & D. M. WATTERSON. 1977. Structural homology between brain modulator protein and muscle TnCs. *In* Calcium-Binding Proteins and Calcium Function. R. H. Wasserman, R. A. Corradino, E. Carafoli, R. H. Kretsinger, D. H. MacLennan & F. L. Siegel, Eds.: 107–116. Elsevier-North Holland Publishing Co., New York, N.Y.
17. VOLLER, A., D. BIDWELL & A. BARTLETT. 1976. Microplate enzyme immunoassays for the immunodiagnosis of virus infections. *In* Protides of the Biological Fluids. H. Peeters, Ed.: 751–758. Pergamon Press. Oxford.
18. KEBABIAN, J. W., G. L. PETZOLD & P. GREENGARD. 1977. Dopamine-sensitive adenylate cyclase in caudate nucleus of rat brain and its similarity to the dopamine receptor. Proc. Natl. Acad. Sci. USA **69:** 2145–2149.
19. MIYAMOTO, E., J. F. KUO & P. GREENGARD. 1969. Cyclic nucleotide-dependent protein kinases. III. Purification and properties of adenosine 3′,5′-monophosphate-dependent protein kinase from bovine brain. J. Biol. Chem. **244:** 6395–6402.

20. REVUELTA, A., P. UZUNOV & E. COSTA. 1976. Release of phosphodiesterase activator from particulate fractions of cerebellum and striatum by putative neurotransmitters. Neurochem. Res. **1:** 217–228.
21. THOMPSON, W. J. & M. M. APPLEMAN. 1971. Characterization of cyclic nucleotide phosphodiesterase of rat tissues. J. Biol. Chem. **246:** 3145–3150.
22. UZUNOV, P. & B. WEISS. 1972. Separation of multiple forms of cyclic nucleotide phosphosdiesterase in rat cerebellum by polyacrylamide gel electrophoresis. Biochem. Biophys. Acta **284:** 220–226.
23. RANDRUP, A. & I. MUNKVAD. 1970. Biochemical, anatomical and psychological investigations of sterotyped behavior induced by amphetamines. *In* Amphetamines and Related Compounds. E. COSTA & S. GARATTINI, Eds.: 695–713. Raven Press. New York, N.Y.
24. COSTA, E., A. GROPETTI & M. K. NAIMZADA. 1972. Effects of amphetamine on the turnover rate of brain catecholamines and motor activity. Brit. J. Pharmacol. **44:** 742–751.
25. CARENZI, A., D. L. CHENEY, E. COSTA, A. GUIDOTTI & G. RACAGNI. 1975. Action of opiates, antipsychotics, amphetamines and apomorphine on dopamine receptors in rat striatum: *in vivo* changes of 3′,5′-cAMP content and acetylcholine turnover rate. Neuropharmacology **14:** 927–940.
26. CREESE, I., D. D. BURT & S. H. SNYDER. 1977. Dopamine receptor binding enhancement accompanies lesion-induced behavioral supersensitivity. Science **197:** 596–598.
27. GNEGY, M. E., P. UZUNOV & E. COSTA. 1977. Participation of an endogenous $Ca^{2+}$-binding protein activator in the development of drug-induced supersensitivity of striatal dopamine receptors. J. Pharmacol. Exp. Ther. **202:** 558–564.
28. LUCCHELLI, A., A. GUIDOTTI & E. COSTA. 1978. Striatal content of $Ca^{2+}$-dependent regulator protein and dopaminergic receptor function. Brain Res. **155:** 130–135.

## DISCUSSION OF THE PAPER

DR. H. RASMUSSEN (*Yale University, New Haven, CT*): In a simple cell like the rat erythrocyte, using catecholamines as the agonist, we can show an opposite kind of an effect. Namely, when the catecholamine binds to its receptor there is a shift of calmodulin from the cytosol to the membrane and that activates a membrane bound phosphodiesterase. So, there is a symmetry in nature.

DR. R. CARLIN: When you are looking at this compartmentalization between membranes and cytosol, you might also be interested in looking at the postsynaptic density because we find all but two calmodulin binding proteins within the postsynaptic density. In relation to this, have you looked at phosphodiesterase activity in the membranes or in the postsynaptic membrane which is part of the membrane?

DR. HANBAUER: No. The phosphodiesterase is not extracted from membrane. We prepared synaptosomal pellet and cytosol fraction with a sucrose gradient and, therefore, there is no extraction of the enzyme from membranes. This is only a cytosolic enzyme.

DR. M. T. PIASCIK: In the animals where you pre-treated with haloperidol and concomitantly injected cyclohexamide, was there supersensitivity in those receptors?

DR. HANBAUER: Yes.

DR. PIASCIK: How do you reconcile these data with the increase in receptor sites as monitored by binding? In other words, you are saying it is the calmodulin in the membranes whereas others have reported an increase in receptors.

DR. HANBAUER: Well, there are some states, such as six hydroxy dopamine administration, chronic injection of reserpine, or denervation, which have been shown to involve supersensitivity. In these studies there was an increase in the number of binding sites and no change in the affinity. But, it has not yet been explained what triggers the increase in the number of receptor sites.

# ROLE OF CALMODULIN IN STATES OF ALTERED CATECHOLAMINE SENSITIVITY

M. E. Gnegy, Y. S. Lau, and G. Treisman

*Department of Pharmacology*
*University of Michigan Medical School*
*Ann Arbor, Michigan 48109*

## INTRODUCTION

Calmodulin, an endogenous calcium-binding protein, can modulate the intracellular concentration of adenosine 3′,5′-cyclic monophosphate (cyclic AMP) by stimulating membrane-bound adenylate cyclase activity and soluble phosphodiesterase (PDE) activity.[1] There is increasing evidence that calmodulin can modulate the effects of calcium at both pre- and postsynaptic sites in some areas of brain. Calmodulin is enriched in rat brain synaptic membranes[2] and has been located in postsynaptic densities in mouse basal ganglia[3] and canine cerebral cortex.[4] Previous studies have suggested that calmodulin is involved with dopamine (DA)-sensitive adenylate cyclase activity in rat striatum. Gnegy *et al.*[5] found that depletion of calmodulin from striatal membranes results in a decreased ability of DA to stimulate adenylate cyclase activity. Furthermore, the calmodulin content in the striatum was increased under conditions of dopamine supersensitivity.[6] Rats treated chronically with cataleptogenic antipsychotic drugs and then withdrawn from the drugs had increased calmodulin content in their striatal membranes. These animals exhibited behavioral supersensitivity to apomorphine and their striatal adenylate cyclase had an increased affinity for dopamine as demonstrated by a 3- to 4-fold increase in dopamine sensitivity. The increase in calmodulin could positively affect DA-sensitive adenylate cyclase activity and thus contribute to dopaminergic supersensitivity.

The increased calmodulin in the striatal membranes of the supersensitive rats seems to be more tightly bound to its membrane binding proteins.[7] In the brain, calmodulin can be bound to membrane proteins in a calcium-dependent and calcium-independent manner.[8] Calmodulin has been shown to bind to $Ca^{2+}$-dependent PDE[9] and $Ca^{2+}$-sensitive adenylate cyclase[10] as well as other proteins. Vandermeers *et al.*[11] have demonstrated a correlation between increased [$^{125}$I]calmodulin binding and activation of adenylate cyclase activity in guinea pig membranes.

In this work we investigate more directly whether calmodulin can affect DA-sensitive activity in rat striatal membranes. We found that calmodulin could affect both the maximal velocity of the reaction and the sensitivity to DA. We studied the binding of calmodulin to striatal membranes in order to further investigate the relationship between calmodulin interaction with adenylate cyclase activity. We showed specific binding of [$^{125}$I]calmodulin to striatal membranes that was competitively inhibited by trifluoperazine, an antipsychotic drug that binds to calmodulin and inhibits its action.[12] We have also shown that the ability of the calmodulin content to increase could be specific for dopaminergic supersensitivity in the striatum.



0077-8923/80/0356-0304 $01.75/0 © 1980, NYAS

## METHODS

### *Membrane Preparation*

Striatal membranes were depleted of calmodulin by a modification of the method of Brostrom *et al.*[13] Male, Sprague-Dawley rats (150–200 g) were sacrificed by decapitation; the striata were removed and homogenized in 9 vol of 10 mM Tris-maleate buffer, pH 7.5, containing 1 mM $MgSO_4$, 1.2 mM EGTA, and 10 $\mu$M guanosine triphosphate (GTP). The homogenate was centrifuged at 27,000 × g for 20 min, resuspended in the same buffer, and centrifuged a second time at 27,000 × g. The pellet resulting from the final centrifugation was resuspended in 10 mM Tris-maleate buffer, pH 7.5, containing 1.2 mM EGTA and 1 mM $MgSO_4$.

### *Adenylate Cyclase Assay*

Adenylate cyclase activity was measured in an assay (200 $\mu$l vol) containing: 80 mM Tris-maleate buffer, pH 7.5; 5 mM $MgSO_4$; 2 mM cyclic AMP, 4 mM phosphoenolpyruvate, 20 $\mu$g pyruvate kinase, 0.12 mM isobutylmethylxanthine, 100–150 $\mu$g membrane protein, 0.15 mM EGTA (carried over from the membrane preparation), 1 mM [*alpha*-$^{32}$P]ATP (1–2 × $10^6$ cpm/assay), with or without additions, such as, 10 $\mu$M guanyl-5′-yl-imidodiphosphate (Gpp(NH)p), 10 $\mu$M GTP, 125 $\mu$M $CaCl_2$, and 500 ng (28 pmol) of highly purified calmodulin prepared from bovine brain according to Klee.[14] Assays were incubated for 3 min, stopped by heating 1 min at 95°C, and 200 $\mu$l of a solution containing 20 mM ATP and 0.7 mM cyclic AMP was added to the tubes. The membranes were centrifuged, and the [$^{32}$P]-cyclic AMP in the supernatant was determined by the method of Krishna *et al.*[15] Recovery of the cyclic AMP was measured using [$^3$H]-cyclic AMP and was 65–70%.

### *Determination of Calmodulin Content*

Calmodulin was assayed by its ability to stimulate calmodulin-deficient PDE activity[9] in the PDE assay described previously.[2] The calmodulin content in nanograms was determined from a standard curve using highly purified calmodulin.

### *Binding Assay*

Calmodulin was iodinated using [$^{125}$I]N-succinimidyl-3-(4-hydroxyphenol) propionate (Bolton-Hunter reagent, Amersham) as described by Chafouleas *et al.*[16] and had a specific radioactivity of 428 Ci/mmol. Calmodulin binding studies were performed as described by Vandermeers *et al.*[10] with some modifications. The assay medium (in final volume of 0.2 ml) contained 80 mM Tris-maleate buffer, pH 7.5, 5 mM $MgSO_4$, 0.2% bovine serum albumin, and various testing agents as indicated. Calmodulin ($^{125}$I-labeled and non-labeled) was added to each assay so that the specific activity remained constant at 8 nCi/pmol of calmodulin. The reaction was initiated by the addition of the calcium- and calmodulin-depleted striatal membrane preparation (60–80 $\mu$g protein) and the reaction mixture was incubated at 37°C for 2 minutes, unless otherwise indicated. To terminate the reaction, the sample was diluted with 2 ml of the ice-cold incubation buffer and filtered immediately under vacuum through a EHWP 0.5 μm Millipore filter (Millipore Co., Bedford, Mass.) that had been prewashed with the buffer. The filters were rapidly washed twice with 2 ml of the same ice-cold buffer, dried, and counted in an Auto-gamma scintillation counter (Packard Model 5220). The nonspecific binding, determined by adding a 100-fold excess of nonlabeled calmodulin to the assay showed no significant difference from the assay blank (without membrane). Both the blank and the nonspecific binding represented less than 0.7% of the total radioactivity. Trifluoperazine, when used, was preincubated in the medium for 20 minutes at 37°C before the membranes were added.

### *Membrane Calcium Determination*

Striatal membrane calcium content was measured using the method as described by Schmidt and Way[17] with some modifications. An aliquot (0.5 ml) of membrane fractions (4–5 mg protein) was treated with 0.5 ml of 10% trichloroacetic acid (TCA). The mixture was equilibrated for 30 minutes and centrifuged at 5,000 × g for 15 minutes. The pellet was used to determine protein content[18] while the clear supernatant of the sample was mixed with 1 ml of 1% $La^{3+}$ in 5% HCl. Sample blanks and standards were similarly treated. Calcium concentrations were assayed using an atomic absorption spectrophotometer (Varian Model AA 375). Four readings per sample were taken and averaged. The limit of detection by this spectrophotometric method was 5 nmol calcium/0.5 ml sample.

The free calcium concentration reported in the binding studies was calculated according to Nanninga and Kempen[19] where the concentrations of EGTA and $Mg^{2+}$ and the buffer pH were taken into consideration. All data were calculated in terms of mean ± S.E.M. Statistical analysis was performed using the *t* test either for unpaired or for paired comparison whenever appropriate. A $P < 0.05$ was considered significant.

## Results

### *Effect of Calmodulin and Calcium on Basal Adenylate Cyclase Activity*

Depletion of most of the calmodulin content from the striatal membranes was necessary to observe the stimulatory effects of this protein on adenylate cyclase

Table 1

Calmodulin Content, Calcium Content, and Adenylate Cyclase Activity in Rat Striatal Membranes Before and After EGTA Treatment

| Conditions | Membrane Calmodulin Content (μg/mg protein) | Membrane Calcium Content (nmol/mg protein) | Adenylate Cyclase Activity (pmol/min/mg protein) |
|---|---|---|---|
| Buffer | 4.9 ± 0.6 (3) | 12.3 ± 0.8 (5) | 175 ± 7 (3) |
| Buffer + EGTA | 2.9 ± 0.3 (3) | not detectable (10) | 105 ± 17 (3) |

TABLE 2

EFFECT OF CALCIUM AND CALMODULIN ON BASAL ADENYLATE CYCLASE ACTIVITY IN RAT STRIATAL MEMBRANES

| Calcium added to Assay ($\mu$M) | Adenylate Cyclase Activity (pmol/min/mg protein ± S.E.M.) | |
|---|---|---|
| | − Calmodulin | + Calmodulin |
| 0 | 96 ± 6 (4) | 76 ± 11 (4) |
| 50 | 97 ± 12 | 78 ± 10 |
| 100 | 94 ± 16 | 135 ± 8 |
| 125 | 86 ± 10 | 145 ± 18 |
| 200 | 80 ± 14 | 105 ± 16 |

activity. The EGTA treatment depleted most, but not all, of the calmodulin from the membrane fraction. As shown in TABLE 1, the calmodulin content in the membranes was depleted by 41% after EGTA treatment. Originally, nearly 50% of the calmodulin content in the brain cell is located in the membranes. EGTA washing similarly decreased the specific activity of the basal adenylate cyclase activity in the membrane fraction by 41% (TABLE 1). The calcium content in EGTA-washed membranes was below the level of detection.

Neither calcium nor calmodulin individually were able to stimulate basal adenylate cyclase activity in the striatum. Calcium itself did not really affect basal activity until high concentrations were added (>200 $\mu$M). Calmodulin and calcium added together had a biphasic action on adenylate cyclase activity (TABLE 2). In the presence of 500 ng of calmodulin ($1.4 \times 10^{-7}$ M), concentrations of calcium added between 50 $\mu$M and 150 $\mu$M elicited stimulation of adenylate cyclase activity with a maximum effect at 125 $\mu$M calcium. Adenylate cyclase activity was inhibited at concentrations of calcium of less than 50 $\mu$M or greater than 200 $\mu$M in the presence of calmodulin. The effective concentrations of calcium providing stimulation was actually very low because the assay contained 150 $\mu$M EGTA. Using a $K_D$ for Ca-EGTA calculated according to Nanninga and Kempen,[19] addition of 125 $\mu$M calcium will result in an effective concentration of 0.14 $\mu$M calcium.

As shown in FIGURE 1, guanyl nucleotides were able to produce a greater maximal stimulation of adenylate cyclase than did calmodulin. 10 $\mu$M Gpp(NH)p stimulated the adenylate cyclase activity by 200%. Calcium significantly inhibited the Gpp(NH)p-stimulated adenylate cyclase activity by 60%. Addition of calmodulin reversed the calcium inhibition with the result that the stimulation by Gpp(NH)p, calcium, and calmodulin was the same as that of Gpp(NH)p alone. The interaction of calcium and calmodulin with GTP stimulation of basal activity qualitatively differed from that observed with Gpp(NH)p. GTP-stimulated adenylate cyclase activity was slightly stimulated by both calcium and calmodulin. Calcium will inhibit GTP-stimulated adenylate cyclase activity but much greater concentrations of calcium are required (200 $\mu$M added calcium, which is >2 $\mu$M effective calcium). Stimulation of adenylate cyclase activity by calmodulin was specific because troponin C, calcium-binding protein from muscle, did not stimulate basal adenylate cyclase activity even at 10-fold higher concentrations. Furthermore, troponin C did not reverse the effects of calcium on Gpp(NH)p stimulation as did calmodulin.

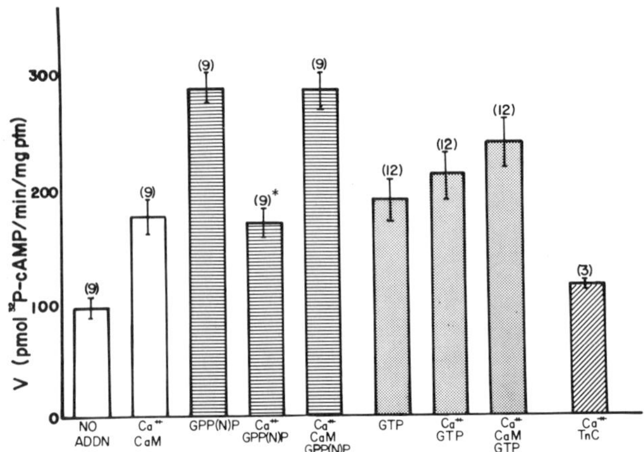


FIGURE 1. Stimulation of adenylate cyclase activity in rat striatum by calmodulin and guanyl nucleotides. Adenylate cyclase assays were performed as described in METHODS with the following additions: 10 μM Gpp(NH)p (GPP(N)P) or GTP; 500 ng of calmodulin (CaM) or 125 μM $CaCl_2$. The number of experiments is given in parentheses. *$P < 0.01$ as compared to the value for Gpp(NH)p alone.

### *Effect of Calmodulin on DA-Sensitive Adenylate Cyclase Activity*

Adenylate cyclase activity in striatal membranes that had been washed repeatedly with EGTA could not be stimulated by DA. Addition of calcium, calmodulin, or both agents together was insufficient to restore DA sensitivity. Inclusion of 10 μM Gpp(NH)p or GTP in the assay was necessary to achieve a dose-dependent response to DA in the formation of cyclic AMP. Adenylate cyclase activity was stimulated 1.4–1.5 fold by DA in the presence of 10 μM Gpp(NH)p. This actually represented a large stimulation over the original basal activity (TABLE 1). The increase in adenylate cyclase activity in the presence of 10 μM Gpp(NH)p reached a maximum at 5 μM DA. The $K_{act}$ for DA in the presence of Gpp(NH)p was 1.1 μM and the $V_{max}$ was 100 pmol/min/mg protein (TABLE 3). Addition of 125 μM calcium brought about a 50% decrease in $V_{max}$ with no

TABLE 3

KINETIC CONSTANTS FOR STIMULATION OF STRIATAL DOPAMINE-SENSITIVE ADENYLATE CYCLASE BY GPP(NH)P, CALCIUM, AND CALMODULIN

| Additions to Assay | $K_{act}$ (μM ± S.E.M.) | $V_{max}$ (pmol/min/mg protein ± S.E.M.) |
|---|---|---|
| Gpp(NH)p (5) | 1.1 ± 0.2 | 100 ± 15 |
| Gpp(NH)p + $Ca^{2+}$ (5) | 0.83 ± 0.1 | 45 ± 5** |
| Gpp(NH)p + $Ca^{2+}$ (5) + calmodulin | 0.56 ± 0.1* | 77 ± 12 |

*$P < 0.005$ as compared to value with Gpp(NH)p alone or Gpp(NH)p + $Ca^{2+}$.
**$P < 0.02$ as compared to value with Gpp(NH)p alone or Gpp(NH)p + $Ca^{2+}$ + calmodulin.

TABLE 4

KINETIC CONSTANTS FOR STIMULATION OF STRIATAL DOPAMINE-SENSITIVE ADENYLATE CYCLASE ACTIVITY IN THE PRESENCE OF GTP, CALCIUM AND CALMODULIN

| Additions to Assay | $K_{act}$ ($\mu$M ± S.E.M.) | $V_{max}$ (pmol/min/mg protein ± S.E.M.) |
|---|---|---|
| GTP (8) | 3.8 ± 0.7 | 122 ± 9 |
| GTP + $Ca^{2+}$ (5) | 5.2 ± 0.1 | 68 ± 8** |
| GTP + $Ca^{2+}$ (8) + Calmodulin | 1.8 ± 0.4* | 112 ± 10 |

*$P < 0.04$ as compared to value with GTP alone or GTP + $Ca^{2+}$.
**$P < 0.001$ as compared to value with GTP alone or GTP + $Ca^{2+}$ + calmodulin.

change in sensitivity for DA. When 500 ng of calmodulin was added with the calcium, the maximal velocity was restored to 80% of the value found with Gpp(NH)p alone and the dose response curve for DA was shifted in a parallel manner to the left. The $K_{act}$ for DA was decreased 2-fold to 0.56 $\mu$M (TABLE 3). The average kinetic parameters of 5 separate experiments for DA stimulation of adenylate cyclase in the presence of Gpp(NH)p are shown in TABLE 3.

The same pattern was seen for the effects of calcium and calmodulin on DA-stimulated activity in the presence of 10 $\mu$M GTP. Adenylate cyclase activity was stimulated 1.7–1.8 fold by DA in the presence of 10 $\mu$M GTP. The $K_{act}$ for DA stimulation of adenylate cyclase was 3.8 $\mu$M (TABLE 4), nearly 3 times greater than that found in the presence of Gpp(NH)p. The $V_{max}$ of the reaction (122 pmol/min/mg protein) was slightly greater, however, than that obtained with Gpp(NH)p. Addition of 125 $\mu$M calcium to the assay again decreased DA stimulation, in spite of the fact that calcium did not decrease GTP stimulation of

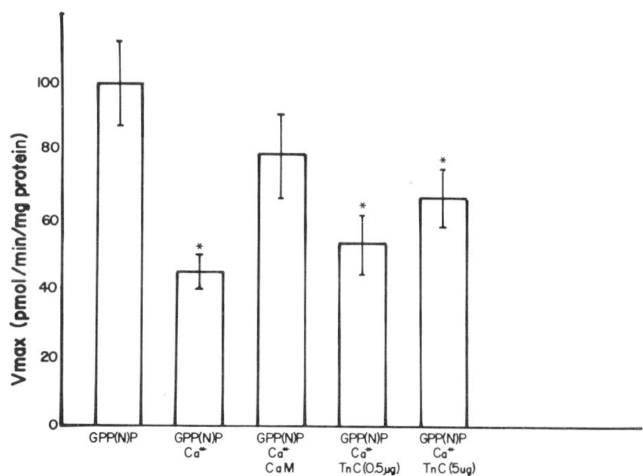


FIGURE 2. Effect of troponin C on maximal velocity of dopamine-sensitive adenylate cyclase activity in rat striatum. The $V_{max}$ of the dopamine-sensitive adenylate cyclase reaction was calculated from Lineweaver Burk plots from 3 different experiments. Concentrations of additions are: 10$\mu$M Gpp(NH)p (GPP(N)P), 500 ng calmodulin (CaM), 500 ng or 5 $\mu$g troponin C (TnC), or 125 $\mu$M $CaCl_2$. *$P < 0.05$ as compared to the value for Gpp(NH)p alone.

basal activity. In this case calcium reduced the $V_{max}$ 2-fold with no significant change in the $K_{act}$ for DA. Addition of calmodulin with the calcium elicited an increase in $V_{max}$ to the level found with GTP alone and resulted in a 2-fold sensitivity for DA ($K_{act}$ for DA = 1.8 $\mu$M). Kinetic constants for DA-stimulation of adenylate cyclase in the presence of GTP for 8 different experiments are shown in TABLE 4.

The effect of calmodulin on DA-sensitive adenylate cyclase activity was specific for calmodulin in the capacity of a calcium-binding protein. Troponin C could not completely substitute for the effects of calmodulin on the maximal velocity of DA-stimulated activity even at nearly 10-fold higher concentration than calmodulin (FIGURE 3). The molecular weight of troponin C is 17,800 daltons.[20] Troponin C did not increase the affinity for DA.

### *Binding of [$^{125}$I] Calmodulin to Striatal Membranes*

Although the experiments using troponin C indicated a specific interaction of calmodulin with adenylate cyclase, we wished to show that calmodulin was able to bind specifically to striatal membranes under the conditions of the adenylate cyclase assay. Therefore, we developed an assay to measure the binding of [$^{125}$I]calmodulin to striatal membranes depleted of calcium and calmodulin. The dependence of calmodulin binding upon time and temperature was determined in the presence of $1.25 \times 10^{-4}$ M added calcium (0.14 $\mu$M) in membranes previously depleted of calcium and calmodulin. Binding of [$^{125}$I] calmodulin was

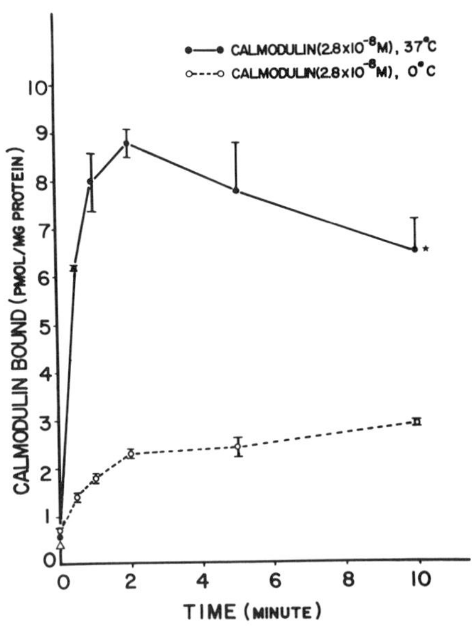


FIGURE 3. Effects of time and temperature on [$^{125}$I]calmodulin binding in rat striatal membranes. The membranes were depleted of calcium and calmodulin as described in the METHODS. The assays contained $1.25 \times 10^{-4}$ M $CaCl_2$. *Value significantly lower than the peak binding response at 2 min. $N = 4$.

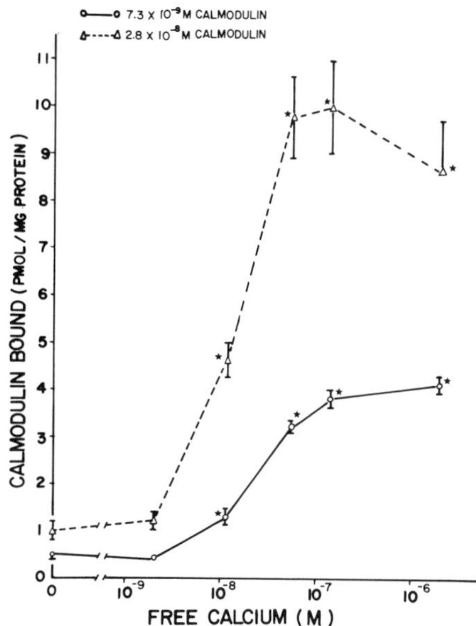


FIGURE 4. Effect of calcium on [$^{125}$I]calmodulin binding in rat striatal membranes. Free calcium concentrations in the assay were calculated according to Nanninga and Kempen.[19] The assays were incubated at 37°C for 2 minutes. $N = 3$. *Significantly different from the value when no calcium was added.

found to be rapid and maximal binding was reached after 2 minutes of incubation. However, the binding decreased when incubation was continued at 37°C (FIGURE 3). The extent of calmodulin binding was markedly decreased when incubation was performed at 0°C (FIGURE 3) but the time-dependent decrease was not seen. A protease could be destroying the calmodulin at the higher temperature.

Calmodulin binding to calcium- and calmodulin-depleted membranes was dependent on the concentration of calcium added into the assay. Calmodulin binding was substantially enhanced by increasing concentrations of calcium (FIGURE 4). $1.25 \times 10^{-4}$ M calcium ($1.4 \times 10^{-7}$ M calculated free $Ca^{2+}$) exhibited a maximal binding of calmodulin to the membranes, while calcium less than $10^{-5}$ M ($2 \times 10^{-9}$ M free $Ca^{2+}$) did not affect the binding significantly, as compared to the binding with no calcium added (FIGURE 4).

Calmodulin binding to the striatal membranes was saturable; saturation being reached at 500 ng of calmodulin ($1.4 \times 10^{-7}$ M) in the presence of $1.25 \times 10^{-4}$ M calcium. Trifluoperazine ($10^{-4}$ M) produced a significant shift of the control calmodulin-binding to the right (FIGURE 5). In four experiments, trifluoperazine at $10^{-5}$ M did not antagonize calmodulin binding significantly and at $10^{-3}$ M the drug did not produce further inhibition of calmodulin binding than at $10^{-4}$ M (data not shown).

The above results (FIGURE 5) were plotted according to the Scatchard analysis. Saturation of binding occurred with a maximal number of binding sites of 11.8 ± 1.2 pmol/mg protein (determined by extrapolation from 11 experiments) (TABLE 5). The Scatchard plot, as well as the displacement curve, suggests that there are

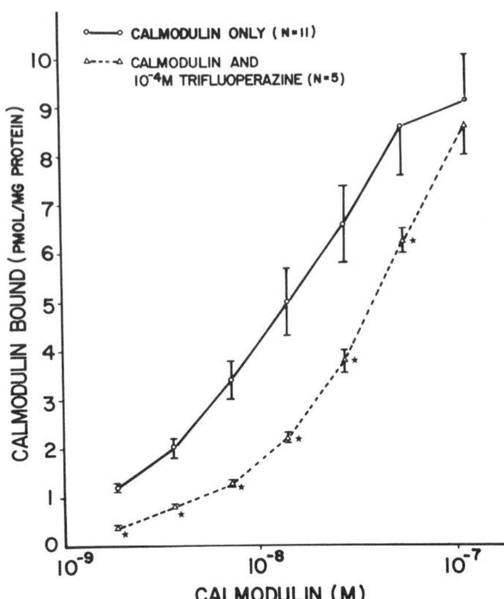


FIGURE 5. Effect of trifluoperazine ($10^{-4}$ M) on the dose-response curve of [$^{125}$I]calmodulin binding in rat striatal membranes. Trifluoperazine was preincubated for 20 minutes at 37°C in the presence of $1.25 \times 10^{-4}$ M $CaCl_2$ before the membranes were added. The binding reaction was carried out at 37°C for 2 minutes. *Significantly different from control value.

two populations of calmodulin binding sites: a higher-affinity binding site with a dissociation constant ($K_D$) of $1.3 \times 10^{-7}$ M and a lower-affinity binding site with a $K_D$ of $2.9 \times 10^{-7}$ M (TABLE 5). When the calmodulin binding curve was evaluated by Hill analysis[21] to determine whether there were cooperative interactions among the binding sites, a linear Hill plot with a coefficient of 0.90 ± 0.14 (95% confidence limit) was obtained. This may suggest that there is no appreciable cooperativity among calmodulin binding sites.

The effect of trifluoperazine ($10^{-4}$ M) on calmodulin binding was also analyzed by the Scatchard method. The analysis revealed that the higher-affinity

TABLE 5

EFFECT OF $10^{-4}$ M TRIFLUOPERAZINE ON [$^{125}$I]CALMODULIN BINDING IN RAT STRIATAL MEMBRANES*

| | Control (11)† | +$10^{-4}$ M Trifluoperazine (5)† |
|---|---|---|
| Higher affinity binding | | |
| Number of sites | 8.5 ± 0.8 pmol/mg protein | 3.3 ± 0.4 pmol/mg protein‡ |
| App. $K_D$ | $1.3 \times 10^{-7}$ M | $1.4 \times 10^{-7}$ M |
| Lower affinity binding | | |
| Number of sites | 11.8 ± 1.2 pmol/mg protein | 13.3 ± 1.5 pmol/mg protein |
| App. $K_D$ | $2.9 \times 10^{-7}$ M | $1.3 \times 10^{-6}$ M‡ |

*All values were determined from Scatchard plot.
†Number of experiments.
‡Significantly different from control values ($P < 0.05$).

binding component for calmodulin was significantly inhibited by trifluoperazine ($10^{-4}$ M), while the $K_D$ for lower-affinity binding component was increased (TABLE 5).

### *Alterations in Calmodulin Content During Supersensitivity to Catecholamines in Brain*

We have shown previously that calmodulin content was increased in rat striatal membranes after chronic treatment with cataleptogenic antipsychotic drugs.[5] Furthermore, Lucchelli *et al.*[22] have shown increases in striatal content of calmodulin in rat striatum after hemisection, indicating that increases in calmodulin will occur after supersensitivity to dopamine is produced by a variety of means. We wished to investigate whether other dopaminergic areas, such as the hypothalamus and the limbic system, had increased calmodulin content as well. The calmodulin content was measured in the membranes (27,000 × g fraction) of

TABLE 6

CALMODULIN CONCENTRATION IN MEMBRANES OF SEVERAL BRAIN AREAS FROM RATS TREATED CHRONICALLY WITH SALINE AND HALOPERIDOL*

| Area | N | Calmodulin Saline | Calmodulin Haloperidol |
|---|---|---|---|
| | | (μg/mg membrane protein ± S.E.M.) | |
| Striatum | 8 | 4.0 ± 0.2 | 5.1 ± 0.1† |
| Hypothalamus | 13 | 3.5 ± 0.2 | 2.9 ± 0.2‡ |
| Nucleus Accumbens | 5 | 3.3 ± 0.1 | 3.2 ± 0.1 |

*Male Sprague Dawley rats were injected daily for 20 days with 2.5 μmol/kg (s.c.) haloperidol or 0.1 ml saline. One week after stopping chronic injections the rats were sacrificed and calmodulin content was measured as described in METHODS.

†$P < 0.05$ as compared to value for saline-treated animals.

‡$P < 0.001$ as compared to value for saline-treated animals.

the hypothalamus and nucleus accumbens after chronic haloperiodol treatment in rats. The rats were treated as described in TABLE 6. As TABLE 6 illustrates, the only area showing an increase in membrane calmodulin after chronic haloperidol treatment was the striatum. A decrease in calmodulin was found in the membrane fraction of the hypothalamus but the change was quite small. At the present time the significance of this is obscure.

The ability of calmodulin content to be altered under conditions of noradrenergic supersensitivity was examined. Treatment of rats with reserpine[23] or 6-hydroxydopamine[24] (6-OHDA) can result in supersensitivity of noradrenergic receptors in the cerebral cortex. In order to examine whether calmodulin could be involved in modulating NE responses in the brain, supersensitivity to NE was developed by these methods. One group of rats was treated with reserpine (2.5 mg/kg i.p.) for four days and sacrificed 4 hours later. The calmodulin content in the membranes (27,000 × g fraction) of the cerebral cortex was measured. In another experiment rats were injected intraventricularly on two successive days with 6-OHDA (200 μg) as described by Kalisker *et al.*[24] Two weeks later the animals were sacrificed and calmodulin was measured in the membrane fractions (27,000 × g) from the cerebral cortex and striatum. The results are shown in TABLE 7. The calmodulin content did not change in the cerebral cortex of rats subjected to either drug treatment. The calmodulin content was, however, significantly increased in the striatal membranes of 6-OHDA rats. This is not too surprising since intraventricularly administered 6-OHDA can cause extensive depopulation in the zona compacta of the substantia nigra. Although NE-sensitive adenylate cyclase activity was not measured, the appearance of noradrenergic supersensitivity in the cortex after these treatments is well documented.

## DISCUSSION

The participation of calcium in catecholamine-stimulated adenylate cyclase activity in the brain is not well understood. Calcium has generally been consid-

TABLE 7

CALMODULIN CONTENT IN MEMBRANES OF RAT CEREBRAL CORTEX AND STRIATUM AFTER 6-OHDA OR RESERPINE TREATMENT

| | Calmodulin Concentration | |
|---|---|---|
| | Striatum | Cerebral Cortex |
| Treatment | (μg/mg protein ± S.E.M.) | |
| Saline (5) | 5.0 ± 0.2 | 5.2 ± 0.2 |
| 6-OHDA (5)* | 6.2 ± 0.1‡ | 5.5 ± 0.4 |
| Saline (8) | – | 5.0 ± 0.3 |
| Reserpine (8) | – | 5.1 ± 0.3 |

*6-OHDA-treated rats were injected intraventricularly using stereotoxic coordinates with 200 μg of 6-OHDA on each of 2 successive days. Animals were sacrificed 2 weeks after last injection.

†Reserpine-treated rats were treated daily with reserpine (2.5 mg/kg i.p.) for 4 days and sacrificed 4 hours after last injection. Number of animals is given in parentheses.

‡$P < 0.001$ when compared to value for saline-treated animals.

ered to be inhibitory to DA-sensitive adenylate cyclase activity in the striatum because stimulation in homogenates can only be detected in the presence of EGTA, a calcium chelating agent. This effect occurred because EGTA decreased basal adenylate cyclase activity and permitted expression of DA sensitivity.[25] In other systems, however, calcium has been shown to enhance catecholamine stimulation of cyclic AMP production. For example, extracellular calcium is required for the increased production of cyclic AMP by α-adrenergic agonists in cerebral cortical slices.[26] A stronger role for a stimulatory effect of intracellular calcium in NE-stimulation of cyclic AMP production was shown by Brostrom *et al.*[27] using C6 glioma cells. The ability of C6 glioma cells to accumulate cyclic AMP in response to NE was reduced 60–70% following nearly total calcium depletion of the cells in culture by EGTA.

Since calmodulin is so integrally involved with the effects of calcium on cyclic nucleotide metabolism in the brain, one cannot properly determine the effects of calcium without consideration of calmodulin. Calmodulin is present in brain membranes and is enriched in synaptic membranes.[2] Therefore, we studied the effects of calcium and calmodulin on basal and DA-stimulated adenylate cyclase activity in calcium and calmodulin depleted rat striatal membranes.

We found that calmodulin stimulated basal adenylate cyclase activity in a biphasic manner that was dependent upon the concentration of calcium. Stimulation of adenylate cyclase by calmodulin was maximal at an effective calcium concentration of 0.14 $\mu$M which is within resting levels of calcium in nerve membranes.[28] Calcium and calmodulin inhibited adenylate cyclase activity at higher concentrations of calcium. The effects of calmodulin on basal adenylate cyclase activity would conceivably depend on calcium flux in the membrane. At resting levels of calcium, adenylate cyclase activity would be increased by calmodulin but as calcium levels rise the system would become inhibited.

DA sensitivity was abolished in calcium and calmodulin-depleted membranes and could not be restored by addition of either agent. A dose dependent stimulation of adenylate cyclase by DA was found only when guanyl nucleotides were included in the assay at concentrations of at least 10 $\mu$M. Similar results were found by Sulakhe *et al.*[29] Addition of 125 $\mu$M calcium inhibited DA-stimulated adenylate cyclase activity in the presence of either guanyl nucleotide. When calmodulin was present, however, the maximal velocity of the reaction was somewhat restored and the sensitivity to DA was increased. Thus calmodulin affected both the $K_a$ for DA and $V_{max}$ of the DA-sensitive adenylate cyclase activity. This suggests that at physiological, resting concentrations of calcium, DA will not greatly stimulate adenylate cyclase activity unless calmodulin is present.

The exact molecular mechanism in which calmodulin affects DA-sensitive adenylate cyclase activity is not yet known. Calmodulin could be simply acting as a calcium scavenger, preventing it from inhibiting DA-sensitive adenylate cyclase. There are several findings that do not support this explanation. First, calmodulin actually stimulated basal adenylate cyclase activity in the presence of calcium. Calcium alone had little effect on basal adenylate cyclase activity until very high concentrations. Second, troponin C, a calcium-binding protein with some sequence homology to calmodulin and a similar molecular weight, did not mimic the action of calmodulin on basal or DA-stimulated adenylate cyclase activity, even at concentrations ten times that of calmodulin. Troponin C actually binds calcium with an affinity similar to that of calmodulin. Third, we have shown that under conditions used in this assay, [$^{125}$I]calmodulin will bind to the striatal membranes in a calcium-dependent manner. The binding was specific for calmodulin and was inhibited by trifluoperazine, an antipsychotic drug known to bind calmodulin. We found that there were at least two classes of binding sites for calmodulin in the membranes. The fact that trifluoperazine mainly affected the higher affinity binding site suggests that this site in the striatum could play an important role in mediating the action of trifluoperazine.

It is highly likely that there is an interaction between calcium, calmodulin, and the guanyl nucleotide binding protein in brain membranes. Wescott *et al.*[10] found that a guanyl nucleotide-sensitive fraction was required for calmodulin stimulation in a purified calmodulin-dependent adenylate cyclase preparation from cerebral cortex. We found, as have others,[29] that both GTP and Gpp(NH)p stimulated basal adenylate cyclase activity in our membrane preparation and were required for DA stimulation. The effects of guanyl nucleotides on basal and DA-stimulated adenylate cyclase activity are variable and may depend on the type of membrane or homogenate preparation.[30] In our preparation, Gpp(NH)p stimulation of adenylate cyclase activity was much more sensitive to calcium inhibition than was GTP stimulation. Calcium, did, however, decrease DA stimulation of adenylate cyclase activity in the presence of both Gpp(NH)p and GTP. The molecular association induced by Gpp(NH)p in striatal membranes may be similar to that achieved in the presence of GTP and hormone. In this respect the role of guanyl nucleotides in dopamine receptor activation could parallel their mechanism of activation of the $\beta$-adrenergic receptor in erythrocytes.[31] Our data suggest that it is this association or "activation state" achieved by Gpp(NH)p or DA that can be inhibited by calcium and with which calmodulin can interact.

Our present findings correlate with our previous work concerning the relationship between calmodulin and dopaminergic supersensitivity in rat striatum.[5,32] Animals chronically treated with antipsychotic drugs had increased calmodulin content in their striatal membranes. The adenylate cyclase in the striatal membranes of the drug-treated animals showed an increased sensitivity to DA as shown by a 3-fold decrease in the apparent $K_a$ for DA. This further supports a role for calmodulin in dopaminergic activity and even supersensitivity to DA in the rat striatum.

Our results indicate that calmodulin is mainly important for dopaminergic supersensitivity in the striatum, although we did not do experiments validating that there was dopaminergic supersensitivity occurring in the other areas. The strong change of calmodulin seen in the striatum may be due to the feedback regulation mechanism of the nigrostriatal system. The difference in regulation of the striatal receptors could account for these results. Similarly, the amount of calmodulin did not change a membrane fraction from the cerebral cortex of rats made supersensitive to NE. There is evidence that calmodulin is important for adrenergic stimulation of cyclic AMP in the brain[27,33] so its possible participation in adrenergic supersensitivity may occur in a more subtle manner. We can conclude that in rat striatum, during normal synaptic activity or under conditions of greatly altered synaptic input, calmodulin could be vital in attempting to maintain normal dopaminergic function through stimulatory and inhibitory effects.

## Acknowledgments

We would like to thank Joe Romson for helping with the drug treatment studies. We are grateful to Dr. James D. Potter, Department of Pharmacology, University of Cincinnati for graciously supplying us with troponin C.

## References

1. Cheung, W. Y. 1980. Calmodulin plays a pivotal role in cellular regulation. Science **207:** 19–27.
2. Gnegy, M. E., J. A. Nathanson & P. Uzunov. 1977. Release of the phosphodiesterase activator by cyclic AMP-dependent ATP: protein phosphotransferase from subcellular fractions of rat brain. Biochim. Biophys. Acta **497:** 75–85.
3. Wood, J. G., R. W. Wallace, J. N. Whitaker & W. Y. Cheung. 1980. Immunocytochemical localization of calmodulin and a heat-labile calmodulin-binding protein (CaM-$BP_{80}$) in basal ganglia of mouse brain. J. Cell Biol. **84:** 66–76.
4. Grab, D. J., K. Berzins, R. S. Cohen & P. Siekevitz. 1979. Presence of calmodulin in postsynaptic densities isolated from canine cerebral cortex. J. Biol. Chem. **254:** 8690–8696.

5. GNEGY, M. E., P. UZUNOV & E. COSTA. 1976. Regulation of dopamine stimulation of striatal adenylate cyclase by an endogenous $Ca^{++}$-binding protein. Proc. Natl. Acad. Sci. USA **73:** 3887–3890.
6. GNEGY, M., P. UZUNOV & E. COSTA. 1977. Participation of an endogenous $Ca^{++}$-binding protein activator in the development of drug-induced supersensitivity of striatal dopamine receptors. J. Pharm. Exptl. Therap. **202:** 558–564.
7. GNEGY, M. E. & Y. S. LAU. 1980. Effects of chronic and acute treatment of antipsychotic drugs on calmodulin release from rat striatal membranes. Neuropharmacology **19:** 319–323.
8. TESHIMA, Y. & S. KAKIUCHI. 1978. Membrane-bound forms of $Ca^{+2}$-dependent protein modulator: $Ca^{+2}$-dependent and independent binding of modulator protein to the particulate fraction from rat brain. J. Cyclic Nucleotide Res. **4:** 219–231.
9. KLEE, C. B. & M. H. KRINKS. 1978. Purification of cyclic 3′,5′-nucleotide phosphodiesterase inhibitory protein by affinity chromatography on activator protein coupled to Sepharose. Biochemistry **17:** 120–126.
10. WESCOTT, K. R., D. C. LAPORTE & D. R. STORM. 1979. Resolution of adenylate cyclase sensitive and insensitive to $Ca^{++}$ and calcium-dependent regulatory protein (CDR) by CDR-Sepharose affinity chromatography. Proc. Natl. Acad. Sci. USA **76:** 204–208.
11. VANDERMEERS, A., P. ROBBERECHT, M.-C. VANDERMEERS-PIRET, J. RATHE & J. CHRISTOPHE. 1978. Specific binding of the calcium-dependent regulator protein to brain membranes from the guinea pig. Biochem. Biophys. Res. Commun. **84:** 1076–1081.
12. LEVIN, R. M. & B. WEISS. 1977. Binding of trifluoperazine to the calcium-dependent activator of cyclic nucleotide phosphodiesterase. Molec. Pharmacol. **13:** 690–697.
13. BROSTROM, M. A., C. O. BROSTROM & D. J. WOLFF. 1978. Calcium-dependent adenylate cyclase from rat cerebral cortex: activation by guanine nucleotides. Arch. Biochem. Biophys. **191:** 341–350.
14. KLEE, C. B. 1977. Conformation transition accompanying the binding of $Ca^{++}$ to the protein activator of 3′,5′-adenosine monophosphate phosphodiesterase. Biochemistry **16:** 1017–1024.
15. KRISHNA, G., B. WEISS & B. B. BRODIE. 1968. A simple sensitive method for the assay of adenyl cyclase. J. Pharmac. Exptl. Therap. **163:** 379–385.
16. CHAFOULEAS, J. G., J. R. DEDMAN, R. P. MUNJAAL & A. R. MEANS. 1979. Calmodulin. Development and application of a sensitive radioimmunoassay. J. Biol. Chem. **254:** 10262–10267.
17. SCHMIDT, W. K. & E. L. WAY. 1979. Direct assay for calcium in brain homogenates and synaptosomal pellets. J. Neurochemistry **32:** 1095–1098.
18. LOWRY, O. H., N. J. ROSEBROUGH, A. L. FARR & R. J. RANDALL. 1951. Protein measurement with the Folin phenol reagent. J. Biol. Chem. **193:** 265–275.
19. NANNINGA, B. & R. KEMPEN. 1971. Role of magnesium and calcium in the first and second contraction of glycerin-extracted muscle fibers. Biochemistry **10:** 2449–2456.
20. STEVENS, F. C., M. WALSH, H. C. HO, T. S. TEO & J. H. WANG. 1976. Comparison of calcium-binding proteins. Bovine heart and brain protein activators of cyclic nucleotide phosphodiesterase and rabbit skeletal muscle troponin C. J. Biol. Chem. **251:** 4495–4500.
21. HILL, A. V. 1910. The possible effects of aggregation of the molecular of haemoglobin on its dissociation curves. J. Physiol. **40:** iv–vii.
22. LUCCHELLI, A., A. GUIDOTTI & E. COSTA. 1978. Striatal content of a $Ca^{++}$-dependent regulator protein and dopaminergic receptor function. Brain Res. **155:** 130–135.
23. PALMER, G. C., H. R. WAGNER & R. W. PUTNAM. 1976. Neuronal localization of the enhanced adenylate cyclase responsiveness to catecholamines in the rat cerebral cortex following reserpine injections. Neuropharmacol. **15:** 695–702.
24. KALISKER, A., C. O. RUTLEDGE & J. P. PERKINS. 1973. Effect of nerve degenerations by 6-hydroxydopamine on catecholamine-stimulated adenosine 3′,5′-monophosphate formation in rat cerebral cortex. Molec. Pharmacol. **9:** 619–629.
25. CLEMENT-CORMIER, Y. C., R. G. PARRISH, G. L. PETZOLD, J. W. KEBABIAN & P. GREENGARD. 1975. Characterization of a dopamine-sensitive adenylate cyclase in rat caudate nucleus. J. Neurochem. **25:** 143–149.

26. SCHWABE, U., Y. OHGA & J. W. DALY. 1978. The role of calcium in the regulation of cyclic nucleotide levels in brain slices of rat and guinea pig. Naunyn-Schmied. Arch. Pharmacol. **302:** 141–151.
27. BROSTROM, M. A., C. O. BROSTROM & D. J. WOLFF. 1979. Calcium dependence of hormone-stimulated cAMP accumulation in intact glial tumor cells. J. Biol. Chem. **254:** 7548–7557.
28. KRETSINGER, R. H. 1979. The informational role of calcium in the cytosol. Adv. Cyclic Nucleotide Res. **11:** 1–26.
29. SULAKHE, P. V., N. L. LEUNG, A. T. ARBUS, S. J. SULAKHE, S.-H. JAN & N. NARAYARAN. 1977. Catecholamine-sensitive adenylate cyclase of caudate nucleus and cerebral cortex: Effects of guanine nucleotides. Biochem. J. **164:** 67–74.
30. ROUFOGALIS, B. D., M. THORNTON & D. N. WADE. 1976. Nucleotide requirement of dopamine sensitive adenylate cyclase in synaptosomal membranes from the straitum of rat brain. J. Neurochem. **27:** 1533–1535.
31. LIMBIRD, L. E., D. M. GILL & R. J. LEFKOWITZ. 1980. Agonist-promoted coupling of the β-adrenergic receptor with the guanine nucleotide regulatory protein of the adenylate cyclase system. Proc. Natl. Acad. Sci. USA **77:** 775–779.
32. GNEGY, M. E., A. LUCCHELLI & E. COSTA. 1977. Correlation between drug-induced supersensitivity of dopamine dependent striatal mechanisms and the increase in striatal content of the $Ca^{++}$-regulated protein activator of cAMP phosphodiesterase. Naunyn-Schmied. Arch. Pharmacol. **301:** 121–127.
33. GNEGY, M. E., T. HULTIN & G. TREISMAN. 1980. Effect of calmodulin on catecholamine-linked adenylate cyclase in rat striatum and cerebral cortex. Adv. Biochemical Psychopharmacology **21:** 125–131.

## DISCUSSION OF THE PAPER

DR. D. FLOCKHART (*Vanderbilt University, Nashville, TN*): Do you consider it possible that your two calmodulin affinities are actually due to two different populations of iodonated calmodulin?

DR. N. GNEGY: We put iodonated calmodulin over gels. It seems to be a pure species.

DR. FLOCKHART: Yes, but, you can have a mono-, di- or tri-iodonated calmodulin which would have altered it.

DR. GNEGY: Well, our stoichiometry doesn't seem to support that but it is always possible.

DR. WOLFE (*National Institutes of Health, Bethesda, MD*): If the calmodulin interacts with the G subunit you should be able to affect ATP ribosylation with cholera toxin and NAD. Have you tried such experiments?

DR. GNEGY: No. But I think due to the experiments of Moss and Vaughn I would say it seems a very likely possibility because they showed that calmodulin can affect cholera toxin action in the brain. Another thing I had to take into account going along with the work shown by the Brostroms is that there could be two forms of cyclase in the membrane, one calcium sensitive, one not, and one may be inhibited by calcium and the other stimulated. So, what I may be looking at is an interplay between the two.

# PHARMACOLOGICAL REGULATION OF CALMODULIN*

Benjamin Weiss, Walter Prozialeck, Mauro Cimino,†
Mary Sellinger Barnette, and Thomas L. Wallace

*Department of Pharmacology*
*Medical College of Pennsylvania*
*Philadelphia, Pennsylvania 19129*

## INTRODUCTION

Pharmacologists characteristically respond to the discovery of an endogenous modulator by searching for ways to mimic or inhibit its actions since selective agonists and antagonists of the modulator may greatly aid in uncovering its biological function. Moreover, studying the actions of endogenous modulators often provides a better understanding of how some presently available drugs act and may possibly lead to the development of novel therapeutic agents.

The studies described in this report review some of the evidence that certain pharmacological agents can selectively inhibit the actions of calmodulin, the endogenous calcium binding protein that has been shown to regulate several biochemical and physiological processes. We will show that this inhibition is caused by a selective calcium-dependent binding of the drugs to calmodulin and that, of a large number of compounds studied, those having antipsychotic activity displayed the highest affinity for calmodulin. These studies suggest that many of the biochemical actions of these antipsychotics may be explained by a common mechanism; namely, by selectively binding to and inhibiting the actions of calmodulin. Finally, we will present preliminary evidence that certain endogenous proteins, such as ACTH and beta-endorphin, which have been shown to alter behavior in animals, also inhibit certain actions of calmodulin. Together, these results raise the possibility that calmodulin may be involved in certain behavioral processes and that drugs known to alter behavior may produce their effects by interfering with the actions of calmodulin.

## PHYSICAL AND CHEMICAL PROPERTIES OF CALMODULIN

Calmodulin is an ubiquitous, heat-stable, calcium binding protein having a molecular weight of approximately 17,000 daltons.[1–4] One molecule of calmodulin can bind 4 calcium ions.[5] The binding of calcium alters the tertiary structure of calmodulin, resulting in an increase in its alpha helical content and making it more resistant to boiling[2] and to treatment with trypsin,[6,7] urea,[7] and sodium dodecyl sulfate.[2]

Chemically, calmodulin contains no phosphate, cysteine, or tryptophan, but does contain a large proportion of acidic amino acids, which give the protein an isoelectric point of approximately 4.2.[8,9] Most studies have shown that calmodulins isolated from different species have similar amino acid compositions. However, slight structural differences have been noted,[10,11] and it may be possible

*Supported by the National Institute of Mental Health (Grant MH 30096).

†Present address: Mario Negri Institute of Research, Via Eritrea 62, 20157 Milan, Italy.

to take advantage of these differences with selective pharmacological agents. Modification of lysine or methionine residues results in the loss of biological activity,[12-14] suggesting that new antagonists might be developed that act by interfering with these amino acids.

## Localization of Calmodulin

Calmodulin has been found in most animal[15,16] and plant[11,117] tissues studied. In animal tissues, the cytoplasmic fraction generally contains high amounts and the nuclear, mitochondrial, and microsomal fractions contain lower amounts of calmodulin.[15,17,18,118]

Studies on the distribution of calmodulin in mammalian brain showed highest levels in the corpus striatum and frontal cerebral cortex with intermediate levels in the cerebellum, corpus callosum, hippocampus, and hypothalamus, and lowest levels in the medulla and pons.[17] Interestingly, those areas with the highest content of calmodulin, i.e., corpus striatum, limbic system, and cerebral cortex, are those believed to be the major sites of action of antipsychotic drugs.

Insofar as its subcellular distribution is concerned, calmodulin has been found in synaptic[19] and vesicular membranes,[20] postsynaptic densities,[21] and microtubules.[22] The results of these studies must be interpreted cautiously, however, since the available assays for calmodulin may not accurately reflect the amounts of this protein. For example, the apparent quantity of calmodulin, as determined from its ability to activate phosphodiesterase, will be influenced by other endogenous proteins that have been shown to inhibit calmodulin activity.[23-28] Likewise, the amount of calmodulin measured by radioimmunoassay may be misinterpreted due to cross-reactivity of the antibody with endogenous proteins of similar structure. Various conditions can also alter the distribution of this protein between membrane and cytosolic fractions. The addition of calcium to tissue increases the amount of calmodulin associated with membranes, and removal of calcium with EGTA raises the amount of calmodulin in the cytosol.[29-31] Further, certain putative neurotransmitters, peptides, and drugs have been reported to change the distribution of calmodulin in the cell.[32-36]

## Biological Actions of Calmodulin That Are Inhibited by Phenothiazine Antipsychotics

### *Effects on Phosphodiesterase Activity*

In 1970 Cheung[128] and Kakiuchi *et al.*[129] independently reported the presence in brain of an endogenous, heat-stable, calcium-dependent protein that could activate phosphodiesterase. This protein, which has variously been called calcium-dependent regulator, CDR, and activator, and is now termed calmodulin, was subsequently shown to selectively activate one of the molecular forms of phosphodiesterase present in brain.[37,38] In the studies described below the multiple molecular forms of phosphodiesterase were prepared from brain by subjecting the soluble supernatant fraction of brain homogenates to polyacrylamide gel electrophoresis.[37] Calmodulin was prepared and isolated to homogeneity according to the procedure of Watterson *et al.*,[30] and phosphodiesterase activity was measured by the luciferin-luciferase method.[39]

Figure 1 shows that concentrations of calmodulin that increase the activity of

Peak II phosphodiesterase have little or no effect on the other molecular forms of phosphodiesterase isolated from cerebrum. The slight activation of Peak III phosphodiesterase is probably caused by contamination with Peak II, since Peak III phosphodiesterase isolated from tissues that lack Peak II phosphodiesterase is not activated by calmodulin.

The finding that certain phenothiazine antipsychotics inhibited the same molecular form of phosphodiesterase that was stimulated by calmodulin prompted us to study further the interaction between the drugs and this enzyme system.[38,40,119]

FIGURE 2a shows that the calmodulin-induced activation of Peak II phosphodiesterase was selectively inhibited by trifluoperazine; concentrations of the drug that had little effect on the unactivated form of the enzyme completely blocked the activation of phosphodiesterase elicited by calmodulin. It should be noted here that high concentrations of the phenothiazines will inhibit calmodulin-independent enzymes as well.[38,41,42] Thus, one must be wary of using these drugs,

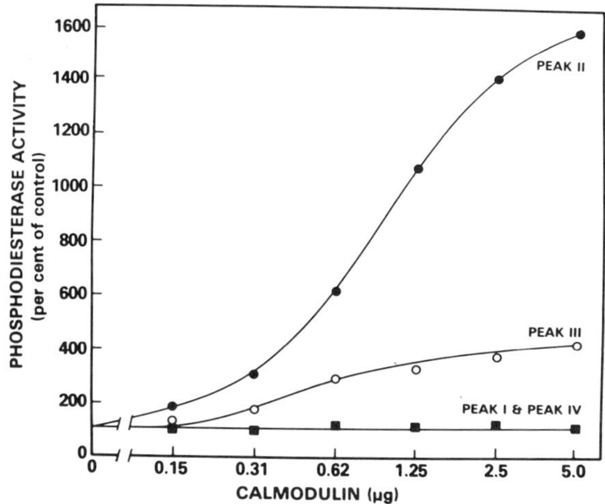


FIGURE 1. Effect of calmodulin on purified peaks of cyclic AMP phosphodiesterase of bovine cerebrum. The soluble supernatant fraction of bovine cerebral homogenates were subjected to polyacrylamide gel electrophoresis, and the cyclic AMP phosphodiesterase activity of peaks I, II, III and IV was determined in the presence and absence of varying quantities of calmodulin. (From Reference 40)

particularly in high concentrations, as the sole criterion for demonstrating a calmodulin-dependent process. Further, it is important to note that the ability of these drugs to inhibit the activation of phosphodiesterase is dependent upon the concentration of calmodulin, since high concentrations of calmodulin can overcome the inhibition of phosphodiesterase induced by phenothiazines.[38,43,44]

Additional studies on the inhibition of calmodulin-activated phosphodiesterase showed that other drugs besides the phenothiazines could block calmodulin's action. FIGURE 2b shows the inhibition of the calmodulin-activated phosphodiesterase by the diphenylbutylpiperidine antipsychotic penfluridol. Like trifluoperazine, penfluridol prevented the activation of phosphodiesterase at concentrations that had little or no effect on the basal enzyme activity.

Studies of a variety of agents that influence central nervous system function

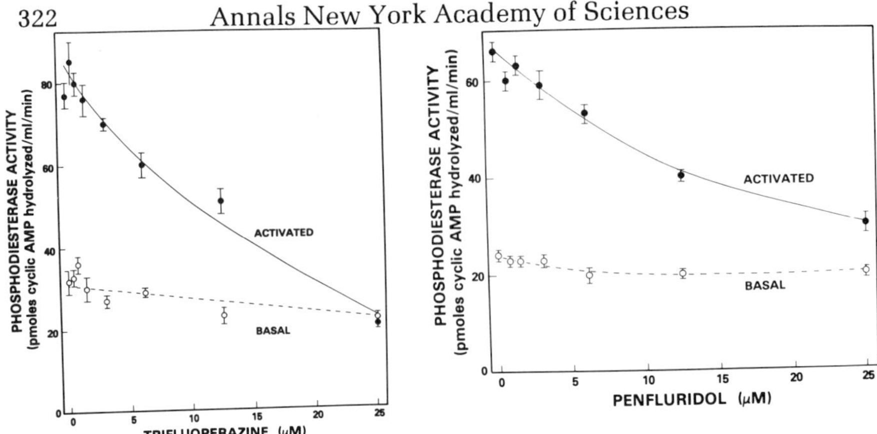


FIGURE 2. Inhibition of the calmodulin-induced activation of phosphodiesterase by trifluoperazine and penfluridol. An activatable Peak II phosphodiesterase was prepared from rat brain and assayed for phosphodiesterase activity in the presence and absence of calmodulin (1.5 units) and varying concentrations of trifluoperazine (Figure 2a) and penfluridol (Figure 2b). Each point represents the mean of 5 experiments. Vertical brackets indicate the standard error.

showed that the calmodulin-sensitive phosphodiesterase was inhibited by antipsychotic drugs of other chemical classes, including thioxanthenes and butyrophenones but was inhibited weakly or not at all by other centrally active compounds devoid of antipsychotic activity.[43] Thus, antidepressant drugs, such as amitriptyline and desipramine, and anxiolytics, such as medazepam and chlordiazepoxide, were much less potent in blocking calmodulin's action. Other phosphodiesterase inhibitors, such as theophylline and papaverine, and several other centrally acting drugs, such as amphetamine, d-lysergic acid diethylamide (d-LSD), pentobarbital, and morphine, showed little specific inhibition of calmodulin-sensitive phosphodiesterase.

Although these early studies showed that clinically effective antipsychotic drugs were the most potent inhibitors of calmodulin-sensitive phosphodiesterase,[38,43,45] subsequent studies using stereoisomers of these compounds raised important questions of the clinical relevance of this interaction.[48] For example, we have shown that the thioxanthene stereoisomers, which have been reported to selectively block the activation of catecholamine-sensitive adenylate cyclase of brain,[45-47] failed to exhibit the same stereospecificity in their ability to interact with calmodulin.[48] There appears to be little or no difference in the potency between the cis and trans isomers of flupenthixol and thiothixene in blocking the calmodulin-induced activation of phosphodiesterase (TABLE 1). A similar lack of stereospecificity was found by Norman *et al.*[49]

Other recent studies show that vinblastine can also inhibit the activation of cyclic AMP phosphodiesterase induced by a calcium-dependent protein activator.[50] In addition, several vascular relaxing agents have been shown to inhibit other calmodulin-activated enzymes, including ($Ca^{2+}+Mg^{2+}$)-ATPase[51,52] and myosin light chain kinase.[52,53] These agents include prenylamine, $N^2$-dansyl-L-arginine-4-*t*-butyl-piperidine amide, and N-(6-amino-hydroxyl)-5-chloro-1-naphthalene-sulfonamide. Thus, it appears that additional classes of pharmacological agents may also inhibit the activity of calmodulin. However, of the compounds studied so far, the clinically effective antipsychotics are still among the most potent drugs that block the calmodulin-induced activation of phosphodiesterase.

### *Effects on Adenylate Cyclase and Guanylate Cyclase Activity*

The observation that calmodulin is present in tissues that lack a calmodulin-dependent phosphodiesterase suggested that this calcium binding protein may have functions other than that of stimulating phosphodiesterase.[130,131] This proved

TABLE 1

PHARMACOLOGICAL INHIBITION OF CALMODULIN-INDUCED ACTIVATION OF PHOSPHODIESTERASE

| Class of Drug | Drug | $IC_{50}$($\mu$M) Activated | Unactivated |
|---|---|---|---|
| Phenothiazine | Trifluoperazine | 10 | >300 |
| | Thioridazine | 18 | >1,000 |
| | Chlorpromazine | 42 | >500 |
| | Promethazine | 340 | >1,000 |
| | Chlorpromazine sulfoxide | 2,500 | 7,500 |
| Diphenylbutylpiperidine | Pimozide | 7 | >100 |
| | Penfluridol | 10 | >50 |
| Thioxanthene | Chlorprothixene | 16 | >200 |
| | Trans flupenthixol | 15 | 50 |
| | Cis flupenthixol | 25 | 100 |
| | Cis thiothixene | 30 | >100 |
| | Trans thiothixene | 40 | >100 |
| Butyrophenone | Benperidol | 58 | >300 |
| | Haloperidol | 60 | >1,000 |
| Dibenzodiazepine | Clozapine | 80 | >200 |
| Benzodiazepine | Medazepam | 150 | >1,000 |
| | Chlordiazepoxide | 320 | 2,500 |
| Dibenzazepine | Desipramine | 125 | >1,000 |
| | Amitriptyline | 130 | >1,000 |
| | Protriptyline | 200 | >1,000 |
| Phosphodiesterase | | | |
| inhibitors | Papaverine | 130 | 210 |
| | Theophylline | 1,500 | 2,500 |
| | Other centrally | | |
| active agents | Amphetamine | >10mM | >10mM |
| | Mescaline | >5mM | >5mM |
| | Pentylenetetrazol | >10mM | >10mM |
| | LSD | >2.5mM | >2.5mM |
| | Pipradrol | >1mM | >1mM |
| | Azacyclonol | >1mM | >1mM |
| | Pentobarbital | >10mM | >10mM |
| | Morphine | >4mM | >4mM |

An activatable preparation of phosphodiesterase was purified from bovine brain and phosphodiesterase activity was measured in the presence and absence of 10 units of calmodulin and various concentrations of the compounds under study. The $IC_{50}$ value is the concentration of drug required to inhibit either the basal phosphodiesterase activity ($IC_{50}$, unactivated) or the calmodulin-induced activation of phosphodiesterase ($IC_{50}$, activated). Taken in part from Reference 43.

to be the case as calmodulin has now been shown to activate a number of enzymes including another brain enzyme involved in cyclic nucleotide metabolism; namely, adenylate cyclase. Originally this was shown in detergent-solubilized preparations[54,55] and, subsequently, in particulate fractions of brain.[56-58] Of great interest was the observation[56,57] that chlorpromazine prevented the calmodulin-induced activation of adenylate cyclase in a manner analogous to that reported for the inhibition by phenothiazines of the calmodulin-induced activation of phosphodiesterase.[38]

Several studies have suggested that activation of adenylate cyclase by guanine nucleotides is dependent, at least in part, on calmodulin, and that chlorpromazine may influence this interaction.[57,59,60] Thus, guanine nucleotides increased the activity of the calmodulin-sensitive adenylate cyclase and reduced the ability of chlorpromazine to inhibit this enzyme.[57] Since guanine nucleotides are also required for the catecholamine-induced activation of adenylate cyclase[61] these results raise the possibility that calmodulin may be involved in the catecholamine-induced stimulation of adenylate cyclase and that blockade of this activation may be explained by an interaction of phenothiazines with calmodulin. This idea is supported by studies showing that activation of adenylate cyclase by dopamine caused a translocation of calmodulin from plasma membrane to cytoplasm, resulting in a diminished responsiveness of adenylate cyclase to dopamine.[62,63] Studies demonstrating that chronic administration of antipsychotic drugs increased the calmodulin content[18,64] and the density of adrenergic receptors[65] in brain also suggest that there may be a functionally important interaction between catecholamine receptors, calmodulin, and antipsychotic drugs.

More recently it was reported that calmodulin can activate guanylate cyclase from tetrahymena.[66] Apparently, at least in some tissues, calmodulin can regulate the biosynthesis and the biodegradation of both cyclic AMP and cyclic GMP. Therefore, the consequence of inhibiting calmodulin on cyclic nucleotide levels would be difficult to predict without knowing the relative proportion of the various activatable enzymes in each tissue.

### *Effects on ATPase Activity*

Antipsychotic drugs can also inhibit certain ATPases.[67] The demonstration that the activity of the $(Ca^{2+}+Mg^{2+})$-ATPase from erythrocyte membranes[68-70] and brain synaptic membranes[19] could be increased by calmodulin led to our observation that antipsychotics inhibit this specific form of ATPase by blocking the actions of calmodulin; trifluoperazine inhibited the calmodulin-induced activation of $(Ca^{2+}+Mg^{2+})$-ATPase at concentrations that failed to inhibit the nonactivated $(Ca^{2+}+Mg^{2+})$-ATPase or the $Mg^{2+}$-ATPase.[42] This selective inhibition of calmodulin-stimulated ATPase has now been confirmed by several investigators.[52,53,71]

### *Effects on Kinases*

Calmodulin can activate a large number of protein kinases, including myosin light chain kinase,[72,73] phosphorylase kinase,[74,75] glycogen synthase kinase,[76,77] NAD kinase,[117] and a calcium-dependent protein kinase located in synaptosomal membranes and synaptic vesicles.[78-81] The protein kinase in synaptic vesicles appears to be involved in the calcium-stimulated release of norepinephrine,[20] an action that is inhibited by phenothiazine antipsychotics.[82]

### *Effects on Other Processes*

In addition to the actions previously described, calmodulin apparently plays a role in processes such as prostaglandin synthesis,[83] smooth muscle contraction,[52] intestinal secretion,[84] insulin secretion,[85,114] dissassembly of microtubules,[86-88] and may possibly be involved in ciliary motility[89] and fast axonal transport.[90] In virtually every case, the phenothiazine antipsychotics have been shown to inhibit these calmodulin-dependent enzymes and processes. A list summarizing those actions of calmodulin that have already been shown to be blocked by phenothiazines is presented in TABLE 2. The specific molecular mechanism by which antipsychotics inhibit these effects of calmodulin is discussed below.

TABLE 2

CALMODULIN-DEPENDENT PROCESSES THAT ARE INHIBITED BY PHENOTHIAZINE ANTIPSYCHOTICS

| Process | Reference |
|---|---|
| Activation of phosphodiesterase | 43 |
| Activation of adenylate cyclase | 56, 57 |
| Activation of $(Ca^{2+}+Mg^{2+})$-ATPase | 42, 52, 53 |
| Activation of myosin light chain kinase | 53, 113 |
| Activation of phosphorylase kinase | 74, 112 |
| Activation of glycogen synthase kinase | 77 |
| Activation of phospholipase A2 | 83 |
| Activation of NAD kinase | 117 |
| Activation of tryptophan hydroxylase | 116 |
| Release of norepinephrine from nerve terminals | 82 |
| Secretion of chloride in intestine | 84 |
| Contraction of smooth muscle | 53 |
| Glucose-stimulated insulin release | 85, 114 |
| Release of trichocyst in paramecium | 115 |

## INHIBITION OF CALMODULIN-INDUCED ACTIVATION OF PHOSPHODIESTERASE BY POLYPEPTIDE HORMONES

Stimulated by the recent studies showing that certain neuropeptides can alter behavior[91,111] and by studies showing that there are proteins in brain capable of inhibiting the calmodulin-induced activation of phosphodiesterase,[23,24,26-28] we have initiated a study to determine whether certain neuropeptides can modify the action of calmodulin. Of the compounds examined thus far, only histone, ACTH, and beta-endorphin, when used at concentrations of 25 $\mu$M, produced more than a 50% inhibition of the calmodulin-induced activation of phosphodiesterase (TABLE 3). These three peptides are highly basic and might be expected to interact with the acidic calmodulin. However, other physical and chemical factors must also be important since neurotensin, substance P, and alpha MSH are equally basic but nevertheless failed to block calmodulin's action at these concentrations.

TABLE 3

EFFECT OF VARIOUS PEPTIDES ON CALMODULIN-INDUCED ACTIVATION OF PHOSPHODIESTERASE ACTIVITY

| Peptide | Percent Inhibition of Phosphodiesterase Activation |
|---|---|
| ACTH | 100 |
| Beta-endorphin | 100 |
| Histone (lysine rich) | 100 |
| Neurotensin | 23 |
| Met-enkephalin | 15 |
| Beta-MSH | 11 |
| Delta sleep inducing factor | 11 |
| Leu-enkephalin | 10 |
| Des-tyr-γ-endorphin | 10 |
| Bombesin | 9 |
| Substance P | 9 |
| Alpha-MSH | 8 |
| Arginine vasopressin | 1 |
| ACTH 4-10 | 0 |

An activatable preparation of phosphodiesterase was purified from rat brain. The influence of various peptides on phosphodiesterase activity was measured in the presence and absence of 1.5 units of calmodulin. The values presented are the percent inhibition of calmodulin-stimulated phosphodiesterase activity produced by the peptide at a final concentration of 25 μM. ACTH was added to a final concentration of 9.6 I.U./ml. None of the peptides examined significantly altered the basal activity of phosphodiesterase.

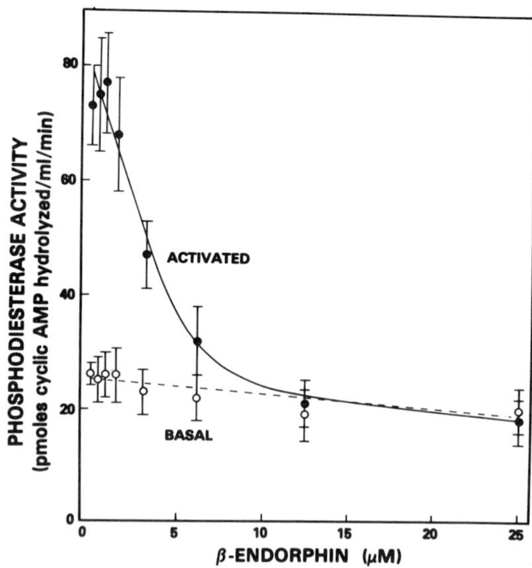


FIGURE 3. Inhibition of the calmodulin-induced activation of phosphodiesterase by beta endorphin. An activatable Peak II phosphodiesterase was prepared from rat brain and assayed for phosphodiesterase activity in the presence and absence of calmodulin (1.5 units) and varying concentrations of beta-endorphin. Each point represents the mean of 5 experiments. Vertical brackets indicate the standard error.

FIGURE 3 shows that beta-endorphin completely inhibits the calmodulin-induced activation of phosphodiesterase at concentrations that have little or no effect on the basal phosphodiesterase activity. The $IC_{50}$ value for blocking calmodulin's action was about 3 $\mu$M, which is lower than that obtained with the most potent antipsychotic drugs we have studied. This effect of beta-endorphin on calmodulin was of particular interest since this compound has been implicated in the pathogenesis of schizophrenia.[132,133]

Although these are preliminary studies, they do raise the possibility that there may be neuropeptides in brain that block the action of calmodulin in the same manner as the antipsychotics do. This could suggest that such peptides may function as endogenous antipsychotics.

## Interaction of Antipsychotic Drugs with Calmodulin

There are at least three different mechanisms by which antipsychotics could inhibit the activation of phosphodiesterase by calmodulin: 1) They might act by chelating calcium since calmodulin requires calcium for its activation; 2) they might interact with phosphodiesterase and thereby prevent its activation; or 3) they may interact with calmodulin. Calcium chelation could not account for this inhibition since adding excess calcium failed to reverse the inhibitory effects of chlorpromazine on the calmodulin-induced activation of phosphodiesterase.[43] Further, direct measurement of the binding of trifluoperazine to purified forms of phosphodiesterase failed to reveal any binding between the drug and this enzyme. However, similar studies on the interaction of trifluoperazine with calmodulin showed that the phenothiazine could bind directly to this calcium binding protein (see below).

### *Characteristics of the Binding of Antipsychotic Drugs to Calmodulin*

The binding of [$^3$H]trifluoperazine to calmodulin was determined by equilibrium dialysis.[92] Initial experiments showed that the binding of [$^3$H]trifluoperazine to calmodulin reached equilibrium within 6 hours at 4°C. The binding was reversible, pH-dependent, and required the presence of a divalent cation. Including EGTA in the dialysis medium freed trifluoperazine from calmodulin, whereas increasing the concentration of calcium to overcome the chelating effects of EGTA restored the binding of trifluoperazine to this protein.[120] A variety of other divalent cations such as strontium, nickel, cobalt, zinc, and manganese (in decreasing order of potency) could replace calcium in promoting the binding of trifluoperazine to calmodulin. Barium and magnesium had no effect on the binding.[92]

Other studies indicated that ionic forces are involved in the interaction of trifluoperazine with calmodulin. For example, an examination of the influence of pH on the binding showed that binding increased markedly above pH 4.2, reached a maximum at pH 5 and remained constant until pH 7.4 (FIGURE 4). At pH 8 and above the binding decreased markedly. Since the isoelectric point for calmodulin is about 4.2[8,9] and the $pK_a$ for trifluoperazine is 8.1,[100] these results suggest an electrostatic interaction between the negatively charged calmodulin and the positively charged trifluoperazine.

To examine the role of hydrophobic interactions in the binding of trifluoperazine to calmodulin, we studied the influence of ethanol. No significant inhibition

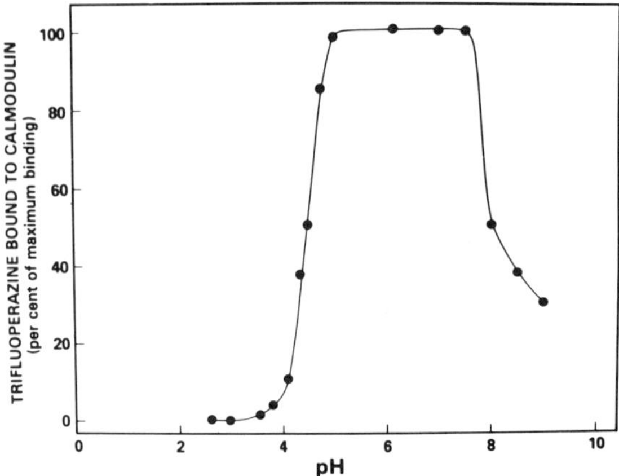


FIGURE 4. Effect of pH on the binding of trifluoperazine to calmodulin. Calmodulin was dialyzed to equilibrium against [$^3$H]-trifluoperazine (1 $\mu$M) in the presence of 100 $\mu$M calcium using either citric acid-phosphate buffer (pH 3 to 7) or Tris-HCl buffer (pH 5 to 9). The points represent the mean of duplicate samples. (Taken in part from Reference 120)

of binding was seen until extremely high concentrations of ethanol were reached, i.e., greater than 1 M. Further, a comparison of the octanol-water partition coefficients of several drugs that bind to calmodulin showed that lipid solubility does not necessarily correlate with the affinity of the drug for the calcium binding protein. For example, diazepam and pentobarbital have a higher

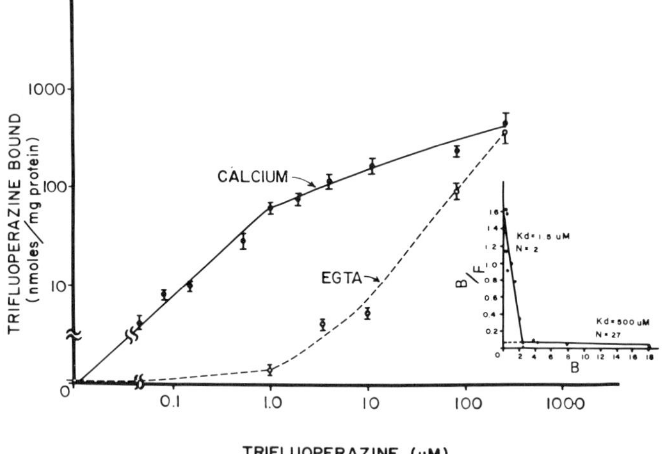


FIGURE 5. Binding of trifluoperazine to calmodulin as a function of the concentration of trifluoperazine. Samples of calmodulin (25 $\mu$gm/ml) were dialyzed to equilibrium against varying concentrations of [$^3$H]-trifluoperazine (0.002 to 500 $\mu$M) in the presence of 100 $\mu$M $Ca^{2+}$ or 300 $\mu$M EGTA. Each point represents the mean value of three samples. Vertical brackets indicate the standard error. Inset: Data plotted according to the Scatchard equation; B, moles of trifluoperazine bound per mole of calmodulin; F, concentration of free trifluoperazine at equilibrium. (From Reference 92)

lipid solubility than trifluoperazine hydrochloride or chlorpromazine hydrochloride[136] and yet the latter two compounds bind to calmodulin to a much greater extent than either diazepam or pentobarbital. Thus, although hydrophobic interactions may be involved to some extent, these results suggest that the binding of phenothiazines to calmodulin may depend more on the ionic attraction between calmodulin and the drugs.

A study of the kinetic properties of the binding of trifluoperazine to calmodulin showed that at low concentrations of drug the binding is calcium-dependent, whereas at high concentrations, the binding does not require calcium (FIGURE 5).[92] Analysis of these data by Scatchard plots revealed two types of binding sites: (1) high affinity, saturable, calcium-dependent binding sites, and (2) low affinity, high capacity, calcium-independent binding sites (FIGURE 5, inset). In the presence of calcium, the calculated dissociation constant ($K_d$) for trifluoperazine was 1 $\mu$M, and at saturation there were approximately two molecules of trifluoperazine specifically bound per molecule of calmodulin. In the absence of calcium, trifluoperazine bound only to the low affinity, high capacity sites.

TABLE 4

CALCIUM-SPECIFIC BINDING OF [$^3$H]-TRIFLUOPERAZINE TO VARIOUS PROTEINS

| Protein | Trifluoperazine Bound (nmoles/mg protein) |
|---|---|
| Bovine brain calmodulin | 19 |
| Chick embryo fibroblast calmodulin | 20 |
| Human brain calmodulin | 15 |
| Rabbit brain calmodulin | 12 |
| Rat brain calmodulin | 16 |
| Aldolase | NS |
| Bovine serum albumin | NS |
| Catalase | NS |
| Chymotrypsinogen | NS |
| Cytochrome C | NS |
| Myosin light chain | NS |
| Phospholipase A | NS |
| Porcine brain S-100 | NS |
| Troponin C | 1.3 |

Various proteins were dialyzed to equilibrium against 0.25 $\mu$M [$^3$H]trifluoperazine in the presence of 100 $\mu$M calcium or 300 $\mu$M EGTA. Calcium-specific binding is the difference between trifluoperazine bound in the presence and absence of calcium. N.S. = no significant binding. (Taken from References 92, 93)

### *Specificity of the Binding of Trifluoperazine to Calmodulin Compared to its Binding to Other Proteins*

To examine the specificity by which trifluoperazine binds to calmodulin, we studied the binding of this phenothiazine to a variety of proteins, including calmodulin prepared from different species and tissues, several other calcium binding proteins, and a number of purified proteins of widely varying molecular weights.[92,93] All calmodulin samples, regardless of the source, displayed calcium-dependent binding of trifluoperazine (TABLE 4). The magnitude of this binding differed somewhat among species but was proportional to the ability of the calmodulin preparation to stimulate phosphodiesterase. Of the other calcium binding proteins, including some that showed substantial structural homology with calmodulin, only troponin C displayed calcium-dependent binding of trifluoperazine. However, at low concentrations of trifluoperazine (0.2 $\mu$M) the calcium-dependent binding to troponin C was less than 10% that seen with

TABLE 5

CALCIUM-SPECIFIC BINDING OF VARIOUS DRUGS TO CALMODULIN

| Compounds | Calcium-specific Binding (nmoles bound/mg protein) |
|---|---|
| *Antipsychotics* | |
| Penfluridol | 105 ± 11 |
| Pimozide | 65 ± 5 |
| Trifluoperazine | 56 ± 6 |
| Chlorpromazine | 23 ± 1 |
| Haloperidol | 13 ± 1 |
| Spiroperidol | 3.8 ± 0.2 |
| *Tricyclic antidepressants* | |
| Amitriptyline | 9.0 ± 0.7 |
| Imipramine | 6.8 ± 0.6 |
| *Anxiolytics* | |
| Chlordiazepoxide | 5.5 ± 0.2 |
| Diazepam | 4.7 ± 0.1 |
| *Miscellaneous* | |
| Dihydroergocryptine | 6.6 ± 0.4 |
| Apomorphine | 2.7 ± 0.3 |
| Dihydroalprenolol | 1.5 ± 0.2 |
| ADTN | NS |
| Dopamine | NS |
| Histamine | NS |
| d-Amphetamine | NS |
| d-LSD | NS |
| Morphine | NS |
| Pentobarbital | NS |

Calmodulin was dialyzed to equilibrium against radiolabeled drug (1 $\mu$M) in the presence of 300 $\mu$M EGTA or 100 $\mu$M $Ca^{2+}$. Calcium specific binding is the difference between drug bound in the presence and absence of calcium. Each value is the mean ± standard error of 3 to 6 experiments.

N.S. = no significant binding.

d-LSD = d-lysergic acid diethylamide

ADTN = amino-6,7-dihydroxy-1,2,3,4 tetrahydro napthalene

(Taken in part from Reference 48)

calmodulin. From these studies it was concluded that the high affinity, calcium-dependent binding of trifluoperazine is a relatively specific property of calmodulin.[92,93]

### *Specificity of the Binding of Various Drugs to Calmodulin*

To determine whether the binding of drugs to calmodulin is associated with a particular pharmacological activity, we have examined the binding of a number of compounds to this protein. Of the compounds studied, antipsychotic agents displayed the highest degree of calcium-dependent binding to calmodulin (TABLE 5).[48] Among the antipsychotics the diphenylbutylpiperidines, penfluridol, and pimozide exhibited the most binding; the phenothiazines, trifluoperazine, and chlorpromazine showed a high degree of calcium-dependent binding; and the butyrophenones, haloperidol, and spiroperidol exhibited less binding. The tricyclic antidepressants and antianxiety agents displayed a low degree of calcium-dependent binding whereas a number of other potent centrally acting compounds that are devoid of antipsychotic activity, including d-LSD, pentobarbital, morphine, and d-amphetamine, did not bind to calmodulin.

A number of neurotransmitters and agents that presumably bind to neurotransmitter receptors also showed little or no calcium-dependent binding to calmodulin. These included the alpha-adrenergic receptor antagonist dihydroergocryptine, the beta-adrenergic receptor antagonist dihydroalprenolol, the dopamine receptor agonist apomorphine, and the dopamine receptor agonist ADTN, as well as the neurotransmitters dopamine and histamine. The results with dopamine were particularly interesting since dopamine and antipsychotics are generally thought to act on the same receptor. Our results showed that, even in the presence of guanine nucleotides, dopamine failed to bind to calmodulin. These results show that some of the biochemical actions of antipsychotic drugs can be explained by an interaction with a protein to which dopamine does not bind.[94]

### *Competition of Penfluridol and Pimozide for the [$^3$H]Trifluoperazine Binding Site*

Since antipsychotic drugs of several different chemical classes displayed high affinity, calcium-dependent binding to calmodulin,[48] the question arose as to whether these compounds were binding to the same sites on the calmodulin molecule. To answer this question, we studied the effect of the diphenylbutylpiperidines, penfluridol and pimozide, on the calcium-dependent binding of trifluoperazine to calmodulin.[48]

Both drugs caused a concentration-dependent shift in the binding curve for [$^3$H]trifluoperazine to calmodulin. At low concentrations of trifluoperazine, penfluridol and pimozide readily displaced trifluoperazine from calmodulin. However, at high concentrations of trifluoperazine, neither drug significantly displaced the phenothiazine. An analysis of a Lineweaver-Burk plot of the binding of trifluoperazine in the presence of different concentrations of penfluridol indicated that penfluridol does not change the maximum binding of trifluoperazine but rather decreases the apparent affinity of trifluoperazine for the binding site on calmodulin, suggesting that penfluridol and trifluoperazine compete for the same high affinity binding sites on the calmodulin molecule. Analysis of the influence of pimozide on the binding of [$^3$H]trifluoperazine demonstrated that pimozide, like penfluridol, also competed for the trifluoperazine binding site on calmodulin.[48]

### *Displacement of Labeled Antipsychotic Compounds by Various Agents*

Relatively few compounds are available in a radiolabeled form, thus precluding direct studies of the binding of many agents to calmodulin. However, the binding of drugs to calmodulin may be determined indirectly by measuring their ability to displace radiolabeled trifluoperazine from this calcium binding protein. In these studies, preparations of calmodulin were dialyzed to equilibrium against [$^3$H]trifluoperazine and various concentrations of the drug being studied. The [$^3$H]trifluoperazine that remained bound to calmodulin was then determined. It should be pointed out that although studies of this sort may provide information on the relative binding affinities of various drugs for the trifluoperazine binding site, they do not rule out the possibility that the nonlabeled drugs can bind to a site on calmodulin other than that to which trifluoperazine binds.

TABLE 6

DISPLACEMENT OF [$^3$H]-TRIFLUOPERAZINE FROM CALMODULIN BY VARIOUS COMPOUNDS

| Compound | Ki (μM) |
|---|---|
| Pimozide | 0.8 |
| Trifluoperazine | 1.0 |
| Trans-flupenthixol | 1.7 |
| Penfluridol | 2.5 |
| Cis-flupenthixol | 2.7 |
| Chlorpromazine | 5.3 |
| Trans-thiothixene | 6.7 |
| Haloperidol | 10 |
| Cis-thiothixene | 11 |
| Clozapine | 13 |
| Trifluoperazine sulfoxide | 30 |
| Promethazine | 40 |
| (+) – Butaclamol | 50 |
| Nortriptyline | 53 |
| Chlordiazepoxide | 60 |
| Amitriptyline | 67 |
| Chlorpromazine sulfoxide | 170 |
| Phentolamine | 230 |
| (–) – Butaclamol | 230 |
| Reserpine | >130 |
| Perlapine | >130 |
| Metoclopromide | >130 |
| Molindone | >130 |
| Papaverine | >670 |
| Theophylline | >670 |
| Methacholine | >670 |

Calmodulin was dialyzed to equilibrium against [$^3$H]trifluoperazine (0.5 μM) and various concentrations of the drug under study, and the concentration of unlabeled drug which displaced 50% of the [$^3$H]trifluoperazine was calculated ($IC_{50}$). Ki values were calculated from the equation $Ki = IC_{50}/1 + (L)/Kd$, where L = the concentration of trifluoperazine and Kd = the dissociation constant for the binding of trifluoperazine to calmodulin. (data taken in part from Reference 48)

The results of these studies showed that the effectiveness of phenothiazine derivatives in displacing [$^3$H]trifluoperazine from calmodulin correlated with their clinical antipsychotic activity.[48] TABLE 6 shows the $K_i$ values for a number of phenothiazines and other agents for displacing labeled trifluoperazine from calmodulin. As can be seen, the clinically active phenothiazines, trifluoperazine and chlorpromazine, displaced [$^3$H]trifluoperazine from calmodulin whereas phenothiazines having little or no antipsychotic activity, such as promethazine, trifluoperazine sulfoxide, and chlorpromazine sulfoxide, were significantly less potent in displacing [$^3$H]trifluoperazine. Other antipsychotic agents besides the phenothiazines, such as clozapine, pimozide, and penfluridol, also displaced [$^3$H]trifluoperazine from calmodulin.

The antianxiety agent, chlordiazepoxide and the antidepressants, amitriptyline and nortriptyline, displayed moderate potency at displacing [$^3$H]trifluoperazine. Agents that were relatively ineffective in displacing [$^3$H]trifluoperazine from calmodulin included the alpha-adrenergic blocking agent, phentolamine, the phosphodiesterase inhibitors, theophylline and papaverine, and the cholinomimetic agent methacholine.

Of particular interest was the lack of effect of perlapine and metoclopramide on the binding of [$^3$H]trifluoperazine. These two compounds have been reported to exert pharmacological actions suggesting that they block dopamine receptors[95,96] and were, therefore, thought to be antipsychotic. However, clinical trials of these compounds failed to demonstrate antipsychotic activity.[97,98]

Although we have not as yet identified the particular chemical or physical property of drugs that enable them to interact with calmodulin, an examination of their physical constants showed that the ability of compounds to displace trifluoperazine from calmodulin is not related either to their surface activity[99] or to their ionization constants.[100]

A notable exception to the general observation that there was a relationship between displacement of trifluoperazine from calmodulin and clinical antipsychotic activity was seen with the cis and trans isomers of the thioxanthenes. Although the cis isomers of these compounds have been reported to block the norepinephrine-induced activation of adenylate cyclase,[45] the dopamine-induced activation of adenylate cyclase,[46,47] and the binding of various ligands to brain membranes,[101] and have also been reported to have greater antipsychotic activity than the trans isomers,[120] we did not find any marked differences in the binding of the two isomers to calmodulin.

Despite the fact that the clinical data comparing the antipsychotic potencies of the stereoisomers are not well documented,[103] neither can they be ignored, and explanations for our inability to demonstrate stereospecific binding of drugs to calmodulin must be sought. One possibility is that during purification of calmodulin, the protein may have undergone sufficient conformational changes to lose its stereospecificity. Alternatively, perhaps stereospecific binding of antipsychotics to calmodulin is confered by cellular elements *in vivo*. Another possibility is that there may be a stereospecific transport mechanism for these drugs in brain. A selective transport of the antipsychotic haloperidol has recently been proposed.[135] And finally, there is evidence that the active cis isomer of flupenthixol can convert to the trans isomer under certain conditions.[104] Such conversion would greatly complicate the interpretation of studies on the pharmacological effects of cis flupenthixol. We believe these questions may best be answered by studying the binding of drugs to calmodulin in an *in vivo* preparation or in tissue slices. Since these studies could be more readily performed if the drugs were bound irreversibly to calmodulin, we have attempted to develop a technique to bind phenothiazine antipsychotics irreversibly to this protein (see below).

Regardless of the outcome of these *in vivo* studies, our *in vitro* results demonstrate a clear relationship between two biochemical events: the binding of drugs to calmodulin and their inhibition of the calmodulin-dependent activation of certain enzymes. This relationship, which has been demonstrated using a wide variety of drugs and even extends to the stereoisomers of antipsychotics, can be demonstrated graphically by a regression analysis of the data (FIGURE 6): a positive correlation ($R=0.99$) was found between the magnitude of calcium-specific binding to calmodulin and the ability of these agents to inhibit phosphodiesterase activation.[48]

## IRREVERSIBLE BINDING OF PHENOTHIAZINE ANTIPSYCHOTICS TO CALMODULIN

### *Photoaffinity Labeling with Ultraviolet Irradiation*

Previous studies showed that the binding of antipsychotic drugs to calmodulin could be easily reversed by dialyzing against EGTA or by dialyzing against an excess of competing antipsychotic drug.[92,120] However, there is evidence that halogenated phenothiazine antipsychotics could undergo photochemical activation by ultraviolet (UV) light, generating free radicals that could interact irreversibly with proteins such as ($Na^{+}+K^{+}$)-ATPase[105] and serum albumin (Colin Chignell, personal communication). Accordingly, we have attempted to use UV irradiation to induce the irreversible binding of phenothiazine antipsychotics to calmodulin.[126,134,137]

FIGURE 7 shows that UV irradiation causes the irreversible binding of trifluoperazine to calmodulin. When solutions of calmodulin and [$^{3}$H]trifluoperazine were not irradiated, about 1 nmole of [$^{3}$H]trifluoperazine per mg of calmodulin remained bound after dialyzing against $Ca^{2+}$. This binding could be reduced by 90% or more by dialyzing against EGTA or against an excess of nonradiolabeled trifluoperazine. By contrast, if solutions of trifluoperazine and calmodulin were irradiated by UV light, about 4 times more trifluoperazine remained bound after dialyzing against $Ca^{2+}$, and more importantly, this binding could not be reduced by dialyzing against either EGTA or nonlabeled trifluoperazine. In more

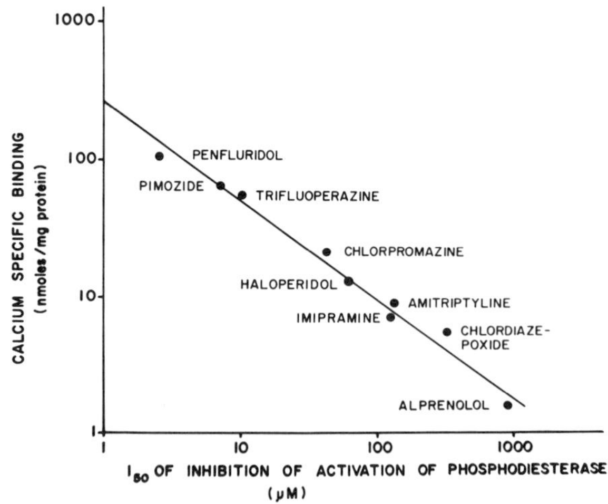


FIGURE 6. Correlation of the calcium-specific binding of neurotropic agents to calmodulin with their ability to inhibit the activation of phosphodiesterase. The line of best fit was determined by regression analysis after logarithmic transformation. The correlation coefficient was 0.99. (From Reference 48.)

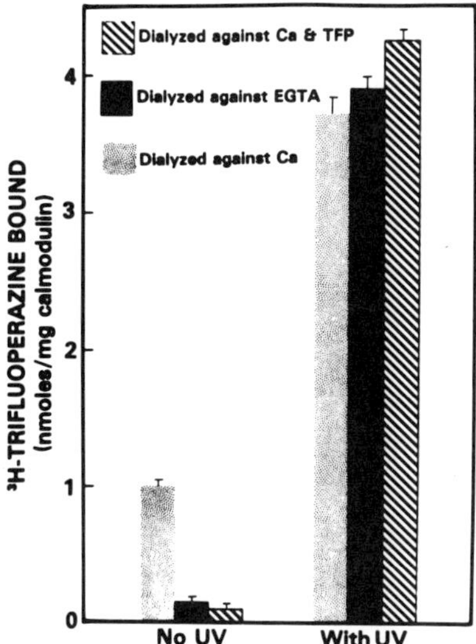


FIGURE 7. Irreversible binding of trifluoperazine to calmodulin following UV irradiation. Calmodulin (40 $\mu$gm/ml) and [$^3$H]-trifluoperazine (1$\mu$M) were incubated in 5 mM Tris-HCl buffer, pH 7.0, containing 1 mM $MgCl_2$ and 100 $\mu$M $CaCl_2$ at 4°C for 30 min in presence or absence of UV light (254 nm). Samples were then dialyzed for 16 hr against Tris-HCl buffer containing either: a) 100 $\mu$M $CaCl_2$; b) 300 $\mu$M EGTA; or c) 100 $\mu$M $CaCl_2$ plus 100 $\mu$M nonlabeled trifluoperazine. The radioactivity that remained bound to calmodulin was then determined. Each value is the mean ± standard error of 4 replicates. (From Reference 134)

recent experiments, we have examined the ability of denaturing agents, such as urea and hexadecyltrimethylammonium bromide, to disrupt the bond between trifluoperazine and calmodulin. These agents, which alter the ionic characteristics and tertiary structure of proteins, failed to change the amount of trifluoperazine bound to calmodulin, suggesting that UV irradiation may induce the formation of a covalent bond between the phenothiazines and calmodulin.[137]

The properties of the irreversible binding of photoactivated trifluoperazine to calmodulin were similar to those that had previously been observed in studies of the reversible binding of the antipsychotic drugs to this protein. FIGURE 8 shows that the binding of trifluoperazine to calmodulin increased with increasing concentrations of trifluoperazine and that like the reversible binding, the irreversible binding of trifluoperazine to calmodulin was greatly enhanced by calcium. This calcium-dependent binding (defined as the amount of trifluoperazine bound in the presence of calcium minus the amount bound in the presence of EGTA) was saturable; the maximum binding was approximately 60 nmoles of trifluoperazine bound per mg of calmodulin, or approximately 1 binding site per molecule of calmodulin, and one-half maximum binding ($EC_{50}$) occurring at a trifluoperazine concentration of about 5 $\mu$M.

A study of the specificity by which phenothiazine antipsychotics bind to calmodulin showed that there was little or no calcium-specific binding of these drugs to hemoglobin, trypsin, bovine serum albumin or histone.[137]

Earlier experiments on the reversible binding showed that antipsychotics of

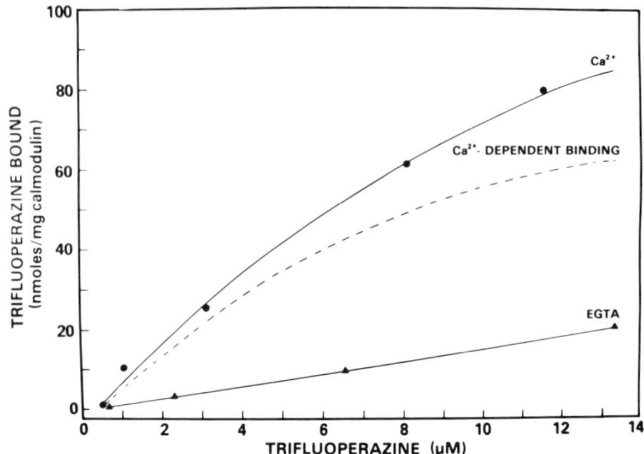


FIGURE 8. Irreversible calcium-dependent binding of trifluoperazine to calmodulin as a function of trifluoperazine concentration. Calmodulin (30 $\mu$g) in 1 ml 5 mM Tris buffer, pH 7.0, containing 1 mM $MgCl_2$ and either 100 $\mu$M $CaCl_2$ or 300 $\mu$M EGTA was irradiated for 30 min in the presence of varying concentrations of [$^3$H]-trifluoperazine. Following irradiation, samples were dialyzed for 16 hr against Tris-HCl buffer containing either 100 $\mu$M $Ca^{2+}$ or 300 $\mu$M EGTA and 1 mM non-labeled trifluoperazine. The dotted line represents the calculated calcium-dependent binding (difference between binding in presence of $Ca^{2+}$ and binding in presence of EGTA). Each point is the mean of 4 replicates.

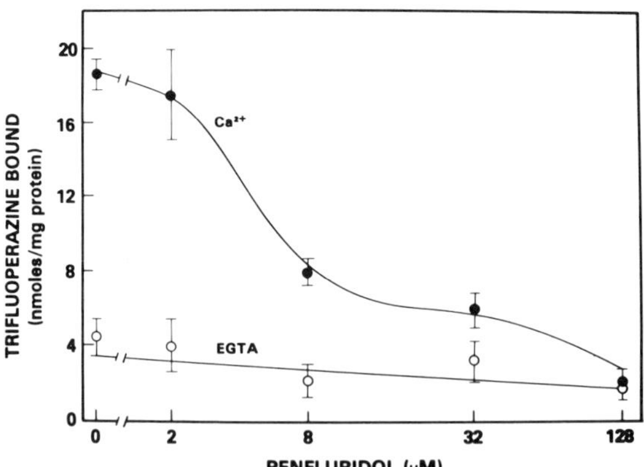


FIGURE 9. Inhibition by penfluridol of the irreversible binding of trifluoperazine to calmodulin. Calmodulin (30 $\mu$gm) and varying concentrations of penfluridol were incubated at 4°C for 15 min in 1 ml of 5 mM Tris-HCl buffer (pH 7.0) containing either 100 $\mu$M $Ca^{2+}$ or 300 $\mu$M EGTA. [$^3$H]-Trifluoperazine (final concentration 10 $\mu$M) was added, and the samples were irradiated with UV light. Samples were then dialyzed for 20 hr and the radioactivity that remained bound to calmodulin was determined. Each point represents the mean ± standard error of 4 replicates.

different chemical classes competed for the same high affinity binding site on calmodulin.[48,120] Similarly, antipsychotics of different chemical classes also inhibited the irreversible binding of trifluoperazine to calmodulin. Results in FIGURE 9 show the irreversible binding of [$^3$H]trifluoperazine to calmodulin in the presence of increasing concentrations of penfluridol. As can be seen, penfluridol had little effect on the nonspecific binding (i.e., binding in the presence of EGTA) but greatly reduced the calcium-dependent binding. Studies of the ability of a number of other psychoactive substances to inhibit the irreversible, calcium-dependent binding of trifluoperazine to calmodulin showed that the most potent compounds examined were the antipsychotic drugs penfluridol and chlorpromazine; chlorpromazine sulfoxide and diazepam were 5–10 times less potent in preventing the irreversible binding of trifluoperazine to calmodulin.

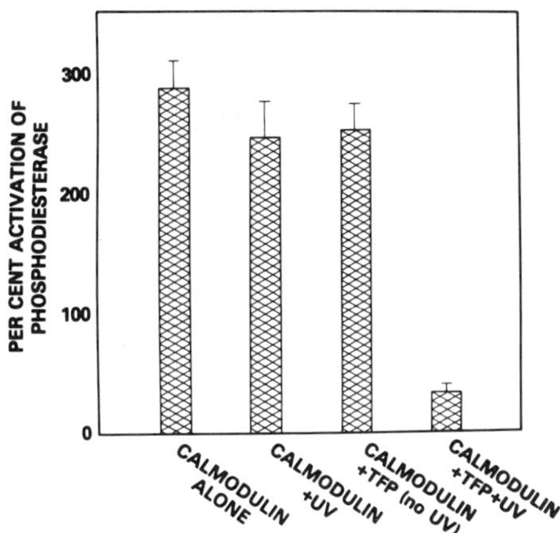


FIGURE 10. Irreversible inhibition of the calmodulin-induced activation of phosphodiesterase. Calmodulin (40 μgm/ml) was incubated in the presence and absence of 12.5 μM non-labeled trifluoperazine for 30 min at 4°C with and without UV irradiation. Samples (1 ml each) were then dialyzed for 20 hr against 2 liters of 5 mM Tris-HCl buffer, pH 7.0 containing 1 mM $MgCl_2$ and 100 μM $CaCl_2$. The dialysis buffer was changed 3 times. Following dialysis the ability of these samples to activate a purified calmodulin-deficient phosphodiesterase was determined using 400 μM cyclic AMP as substrate.

To determine whether calmodulin that had been irradiated in the presence of trifluoperazine had irreversibly lost its biological activity, samples of calmodulin and trifluoperazine were irradiated and examined for their ability to activate a calmodulin-dependent phosphodiesterase. Results presented in FIGURE 10 show that UV irradiation of calmodulin did not effect its ability to activate phosphodiesterase. By contrast, UV irradiated samples of calmodulin and trifluoperazine failed to activate phosphodiesterase even after extensive dialysis although similarly dialyzed preparations of calmodulin that had not been irradiated increased phosphodiesterase activity about 2.5-fold. Thus, it appears that the irreversible binding of the phenothiazine antipsychotics to calmodulin results in the irreversible inactivation of calmodulin.

If the irreversible binding of phenothiazines to calmodulin occurs *in vivo* as well as *in vitro*, it may explain the well known observation that metabolites of the

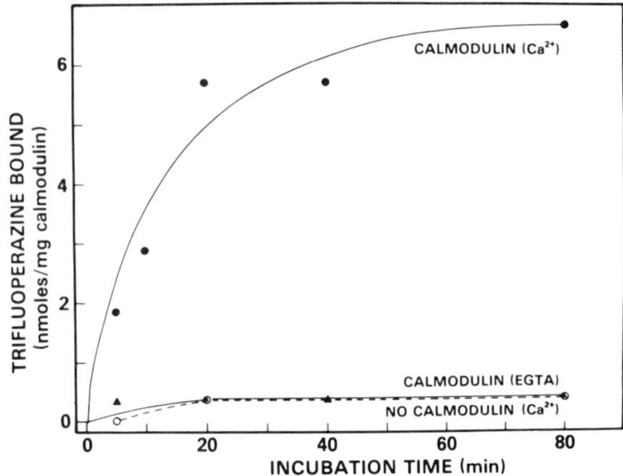


FIGURE 11. Irreversible binding of trifluoperazine to calmodulin induced by a peroxidase-hydrogen peroxide system. Calmodulin (70 $\mu$gm) and horseradish peroxidase (2 $\mu$gm) were incubated at 37°C for varying periods of time in 2 ml of 5 mM Tris buffer (pH 7.0) containing 150 $\mu$M hydrogen peroxide and either 100 $\mu$M $CaCl_2$ or 300 $\mu$M EGTA. Samples were then dialyzed against 1 mM non-labeled trifluoperazine and counted for radioactivity. Points represent the mean value of 3 samples.

phenothiazine antipsychotics remain in the body for extremely long periods of time following the cessation of drug treatment.[106] Perhaps, *in vivo* the phenothiazines are metabolically activated to yield free radicals[108] which might bind irreversibly to calmodulin.

In a preliminary effort to examine this question, we have used a hydrogen peroxide-horseradish peroxidase system to enzymatically form free radicals of trifluoperazine.[107] In these experiments, samples of calmodulin and [$^3$H]trifluoperazine that had been incubated in the presence of hydrogen peroxide and peroxidase, were dialyzed to remove any trifluoperazine that was not bound irreversibly. As can be seen in FIGURE 11, when no calmodulin was present, little or no trifluoperazine remained after dialysis, indicating that trifluoperazine does not bind irreversibly to the horseradish peroxidase. Also, when calmodulin and the peroxidase system were incubated in the presence of EGTA, no irreversible binding of trifluoperazine occurred. However, when solutions of calmodulin and the peroxidase system were incubated in the presence of calcium, trifluoperazine remained bound to calmodulin even after dialysis, indicating an irreversible binding of the drug to the calcium binding protein. Interestingly, the amount of trifluoperazine bound to calmodulin using this enzyme system was similar to that bound following UV irradiation (FIGURE 7).

Since the peroxidase system converts the phenothiazines to sulfoxide analogues,[107,109] which are similar to the end products formed from the oxidative metabolism of phenothiazines *in vivo*,[110] one might predict that during their metabolism, phenothiazines might combine irreversibly with calmodulin and accumulate in tissue. We are currently addressing this problem by determining whether certain microsomal preparations might also be able to induce the irreversible binding of the phenothiazines to calmodulin.

## BINDING OF PHENOTHIAZINE ANTIPSYCHOTICS TO CALMODULIN IN BRAIN

Although we have demonstrated that phenothiazine antipsychotics could bind to purified calmodulin, it is important to know whether the drugs will still bind to the calcium-binding protein in the presence of other competing brain proteins. Accordingly, we incubated [$^3$H]chlorpromazine with the soluble fraction of rat cerebral cortex, irradiated the samples with UV light to cause an irreversible binding of the drug to calmodulin, dialyzed the sample to remove unbound [$^3$H]chlorpromazine, and then subjected the material to disc gel electrophoresis. Following electrophoresis, the gels were cut into 2 mm sections and counted for radioactivity. These results showed a peak of radioactivity (designated as Peak 1, FIGURE 12) which migrated with the same $R_f$ value as purified calmodulin. Another peak (Peak II) with a higher $R_f$ value than calmodulin was also detected and may represent an additional binding site for the drug. Coomassie blue staining did not reveal any stained proteins associated with the smaller peaks of radioactivity. A significant amount of radioactivity was found at the origin, suggesting that some materials that bound chlorpromazine did not enter the gel.

These findings indicate that a relatively large proportion of chlorpromazine binds to calmodulin even in the presence of other brain proteins. Although the pharmacological and clinical significance of this binding has yet to be determined, there is a large body of evidence suggesting that many of the biochemical

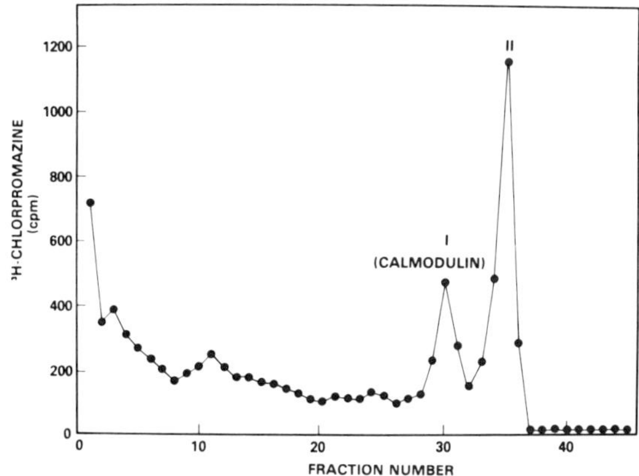


FIGURE 12. Binding of phenothiazine antipsychotics to calmodulin in brain. Cerebral cortices of rats were homogenized in 4 volumes of Tris-HCl buffer, pH 7.0, containing 3 mM EGTA and centrifuged at 100,000×g for 60 min. Homogenization in EGTA causes the dissociation of calmodulin from calcium-dependent membrane binding sites, thereby increasing its concentration in the soluble fraction.[29,31] After dialyzing to remove the EGTA, the 100,000×g soluble fraction was dialyzed against Tris buffer containing 100 $\mu$M $CaCl_2$. [$^3$H]-Chlorpromazine (2 $\mu$M; 26.5 $\mu$Ci) was then added, the samples were irradiated with ultraviolet light and then dialyzed against Tris buffer containing 1 mM unlabeled chlorpromazine to remove unbound [$^3$H]-chlorpromazine. The dialyzed sample was subjected to disc electrophoresis according to the method of Davis.[123] After electrophoresis, gels were either stained for protein in 0.25% Coomassie blue or were cut into 2 mm sections and counted for radioactivity.

effects of these drugs can be explained by their binding to and inhibition of calmodulin. A scheme depicting our proposed mechanism by which antipsychotic drugs inhibit calmodulin's actions is illustrated in FIGURE 13.

## SUMMARY

A number of psychotropic drugs, particularly the phenothiazines and related antipsychotic compounds, inhibit a variety of calmodulin-dependent enzymes.

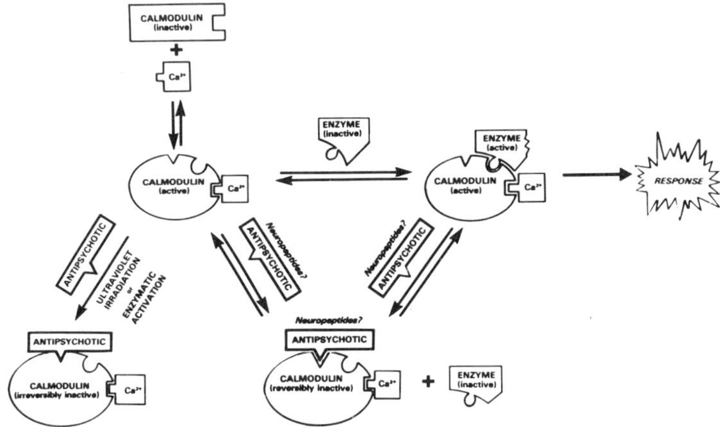


FIGURE 13. Mechanism by which antipsychotic agents inhibit the action of calmodulin. The scheme describes our hypothesis for the mechanism by which antipsychotic agents block the calmodulin-induced activation of enzymes. The first step involves the interaction of calcium with calmodulin. Calcium binds to calmodulin (up to four calcium ions to one molecule of calmodulin[127]). This results in a conformational change in calmodulin.[6,12] The calcium-calmodulin complex then interacts with a specific calmodulin-sensitive form of the enzyme, causing the formation of the active calmodulin-calcium-enzyme complex[122,124,125] and, subsequently, the biochemical response. The calcium-calmodulin complex may also bind to antipsychotic agents (or neuropeptides) resulting in a calmodulin-calcium-antipsychotic complex.[92] In the absence of calcium, there is no interaction of calmodulin with either the enzyme or the antipsychotics.[38,42,92] In purified preparations under in vitro conditions, the calcium-dependent binding between antipsychotic (or neuropeptide) drugs and calmodulin is reversible.[92] However, if irradiated with UV light[126] or activated by hydrogen peroxide-peroxidase, the binding of phenothiazine antipsychotics becomes irreversible. In either case, this complex cannot activate the calmodulin-sensitive form of enzymes.[38,43] The addition of antipsychotics to an enzyme that is already activated will reverse its activation. This inhibition cannot be overcome by adding more calcium or by adding more enzyme.[43] We suggest that a scheme similar to the one described here explains the mechanism by which antipsychotics inhibit many calmodulin-sensitive enzymes and processes and may even explain how antipsychotics exert some of their diverse pharmacological actions (From Ref. 21).

The mechanism by which these compounds inhibit the activity of calmodulin is through a selective calcium-dependent binding to this protein.

With the notable exception of certain stereoisomers, compounds that are clinically effective antipsychotic agents showed the greatest degree of binding to calmodulin. Other classes of pharmacological agents, including aminergic agonists and antagonists, and nonspecific central nervous system depressants and stimulants, showed little or no binding to calmodulin. In fact, the specificity with which antipsychotic drugs bind to calmodulin suggests the possibility of screening for new and clinically more effective antipsychotic agents based on their selective binding to calmodulin.

Certain neuropeptides that produce behavioral effects in animals also were found to inhibit the activity of calmodulin, suggesting that there may be endogenous psychotogens or antipsychotic peptides that interact with calmodulin.

Although under ordinary conditions the binding of antipsychotics to calmodulin is reversible, the binding of phenothiazine antipsychotics to calmodulin can be made irreversible either photochemically by ultraviolet irradiation, or enzymatically by a hydrogen peroxide–peroxidase system. Such a labeling technique should prove to be a useful tool to study the localization and turnover of calmodulin.

These results indicate that several of the diverse biochemical actions of antipsychotic agents can be explained by a common mechanism, namely, by their binding to and inhibition of calmodulin, and raise the possibility that calmodulin may serve as one of the cellular receptors for certain antipsychotic compounds. However, further studies must be completed before we can state with any degree of certainty that these *in vitro* biochemical findings can explain the pharmacological and clinical actions of the antipsychotics.

## Acknowledgments

We wish to thank Ms. Becky Simon for her excellent technical assistance and Ms. Jose Weycis for her help with the manuscript.

Supported by Grant #MH30096 awarded by the National Institute of Mental Health, DHEW, and by a fellowship awarded to T.L.W. by the Scottish Rite Schizophrenia Research Program.

## References

1. Teo, T. S., T. H. Wang & J. H. Wang. 1973. J. Biol. Chem. **248:** 588–595.
2. Dedman, J. R., J. D. Potter, R. L. Jackson, J. D. Johnson & A. R. Means. 1977. J. Biol. Chem. **252:** 8415–8422.
3. Jarrett, H. W. & J. T. Penniston. 1978. J. Biol. Chem. **253:** 4676–4682.
4. Watterson, D. M., F. Sharief & T. C. Vanaman. 1980. J. Biol. Chem. **255:** 962–975.
5. Wolff, D. J., P. G. Poirier, C. O. Brostrom & M. A. Brostrom. 1977. J. Biol. Chem. **252:** 4108–4117.
6. Klee, C. B. 1977. Biochemistry **16:** 1017–1024.
7. Walsh, M., F. C. Stevens, K. Oikawa & C. M. Kay. 1979. Can. J. Biochem. **57:** 267–278.
8. Lin, Y. M., Y. P. Liu & W. Y. Cheung. 1974. J. Biol. Chem. **249:** 4943–4954.
9. Cheung, W. Y., T. J. Lynch & R. W. Wallace. 1978. Adv. Cyclic Nucleotide Res. **9:** 233–251.
10. Wolff, D. J. & C. O. Brostrom. 1979. Adv. Cyclic Nucleotide Res. **11:** 27–88.
11. Grand, R. J. A., A. C. Nairn & S. V. Perry. 1980. Biochem. J. **185:** 755–760.
12. Liu, Y. P. & W. Y. Cheung. 1976. J. Biol. Chem. **251:** 4193–4198.
13. Walsh, M. & F. C. Stevens. 1977. Biochemistry **16:** 2742–2749.
14. Luthra, M. G. & H. D. Kim. 1979. Life Sciences **24:** 2441–2448.
15. Cheung, W. Y., Y. M. Lin, Y. P. Liu & J. A. Smoake. 1975. *In* Cyclic Nucleotides in Disease. B. Weiss, Ed.:321–350. University Park Press. Baltimore, Md.

16. WAISMAN, D., F. C. STEVENS & J. H. WANG. 1975. Biochem. Biophys. Res. Commun. **65:** 975–982.
17. EGRIE, J. C., J. A. CAMPBELL, A. L. FLANGAS & F. L. SIEGEL. 1977. J. Neurochem. **28:** 1207–1213.
18. GNEGY, M. E., J. A. NATHANSON & P. UZUNOV. 1977. Biochim. Biophys. Acta. **497:** 75–85.
19. SOBUE, K., S. ICHIDA, H. YOSHIDA, R. YAMAZAKI & S. KAKIUCHI. 1979. FEBS Lett. **99:** 199–202.
20. DELORENZO, R. J., S. D. FREEDMAN, W. B. YOHE & S. C. MAURER. 1979. Proc. Natl. Acad. Sci. USA **76:** 1838–1842.
21. GRAB, D. J., K. BERZINS, R. S. COHEN & P. SIEKEVITZ. 1979. J. Biol. Chem. **254:** 8690–8696.
22. RUNGE, M. S., P. B. HEWGLEY, D. PUETT & R. C. WILLIAMS, JR. 1979. Proc. Natl. Acad. Sci. USA **76:** 2561–2565.
23. WANG, J. H. & R. DESAI. 1977. J. Biol. Chem. **252:** 4175–4184.
24. KLEE, C. B. & M. H. KRINKS. 1978. Biochemistry **17:** 120–126.
25. WALLACE, R. W., T. J. LYNCH, E. A. TALLANT & W. Y. CHEUNG. 1978. J. Biol. Chem. **254:** 377–382.
26. SHARMA, R. K., R. DESAI, T. R. THOMPSON & J. H. WANG. 1978. Can. J. Biochem. **56:** 598–604.
27. SHARMA, R. K., E. WIRCH & J. H. WANG. 1978. J. Biol. Chem. **253:** 3575–3580.
28. KLEE, C. B., T. H. CROUCH & M. H. KRINKS. 1979. Proc. Natl. Acad. Sci. USA **76:** 6270–6273.
29. VANAMAN, T. C., F. SHARIEF, J. L. AWRAMIK, P. A. MENDEL & D. M. WATTERSON. 1976. *In* Contractile Systems in Non-Muscle Tissues. S. V. Perry, Ed.:165–176. Elsevier/North Holland Biomedical Press. New York, N.Y.
30. WATTERSON, D. M., W. G. HARRELSON, P. M. KELLER, F. SHARIEF & T. C. VANAMAN. 1976. J. Biol. Chem. **251:** 4501–4513.
31. TESHIMA, Y. & S. KAKIUCHI. 1978. J. Cyclic Nucl. Res. **4:** 219–231.
32. REVUELTA, A., P. UZUNOV & E. COSTA. 1976. Neurochem. Res. **1:** 217–227.
33. HANBAUER, I., J. GIMBLE & W. LOVENBERG. 1979. Neuropharmacology **18:** 851–857.
34. HANBAUER, I., J. GIMBLE, K. SANKARAN & R. SHERARD. 1979. Neuropharmacology **18:** 859–864.
35. HARPER, J. F., W. Y. CHEUNG, R. W. WALLACE & H.-L. HUANG. 1980. Proc. Natl. Acad. Sci. USA **77:** 366–370.
36. GNEGY, M. E. & Y. S. LAU. 1980. Neuropharmacology **19:** 319–323.
37. UZUNOV, P. & B. WEISS. 1972. Biochim. Biophys. Acta **284:** 220–226.
38. WEISS, B., R. FERTEL, R. FIGLIN & P. UZUNOV. 1974. Mol. Pharmacol. **10:** 615–625.
39. WEISS, B., R. LEHNE & S. J. STRADA. 1972. Anal. Biochem. **45:** 222–235.
40. WEISS, B. 1975. Adv. Cyclic Nucleotides Res. **5:** 195–211.
41. WALLACE, T. L., R. M. LEVIN & B. WEISS. 1978. Pharmacologist **20:** 232.
42. LEVIN, R. M. & B. WEISS. 1980. Neuropharmacology **19:** 169–174.
43. LEVIN, R. M. & B. WEISS. 1976. Mol. Pharmacol. **12:** 581–589.
44. FILBURN, C. R., F. T. COLPO & B. SACKTOR. 1979. Mol. Pharmacol. **15:** 257–262.
45. UZUNOV, P. & B. WEISS. 1971. Mol. Pharmacol. **10:** 697–708.
46. MILLER, R. J., A. S. HORN & L. L. IVERSON. 1974. Mol. Pharmacol. **10:** 759–766.
47. MILLER, R. J. 1976. Biochem. Pharmacol. **25:** 537–541.
48. LEVIN, R. M. & B. WEISS. 1979. J. Pharmacol. Exptl. Ther. **208:** 454–459.
49. NORMAN, J. A., A. H. DRUMMOND & P. MOSER. 1979. Mol. Pharmacol. **16:** 1089–1094.
50. WATANABE, K., E. F. WILLIAMS, J. S. LAW & W. L. WEST. 1980. Experientia (In press.)
51. KOBAYSHI, R., M. TAWATA & H. HIDAKA. 1979. Biochem. Biophys. Res. Commun. **88:** 1037–1045.
52. HIDAKA, H., T. YAMAKI, M. NAKA, T. TANAKA, H. HAYASHI & R. KOBAYASHI. 1980. Mol. Pharmacol. **17:** 66–72.
53. HIDAKA, H., M. NAKA & T. YAMAKI. 1979. Biochem. Biophys. Res. Commun. **90:** 694–699.
54. CHEUNG, W. Y., L. S. BRADHAM, T. J. LYNCH, Y. M. LIN & E. A. TALLANT. 1975. Biochem. Biophys. Res. Commun. **65:** 1055–1062.

55. BROSTROM, C. O., Y.-C. HUANG, B. M. BRECKENRIDGE & D. J. WOLFF. 1975. Proc. Natl. Acad. Sci. USA **72:** 64–68.
56. BROSTROM, C. O., M. A. BROSTROM & D. J. WOLFF. 1977. J. Biol. Chem. **252:** 5677–5685.
57. BROSTROM, M. A., C. O. BROSTROM, B. M. BRECKENRIDGE & D. J. WOLFF. 1978. Adv. Cyclic Nucleotide Res. **9:** 85–99.
58. LYNCH, T. J., E. A. TALLANT & W. Y. CHEUNG. 1976. Biochem. Biophys. Res. Commun. **68:** 616–625.
59. MOSS, J. & M. VAUGHN. 1977. Proc. Natl. Acad. Sci. USA **74:** 4396–4400.
60. TOSCANO, W. A., JR., K. R. WESTCOTT, D. C. LAPORTE & D. R. STORM. 1979. Proc. Natl. Acad. Sci. USA **76:** 5582–5586.
61. LEVITZKI, A. & E. J. M. HELMREICH. 1979. FEBS Lett. **101:** 213–219.
62. GNEGY, M. E., P. UZUNOV & E. COSTA. 1976. Proc. Natl. Acad. Sci. USA **73:** 3887–3890.
63. COSTA, E., M. GNEGY, A. REVUELTA & P. UZUNOV. 1977. *In* Advances in Biochemical Psychopharmacology. E. Costa & G. L. Gessa, Eds. Vol. **16:** 403–408. Raven Press. New York, N.Y.
64. GNEGY, M. E., A. LUCCHELLI & E. COSTA. 1977. Naunyn-Schmiedeberg's Arch. Pharmacol. **301:** 121–127.
65. GREENBERG, L. H. & B. WEISS. 1978. *In* Recent Advances in Pharmacology of Adrenoceptors. E. SZABADI, C. M. BRADSHAW & P. BEVAN, Eds.: 241–250. Elsevier/North Holland Biomedical Press. New York, N.Y.
66. NAGAO, S., Y. SUZUKI, Y. WATANABE & Y. NOZAWA. 1979. Biochem. Biophys. Res. Commun. **90:** 261–268.
67. DAVIS, P. W. & T. M. BRODY. 1966. Biochem. Pharmacol. **15:** 703–710.
68. BOND, G. H. & D. L. CLOUGH. 1973. Biochim. Biophys. Acta **323:** 592–599.
69. JARRETT, H. W. & J. T. PENNISTON. 1977. Biochem. Biophys. Res. Commun. **77:** 1210–1216.
70. GOPINATH, R. M. & F. F. VINCENZI. 1977. Biochem. Biophys. Res. Commun. **77:** 1203–1209.
71. VINCENZI, F. F., T. R. HINDS & B. RAESS. 1980. Ann. N.Y. Acad. Sci. (This volume.)
72. DABROWSKA, R., D. AROMATORIO, J. M. F. SHERRY & D. J. HARTSHORNE. 1977. Biochem. Biophys. Res. Commun. **78:** 1263–1272.
73. YAGI, K., M. YAZAWA, S. KAKIUCHI, M. OHSHIMA & K. UENISHI. 1978. J. Biol. Chem. **253:** 1338–1340.
74. COHEN, P., A. BURCHELL, J. G. FOULKES, P. T. W. COHEN, T. C. VANAMAN & A. C. NAIRN. 1978. FEBS Lett. **92:** 287–293.
75. WAISMAN, D. M., T. J. SINGH & J. H. WANG. 1978. J. Biol. Chem. **253:** 3387–3390.
76. RYLATT, D. B., N. EMBI & P. COHEN. 1979. FEBS Lett. **98:** 76–80.
77. SRIVASTAVA, A. K., D. M. WAISMAN, C. O. BROSTROM & T. R. SODERLING. 1979. J. Biol. Chem. **254:** 583–586.
78. SCHULMAN, H. & P. GREENGARD. 1978. Nature **271:** 478–479.
79. SCHULMAN, H. & P. GREENGARD. 1978. Proc. Natl. Acad. Sci. USA **75:** 5432–5436.
80. SIEGHART, W., H. SCHULMAN & P. GREENGARD. 1980. J. Neurochem. **34:** 548–553.
81. KENNEDY, M. B. & P. GREENGARD. 1980. Ann. N.Y. Acad. Sci. (This volume.)
82. DELORENZO, R. J. 1980. Ann. N.Y. Acad. Sci. (This volume.)
83. WONG, P. Y-K. & W. Y. CHEUNG. 1979. Biochem. Biophys. Res. Commun. **90:** 473–480.
84. ILUNDAIN, A. & R. J. NAFTALIN. 1979. Nature **279:** 446–448.
85. SUGDEN, M. C., M. R. CHRISTIE & S. J. H. ASHCROFT. 1979. FEBS Lett. **105:** 95–100.
86. MARCUM, J. M., J. R. DEDMAN, B. R. BRINKLEY & A. R. MEANS. 1978. Proc. Natl. Acad. Sci. USA **75:** 3771–3775.
87. NISHIDA, E., H. KUMAGAI, I. OHTSUKI & H. SAKAI. 1979. J. Biochem. **85:** 1257–1266.
88. KUMAGAI, H. & E. NISHIDA. 1979. J. Biochem. **85:** 1267–1274.
89. JAMIESON, G. A., JR., T. C. VANAMAN & J. J. BLUM. 1979. Proc. Natl. Acad. Sci. USA **76:** 6471–6475.
90. IQBAL, Z. & S. OCHS. 1980. J. Neurobiol. **11:** 311–318.
91. KASTIN, A. J., R. D. OLSON, A. V. SCHALLY & D. H. COY. 1979. Life Sciences **25:** 401–414.
92. LEVIN, R. M. & B. WEISS. 1977. Mol. Pharmacol. **13:** 690–697.

93. LEVIN, R. M. & B. WEISS. 1978. Biochim. Biophys. Acta **540:** 197–204.
94. WEISS, B., R. M. LEVIN & L. H. GREENBERG. 1979. *In* Catecholamines: Basic and Clinical Frontiers. E. USDIN, I. J. KOPIN & J. BARCHAS, Eds.: 529–531. Pergamon Press. New York, N.Y.
95. BURKI, H. R., W. RUCH & H. ASPER. 1975. Psychopharmacologia **41:** 27–33.
96. PERINGER, T., P. JENNER, P. I. M. DONALDSON & C. D. MARSDEN. 1976. Neuropharmacology **15:** 463–469.
97. STILLE, G., A. C. SAYERS, H. LAVENER & E. EICHENBERGER. 1973. Psychopharmacologia **28:** 325–337.
98. NAKRA, B. R. S., A. J. BOND & M. H. LADER. 1975. J. Clin. Pharmacol. **15:** 449–454.
99. SEEMAN, P. M. & H. S. BIALY. 1963. Biochem. Pharmacol. **12:** 1181–1191.
100. GREEN, A. L. 1967. J. Pharm. Pharmacol. **19:** 10–16.
101. ENNA, S. J., J. P. BENNET, D. R. BURT, I. CREESE & S. H. SNYDER. 1976. Nature **263:** 338–341.
102. JOHNSTONE, E. C., C. D. FRITH, T. J. CROW, M. W. P. CARNEY & J. S. PRICE. 1978. Lancet **1:** 848–851.
103. CROW, T. J., J. W. F. DEAKIN & E. C. JOHNSTONE. 1977. Nature **267:** 183–184.
104. PO, A. L. W. & W. J. IRWIN. 1980. J. Pharm. Pharmacol. **32:** 25–29.
105. AKERA, T. & T. M. BRODY. 1969. Mol. Pharmacol. **5:** 605–614.
106. BREYER, V. & H. J. GAERTNER. 1974. Adv. Biochem. Psychopharmacol. **9:** 167–173.
107. PIETTE, L. H., G. BULOW & I. YAMAZAKI. 1964. Biochem. Biophys. Acta **88:** 120–129.
108. FORREST, I. S., F. M. FORREST & M. BERGER. 1958. Biochim. Biophys. Acta **29:** 441–442.
109. CAVANAUGH, D. J. 1957. Science **125:** 1040–1041.
110. CHAN, T. L., G. SAKALIS & S. GERSHON. 1974. Adv. Biochem. Psychopharmacol. **9:** 323–333.
111. PRANGE, A. J., P. T. LOOSEN & G. B. NEMEROFF. 1979. *In* New Frontiers of Psychotropic Drug Research. S. FIELDING, Ed.: 117–189. Future Publishing Co. New York, N.Y.
112. SHENOLIKAR, S., P. T. W. COHEN, P. COHEN, A. C. NAIRN & S. V. PERRY. 1979. Europ. J. Biochem. **100:** 329–337.
113. HARTSHORNE, D. 1980. Ann. N.Y. Acad. Sci. (This volume.)
114. SCHUBART, V. K., J. ERLICHMAN & N. FLEISCHER. 1980. J. Biol. Chem. **255:** 4120–4124.
115. SATIR, B. H., R. S. GAROFALO, N. J. MAIHLE & D. M. GILLIGAN. 1980. Ann. N.Y. Acad. Sci. (This Volume.)
116. KUHN, D. M., J. P. O'CALLAGHAN & W. LOVENBERG. 1980. Ann. N.Y. Acad. Sci. (This Volume.)
117. ANDERSON, J. M. & M. J. CORMIER. 1978. Biochem. Biophys. Res. Commun. **84:** 595–602.
118. BEALE, E. G., J. R. DEDMAN & A. R. MEANS. 1977. Endocrinology **101:** 1621–1634.
119. UZUNOV, P., H. M. SHEIN & B. WEISS. 1974. Neuropharmacology **13:** 377–391.
120. WEISS, B. & R. M. LEVIN. 1978. Adv. Cyclic Nucleotide Res. **9:** 285–303.
121. WEISS, B. & T. L. WALLACE. 1980. *In* Calcium Binding Proteins and Calcium Function. W. Y. CHEUNG, Ed. Academic Press, Inc. New York, N.Y. (In press.)
122. TESHIMA, Y. & S. KAKIUCHI. 1974. Biochem. Biophys. Res. Commun. **56:** 489–495.
123. DAVIS, B. J. 1964. Ann. N.Y. Acad. Sci. **121:** 404–427.
124. LIN, Y. M., Y. P. LIU & W. Y. CHEUNG. 1975. FEBS Lett. **49:** 356–360.
125. BROSTROM, C. O. & D. J. WOLFF. 1974. Arch. Biochem. Biophys. **165:** 715–727.
126. CIMINO, M., W. PROZIALECK & B. WEISS. 1979. Pharmacologist **21:** 240.
127. WOLFF, D. J., M. A. BROSTROM & C. O. BROSTROM. 1977. *In* Calcium Binding Proteins and Calcium Function. R. H. WASSERMAN, R. A. CORRADINO, E. CARAFOLI, R. H. KRETSINGER, D. H. MACLENNAN & F. L. SIEGEL, Eds.: 97–106. North Holland. New York, N.Y.
128. CHEUNG, W. Y. 1970. Biochem. Biophys. Res. Commun. **38:** 533–538.
129. KAKIUCHI, S. & R. YAMAZAKI. 1970. Biochem. Biophys. Res. Commun. **41:** 1104–1110.
130. FERTEL, R. & B. WEISS. 1976. Mol. Pharmacol. **12:** 678–687.
131. HAIT, W. N. & B. WEISS. 1976. Nature **259:** 321–323.
132. WATSON, S. J., H. AKIL, P. A. BERGER & J. D. BARCHAS. 1979. Arch. Gen. Psych. **36:** 35–41.

133. LEHMANN, H., N. P. VASAVAN & N. S. KLINE. 1979. Am. J. Psych. **136:** 762–766.
134. WEISS, B., W. PROZIALECK & M. CIMINO. 1980. Adv. Cyclic Nucleotide Res. **12:** 213–225.
135. JEWSON, T. J., M. E. RAICHLE & M. J. WELCH. 1980. Brain Res. **192:** 291–295.
136. LEO, A., C. HANSCH & D. ELKINS. 1971. Chem. Rev. **71:** 525–616.
137. PROZIALECK, W. C., M. CIMINO & B. WEISS. 1981. Mol. Pharmacol. (In press.)

## DISCUSSION OF THE PAPER

DR. P. EPSTEIN: You showed that there is binding of trifluoperazine to calmodulin. Is it definitive that that really is the mechanism by which it inhibits the activation? In other words, do you have any data that would rule out that binding of calmodulin to another enzyme would open a site on the enzyme to which trifluoperazine might bind?

DR. B. WEISS: There is no question that trifluoperazine binds directly to calmodulin and that the binding of this compound as well as the binding of several other compounds to calmodulin is directly related to their ability to inhibit the activation of phosphodiesterase. We have no evidence whatever that the drugs bind to phosphodiesterase either in the presence or absence of calmodulin.

DR. MARSHAK: In the control, when you UV irradiate calmodulin without drug present, does the calmodulin break down at all? Is there any physical property of calmodulin that is different after irradiation?

DR. WEISS: The irradiated calmodulin shows, as far as we can tell, exactly the same biological properties and it activates phosphodiesterase exactly as it did before it was irradiated.

DR. MARSHAK: What does it look like on a gel? Is it still a single band?

DR. WEISS: After it is irradiated, we find that it's a single band provided the samples are fresh. But, as others also found, if calmodulin is stored for long periods of time we get multiple bands.

DR. MARSHAK: What is the molar ratio of the drug bound to the calmodulin after the UV irradiation. That is, how many moles of drug are covalently incorporated per mole of calmodulin?

DR. WEISS: With UV irradiation we get one mole of trifluoperazine bound for every mole of calmodulin.

DR. MARSHAK: Yet, one of your slides indicated that there may be two sites.

DR. B. WEISS: That's right. This may be related to experimental variability or to the difference in the design of the studies. In our experiments of the reversible binding, our calculations show approximately two drug binding sites on calmodulin. The studies on the irreversible binding show one site.

DR. MARSHAK: In the covalently modified calmodulin was there still no phosphodiesterase activating activity?

DR. WEISS: If you just irradiate calmodulin you find it still retains biological activity. If you irradiate calmodulin with trifluoperazine and then dialyze it, it no longer has biological activity. The irreversibly bound complex is irreversibly inactivated.

# CALCIUM AND cAMP IN STIMULUS-RESPONSE COUPLING*

Howard Rasmussen

*Departments of Cell Biology and Internal Medicine*
*Yale University School of Medicine*
*New Haven, Connecticut 06511*

In the preceding reports in this symposium the emphasis has been focused upon the role of calcium and calmodulin in the regulation of cell function. In closing this meeting, I should like to put the preceding discussion into a broader context of the regulation of cell function by neurotransmitters and peptide hormone. Within this context I should like to consider two related questions. The first, whether the distinction between excitable and non-excitable tissues means that the systems which couple stimulus to response in them differ. The second whether the distinction between calcium-regulated and cAMP-regulated cellular responses is as sharp as has been suggested.

From the ensuing discussions, I shall draw the conclusion that stimulus-response coupling in excitable and non-excitable cells is achieved by the same messenger system, and that in nearly all cells regulation of cellular function by these classes of extracellular messages involves the integrated actions of cAMP and calcium (see Rasmussen and Goodman[1] for review of experimental data). In order to emphasize the universality of their role in stimulus-response coupling, I suggest that this concept of dual universal interrelated messengers be identified as the synarchy to indicate that these two messengers or heralds within the cell act together in one of several ways to regulate and direct cell responses.

I use the term stimulus-response coupling in a broader context than that usually employed. I intend that the specialized response of any differentiated tissue to a specific class of extracellular stimuli be included in this term. This class is defined as agents that possess two attributes: 1) they interact predominantly, if not exclusively, with appropriate receptors on the cell surface; and 2) they bring forth, as a consequence of this initial recognition event, the specialized response of the differentiated cell. This would mean muscle contraction, nerve conduction, gland secretion, a change in an epithelial transport process, or activation of a specific metabolic response, e.g., gluconeogenesis. Excluded from this class would be hormones like insulin, epidermal growth factor, nerve growth factor, and the somatomedins which are energy-storage and growth-promoting hormones. Even though they too bring about a cellular response by an initial interaction with a cell surface receptor, they do not normally initiate the unique physiological response by which that particular tissue contributes to the integrated physiological behavior of the total organism.

The corollary to this broader use of the definition of stimulus-response coupling is that of understanding the distinction between excitable and non-excitable tissues. Excitability, in its restricted definition, has become the province of the neuro- and muscular physiologist and has come to be restricted to those tissues in which regenerative ion currents lead to action potentials that are

*Supported by the National Institutes of Health (Grant AM 198103).



0077-8923/80/0356-0346 $01.75/0 © 1980, NYAS

propagated across the cell surface from the site of initial stimulus. The tissues that most clearly possess this attribute are nerve and muscle. Hence, by convention these are known as excitable in contrast to other so-called non-excitable tissues like liver, parotid, or adipose tissue in which stimulation does not involve initiation of action potentials.

This distinction was further reinforced by the discoveries that calcium ion was the factor that coupled excitation to neurosecretion and excitation to contraction,[2,3] but that cAMP was the second messenger in epinephrine action upon hepatic glycogenolysis.[4] Hence, it became convenient to consider that stimulus-response coupling in excitable tissues involved ionic messengers, and hormonal (or chemical) activation of non-excitable tissues involved chemical intracellular messenger (FIGURE 1).

The major point to be made in the ensuing discussion is that this distinction is invalid. If we expand the definition of excitability to include mechanical,

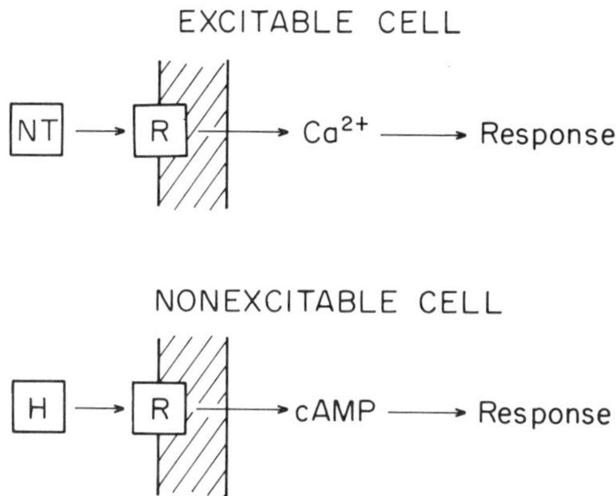


FIGURE 1. A schematic representation of the concept of calcium ion as coupling factor in excitable tissues (upper); and of cAMP as second messenger in the action of peptide and amine hormones on non-excitable cells.

thermal, chemical, and osmotic as well as electrical excitability then liver is an excitable tissue just as is nerve. Furthermore, if we analyze the properties of certain slow muscle (those with a slow response time) we find that they are not excitable in the classic, restricted sense because a stimulus sufficient to cause them to contract does not induce action potentials in their surface membranes. Nevertheless, the basic aspects of stimulus-response coupling in these tissues are otherwise the same as in 'excitable' muscle. On the other hand, the beta cell of the endocrine pancreas is an excitable cell in the restricted definition of this term. Glucose induces calcium-dependent trains of action potentials in the plasma membrane of these cells.[5] In contrast, adrenal medullary cells which are also endocrine secretory cells derived from neural tissue are not excitable by this criterion even though they are stimulated to secrete by the classic neurotransmitter, acetylcholine. If, on the other hand, one analyzes the events in each of these tissues in terms of the intracellular messengers coupling stimulus to response a common and universal theme emerges. Calcium ion and cAMP serve as dual, interrelated messengers in determining cellular response. Their interrelatedness is the prime basis for applying the term *synarchic* regulation to the universal system in question.

Encompassed by this term *synarchic* regulation is the dual meaning that these two messengers nearly always serve together, but that their relationship may vary from one cell type to another. One can recognize within this universal theme a number of particular variations. Those to be discussed will include: 1) coordinate control; 2) hierarchical control; 3) redundant control; 4) sequential control; and 5) antagonistic control. Regardless of which operates, the synarchic system possesses a number of universal attributes. These include: 1) there are receptor proteins for both messengers within the cell; 2) cAMP regulates and by its actions can either expand or contrast the temporal and/or spatial domain of the calcium message; 3) calcium ion regulates cAMP metabolism and by its actions can either expand or contract the temporal and/or spatial domain of the cAMP message; 4) cAMP can alter the sensitivity of calcium response elements to calcium; 5) calcium can alter the sensitivity of cAMP response elements to cAMP; 6) a major mechanism by which both calcium and cAMP control cell function is by controlling the activity of a class of enzymes known as protein kinases; 7) the calcium- and cAMP-dependent kinases are different enzymes but may nevertheless act upon the same or different protein substrates; and 8) calcium mediates a number of its effects by controlling the functions of proteins other than protein kinases whereas there is little evidence that cAMP exerts effects other than those controlled by protein kinases.[1,6,7]

## Coordinate Synarchic Regulation

Having discovered with Naokazu Nagata that both calcium and cAMP served messenger functions in the action of parathyroid hormone on renal gluconeogenesis, I proposed in 1970 that a relationship between these two messengers was probably of widespread occurrence. I was fortunate enough to join Michael Berridge and William Prince of Cambridge University in demonstrating that both messengers functioned in coupling stimulus to response in the action of 5HT on fluid secretion in the fly salivary gland. The critical data in the development of this evidence have been discussed in detail.[1,6] Of note is the fact that the action of 5HT met all the criteria Sutherland and coworkers suggested were necessary to establish cAMP as the messenger in hormone action: 1) 5HT caused a rise in tissue cAMP; 2) phosphodiesterase inhibitors enhanced the effect of 5HT; 3) a 5HT-sensitive adenylate cyclase was found in a particulate fraction of the gland; 4) exogenous cAMP mimicked the effect of 5HT on fluid secretion; and 5) a cAMP-dependent protein kinase was found in the tissue. However, when cAMP stimulated fluid secretion the change in transepithelial membrane potential was exactly opposite to that seen after 5HT addition. Also, either the calcium ionophore, A23187, or an elevation of external $K^+$ led to a calcium-dependent increase in fluid secretion under circumstances in which cAMP content either did not change or actually fell. An analysis of the effects of 5HT and cAMP on tissue calcium metabolism showed that although both agents stimulated calcium efflux from prelabeled cells, only 5HT stimulated calcium uptake. Subsequent work by Fain and Berridge showed that 5HT but not cAMP increased phosphoinositide turnover which is thought to be a biochemical component of calcium gating in the plasma membrane.[8,9]

On the basis of the data available, a model of 5HT action can be constructed in which 5HT has two simultaneous and independent effects on the function of the plasma membrane: a) it activates adenylate cyclase; and b) it opens a calcium channel in the membrane. As a result, both the cAMP and calcium ion content of the cell cytosol increases. The two messengers act in a coordinated fashion on the major response element within the cell: a) cAMP activates a $K^+$ -pump in the luminal membrane; and b) calcium increases the permeability of the membrane to $Cl^-$. As a consequence, KCl secretion increases and this leads to an increase in fluid secretion. In addition, the two messengers interact by controlling the metabolism of the other. Calcium acts as a negative feedback modulator of cAMP concentration; cAMP as a positive feedforward modulator of cytosolic calcium ion concentration. The fundamental feature, however, is that the actions of these two messengers are coordinated in achieving cell response (FIGURE 2). This type of coordinate synarchic regulation is also seen in the control of renal gluconeogenesis by parathyroid hormones. It probably operates in other cell types as well.

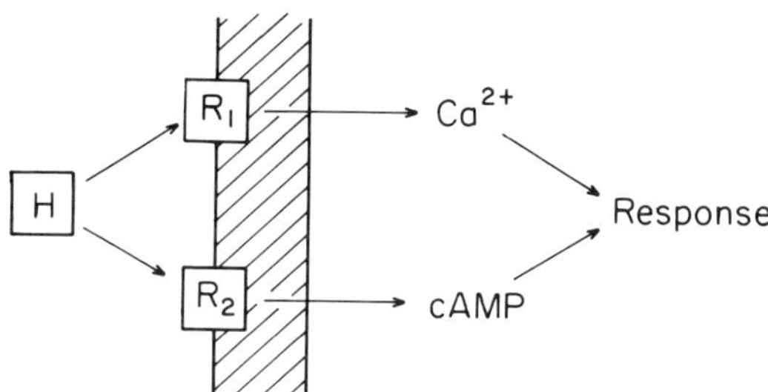


FIGURE 2. Schematic representation of coordinate synarchic regulation of cell function by calcium and cAMP.

## REDUNDANT SYNARCHIC REGULATION

Although the control of hepatic glycogenolysis and gluconeogenesis by catecholamines and glucagon were the systems in which the second messenger function of cAMP was discovered, newer findings in the past decade have shown beyond any doubt that hepatic glucose release is controlled by some agents (glucagon and $\beta$-adrenergic agonists) utilizing cAMP as messenger;[10] and by others (angiotensin and $\alpha$-adrenergic agonists) utilizing calcium as messenger[11,12] (FIGURE 3). In this circumstance either messenger alone is capable of inducing response, each is generated in response to different extracellular stimuli; but each controls the activity of the same key enzymes in the metabolic pathways involved.[13] However, because the calcium-dependent and cAMP-dependent protein kinases catalyze the phosphorylation of different sites on the regulated enzymes, there is a quantitative difference in the flux of materials through the pathways.

Comparison of the operational feature of this type of redundant synarchic control with the coordinate type illustrates on the one hand the commonality of the two intracellular messengers and the mechanisms by which they are gener-

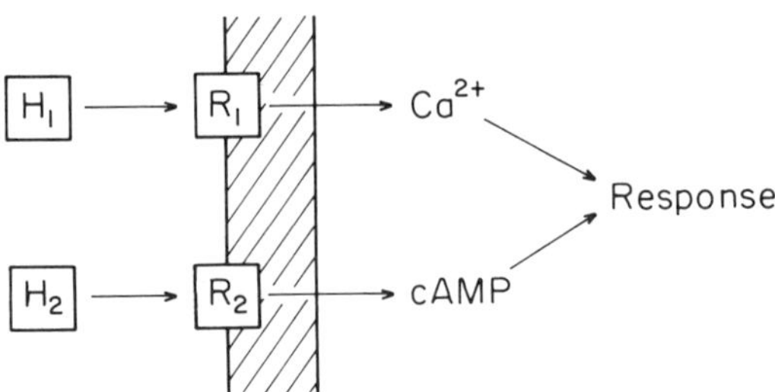


FIGURE 3. Schematic representation of redundant synarchic regulation of cell function by calcium and cAMP.

ated, but shows that separate, rather than the same, extracellular messengers control the separate messenger generators. In addition to the liver, redundant control of this type is also evident in the ionic and hormonal control of aldosterone secretion in the zona glomerulosa of the adrenal cortex.

## SEQUENTIAL SYNARCHIC REGULATION

A particularly interesting type of synarchic regulation is one in which one of the two messengers is generated in response to a specific extracellular stimulus, and the increase in concentration of this first intracellular messenger is responsible for the subsequent increase in the concentration of the other messenger. In the case of the stimulation of catecholamine secretion from the adrenal medulla by acetylcholine, the initial intracellular messenger is calcium.[3] As a result of its increase within the cytosol a calmodulin-dependent activation of adenylate cyclase is found.[14] In contrast, when epinephrine acts on the heart an initial increase in cAMP content prolongs the influx of calcium into the cell via the voltage-dependent calcium channel thereby enhancing the calcium message. Thus, synarchic regulation can involve the sequential appearance of one second messenger in response to an initial increase in the concentration of the other.

## HIERARCHICAL SYNARCHIC REGULATION

A host of tissues demonstrate hierarchical synarchic regulation in which the calcium message is generated in response to a primary extracellular stimulus, and the cAMP message is generated by a second extracellular message and serves to extend the temporal and/or spatial domains of the calcium message (FIGURE 4). One of the most elegantly simple of these systems is that described by Kupfermann and coworkers involving the control of the strength of contraction of the accessory radula closer (ARC) muscle of *Aplysia*.[15,16] This is a particularly interesting example because although it is a muscle and hence would be considered an excitable tissue when stimulated to contract, no action potentials are seen in its plasma membrane; only a depolarization is observed. Contraction is induced by cholinergic neurons just as in fast mammalian skeletal muscles which exhibit action potentials. Also, calcium couples stimulus to response as it does in other muscles. The feature of greatest interest to our present discussion is that the muscle cells are also innervated by serotonergic neurons. Stimulation of these alone cause no change in membrane potential nor do they induce muscle contraction. However, if they are stimulated just prior to the stimulation of the motor neuron, the subsequent contractile response to a standard stimulus is increased. This increase in response is not accompanied by any change in the degree of membrane depolarization induced by a standard motor neuron stimulus. This implies that the action of the serotonergic neuron is that of enhancing the intracellular response to a standard stimulus. Stimulation of this serotonergic neuron leads to a rise in the cAMP content of the muscle, and application of exogenous cAMP induces a similar enhancement of contractile response. Thus, the generation of cAMP acts either to extend the spatial and/or temporal domain of calcium ion and/or changes the sensitivity of a calcium response element to activation by calcium ion.

Several examples of presynaptic facilitation by cAMP have also been described.[17,18] These share the operational features outlined in FIGURE 4. The primary stimulus acts to cause an increase in the calcium message, a facilitatory stimulus by initiating the cAMP message causes an augmentation of cellular response.

## ANTAGONISTIC SYNARCHIC REGULATION

In the preceding examples, even though negative feedback effects of one messenger upon the other can be identified, the totality of their relationship is that of acting in a concerted, coordinate, or supplementary way to regulate response. However, given that within these synarchic systems there are instances in which calcium ion acts to either increase or decrease the domain of the cAMP message, the same is likely to be true in regard to the control of cAMP domain by calcium. Instances, such as presynaptic facilitation, have already been discussed in which cAMP extends this domain. The converse is also frequently seen. A prime example is the regulation of the contraction of a number of different types of smooth muscle.[19,20] In these instances, the muscle is stimulated to contract by either $\alpha$-adrenergic or cholinergic nerve stimulation. The primary message is calcium, which may come either from an intracellular (microsomal) pool and/or from the extracellular pool. Relaxation or an inhibition of contraction is induced

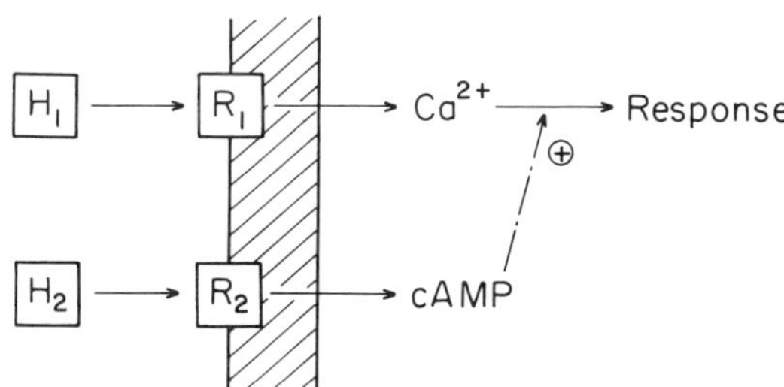


FIGURE 4. Schematic representation of hierarchical synarchic, regulation of cell function by calcium and cAMP.

by β-adrenergic nerve stimulation. This causes a rise in cAMP concentration within the muscle. The increase in cellular cAMP acts to inhibit or attenuate the calcium message by several different effects: It causes an increase in the rate of reaccumulation of calcium by the endoplasmic reticulum, and an increase in rate of calcium efflux from the cell. It also decreases the sensitivity of the calcium response element, the calmodulin-regulated myosin light chain kinase, to activation by calcium.[21]

Operationally, antagonistic synarchic regulation (FIGURE 5) has many basic similarities to the hierarchical system (FIGURE 3), the only difference being that a rise in cAMP extends the calcium messenger domain in hierarchical systems, but restricts it in the antagonistic system. This type of antagonistic synarchic regulation is widespread, being seen in the control of histamine release from the mast cell,[22] in the platelet release reaction, and in the phagocytic response of the leucocyte.

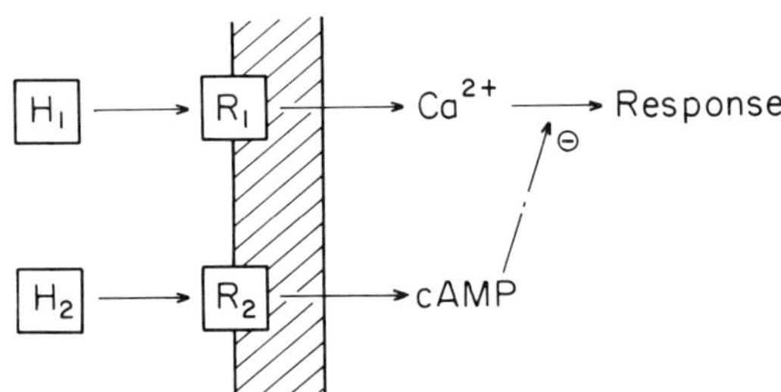


FIGURE 5. Schematic representation of antagonistic synarchic regulation of cell function by calcium and cAMP.

## CONCLUSION

These models are presented as a way of emphasizing that synarchic regulation by calcium and cAMP is the rule rather than the exception, and that this dual messenger system is the major one involved in coupling stimulus to response in most differentiated cells when these cells are called upon to perform their unique physiologic function. However, the models presented do not exhaust the possibilities. For example, a system in which dual messengers are induced by the same extracellular messenger but act in an antagonistic fashion is possible, and may operate in the case of histamine release from the mast cell. Likewise, by the proper organization of feedback interactions synarchic systems could result in initiating and sustaining a variety of oscillatory cellular responses.

The primary purpose of this review has been to call attention to the fact that it is no longer reasonable to arbitrarily divide cells into excitable and non-excitable and to restrict the concept of stimulus-response coupling to the behavior of excitable tissues. It is also no longer valid to emphasize the messenger function of calcium in excitable (neural) tissues and of cAMP in non-excitable (hormonally responsive) tissues. Rather, what has emerged is a picture in which synarchic regulation of cell function by the two messengers, cAMP and calcium, is the rule rather than the exception when one studies the phenomenon of stimulus-response coupling in highly differentiated tissues, and defines stimulus in the broadest possible terms and response in the restricted term of being the unique specialized response of that particular cell type of tissue.

The value of considering the problems of stimulus-response coupling in terms of a synarchic control system lies in broadening the context within which the roles of both calcium and calmodulin are considered. From the philosophic point of view, it emphasizes the fact that the system of stimulus-response coupling is as universal as glycolysis and the Krebs cycle. It extends to another level, that of control, the essential unity of cellular biochemical pathways.

## References

1. Rasmussen, H. & D. B. P. Goodman. 1977. Physiol. Rev. **57:** 421–509.
2. Katz, B. 1966. Nerve, Muscle and Synapse. McGraw-Hill. New York, N.Y.
3. Douglas, W. W. & Rubin. 1961. J. Physiol. **159:** 40–57.
4. Sutherland, E. W. & T. W. Rall. 1958. J. Biol. Chem. **232:** 1065–1078.
5. Dean, P. M. & E. K. Matthews. 1970. J. Physiol. **210:** 255–264.
6. Berridge, M. J. 1976. Adv. Cyclic Nucleotide Res. **6:** 1–96.
7. Cheung, W. T. 1979. Science **207:** 19–27.
8. Berridge, M. J. & J. N. Fain. 1979. Biochem. J. 59–72.
9. Fain, J. N. & M. J. Berridge. 1979. Biochem. J. 45–58.
10. Birnbaum, M. J. & J. N. Fain. 1977. J. Biol. Chem. **252:** 528–535.
11. Van der Werve, G., L. Hue & H. Heis. 1977. Biochem. J. **162:** 135–143.
12. Keppens, S. & N. de Wulf. 1979. Biochem. Biophys. Acta **588:** 63–74.
13. Garrison, J. C. 1978. J. Biol. Chem. **253:** 7091-7103.
14. Guidotti, A., I. Hanbauer & E. Costa. 1975. Adv. Cyclic Nuc. Res. pg. 619.
15. Weiss, K. R., J. L. Cohen & I. Kupfermann. 1978. Neurophysiol. **41:** 181-203.
16. Cohen, J. L., K. R. Weiss & I. Kupfermann. 1978. J. Neurophysiol. **41:** 151–180.
17. Sandaert, G. F. & K. L. Gretchen. 1978. Fed. Proc. **38:** 2183–2192.
18. Langer, S. Z. 1977. Br. J. Pharm. **62:** 481-497.
19. Bolton, T. B. 1979. Physiol. Rev. **59:** 607-718.
20. Mueller, E. & C. Van Breeman. 1979. Nature **281:** 682–683.
21. Conti, M. A. & R. S. Adelstein. 1980. Fed. Proc. **39:** 1569-1573.
22. Foreman, J. C., L. G. Garland & J. L. Mongar. 1976. SEB Symposia XXX: 193–218.

## Discussion of the Paper

Dr. Spitzer (*Louisiana State University, Baton Rouge, LA*): I would like to ask your thoughts about situations regarding other hormones, for example, parathyroid hormone and calcitonin where both cyclic AMP and calcium are implicated as second messengers. How would you categorize those situations?

Dr. H. Rasmussen: It is a very complicated situation and the problem is in getting appropriate bone cells. Both parathyroid hormone and calcitonin activate several different bone cell types. So, the issue is still not resolved as to whether in fact these two hormones both activate adenylate cyclase in the same cell or whether they are activating different cell types. Available data suggest they do and that they have opposite effects on the calcium messenger system and that's why the physiologic responses come out the way they do.

# BINDING OF $Ca^{2+}$ AND $Tb^{3+}$ TO CALMODULIN*

Robert R. Aquaron and Chih-Lueh A. Wang

*Department of Muscle Research*
*Boston Biomedical Research Institute*
*Boston, Massachusetts 02114*

## INTRODUCTION AND METHODS

Calmodulin(CaM), like troponin-C(TnC), has 4 $Ca^{2+}$ binding sites. So far no general agreement has been reached with regard to the binding constants of these sites.[1] Here we present studies on the metal binding properties of CaM using both $Ca^{2+}$ and a luminescent lanthanide ion, $Tb^{3+}$, as an analog of $Ca^{2+}$. $Ca^{2+}$ binding to CaM was monitored by the change in tyrosine fluorescence($\lambda_{exc}$ = 280 nm, $\lambda_{em}$ = 300 nm), and $Tb^{3+}$ binding by its enhanced luminescence($\lambda_{exc}$ = 280 nm, $\lambda_{em}$ = 543 nm). Exept for $Ca^{2+}$ titration in a system buffered by EGTA and NTA (nitrilotriacetic acid), the protein was treated with 1 mM EDTA to remove $Ca^{2+}$ followed by dialysis against buffer containing 25 mM PIPES and 100 mM KCl, pH 7.0. Buffer solutions were treated by passage through a Chelex-100 column.

## RESULTS AND DISCUSSION

Addition of $Ca^{2+}$ to CaM results in a 3-fold enhancement of tyrosine fluorescence with a transition midpoint of 0.2 $\mu$M $Ca^{2+}$. The complete fluorescence change requires the addition of 2 $Ca^{2+}$ per CaM. In view of the location of the two tyrosine residues in the primary structure of CaM (Tyr-99 and Tyr-138), these results suggest that the spectral changes are induced by $Ca^{2+}$ binding to sites III and IV.

Addition of $Tb^{3+}$ to CaM up to 2 $Tb^{3+}$ per CaM produces only small enhancement of metal luminescence which is followed by a large enhancement between 2 and 4 $Tb^{3+}$ per CaM (FIGURE 1, A). This is in contrast to the behavior of TnC where there is no lag in the luminescence enhancement.[2] This suggests that $Tb^{3+}$ binds preferentially to sites where no energy transfer can take place, presumably sites I and II. This is supported by $Tb^{3+}$ titration of CaM preloaded with 2 mol of another lanthanide, $Eu^{3+}$; in this case $Tb^{3+}$ luminescence increases without any lag.

When CaM is initially saturated with $Ca^{2+}$, the $Ca^{2+}$-enhanced tyrosine fluorescence is unaffected by the addition of the first 2 $Tb^{3+}$, but decreases as the third and the fourth $Tb^{3+}$ ions are added (FIGURE 1, C), suggesting that the second pair of added $Tb^{3+}$ displaces $Ca^{2+}$ from sites III and IV. When no $Ca^{2+}$ is initially added to the system, however, the tyrosine fluorescence increases upon addition of the first pair of $Tb^{3+}$ (FIGURE 1, B). The simplest interpretation of these results is that $Ca^{2+}$ bound to sites I and II are displaced to sites III and IV upon addition of the first 2 $Tb^{3+}$ ions, and displaced from them by the second pair of $Tb^{3+}$. An alternate interpretation would be that $Tb^{3+}$ binding to sites I and II induces

*Supported by the National Institutes of Health and Muscular Dystrophy Association.



0077-8923/80/0356-0354 $01.75/0 © 1980, NYAS

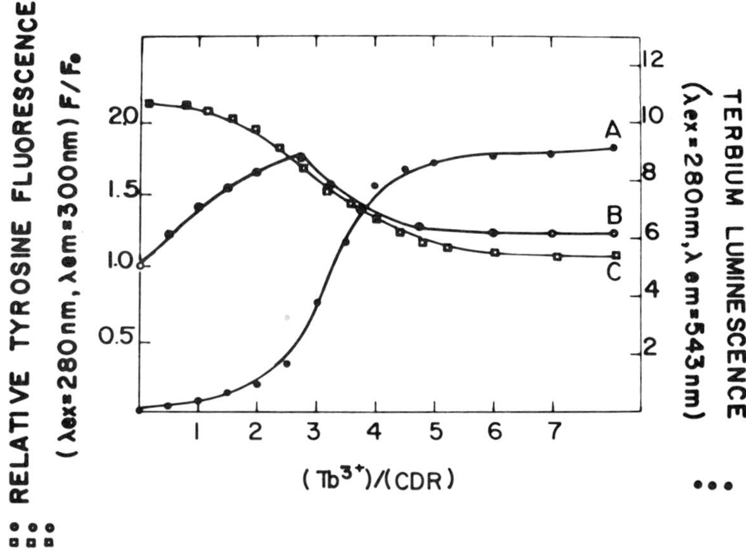


FIGURE 1. Binding of $Tb^{3+}$ to CaM monitored by $Tb^{3+}$ luminescence (A) and by tyrosine fluorescence (B and C).

conformational changes that affect the fluorescence of tyrosine residues in sites III and IV.

## CONCLUSION

Our report presents evidence that differences exist between the two homologous proteins, CaM and TnC, with respect to their metal binding properties. Sites III and IV are the high affinity sites for both $Ca^{2+}$ and $Tb^{3+}$ in TnC, whereas in CaM they are the low affinity sites for $Tb^{3+}$. The fact that $Tb^{3+}$ can discriminate between the low and the high affinity sites in CaM, may prove useful in studies of this protein.

## REFERENCES


1. WOLFF, D. J. AND C. O. BROSTROM. 1979. Adv. Cyclic Nucleotide Res. **11:** 27.
2. LEAVIS, P. C., B. NAGY, S. S. LEHRER, H. BIALKOWSKA, AND J. GERGELY. 1980. Arch. Biochem. Biophys. **200:** 17.

# CALMODULIN AND CALMODULIN-BINDING PROTEINS IN CANINE PANCREAS*

D. C. Bartelt and G. A. Scheele

*The Rockefeller University*
*New York, New York 10021*

The events that couple hormonal stimulation to the discharge of secretory protein in the exocrine pancreas have been studied in our laboratory. Using an *in vitro* system of guinea pig pancreatic lobules,[1] it has been shown that free calcium, recruited from both intracellular and extracellular stores, mediates in its entirety the secretagogue-stimulated discharge of secretory protein.[2] The role of calcium in this process may involve cellular calmodulin and calmodulin binding proteins.

We have isolated calmodulin from the canine pancreas by the following procedure: proteins contained in a postmicrosomal supernate (PMS) fraction and soluble at 60 percent saturated $(NH_4)_2SO_4$ were treated at pH 4.0 for 2 hours at 0°C. Proteins contained in the resulting precipitate were then solubilized, dialyzed against 10 mM Tris·HCl, 300 mM NaCl, 1 mM EGTA, 1 mM $\beta$-mercaptoethanol, and 0.1 mM diisopropylfluorophosphate, and separated by gradient elution (350–600 mM NaCl) ion exchange chromatography on DEAE Sephadex. Effluent containing calmodulin was chromatographed on 2-Cl-10-(3 aminopropyl)-phenothiazine (CAPP) Sepharose. Calmodulin that was eluted from CAPP-Sepharose in buffer containing 200 mM NaCl and 2 mM EGTA, migrated as a single band during electrophoresis in both nondenaturing and denaturing (SDS) polyacrylamide gels. In addition canine pancreatic calmodulin comigrated with calmodulin isolated from bovine brain tissue in both electrophoretic systems and showed the same calcium-dependent shift in mobility during SDS polyacrylamide gel electrophoresis.

Canine pancreatic calmodulin stimulated the activity of a partially purified preparation of bovine brain phosphodiesterase to the same extent as bovine brain calmodulin. The specific activity of both proteins was calculated to be 138 units/mg. (One unit is defined as the amount of calmodulin required for half-maximal stimulation of bovine brain phosphodiesterase based on an amount of phosphodiesterase which, when maximally stimulated by calmodulin, gives 1.0 $\mu$mol/min phosphodiesterase activity.)

The presence of calmodulin-binding proteins in a postmicrosomal supernatant fraction of canine pancreas was detected using Calmodulin-Sepharose as an affinity probe. Proteins precipitated from the PMS between 30 and 60 percent saturation of $(NH_4)_2SO_4$ were subjected to equilibrium ion-exchange chromatography on DEAE Sephadex to remove free calmodulin. Proteins not bound to the resin were chromatographed on Calmodulin-Sepharose in buffer containing 100 mM KCl and 1 mM $CaCl_2$ at pH 7.8. After washing the column with buffer containing 300 mM KCl and 1 mM $CaCl_2$, those proteins bound to the Calmodulin-Sepharose in a calcium-dependent manner were eluted in buffer containing 300 mM KCl and 2 mM EGTA. Among those proteins eluted in 2 mM EGTA,

*Supported by a grant from the Cystic Fibrosis Foundation.



0077-8923/80/0356-0356$01.75/0 © 1980, NYAS

were five major species as judged by electrophoresis on SDS polyacrylamide gels. The apparent molecular weights of these proteins were 75,000, 36,000, 35,500, 14,000, and 12,500. Current research is directed toward the identification of these proteins by their biological activity.

## References

1. Scheele, G. A. & G. E. Palade. 1975. J. Biol. Chem. **250:** 2660–2670.
2. Scheele, G. & A. Haymovits. 1979. J. Biol. Chem. **254:** 10346–10353.

# CALMODULIN FROM *DICTYOSTELIUM DISCOIDEUM**

Wendy Bazari and Margaret Clarke

*Department of Molecular Biology*
*Albert Einstein College of Medicine*
*Bronx, New York 10461*

We are investigating the molecular basis of motility in the eukaryotic microorganism *Dictyostelium discoideum*. Calmodulin has been implicated in the regulation of both actin-myosin interaction and cyclic nucleotide metabolism in a variety of eukaryotic cells. Since motile amoebae of *Dictyostelium* contain actin and myosin similar to that of higher organisms, and the cells carry out an aggregation response during development that is dependent on cyclic AMP signals, we examined the possibility that calmodulin might play similar roles in this system. We found that heated extracts of *Dictyostelium* amoebae fully activated brain cyclic nucleotide phosphodiesterase in a $Ca^{2+}$-dependent reaction. Using this activity as our assay, we purified a protein that shares many properties with mammalian calmodulins (Clarke, Bazari & Kayman. 1980. *J. Bact.* **141:** 397). More detailed characterization has shown that this protein also has several unique features.

We have compared *Dictyostelium* calmodulin to calmodulin purified from bovine brain. Both are heat-stable proteins similar in size and shape, as indicated by chromatography on gel filtration columns. They each require high salt (0.3 M KCl) for elution from a DEAE-cellulose column. Their isoelectric points are virtually identical (pI = 4.3). The *Dictyostelium* protein is capable of fully activating brain cyclic nucleotide phosphodiesterase, although three times more *Dictyostelium* than brain calmodulin is needed to provide the same level of activation. This activation is completely inhibited by the addition of 30 $\mu$M trifluoperazine; sensitivity to trifluoperazine appears to be identical for the two proteins. Both proteins exhibit a shift in mobility on SDS polyacrylamide gels depending on whether $Ca^{2+}$ or EGTA is present in the sample buffer; in the presence of $Ca^{2+}$ they comigrate. All of these properties suggest close structural and functional resemblances between the two proteins, and led to our classification of the *Dictyostelium* protein as a calmodulin.

Despite these similarities, *Dictyostelium* calmodulin has several properties that are different from any previously described calmodulin. *Dictyostelium* calmodulin lacks the amino acid trimethyllysine, as well as tryptophan and cysteine. Its tyrosine content is higher than that of brain, yielding different spectral properties ($\epsilon^{1\%}_{280nm} = 4.3$). The content of glutamic acid and glutamine is also remarkably high; these amino acids make up 30% of the residues. The molecular weight of *Dictyostelium* calmodulin, calculated from its amino acid composition, is about 14,500; this figure is consistent with the relative mobilities of *Dictyostelium* and brain calmodulin on SDS gels in the absence of $Ca^{2+}$.

We have also compared the two proteins immunologically, using anti-serum raised against bovine brain calmodulin by Dr. Tom Sturgill (see Speaker *et al.*,

*Supported by grants from the American Cancer Society (VC-274) and the American Heart Association (77-205). M.C. is an Established Investigator of the American Heart Association.



0077-8923/80/0356-0358 $01.75/0 © 1980, NYAS

this volume). In a competition radioimmune assay *Dictyostelium* calmodulin clearly cross-reacts with antibodies to the brain protein, substantiating the basic similarity of the two proteins. However, it exhibits a significantly lower affinity for most of the antibody population, and apparently lacks altogether a small number of antigenic determinants of brain calmodulin.

It is particularly interesting that *Dictyostelium* calmodulin appears so different from brain calmodulin at the level of primary structure, yet retains sufficient functional similarity to activate brain phosphodiesterase and to cross-react with antibodies to brain calmodulin. The unusual features of *Dictyostelium* calmodulin, such as the lack of trimethyllysine, offer a unique opportunity for studying structure-function relationships of calmodulins.

# ACTIVATION OF PROKARYOTIC ADENYLATE CYCLASE BY CALMODULIN

S. A. Berkowitz, A. R. Goldhammer, E. L. Hewlett, and J. Wolff

*National Institute of Arthritis, Metabolism, and Digestive Diseases*
*Bethesda, Maryland 20205*

Adenylate cyclase of *Bordetella pertussis* organisms has an approximate molecular weight of 70,000, and differs from bacterial and eukaryotic adenylate cyclases in that it is not regulated by nucleoside triphosphates, is unaffected by $\alpha$-ketoacids or glucose, and is inhibited by fluoride. Of the enzyme associated with intact bacteria, a large proportion is extracytoplasmic, while only approximately 10% of the activity resides in an intracellular compartment. In a search for alternative regulatory systems for the enzyme, we found that this cyclase is stimulated 100–1,000-fold in a dose-dependent manner by calf brain calmodulin. Half-maximal activation occurs at ~ 10 nM calmodulin, and significant stimulation can be observed at 30 pM. The system has the following properties: 1) The activation is prevented by EGTA and restored by calmodulin. However, once activated by calmodulin, the stimulation is poorly reversed by EGTA. 2) Oxidation of the methionine residues of calmodulin abolishes the ability to activate the cyclase. 3) Trifluoperazine inhibits calmodulin-activated cyclase. 4) Troponin C preparations stimulate the *B. pertussis* cyclase $<$ 0.01 times the potency of calmodulin. This is the first example of a calmodulin effect in a prokaryote.

We had previously shown that the increased activity of this cyclase in organisms grown on blood agar is due to the presence of an activator in the blood. Activators were also found in diverse purified protein preparations from different sources including rabbit erythrocyte lysates, catalase, peroxidase, creatine phophokinase, and lima bean trypsin inhibitor. These activators have in common an absolute $Ca^{2+}$-dependence, inhibition by trifluoperazine and a remarkable heat stability. In addition, they can be shown to be activators of cyclic nucleotide phosphodiesterase. Gel filtration experiments of samples boiled prior to the application to the column have shown that these activators are eluted in a common position close to calmodulin. Since calmodulin enjoys a very wide distribution in eukaryotic tissues, we propose that a number of the activators are, in fact, calmodulin. These amounts of calmodulin contaminating other protein preparations are sufficient to obscure calmodulin requirements when they are added to reaction mixtures, e.g. creatine kinase as an ATP-regenerating system in adenylate cyclase assays. Finally, the high sensitivity of the *B. pertussis* system makes it useful for assay purposes.



0077-8923/80/0356-0360 $01.75/0 © 1980, NYAS

# CALMODULIN IN AXONAL TRANSPORT

Scott T. Brady, Michael Tytell, Kirk Heriot,
and Raymond J. Lasek

*Department of Anatomy*
*Case Western Reserve University*
*Cleveland, Ohio 44106*

Calmodulin is a soluble, heat-stable protein that has been shown to modulate the activities of both membrane-bound and soluble enzymes,[1] but relatively little is known about its intracellular transport and *in vivo* distribution in the neuron. The wide variety of activities suggested for calmodulin make its intracellular localization of particular importance. The unique morphology and requirements of the neuron permit analysis of the specific intracellular transport processes of the axon.[2,3] Subcellular fractionation and identification of some of the proteins associated with the major rate components of axonal transport have led to the suggestion that each rate component constitutes a distinct cellular structural entity.[4,5] Assignment of a protein to a specific rate component constitutes identification of a long term association of that protein with the other proteins of that rate component. Thus analysis of the axonal transport of a protein represents a valuable complement to traditional methods of cytological localization providing evidence of both physical and functional association. Transported proteins were pulse-labeled by injection of [$^{35}$S]methionine into the posterior chamber of guinea pig eyes and the specific rate components of axonal transport isolated in the optic nerve by sacrifice of the animal at appropriate time intervals postinjection. A 17,000 dalton heat-stable protein was found to move in the axon as part of the complex of proteins known as Slow Component b (SCb; 2–4 mm/day) which includes actin[6] and soluble enzymes of intermediary metabolism. This heat-stable labeled protein comigrated with authentic bovine brain calmodulin (a gift of Drs. D. Jemiolo & W. Burgess, University of Virginia) on one-dimensional gel electrophoresis (PAGE) and both labeled and stained proteins migrate anomalously on PAGE in the presence of EGTA. Upon two-dimensional electrophoresis (modified from O'Farrell[7]) of brain purified calmodulin with radioactively labeled SCb, it was found that the heat-stable SCb protein was identical to calmodulin in pI, SDS molecular weight, and spot morphology. In two-dimensional PAGE of the other two major rate components of axonal transport, no radioactivity could be detected comigrating with calmodulin in either Fast Component (FC: 200–400 mm/day, membrane associated proteins) or Slow Component a (SCa; 0.2–1 mm/day, microtubule-neurofilament network). The methods used would have detected calmodulin at the level of 0.002% of the total material in FC and 0.01% of the total material in SCa. We conclude that calmodulin is transported down the axon solely as part of the SCb complex of proteins—the axoplasmic matrix. Previous suggestions that calmodulin was moved in FC[8] were based on inadequate biochemical analysis which would not provide unequivocal identification of the FC calcium-binding protein. Erickson *et al.*,[9] using a different approach, also find little or no calmodulin in FC. The identification of a small molecular weight calcium binding protein in FC remains of interest, since it may provide an insight into the functions of other neuronal calcium binding proteins such as calcineurin. The primary long term association



0077-8923/80/0356-0361 $01.75/0 © 1980, NYAS

of calmodulin in the axon appears therefore to be with the actin-containing cytoskeletal network of the axon. It moves independently of the movements of the microtubules (SCa) and of membranes (FC). The absence of long term associations of calmodulin with these structures suggests that the effects of calmodulin on proteins associated with FC or SCa in the axon may be due to short term or transient interactions between groups of proteins moving in axonal transport at different rates.

## References


1. Cheung, W. Y. 1979. Science **207:** 19.
2. Willard, M., W. M. Cowan & P. R. Vagelos. 1974. Proc. Natl. Acad. Sci. USA **71:** 2183.
3. Hoffman, P. & R. J. Lasek. 1975. J. Cell. Biol. **66:** 351.
4. Black, M. M. & R. J. Lasek. 1980. J. Cell Biol. 1980. **86:** 616.
5. Lasek, R. J. 1980. Trends Neurosci. **3:** 87.
6. Black, M. M. & R. J. Lasek. 1979. Brain Res. **171:** 401.
7. O'Farrell, P. 1975. J. Biol. Chem. **250:** 4007.
8. Iqbal, Z. & S. Ochs. 1978. J. Neurochem. **31:** 409.
9. Erickson, P., K. B. Seamon, B. W. Moore, R. S. Lasher & L. N. Minier. 1980. J. Neurochem. **35:** 242.

# CALMODULIN REGULATION OF $^{45}Ca^{2+}$ UPTAKE BY SUBCELLULAR FRACTIONS IN THE NEUROHYPOPHYSIS

Arthur D. Conigrave,* Marek Treiman, Torben Særmark, and Niels A. Thorn

*Institute of Medical Physiology C*
*University of Copenhagen*
*DK 2200 Copenhagen N*
*Denmark*

In the neurohypophysis $Ca^{2+}$- and $Mg^{2+}$-dependent ATPases are thought to regulate intracellular free calcium concentration. "Microsomes," mitochondria, and coated microvesicles contain $(Ca^{2+}+Mg^{2+})$-ATPase activity and take up calcium by ATP-dependent processes.[1,2] However, the integration of the function of these different $Ca^{2+}$-sequestering organelles remains an unsolved problem. Calmodulin has been isolated from the neurohypophysis.[3] This protein could act as a regulator of the relative contribution of the different $Ca^{2+}$-sequestering organelles. We have studied the effect of calmodulin upon the ATP-dependent $Ca^{2+}$ accumulation by a microsome fraction, a plasma membrane-enriched fraction, and by coated microvesicles.

Microsomes were assayed for $Ca^{2+}$ uptake and $Ca^{2+}$-dependent ATPase activity in the presence of ouabain. They showed $Ca^{2+}$ uptake in the presence of ATP and $Mg^{2+}$ (8 nmol per mg of membrane protein ± 2 (mean ± SEM, 4 experiments) at a free $Ca^{2+}$ concentration of $10^{-6}$ M). The maximal amount of $Ca^{2+}$ taken up in the presence of calmodulin was 14 nmoles per mg of membrane protein ± 1.5 (4 experiments) at a free $Ca^{2+}$ concentration of $10^{-6}$ M. The dose for half-maximal stimulation was 20 $\mu$g of calmodulin per mg of membrane protein at a free $Ca^{2+}$ concentration of $10^{-7}$ M. Total ATPase activity was increased by 20%.

A plasma membrane-enriched fraction was prepared from secretosomes on the basis of the method of Vilhardt *et al.*[4] Isolated secretosomes were lysed in a buffer containing 5 mM 2-{[2-hydroxy-1,1-bis(hydroxymethyl)ethyl]amino} ethanesulfonic acid (TES) and 5 mM EGTA, pH 8.0. After lysis they were placed on top of a sucrose gradient containing 0.6 M sucrose, 20 mM TES, pH 7.0 and 0.95 M sucrose, TES 20 mM, pH 7.0. After centrifugation for 90 min at 62000 $g_{av}$, the plasma membrane-enriched fraction was recovered from the interphase. The membranes were diluted 4 times in 20 mM TES, 5 mM EDTA, 5 mM EGTA, pH 7.0 and centrifuged at 140,000 $g_{av}$ for 30 min. The pellet was resuspended in 130 mM KCl, 20 mM TES, pH 7.0. The $Ca^{2+}$ uptake in this preparation was comparable to that of the microsome fraction. $Ca^{2+}$ uptake by this fraction was stimulated by a factor of two to three by calmodulin. This was not due to an effect on $(Na^{+}+K^{+})$-ATPase since it was unaffected by addition of ouabain. The activation by calmodulin was maximal at $10^{-6}$ M free $Ca^{2+}$. The effect on uptake could be correlated to $(Ca^{2+}+Mg^{2+})$-ATPase activation by calmodulin.

Microvesicles were purified as described[2] from both neurohypophysis and

*Present address: Department of Physiology, University of Sydney, New South Wales 2006, Australia.



0077-8923/80/0356-0363 $01.75/0 © 1980, NYAS

brain cortex. Neither $Ca^{2+}$-dependent ATPase activity nor $Ca^{2+}$ uptake was influenced by calmodulin (max. conc. 100 $\mu$g per mg protein) in any of the tissues.

In conclusion, calmodulin was found to have a differential effect on calcium uptake in the three preparations tested. It increased calcium uptake in the microsomal and plasma membrane-enriched fraction in a $Ca^{2+}$-dependent way, but had no effect on the uptake of $Ca^{2+}$ into the microvesicles. A correlation seems to exist between calcium uptake and $Ca^{2+}$-dependent ATPase activity in the plasma membrane fraction. Calmodulin had $Ca^{2+}$-dependent effects on these parameters.

## REFERENCES


1. THORN, N. A., J. T. RUSSELL, C. TORP-PEDERSEN & M. TREIMAN. 1978. Ann. N.Y. Acad. Sci. **307:** 618–639.
2. TORP-PEDERSEN, C., T. SÆRMARK, N. A. THORN & M. BUNDGAARD. 1980. J. Neurochem. **35:** 552–557.
3. RUSSELL, J. T. & N. A. THORN. 1977. Biochim. Biophys. Acta **491:** 398–408.
4. VILHARDT, H., R. BAKER & D. B. HOPE. 1975. Biochem. J. **148:** 57–65.

# CRYSTALLOGRAPHIC STUDIES OF CALMODULIN

William J. Cook* and Charles E. Bugg†

*Department of Pathology, †Department of Biochemistry, Comprehensive Cancer Center, and Institute of Dental Research
University of Alabama in Birmingham
Birmingham, Alabama 35294

John R. Dedman and Anthony R. Means

Department of Cell Biology Baylor College of Medicine
Houston, Texas 77030

We are currently working on the high-resolution crystal structure of calmodulin.

Calmodulin from rat testes was isolated and purified as described by Dedman *et al.*[1] Crystals suitable for X-ray analysis were obtained by vapor diffusion equilibration of 10 $\mu$l droplets hanging from siliconized coverslips inverted on Linbro plates. The droplets consisted of 6 $\mu$l of a solution containing 15 mg of calmodulin/ml of distilled water and 5.0 mM $Ca^{2+}$, plus 4 $\mu$l of a solution containing 55% (vol/vol) 2-methyl-2,4-pentanediol (MPD) in 0.05 M cacodylate buffer (pH 6.0). These droplets were equilibrated against 1 ml of a solution containing 55% (vol/vol) MPD in 0.05 M cacodylate buffer (pH 6.0). After 6–10 days at 4°C platelike crystals with dimensions up to 0.1 × 0.1 × 0.8 mm were obtained (FIGURE 1).

For X-ray studies, crystals were transferred to a stabilizing solution of 65% (vol/vol) MPD in 0.05 M cacodylate buffer (pH 6.0) and mounted in glass capillaries. Crystals were photographed on a precession camera at 22°C using nickel-filtered copper radiation from a Rigaku RU-200 rotating anode generator. X-ray precession photographs were taken at multiple settings; these photographs indicated that the crystals are triclinic (FIGURE 2). This symmetry was corrob-

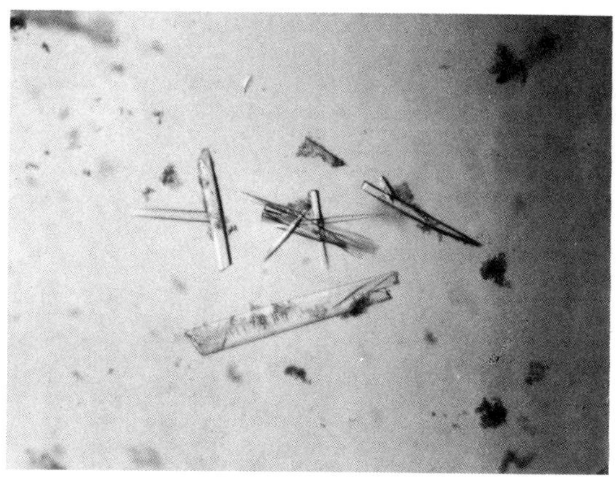

FIGURE 1. Crystals of calmodulin.



0077-8923/80/0356-0365 $01.75/0 © 1980, NYAS

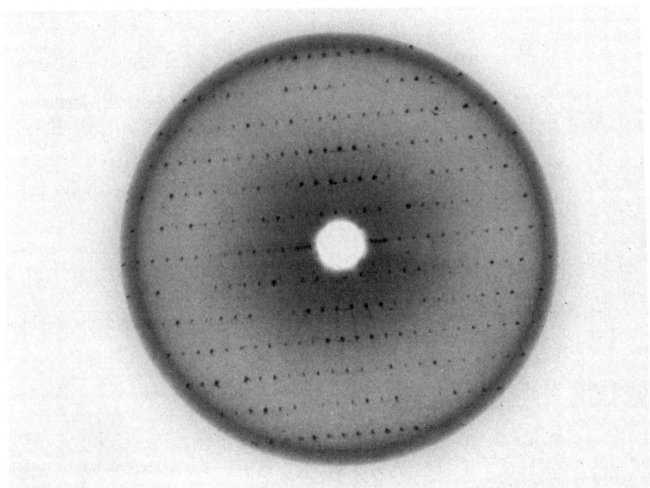

FIGURE 2. X-ray precession photograph of calmodulin. 0k*l* zone, $\bar{\mu}$= 12°.

orated by diffractometer measurements on a Picker FACS-1 diffractometer. Zero level and cone-axis photographs at two different settings revealed a unit cell volume of 39,300 Å$^3$. Thus, the crystals are triclinic, space group P1, with one calmodulin molecule per unit cell. Unit cell parameters for the reduced cell, which were obtained by least-squares refinement of the angular settings for twelve reflections measured on the diffractometer, are $a$ = 29.79(4) Å, $b$ = 53.74(7) Å, $c$ = 24.78(3) Å, $\alpha$ = 93.46(2)°, $\beta$ = 96.98(2)° and $\gamma$ = 89.05(3)°.

The crystals are stable to X-rays at room temperature for at least 10 days and diffract strongly to at least 2.3 Å resolution. Assuming an occupancy of one 16,700 dalton chain per crystallographic asymmetric unit and a partial-specific-volume of 0.74 ml/g, $V_m$ = 2.35 Å$^3$/dalton,[2] and the solvent volume fraction is 48%; these values are within the ranges normally observed for protein crystals. Values calculated assuming two or more molecules per unit cell are far outside those ranges.[2]

Efforts to obtain isomorphous heavy-atom derivatives have indicated several promising candidates.

## REFERENCES

1. DEDMAN, J. R., J. D. POTTER, R. L. JACKSON, J. D. JOHNSON & A. R. MEANS. 1977. J. Biol. Chem. **252:** 8415–8422.
2. MATTHEWS, B. M. 1968. J. Mol Biol. **33:** 491–497.

# EFFECT OF TRIFLUOPERAZINE ON CYCLIC NUCLEOTIDE PHOSPHODIESTERASE ACTIVITY IN CRUDE HOMOGENATES OF BOVINE ADRENAL MEDULLA*

Albert A. Crimaldi, Ingrid C. Y. Kuo, and Carole J. Coffee

*Department of Biochemistry,*
*School of Medicine,*
*University of Pittsburgh,*
*Pittsburgh, Pennsylvania 15261*

The calmodulin regulated phosphodiesterase present in the adrenal medulla has characteristics that are different from those seen for the equivalent enzyme found in other tissues, such as brain and heart. In crude homogenates, as well as in preliminary purification steps, the addition of the calcium chelator EGTA has no effect on the total phosphodiesterase activity. In addition, EGTA does not cause the dissociation of the calmodulin-phosphodiesterase complex. However, using repeated ion-exchange chromatography on DEAE-cellulose, we have been able to partially purify a phosphodiesterase that is totally free of endogenous calmodulin. This phosphodiesterase, which comprises approximately 40% of the total phosphodiesterase activity, can be stimulated by the addition of calcium and calmodulin in a manner identical to that described for the enzyme from other tissues, including the demonstration that the activation and complex formation is completely dependent on calcium and is reversed by EGTA. Other properties of this enzyme include a $K_m$ of 260 $\mu$M for cAMP and 120 $\mu$M for cGMP. It is stimulated to the same extent and with the same specific activity by calmodulin isolated from either brain or adrenal medulla. The calcium concentration required for half-maximal stimulation is 2 $\mu$M.

Using the calmodulin binding drug trifluoperazine, we have been able to directly demonstrate the presence and quantitate the amount of calmodulin-regulated phosphodiesterase in adrenal medulla crude homogenate. Addition of the drug inhibits approximately 90% of the total enzyme activity with an apparent $K_I$ of 0.33 mM. Inhibition was observed only in the presence of calcium. In the absence of calcium, addition of trifluoperazine in concentrations up to 2.5 mM had no effect on the total phosphodiesterase activity. We have also found that the total phosphodiesterase activity is sensitive to fluphenazine (apparent $K_I$ of 0.7 mM) whereas the sensitivity to both promethazine (apparent $K_I$ of 4 mM) and trifluoperazine sulfoxide (apparent $K_I > 10$ mM) is markedly reduced. The addition of calmodulin in an amount equivalent to four times that present in the crude homogenates was not effective in either decreasing the extent of inhibition or increasing the concentration of trifluoperazine required for inhibition. When the DEAE-purified enzyme is assayed in the presence of trifluoperazine, the stimulation by calmodulin is totally removed (apparent $K_I = 10$ $\mu$M), although in the absence of calmodulin the nonstimulated phosphodiesterase activity is unaffected by trifluoperazine. These results indicate that the inhibition observed

*Supported by National Institutes of Health Grant AM 16728.



0077-8923/80/0356-0367 $01.75/0 © 1980, NYAS

in the crude homogenates is due to elimination of calmodulin stimulation and not due to a direct inhibition of the phosphodiesterase enzyme.

These results suggest that most of the phosphodiesterase activity present in the adrenal medulla is activated by calmodulin and that the endogenous enzyme-calmodulin complex is different from that found in other tissues. The complex found in the adrenal medulla is apparently regulated by factors either independent of or in addition to calcium.

# EFFECTS OF $Ca^{2+}$ AND CALMODULIN ON ENZYMES OF CYCLIC NUCLEOTIDE METABOLISM IN OX NEUROHYPOPHYSEAL SECRETOSOMES

Darlene A. Dartt,* Christian Torp-Pedersen, and Niels A. Thorn

*Institute of Medical Physiology C*
*University of Copenhagen*
*DK 2200 Copenhagen N*
*Denmark*

An increase in the intracellular ionized calcium concentration is essential for release of vasopressin from the neurohypophysis. Therefore, to determine whether $Ca^{2+}$ in conjunction with calmodulin could affect neurohypophyseal secretion by regulating the enzymes that metabolize cAMP and cGMP, we measured adenylate cyclase, guanylate cyclase, cAMP phosphodiesterase, and cGMP phosphodiesterase activities in isolated nerve endings (neurosecretosomes) prepared from ox neurohypophyses.

Neurosecretosomes were prepared by mild homogenization and differential centrifugation and lysed by sonication before use. Lysed neurosecretosomes were either used directly or centrifuged at 100,000 × g for 20 min to form a soluble (supernatant) and a membrane (pellet) fraction. Adenylate cyclase activity was determined by incubation of neurosecretosomes or the neurosecretosomal fractions in the presence of EDTA, EGTA, theophylline, IBMX, ATP, and varying $Ca^{2+}$ and $Mg^{2+}$ concentrations. The amount of cAMP formed during the assay was measured by radiolabel displacement from cAMP-binding protein. Guanylate cyclase activity was determined in the presence of EDTA, EGTA, IBMX, GTP, and varying $Ca^{2+}$ and $Mn^{2+}$ concentrations. The amount of cGMP formed during the assay was measured by radioimmunoassay using a Radiochemical Centre (Amersham) kit. The cAMP phosphodiesterase assay was performed in a medium containing EDTA, EGTA, [$^3$H]cAMP, and varying $Ca^{2+}$ and $Mg^{2+}$ concentrations. The assay of cGMP phophodiesterase activity was performed in the same medium as for cAMP phosphodiesterase, except [$^3$H]cGMP replaced [$^3$H]cAMP. The amount of cAMP or cGMP metabolites formed during the assays was measured by thin-layer chromatography on polyethylene-imine sheets.

In the presence of 10 mM $MgCl_2$, 78% of the adenylate cyclase activity was found in the neurosecretosomal membrane fraction. Similarly, with 10 mM $MnCl_2$, 73% of the guanylate cyclase activity was localized to the membrane fraction. Using the membrane fraction, it was shown that in the presence of 3 mM $Mg^{2+}$ increasing the $Ca^{2+}$ concentration from less than $10^{-8}$ M to $10^{-5}$ M, inhibited adenylate cyclase activity. The addition of calmodulin, however, stimulated adenylate cyclase activity 4-fold and 3-fold in the presence of $10^{-7}$ M and $10^{-5}$ M $Ca^{2+}$, respectively. On the other hand, neither varying the $Ca^{2+}$ concentration nor the addition of calmodulin affected the guanylate cyclase activity of the neurosecretosomal membrane fraction. cAMP and cGMP phosphodiesterase activities were measured on both the soluble and membrane

*Present Address: Department of Physiology, Tufts University School of Medicine, Boston, Mass. 02111.



0077-8923/80/0356-0369 $01.75/0 © 1980, NYAS

fractions. In the soluble fraction, both of these enzyme activities were stimulated by increasing the $Ca^{2+}$ concentration from less than $10^{-8}$ M to $10^{-7}$ M or $10^{-5}$ M, but only at $10^{-7}$ M $Ca^{2+}$ were either of the phosphodiesterase activities stimulated by calmodulin. When the membrane fraction was assayed, neither varying the $Ca^{2+}$ concentration from less than $10^{-8}$ M to $10^{-5}$ M nor the presence of calmodulin affected either the cAMP or cGMP phosphodiesterase activities. No attempt, however, was made to distinguish between the several phosphodiesterase isoenzyme activities present in the neurohypophysis.

From the present study it was concluded that calcium ions, a known stimulus of neurohypophyseal secretion, in conjunction with calmodulin, affect the activity of the enzymes that control cAMP metabolism (adenylate cyclase and cAMP phosphodiesterase) and the enzyme that controls cGMP degradation (cGMP phosphodiesterase), without affecting the enzyme that controls cGMP synthesis (guanylate cyclase). It is important to remember, however, that effects of calcium ions and calmodulin *in vivo* may be modified by different cellular locations of enzymes and activation factors as well as differences in the time course of events.

# CALMODULIN-ACTIVATED PLANT MICROSOMAL $Ca^{2+}$ UPTAKE AND PURIFICATION OF PLANT NAD KINASE AND OTHER PROTEINS BY CALMODULIN-SEPHAROSE CHROMATOGRAPHY

Peter Dieter and Dieter Marmé

*Institut für Biologie III*
*D-78 Freiburg*
*Federal Republic of Germany*

Calmodulin from animal (bovine brain) and plant (zucchini squash) sources is partially purified using DEAE-cellulose chromatography. The ATP-dependent $Ca^{2+}$ uptake into a microsomal fraction from plants (zucchini squash) is stimulated by the plant and animal calmodulin (FIGURE 1). This stimulation occurs only when the microsomes are prepared in the presence of EDTA. This is probably due to the removal of endogenous calmodulin from the microsomal fraction. The fact that EDTA-extracted microsomes exhibit much less $Ca^{2+}$ uptake activity than the microsomes extracted without EDTA supports this assumption. The calmodulin-mediated activation of microsomal $Ca^{2+}$ uptake is completely abolished by the antipsychotic drug fluphenazine, while the calmodulin-independent $Ca^{2+}$

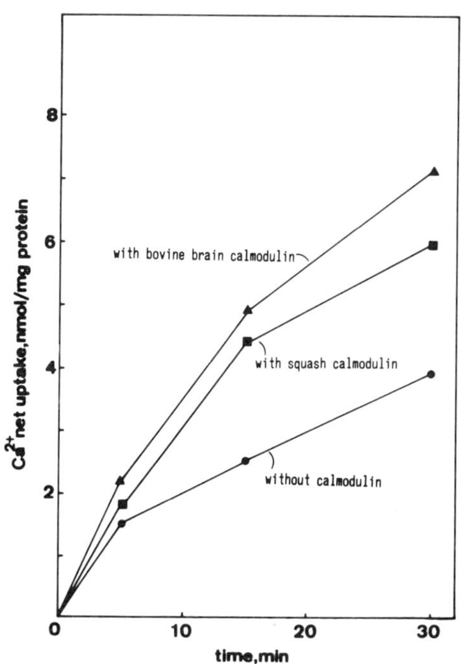


FIGURE 1. Calmodulin-dependent $Ca^{2+}$ net uptake.



0077-8923/80/0356-0371 $01.75/0 © 1980, NYAS

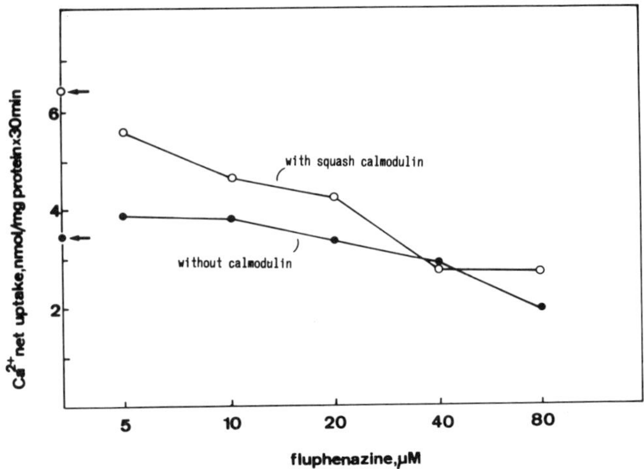


FIGURE 2. Fluphenazine-dependent $Ca^{2+}$ net uptake in the absence and presence of calmodulin.

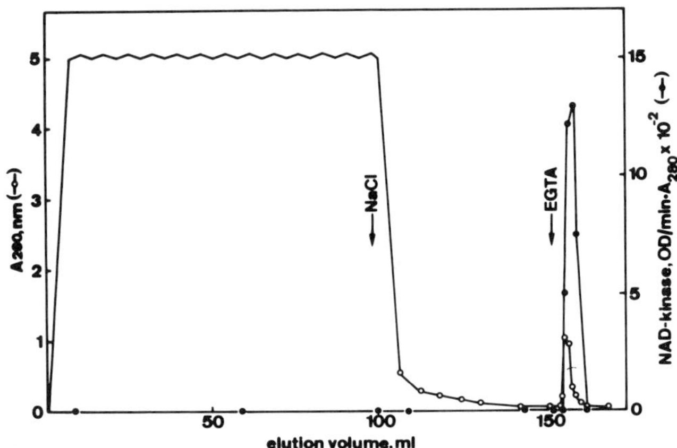


FIGURE 3. Protein and NAD kinase activity elution profiles of a calmodulin-Sepharose chromatography affinity column.

TABLE 1

PURIFICATION OF NAD KINASE BY CALMODULIN-SEPHAROSE AFFINITY COLUMN CHROMATOGRAPHY

| | | NAD-Kinase Activity | | Phosphodiesterase Activity | |
|---|---|---|---|---|---|
| Fraction | Protein (mg) | Total (munits) | Specific (munits/mg protein) | Total (munits) | Specific (munits/mg protein) |
| 50K supernatant | 176.0 | 616.0 | 3.5 | 546.0 | 3.1 |
| NaCl eluate | 140.0 | 55.0 | 0.4 | 414.0 | 3.2 |
| EGTA eluate | 1.9 | 481.0 | 253.0 | 0.4 | 0.25 |

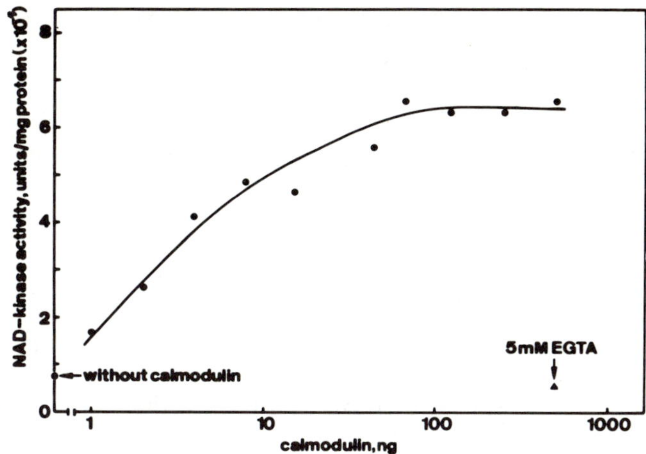


FIGURE 4. Calmodulin activation of plant NAD kinase.

uptake is only slightly decreased (FIGURE 2). Mitochondrial $Ca^{2+}$ accumulation is not affected by calmodulin. Plant calmodulin also activates the bovine brain cyclic nucleotide phosphodiesterase.

Calmodulin from bovine brain has been purified to apparent electrophoretic homogeneity and was coupled to CNBr-activated Sepharose 4B. A crude homogenate prepared from dark-grown squash was loaded onto a calmodulin-Sepharose column in the presence of 0.5 mM $Ca^{2+}$. After washing with 0.5 M NaCl the proteins were eluted with 5 mM EGTA (FIGURE 3). Calmodulin-dependent plant NAD kinase was purified by this procedure at least 70-fold (TABLE 1). Cyclic nucleotide phosphodiesterase from plants is not retained by the calmodulin-Sepharose affinity column (TABLE 1). The apparent molecular weight of the plant NAD kinase is about 50,000 as estimated by Sephadex G-100 chromatography of the EGTA eluate. Addition of 5 ng calmodulin causes 50% stimulation of the enzyme activity (FIGURE 4). SDS polyacrylamide gel electrophoresis of the proteins from the EGTA eluate exhibited at least 7–8 bands that were much intensified on the gel as compared to the protein pattern on the gel of the original homogenate. The identification of those proteins needs further investigation.

# STUDIES ON STRUCTURE AND FUNCTION OF CALMODULIN

W. Drabikowski, H. Brzeska, J. Kuźnicki, and Z. Grabarek

*Department of Biochemistry of Nervous System and Muscle*
*Nencki Institute of Experimental Biology*
*Warsaw, Poland*

Calmodulin contains four $Ca^{2+}$ binding regions, each consisting of two $\alpha$-helical fragments and a $Ca^{2+}$ binding loop in between. In this work the $Ca^{2+}$ induced conformational changes, binding of $Ca^{2+}$ and $Mg^{2+}$, as well as biological function of calmodulin, have been studied with the use of various fragments of calmodulin (called TR-Ca and TR-EDTA when obtained with trypsin in the presence of $Ca^{2+}$ or EDTA, respectively) and consisting of different number of $Ca^{2+}$ binding regions.[1,2]

The present studies included measurement of changes of circular dichroism and intrinsic fluorescence intensity and polarization, caused by the absence or presence of $Ca^{2+}$ and $Mg^{2+}$.

The $TR_2$-Ca peptide shows large enhancement of tyrosine fluorescence intensity upon binding of $Ca^{2+}$, similar to that of the intact molecule. This change is much smaller if the two $Ca^{2+}$ binding regions are separated as a result of cleavage.[3]

Similarly, the removal of $Ca^{2+}$, has only very slight effect on the fluorescence polarization of the whole calmodulin and its $TR_2$-Ca peptide, but it significantly diminishes fluorescence polarization of the $TR_1$-EDTA and $TR_3$-EDTA peptides.[3]

All calmodulin fragments show an increase of ellipticity in the presence of $Ca^{2+}$. $Mg^{2+}$ increases the ellipticity to a smaller extent than $Ca^{2+}$. The CD spectra of both halves of calmodulin ($TR_1$-Ca and $TR_2$-Ca peptides) are additive under all conditions studied.[4]

NMR data indicate that both $TR_1$-Ca and $TR_2$-Ca peptides bind two ions of $Ca^{2+}$. These $Ca^{2+}$ ions can be competitively replaced by two ions of either $Mg^{2+}$ or $Na^+$.

The transition midpoints of the $Ca^{2+}$-induced changes obtained by different methods (fluorescence, NMR, and CD) are of the order of $10^{-7}$–$10^{-6}$ M for all calmodulin peptides, except $TR_3$-EDTA.

In the absence of urea the mobility in PAGE of calmodulin and all of its peptides is higher in the absence of $Ca^{2+}$ than in its presence.[6]

The analysis of $Ca^{2+}$-dependent changes of ellipticity, mobility in PAGE, and the competition between binding of $^{23}Na^+$ and $Ca^{2+}$ shows that the difference between N- and C-terminal halves of calmodulin is much smaller than that in case of troponin C. Both halves of calmodulin molecule are similar to the N-terminal half of troponin C.[4,7,8]

Calmodulin from phylogenically distant species (bovine brain and *Physarum plasmodia*) show biological cross-reactivity in respect to phosphodiesterase activity and substitution for troponin C in the regulatory system of skeletal muscle containing tropomyosin and other subunits of troponin.[9,10]

Contrary to troponin C peptides, however, only the $TR_2$-Ca calmodulin



0077-8923/80/0356-0374 $01.75/0 © 1980, NYAS

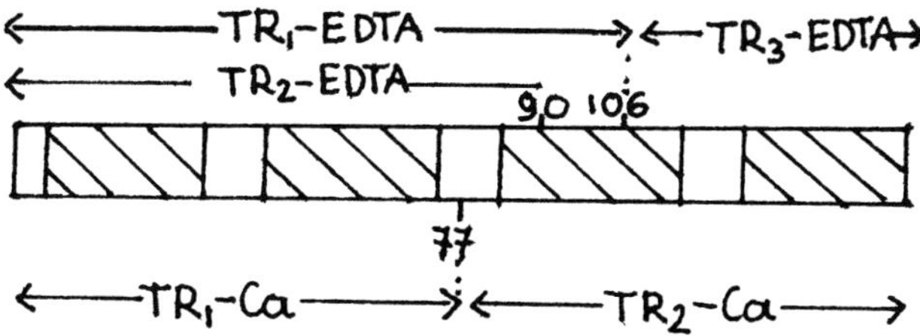


peptide shows significant binding to troponin I in the urea-PAGE,[1] and can only partially substitute for intact calmodulin in the reconstituted actomyosin system.

None of the calmodulin peptides is able to activate myosin light chain kinase and phosphodiesterase to the extent comparable with that of the intact molecule. Only the $TR_1$-EDTA peptide, added in a great excess, can partially activate both enzymes.[2]

## References

1. Drabikowski, W., J. Kuźnicki & Z. Grabarek. 1977. Biochim. Biophys. Acta **485:** 124–133.
2. Walsh, M., F. C. Stevens, J. Kuźnicki & W. Drabikowski. 1977. J. Biol. Chem. **252:** 7440–7443.
3. Grabarek, Z., H. Brzeska, J. Kuźnicki & W. Drabikowski. 1978. 7th Eur. Conf. Muscle and Motility. Warsaw, Poland. p. 61 (abstract).
4. Brzeska, H., S. Venyaminov. Submitted for publication.
5. Deville, E., J. Grandjean, P. Laszlo, Ch. Gerday, H. Brzeska & W. Drabikowski. Submitted for publication.
6. Brzeska, H., Z. Grabarek & W. Drabikowski. 1978. 8th Eur. Conf. Muscle and Motility. Heidelberg, Germany. p. 35.
7. Leavis, P. C., S. S. Rosenfeld, J. Gergely, Z. Grabarek & W. Drabikowski. 1978. J. Biol. Chem. **253:** 5452–5459.
8. Drabikowski, W., H. Brzeska & S. Venyaminov. 1980. Proc. 13th FEBS Mtg. Jerusalem, Israel. (In press.)
9. Kuźnicki, J., L. Kuźnicki & W. Drabikowski. 1979. Cell Biol Int. Rep. **3:** 17–23.
10. Kuźnicki, J. & W. Drabikowski. 1979. Publications of Univ. of Innsbruck. W. Sachsenmaier, Ed. Vol. **120:** 99–105.

# ARE THE SEA URCHIN MITOTIC $Ca^{2+}$-ATPASE AND $Ca^{2+}$-SEQUESTERING ACTIVITIES AFFECTED BY CALMODULIN?

Joan C. Egrie and Barbara W. Nagle

*Department of Zoology*
*University of California*
*Berkeley, California 94720*

Calmodulin has been localized to the mitotic apparatus (MA) of dividing cells. While Marcum *et al.*[1] have provided *in vitro* evidence that calmodulin could interact with microtubules to confer increased calcium sensitivity upon the microtubule disassembly reaction, it is important to look at other possible functions for calmodulin in the MA. One obvious candidate for interaction with calmodulin is the calcium-sequestering system which has been proposed to function in the regulation of free $Ca^{2+}$ levels in the region of the MA. The MA of sea urchin eggs contains a vesicular $Ca^{2+}$-sequestering component[2] and a membrane-bound $Ca^{2+}$-ATPase[3] which could be responsible for the $Ca^{2+}$ pumping activity. Since another $Ca^{2+}$ requiring ATPase, the $(Ca^{2+}+Mg^{2+})$-ATPase of erythrocyte membranes, has been shown to require calmodulin[4,5] we have examined the possible interaction of calmodulin with the mitotic $Ca^{2+}$-ATPase and $Ca^{2+}$-sequestering system.

MAs were isolated by passing *S. purpuratus* eggs suspended in 20 mM MES, 10 mM EGTA, 1 mM $MgCl_2$ (pH 6.5) through a 54 $\mu$m Nitex mesh and centrifuging to pellet the MAs. The $Ca^{2+}$-ATPase obtained by this procedure has a pH optimum of 8.5, a $Ca^{2+}$ optimum of 1.0 mM, and is distinguished from mitochondrial and $(Na^+/K^+)$-ATPases and dynein by its insensitivity to oligomycin, ouabain, and vanadate, respectively. The $Ca^{2+}$-ATPase associated with the MAs was assayed in the presence and absence of porcine brain calmodulin. Neither the total activity of the $Ca^{2+}$-ATPase nor its requirement for millimolar levels of $Ca^{2+}$ were affected by the addition of calmodulin. While the procedures for isolation of MAs should remove most MA-associated calmodulin, it was possible that small amounts of residual calmodulin would be enough to activate the $Ca^{2+}$-ATPase to its maximal level. Therefore, DEAE-cellulose chromatography was used to completely separate the $Ca^{2+}$-ATPase from remaining calmodulin. This separation did not produce an enzyme dependent on calmodulin for maximal activity when assayed at 1 mM $Ca^{2+}$. Furthermore, calmodulin did not change the concentration of $Ca^{2+}$ necessary for maximal activity of the partially purified enzyme.

We have also examined the effect of the drug trifluoperazine (TFP) on the mitotic $Ca^{2+}$-ATPase. Although TFP does produce some inhibition of both crude and DEAE-purified enzyme preparations, this inhibition occurs both in the presence of $Ca^{2+}$ and EGTA, and cannot be overcome by the addition of exogenous calmodulin. Addition of 100 $\mu$M TFP produced at most a 40% inhibition of the $Ca^{2+}$-ATPase. The high concentration of TFP required and the failure of calmodulin to overcome the inhibition by TFP indicate that the inhibition is not related to calmodulin and that calmodulin is not an activator of the ATPase.

Since it has not yet been proven that the mitotic $Ca^{2+}$-ATPase is in fact



0077-8923/80/0356-0376 $01.75/0 © 1980, NYAS

responsible for the $Ca^{2+}$ pumping activity found in MAs, we have looked directly at the effect of calmodulin on the $Ca^{2+}$-sequestering system associated with isolated sea urchin MAs. We have found that calmodulin has no effect on $^{45}Ca^{2+}$ accumulation by the isolated MAs. Similar results were obtained whether or not oxalate was included in the incubation buffer. The accumulation of $^{45}Ca^{2+}$ was dependent on the presence of ATP, and the label could be released by addition of Triton X-100, which destroys vesicle integrity.

## References


1. Marcum, J. M., J. R. Dedman, B. R. Brinkley & A. R. Means. 1978. Proc. Natl. Acad. Sci. USA **75:** 3771.
2. Silver, R. B., R. D. Cole & W. Z. Cande. 1980. Cell **19:** 505.
3. Mazia, D., C. Petzelt, R. O. Williams & I. Meza. 1972. Exptl. Cell Res. **70:** 325.
4. Gopinath, R. M. & F. F. Vincenzi. 1977. Biochem. Biophys. Res. Commun. **77:** 1203.
5. Jarrett, H. W. & J. T. Penniston. 1978. J. Biol. Chem. **253:** 4656.

# SEQUENCE ALIGNMENT OF CALMODULIN DOMAINS BY METRIC ANALYSIS

Bruce W. Erickson, D. Martin Watterson, and Daniel R. Marshak

*The Rockefeller University*
*New York, New York 10021*

The amino acid sequence of bovine brain calmodulin[1] was examined by the mathematical method of metric analysis.[2-4] This method furnishes the evolutionary distance between two amino acid sequences, which is based on the known codons for amino acid residues. This distance is measured in minimum base changes (mbc), which is the minimum number of nucleotide changes and/or deletions needed to convert both sequences into a common sequence.

As previously proposed,[5,6] the four calcium-binding domains of calmodulin probably evolved from an ancestral gene coding for a single calcium-binding

```
                                             Segment 1b
                                           ------------
               1         10        20         30       39
               2:-------------------------------------
Domain 1       ADQLTEEQIAEFKEAFSLFDKDGNGTITTKELGTVMRSL
                 + T+ + +E++EAF++FDKDGNG I++ EL+ VM++L     31 mbc (<0.1%)
Domain 3       M-KDTDSE-EEIREAFRVFDKDGNGYISAAELRHVMTNL
              76  80        90        100        112
                                           ------------
                                            Segment 3b


                    Segment 2a
                   ------------
                  40       50        60        70   75
                  ------------------------------------
Domain 2          GQNPTEAELQDMINEVDADGNGTIDFPEFLTMMARK
                  G+++T++E+ +MI E++ DG+G +++ EF+ MM+ K        29 mbc (<0.1%)
Domain 4          GEKLTDEEVDEMIREANIDGDGEVNYEEFVQMMTAK
                 113    120       130       140    148
                   ------------            ------------
                    Segment 4a              Segment 4b


                        115  120   126
                           ------------
Segment 4a                 KLTDEEVDEMIR
                           ++++EE+++M++           9 mbc (0.2%)
Segment 4b                 EVNYEEFVQMMT
                        135  140   146


                           42      50 53
                           ------------
Segment 2a                 NPTEAELQDMIN
                           + ++AEL+++++          10 mbc (0.2%)
Segment 3b                 YISAAELRHVMT
                           99         110


                        115  120   126
                           ------------
Segment 4a                 KLTDEEVDEMIR
                           ++T +E++ ++R          11 mbc (1.0%)
Segment 1b                 TITTKELGTVMR
                           26  30     37
```

FIGURE. Metric analysis of selected segments of bovine brain calmodulin. The line between the segments indicates identical residues, shows residues that differ by only 1 mbc as "+", and gives the evolutionary distance between the segments and the probability that this distance is due to composition alone.



0077-8923/80/0356-0378 $01.75/0 © 1980, NYAS

domain. Metric analysis has shown that each calmodulin domain (residues 1–39, 40–75, 76–112, 113–148) is significantly similar to the other three domains. But the evolutionary distances between Domains 1 and 3 (31 mbc) and Domains 2 and 4 (29 mbc) are significantly lower than the distances (39–42 mbc) between the other four pairs of domains. Thus Domains 1 and 3 are more similar to one another than to Domains 2 or 4 and, conversely, Domains 2 and 4 are more similar to one another than to Domains 1 or 3. These results are readily explained by duplication of an ancestral one-domain gene followed by divergence and duplication of a two-domain gene. The latter would be a common ancestor of the present gene segments coding for Domains 1 + 2 and 3 + 4.

In addition, metric analysis has provided evidence for the first time that the ancestral one-domain gene itself may have been derived by gene duplication. For example, Domain 4 contains two 12-residue segments denoted Segments 4a and 4b in the FIGURE. The last eight residues of these segments correspond to the conserved positions of the alpha-helical regions of crystalline carp parvalbumin.[5] Segments 4a and 4b are unusually similar because they have identical residues at three positions and residues that differ by only 1 mbc at the other nine positions. The probability that a randomly permuted pair of sequences with amino acid compositions of Segments 4a and 4b would have an evolutionary distance as low as 9 mbc is only one in 500. In the same manner, Segments 2a and 3b and Segments 4a and 1b are also unusually similar (see FIGURE). These three statistically significant comparisons support the possibility that the a and b segment of each domain originated from a half-domain gene coding in part for an alpha-helical region similar to Segment 4a. Thus the present gene segments coding for the eight helical regions of calmodulin may have been derived from a single gene segment coding for a helical half-domain through three successive gene duplications.

## REFERENCES


1. WATTERSON, D. M., F. SHARIEF & T. C. VANAMAN. 1980. J. Biol. Chem. **255:** 962–965.
2. SELLERS, P. H. 1974. Siam J. Appl. Math. **26:** 787–793.
3. ERICKSON, B. W. & P. H. SELLERS. 1980. *In* Time Warps, String Edits, and Macromolecules. D. Sankoff & J. B. Kruskal, Eds. (In press.)
4. SELLERS, P. H. 1980. J. Algorithms (In press.)
5. KRETSINGER, R. H. 1980. C.R.C. Crit. Rev. Biochem. (In press.)
6. BARKER, W. C., L. K. KETCHAM & M. O. DAYHOFF. 1979. *In* Atlas of Protein Sequence and Structure, Vol. **3** (Suppl. 3) M. O. Dayhoff, Ed.: 273–283. National Biomedical Research Foundation. Washington, D.C.

# A COMPARISON OF DIFFERENT PREPARATIONS IF BOVINE BRAIN CALMODULIN

M. P. Esnouf,* R. E. Klevit,† and R. J. P. Williams†

**Nuffield Department of Clinical Biochemistry and*
*†Inorganic Chemistry Laboratory*
*University of Oxford*
*Oxford OX1 3QR England*

During our studies of the conformational changes of calmodulin in solution by $^1$H-NMR we have noticed differences in the spectra of the calcium-bound form of the protein in different preparations. We have compared samples both prepared by us and sent to us from several other laboratories. Although most of the samples, called 'normal', give spectra similar to those published (K. Seamon. 1980. *Biochemistry* **19:** 207.), several are significantly different. The most striking differences appear in the aromatic region of the spectra, suggesting a change in the positions of the aromatic side chains with respect to one another. A change of this kind could also arise from small amounts of contaminating material absorbing in the same region of the spectrum or from binding of an impurity to the protein.

In order to determine the nature of the differences we have observed, we have used techniques other than NMR to compare representative samples: Type A are those which give 'normal' NMR spectra and Type B are those which give 'changed' spectra.

We have noted a difference in the UV spectra of Type B samples. The relative intensities of the absorbance maxima have shifted so that 276 nm is no longer the highest point in the spectrum. Again, this observation could arise from contamination which absorbs in the ultra-violet. The addition of trace amounts of 2-mercaptoethanol, EDTA, or sodium azide, any of which could conceivably be present, to a sample of Type A changes the spectrum but does not generate a Type B spectrum. A difference spectrum of Types A and B gives a maximum at about 260 nm. Interestingly, the peak that elutes directly after calmodulin off DEAE-Sephadex during purification has $\lambda_{max} = 260$ nm. The identity of this material is unknown to us. It does not appear as a protein band on 7.5% polyacrylamide gels.

The two types of calmodulin also have different mobilities when electrophoresed on 12.5% polyacrylamide gels at pH 8.6. In the absence of added $Ca^{+2}$ or EGTA, Type B has a higher mobility than Type A, and its mobility is unchanged by the addition of either EGTA or $Ca^{+2}$ to the sample prior to electrophoresis. The addition of trace amounts of EGTA, sodium azide, 2-mercaptoethanol, or nucleotide to Type A does not generate a sample with Type B mobility.

The isoelectric focusing patterns of Types A and B are quite different. In our hands, even Type A gives multiple bands when 150 μg are focused. We observe 5 bands, with the central band being the major species. Type B also gives 5 bands, but the relative intensities are very different from Type A: there are two major bands of equal intensity that focus on either side of the major Type A band that now appears as a minor band, and two more highly acidic bands.

The electrophoresis and electrofocusing data suggest that the differences observed between Types A and B are due to changes in the protein, rather than



0077-8923/80/0356-0380 $01.75/0 © 1980, NYAS

the presence of a contaminant. Deamidation has been offered as an explanation for the change in mobility during electrophoresis (L. J. Van Eldik & D. M. Watterson. 1979. *J. Biol. Chem.* **254:** 10250.). Preincubation of Type A calmodulin with either glutaminase or asparaginase caused no change in the isoelectric focusing pattern. Further studies are needed to determine the nature of the differences reported here but we note that it is likely that some part of the calmodulin structure is sensitive to small changes of conditions.

**Note added in proof:** Work done subsequent to this conference has shown that Type B calmodulin has undergone limited proteolysis from the C-terminus, eliminating the fourth calcium-binding site. A mixture of cleaved and uncleaved calmodulin has no observable phosphodiesterase-stimulating activity.

# A $(Ca^{2+} + Mg^{2+})$-ATPASE IN EPIDIDYMAL AND EJACULATED SPERM FLAGELLAR PLASMA MEMBRANES: POSSIBLE MODULATION BY EXTRACELLULAR FACTORS*

Ian T. Forrester and Mark P. Bradley

*Department of Biochemistry*
*University of Otago, Dunedin*
*New Zealand*

Mammalian spermatozoa undergo many important functional changes during the migration through the epididymal tract and with ejaculation.[1] The actual control mechanisms associated with many of these maturational changes is still uncertain although evidence indicates that some sperm functions may be modulated by the absorption or integration of specific membrane proteins.[2-4] We have investigated the possibility that seminal plasma contains factors capable of modulating $Ca^{2+}$ transport across the sperm plasmalemma. This concept is supported by the recent finding that human seminal plasma contains a calmodu-

TABLE 1

EFFECTS OF SEMINAL PLASMA EXTRACT ON SPERM PLASMALEMMA $Ca^{2+}$-DEPENDENT $Mg^{2+}$-ATPASE

| PMV Source* | Seminal Plasma Extract† | $Ca^{2+}$-Dependent $Mg^{2+}$ ATPase Activity‡ | Activation (% Change)§ |
|---|---|---|---|
| Epididymal | – | 1.98 | — |
| Epididymal | + | 3.56 | 80 |
| Ejaculated | – | 2.38 | — |
| Ejaculated | + | 2.97 | 25 |

*Plasma membrane vesicles prepared from ram sperm flagella, as previously described.[6]

†Additions: 30 μg protein/assay. Sperm-free human seminal plasma[5] was adjusted to pH 4.0 with HCl, the insoluble proteins recovered by centrifugation and dissolved in 18 mM – 18 mM histidine-imidazole (HI) buffer, pH 7.1. The acid-precipitated protein fraction was centrifuged, 1 hr at 100,000 g. The clear supernatant was removed and dialyzed against the HI buffer, pH 7.1, for 48 hr at 4°C.

‡ATPase assayed as previously described.[5] The $Ca^{2+}$-dependent ATPase is defined as that activity in excess of the $Mg^{2+}$-ATPase produced in the presence of 0.2mM $CaCl_2$. Activity expressed as μmol ADP produced/mg membrane protein/hr.

§Compared to the activity obtained in the absence of seminal plasma extract for each PMV source.

*Supported by grants from the Ford Foundation (790-0658) and the Medical Research Council of New Zealand.



0077-8923/80/0356-0382 $01.75/0 © 1980, NYAS

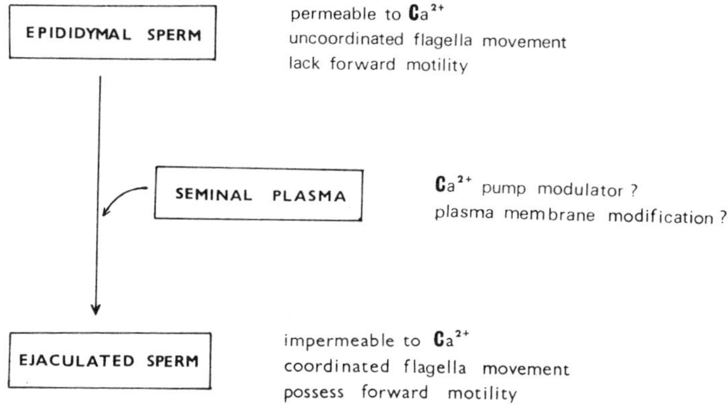


FIGURE 1. Summary of proposed $Ca^{2+}$-linked processes in the maturation of mammalian sperm.

lin-like protein capable of activating the $(Ca^{2+}+Mg^{2+})$-ATPase of erythrocyte plasma membranes.[5]

Plasma membranes were separated from the flagella of ram spermatozoa using osmotic lysis followed by differential centrifugation. These membranes were shown by freeze-fracture electron microscopy to occur as closed vesicles with an average diameter of 400 nm.[6] The plasma membrane vesicles (PMVs) from both epididymal and ejaculated ram sperm contain a $Ca^{2+}$-dependent $Mg^{2+}$-ATPase (TABLE 1). The level of activity is higher in the PMVs of ejaculated sperm and shows limited stimulation (25%) by seminal plasma extract. In comparison the epididymal sperm PMVs have a lower $(Ca^{2+}+Mg^{2+})$-ATPase activity but demonstrate a greater stimulation (80%) by seminal plasma extract. In each case stimulation of the plasmalemma $(Ca^{2+}+Mg^{2+})$-ATPase by seminal plasma extract was abolished by chlorpromazine. The PMVs from both epididymal and ejaculated ram sperm were shown by SDS-polyacrylamide gel electrophoresis to contain at least 30 different protein species.[7] In the PMVs from ejaculated sperm, two protein species (designated proteins A and B) of molecular weight 109,000 and 18,300, respectively, predominate. In comparison, the PMVs obtained from epididymal sperm contain protein A but appear to lack protein B. At this stage the function and origin of protein B is unknown. However, it stains positively with periodate-Schiff's (PAS) reagent and appears to be similar to one of the PAS-positive proteins we have detected in the $(Ca^{2+}+Mg^{2+})$-ATPase activating component of mammalian seminal plasma. Preliminary data suggest that this protein may serve as a calmodulin-like modulator of the sperm plasmalemma $(Ca^{2+}+Mg^{2+})$-ATPase.

These findings could explain some of the functional changes associated with sperm maturation. We propose that the increased impermeability of the ejaculated sperm to $Ca^{2+}$ may be a consequence of seminal-plasma factors interacting with sperm to produce an activation of the plasmalemma $(Ca^{2+}+Mg^{2+})$-ATPase. In this condition intrasperm $Ca^{2+}$ is lowered, flagella movement is coordinated and the sperm acquire forward motility (FIGURE 1).

## References


1. Hamilton, D. W. 1977. *In* Frontiers in Reproduction and Fertility Control. R. O. Greep & M. A. Koblinsky, Eds.: 411–426. The Massachusetts Institute of Technology Press. Cambridge, Mass.
2. Nicolson, G. L. & R. Yanagimachi. 1979. *In* The Spermatozoon. D. W. Fawcett & J. M. Bedford, Eds.: 187–194. Urban and Schwarzenberg, Inc. Baltimore, Md.
3. Acott, T. S., D. J. Johnson, H. Brandt & D. D. Hoskins. 1979. Biol. Reprod. **20:** 247–252.
4. Oliphant, G. & C. A. Singhas. 1979. Biol. Reprod. **21:** 937–944.
5. Forrester, I. T. & M. P. Bradley. 1980. Biochem. Biophys. Res. Commun. **92:** 994–1001.
6. Bradley, M. P. & I. T. Forrester. 1980. Proc. Univ. Otago Med. Sch. **58:** 3–4.
7. Bradley, M. P. & I. T. Forrester. Unpublished observations.

# PROTEIN CARBOXYL-METHYLASE MODIFIES CALMODULIN FUNCTION*

C. Gagnon and S. Kelly

*Department of Pharmacology*
*Laval University*
*Quebec, Canada, G1K 7P4*

V. Manganiello and M. Vaughn

*Laboratory of Cellular Metabolism*
*National Heart, Lung, and Blood Institute*
*Bethesda, Maryland 20205*

W. Strittmatter, A. Hoffman, and F. Hirata

*Laboratory of Clinical Sciences*
*National Institute of Mental Health*
*Bethesda, Maryland 20205*

Posttranslational modification of protein is a mechanism by which a cell can modify protein structure and function. Protein carboxyl-methylase (PCM), a ubiquitous cytosolic enzyme, posttranslationally modifies proteins by methylating their free carboxyl groups. This reaction results in the neutralization of negative charges on the protein substrates.[1] PCM is one of the three elements of the protein carboxyl-methylating system, the other elements being the substrates, methyl acceptor proteins (MAP), and the demethylating enzyme, protein methyl-esterase (PME).[2] Not all proteins are substrate for PCM, and the mere presence of aspartic or glutamic acid residues does not guarantee that the protein will be methylated.[1]

Calmodulin is a ubiquitous calcium binding protein that regulates several enzymatic activities.[3] Since PCM has been shown to be involved in cellular processes requiring calcium such as exocytotic secretion and white blood cell chemotaxis,[2] we investigated the capacity of calmodulin to serve as a substrate for PCM. When compared to various purified proteins, calmodulin was found to be an exceptionally good MAP, even better than the anterior pituitary peptide hormone LH. Analysis by polyacrylamide gel electrophoresis of the carboxyl-methylated calmodulin revealed only one peak of radioactivity that comigrated with calmodulin.

The effect of enzymatic carboxyl-methylation of calmodulin on its capacity to activate calcium-dependent brain cyclic nucleotide phosphodiesterase (PDE) was investigated. The methylation reaction affected calmodulin function. The incorporation of up to 0.5 mole of methyl group per mole of calmodulin produced a parallel decrease in the capacity of calmodulin to activate brain PDE (TABLE 1).

Since the recently described mammalian protein methylesterase[4] can hydrolyze enzymatically synthesized calmodulin-methyl esters (unpublished data), the protein carboxyl-methylation system could, via its action on calmodulin, reversibly regulate the activity of the calcium-dependent cyclic nucleotide phosphodiesterase as well as other enzymes activated by calmodulin.

*Supported by Medical Research Council of Canada and Conseil des Recherches en Santé du Québec.



0077-8923/80/0356-0385 $01.75/0 © 1980, NYAS

TABLE 1

EFFECT OF CARBOXYL-METHYLATION OF CALMODULIN (CAM) ON ITS CAPACITY TO ACTIVATE BRAIN PHOSPHODIESTERASE (PDE)

| Incubation Conditions | PDE Activity (CPM of cGMP Hydrolyzed) | CAM Stimulated PDE (CPM of cGMP Hydrolyzed) |
|---|---|---|
| PDE alone | 690 | |
| PDE + Calmodulin (50 ng/ml) | 1,660 | 970 (100%) |
| PDE + Calmodulin-$CH_3$* (50 ng/ml) | 1,140 | 450 (47%) |

*Up to 0.51 mole of methyl groups were incorporated per mole of calmodulin.

## REFERENCES

1. PAIK, W. K. & S. KIM. 1980. Protein methylation. John Wiley & Sons. New York, N. Y.
2. GAGNON, C. & S. HEISLER. 1979. Biochem. Biophys. Res. Commun. **88:** 847–853.
3. CHEUNG, W. Y. 1980. Science **207:** 19–27.
4. GAGNON, C. 1979. Biochem. Biophys. Res. Commun. **88:** 847–853.

# STIMULATION BY CALMODULIN OF $Ca^{2+}$ TRANSPORT AND ($Ca^{2+}$-$Mg^{2+}$)ATPASE ACTIVITIES IN CANINE TRACHEAL SMOOTH MUSCLE MICROSOMAL PREPARATIONS

G. Kurt Hogaboom and Jeffrey S. Fedan

*Department of Pharmacology and Toxicology*
*West Virginia University Medical Center*
*Morgantown, West Virginia 26505*

The regulation of the cytosolic $Ca^{2+}$ concentration is important in the control of contractile and metabolic processes in many tissues. In smooth muscle, ATP-dependent $Ca^{2+}$ transport has been suggested to regulate the intracellular free $Ca^{2+}$ concentration. In this study, we have investigated various aspects of $Ca^{2+}$ transport ATPase and its interactions with the $Ca^{2+}$ binding protein, calmodulin (CaM), in microsomes of the canine trachealis muscle.

Canine tracheal smooth muscle microsomes were prepared by differential centrifugation and washed with 0.5 mM EDTA and 0.5 mM EGTA during the final step of preparation.

CaM was prepared from rat testis by the method of Dedman *et al.* (*J. Biol. Chem.* 1977. **252:** 8415–8422).

ATP-dependent $^{45}Ca^{2+}$ uptake and $Ca^{2+}$-stimulated, $Mg^{2+}$-dependent ATPase activities were measured in the presence of 40 mM imidazole·HCl, pH 7.0 (37°C), 100 mM KCl, 0.1 mM ouabain, 5 mM $KN_3$, 0.125 mM $CaCl_2$, 25 $\mu$g protein/ml, and where appropriate, $^{45}CaCl_2$ (5 $\mu$Ci/ml), 4 mM MgATP, 1mM EGTA, and 5 mM oxalate. The $[Ca^{2+}]_{free}$ was controlled with the use of a $Ca^{2+}$-EGTA buffer system (pH 7.0, 37°C). Statistical evaluation was made by the Students *t*-test for paired samples and the 0.05 level of probability was accepted as significant.

The ATP-dependent $^{45}Ca^{2+}$ uptake activity (nmoles $Ca^{2+}$/mg/10 min) was dependent upon the $[Ca^{2+}]_{free}$ (1–100 $\mu$M) and was significantly stimulated in a concentration-dependent manner by CaM (30–1500 nM).

The ATP-dependent $^{45}Ca^{2+}$ uptake activity (assayed with 10 $\mu$M $[Ca^{2+}]_{free}$) was significantly stimulated by 150 nM CaM and by oxalate at 0.5, 1, 4, 10, and 30 min. In the presence of both 150 nM CaM and oxalate, $Ca^{2+}$ uptake was not stimulated above the level seen with 150 nM CaM alone, except at 30 min. ATP-independent $^{45}Ca^{2+}$ binding was not affected by 150 nM CaM, by oxalate, or by 150 nM CaM and oxalate.

The $Ca^{2+}$-stimulated, $Mg^{2+}$-dependent ATPase activity ($\mu$moles Pi/mg/30 min) was dependent upon the $[Ca^{2+}]_{free}$ (1–300 $\mu$M) and CaM (30–600 nM) significantly stimulated this activity in a concentration-dependent manner. $Mg^{2+}$-dependent ATPase activity (assayed with 1 mM EGTA) was not affected by the CaM concentrations used (30–600 nM).

The $Ca^{2+}$-stimulated, $Mg^{2+}$-dependent ATPase activity (assayed with $[Ca^{2+}]_{free}$ = 10$\mu$M) was significantly stimulated by 150 nM CaM and by oxalate at 4, 10, and 30 min. The increase in this activity in the presence of both 150 nM CaM and oxalate was additive at all time points measured.



0077-8923/80/0356-0387 $01.75/0 © 1980, NYAS

While the use of a microsomal preparation precludes an evaluation of the interactions of CaM with a single organelle, some conclusions concerning these results can be made.

CaM appears to be interacting with a $Ca^{2+}$ transport ATPase as manifested by the $[Ca^{2+}]_{free}$- and [CaM]-dependence of ATP-dependent $^{45}Ca^{2+}$ uptake and $Ca^{2+}$-stimulated, $Mg^{2+}$-dependent ATPase activities. As CaM is the $Ca^{2+}$-dependent regulator of myosin light chain kinase, the interactions of CaM with ATP-dependent $Ca^{2+}$-pumps located in smooth muscle organelles (i.e., sarcolemma and sarcoplasmic reticulum) suggest another mechanism by which CaM may control contractile processes in canine trachealis muscle.

# FAST AXOPLASMIC TRANSPORT OF CALMODULIN IN MAMMALIAN NERVE: POSSIBLE INVOLVEMENT IN AXOPLASMIC TRANSPORT*

Zafar Iqbal and Sidney Ochs

*Departments of Physiology, Biochemistry and Medical Biophysics*
*Indiana University School of Medicine*
*Indianapolis, Indiana 46223*

Fast axoplasmic transport in peripheral cat nerve was recently observed to be dependent on $Ca^{2+}$.[1,2] A calcium binding protein of 15,000 dalton[3] was found present in mammalian nerve and identified by its molecular weight and isoelectric point as calmodulin.[4]

In this report we show that the calmodulin is transported at a fast rate in mammalian nerve. [$^{35}$S]methionine or [$^{3}$H]leucine injected into the L7 dorsal root ganglion is incorporated into various polypeptide components and these are transported down the nerve fibers. A ligation is placed below the ganglion 2 hr after the injection to stop further transport of labeled components from the ganglion and 3–4 more hrs of downflow of labeled polypeptides is allowed in the isolated portion of nerve before taking 5 mm segments for analysis. The nerve segments were homogenized in 10 mM Tris-HCl buffer at pH 7.5 containing 1 mM each of 2-mercaptoethanol and EGTA to isolate calmodulin.[4] The calmodulin transported in the nerve fibers was identified by SDS-PAGE[5] and isoelectrofocusing on LKB-PAG plates.[4]

The radioactivity present in Coomassie blue-stained protein bands comigrating with calmodulin showed an outflow pattern typical for fast transport. The level of activity in the calmodulin band was much lower compared to total amount of radioactivity transported. The amount of labeled calmodulin in the crest region was higher than in the plateau region which comprises either the activity left behind the fast-transported crest wave or the component which is being transported at a slow rate. The rate as shown by the front of transported calmodulin was at the distance expected of fast transport at the usual fast rate of 410 mm/day.

In the transport filament model proposed for fast axoplasmic transport, energy supplied by the hydrolysis of ATP through the action of a ($Ca^{2+}+Mg^{2+}$)-ATPase is required to move the transport filaments to which various components are bound down along the microtubules.[6] We have shown that the ($Ca^{2+}+Mg^{2+}$)-ATPase present in mammalian nerve[7] is activated by nerve calmodulin at a micromolar level concentration of $Ca^{2+}$.[8] More recently we found the ($Ca^{2+}+Mg^{2+}$)-ATPase associated with tubulin, purified by several cycles of assembly-disassembly, to be calmodulin regulated when $Ca^{2+}$ is present in micromolar levels.[9] Calmodulin is conceived as playing a role in the fast axoplasmic transport of materials in nerve fibers through the regulation of $Ca^{2+}$ and ($Ca^{2+}+Mg^{2+}$)-ATPase, possibly as a part of the transport filament as postulated in the transport filament model.[6]

*Supported by National Institutes of Health Grant RO1 8706-11, National Science Foundation Grant 79-14029, and American Cancer Society Grant AC-IN-46T.



0077-8923/80/0356-0389 $01.75/0 © 1980, NYAS

## REFERENCES


1. OCHS, S., R. M. WORTH & S. Y. CHAN. 1977. Nature **270:** 748–750.
2. LAVOIE, P. A., R. HAMMERSCHLAG & A. TJAN. 1978. Brain Res. **148:** 535–540.
3. IQBAL, Z. & S. OCHS. 1978. J. Neurochem. **31:** 409–418.
4. IQBAL, Z. & S. OCHS. 1980. J. Neurobiol. **11:** 311–318.
5. WEBER, K. & M. OSBORN. 1969. J. Biol. Chem. **244:** 4406–4412.
6. OCHS, S. 1972. Science **176:** 252–260.
7. KHAN, M. A. & S. OCHS. 1974. Brain Res. **91:** 413–426.
8. IQBAL, Z., B. GARG & S. OCHS. 1979. Neuroscience Abst. **5:** 60.
9. OCHS, S. & Z. IQBAL. 1980. Neuroscience, Abst. **6:** 774.

# AFFINITY CHROMATOGRAPHIC ISOLATION OF HIGHLY PURIFIED Ca-CaM SENSITIVE DYNEIN ATPASES FROM *TETRAHYMENA* CILIA*

G.A. Jamieson, Jr. and T.C. Vanaman

*Department of Microbiology and Immunology*

A. Hayes and J.J. Blum

*Department of Physiology*
*Duke University Medical Center*
*Durham, North Carolina 27710*

The key role of $Ca^{2+}$ in the control of ciliary motility has been well documented. Many of the studies on the effects of $Ca^{2+}$ have utilized intact preparations—where $Ca^{2+}$ influx and efflux through the membrane was demonstrated to control the direction and/or frequency of ciliary beating. Several of the effects of $Ca^{2+}$ on functionally demembranated axonemal preparations have also been noted. In no case have these studies revealed any of the molecular mechanisms responsible for the effects observed of $Ca^{2+}$ on the axonemal components. Our recent finding that *Tetrahymena* contain calmodulin (CaM) and that a portion of the CaM is in ciliary axonemes associated with the dynein fraction,[1] suggested that CaM might confer $Ca^{2+}$-sensitivity to dynein—possibly on its ATPase activity.

Preliminary studies showed that the ATPase activity of crude dynein, obtained from a 1 mM Tris-HCl/0.1 mM EDTA, pH 8.2 extraction of demembranated *Tetrahymena* ciliary axonemes, was enhanced by the addition of $Ca^{2+}$, if additional CaM was present. However, the effects observed were small. Doughty[2] recently described a KCl extraction procedure that he used in an effort to observe the effects of low concentrations of $Ca^{2+}$ on the $Mg^{2+}$-dependent ATPase activity of dyneins. After a similar KCl extraction procedure [0.5 M KCl in IMT/6 (8.33 mM Tris-HCl/8.33 mM imidazole/1.25 mM $MgCl_2$/0.067 mM EGTA, pH 7.5)] we investigated the sensitivity of the ATPase activity of the crude dynein obtained to the addition of $Ca^{2+}$ and CaM.

Greater CaM-induced $Ca^{2+}$-sensitivity of dynein ATPase was observed with crude dyneins obtained by KCl extraction. Addition of $Ca^{2+}$ to the KCl extract supernatant produced an 11% increase in ATPase activity, while addition of $Ca^{2+}$ plus 10 $\mu$g of bovine brain CaM caused a 25% increase in ATPase activity. It was then discovered that preincubation of CaM with the dynein extract supernatant fraction (i.e., in the absence of 1 mM ATP) for a short period of time (>1 min) produced a larger activation of the dynein ATPase activity. After fractionation of KCl or Tris-EDTA crude dynein preparations by sucrose density gradient sedimentation, the effect of $Ca^{2+}$ plus CaM on the 14S and 30S dyneins obtained

*Supported by the National Science Foundation (Grant PCM 78-03866) and the National Institutes of Health (Grant NS 10123). G.A.J., Jr. gratefully acknowledges U.S. Public Health Service predoctoral trainee support (Grant 5T32 CA 09111).



0077-8923/80/0356-0391 $01.75/0 © 1980, NYAS

was assessed (ATPase stimulation indexes: 14S-KCl extraction procedure, 7-fold; 14S-EDTA, 1.3-fold; 30S-KCl, 1.9-fold; 30S-EDTA, 1.6-fold).

The data presented above demonstrated that CaM confers $Ca^{2+}$-sensitivity on 14S and 30S dynein ATPases. This implies the presence of a CaM-binding site on these dynein preparations. Direct evidence of such a binding site was sought by subjecting partially purified Tris-EDTA-extracted 14S and 30S dyneins to $Ca^{2+}$-dependent affinity chromatography on CaM-Sepharose 4B.[3] In an effort to remove any aggregate that might be present, and to remove any endogenous CaM present in the samples, the material to be analyzed was first passed sequentially (in $Ca^{2+}$-containing buffer) through Sepharose 4B and phenothiazine-Sepharose 4B[4] before the material reached the CaM-Sepharose 4B column. After sample application (in IMT/6 + 1.5 mM $Ca^{2+}$, pH 7.5) was complete, the columns were washed with $Ca^{2+}$-containing buffer. The columns were then disconnected and eluted separately with buffer containing 1 mM EGTA instead of $Ca^{2+}$.

The 14S and 30S dyneins and their ATPase activities were retained in a $Ca^{2+}$-dependent fashion on the CaM-Sepharose 4B column, thus demonstrating directly the presence of CaM-binding sites on these preparations. No protein or ATPase activity was detected in the CAPP-Sepharose 4B or Sepharose 4B EGTA eluants, or in the material which passed unretarded, in the presence of $Ca^{2+}$, through the set of serial columns. Fractions from this affinity chromatographic procedure were analyzed by SDS-polyacrylamide gel electrophoresis. A majority of the material $Ca^{2+}$-dependently retained on CaM-Sepharose 4B consisted of high molecular weight components. Further analysis revealed that two major ($M_r \sim$ 260K and 253K) and one minor ($M_r \sim$ 270K) components comprised the bulk of the isolated 14S dynein. The 30S dynein was comprised of one major component ($M_r \sim$ 270K), possibly corresponding to the minor component of 14S dynein, and a lesser amount of a 246K ($M_r$) component, which is not found in the 14S dynein.

These results demonstrate that: (1) CaM confers $Ca^{2+}$-sensitivity upon the $Mg^{2+}$-ATPase activity of 14S and 30S dyneins; (2) $Ca^{2+}$-dependent activation of dynein ATPases by CaM requires preincubation of dynein preparations with $Ca^{2+}$ and CaM in the absence of ATP; (3) KCl extraction of ciliary axonemes yield dynein ATPases that are more sensitive to $Ca^{2+}\cdot$CaM-dependent activation than those obtained using Tris-EDTA extraction; (4) 14S dynein ATPase obtained from KCl or Tris-EDTA dynein preparations is more sensitive to $Ca^{2+}\cdot$CaM-dependent activation than 30S ATPase; (5) $Ca^{2+}$-dependent retention of 14S and 30S dyneins on CaM-Sepharose 4B demonstrated the presence of CaM-binding sites on these preparations; and (6) 14S and 30S dyneins consist of different high molecular weight species. An improved $Ca^{2+}$-dependent CaM-Sepharose 4B affinity chromatographic procedure is also presented.

## References

1. Jamieson, G. A., Jr., T. C. Vanaman & J. J. Blum. 1979. Proc. Natl. Acad. Sci. USA **76:** 6471.
2. Doughty, M. J. 1979. Comp. Biochem. Physiol. **64**B: 255.
3. Watterson, D. M. & T. C. Vanaman. 1976. Biochem. Biophys. Res. Comm. **73:** 40.
4. Jamieson, G. A., Jr. & T. C. Vanaman. 1979. Biochem. Biophys. Res. Comm. **90:** 1048.

# LOCALIZATION OF CALMODULIN IN MAMMALIAN SPERMATOZOA BY IMMUNOFLUORESCENCE METHODS*

Harold P. Jones,[†] Richard W. Lenz, Barry A. Palevitz,[‡] and Milton J. Cormier

*†Department of Biochemistry*
*University of South Alabama*
*Mobile, Alabama 36688*
*Departments of Biochemistry and ‡Botany*
*University of Georgia*
*Athens, Georgia 30602*

Prior to fertilization the mammalian spermatozoon undergoes an activation process comprised of two distinct events. The first of these processes, capacitation, results in increased permeability of the plasma membrane to metal ions including calcium. In response to this influx of calcium ions, a second process termed the acrosome reaction is initiated. This reaction results in the vesiculation and eventual loss of plasma and outer acrosomal membranes terminating in the release of the hydrolytic contents of the acrosome.

The possible role of calmodulin as a coupling factor in these processes has been explored by surveying the content of calmodulin in sperm isolated from a variety of species and by examining the localization of calmodulin in mammalian sperm by indirect immunofluorescence techniques. Calmodulin, as measured by its ability to activate porcine brain phosphodiesterase, was found to be present in surprisingly high levels in human, rabbit, hamster, rooster, and sea urchin spermatozoa. Subcellular fractionation of rabbit spermatozoa and subsequent assay of these fractions demonstrated that greater than 99% of the measurable calmodulin activity was associated with the sperm head, while no detectable activity was present within the tail and midpiece fraction.

In order to more properly define the subcellular localization of calmodulin in mammalian sperm, sperm isolated from rabbits and guinea pigs were treated with goat or sheep antibody directed against rat testes calmodulin followed by fluorescein or rhodamine-tagged second antibody, and viewed with video-enhanced fluorescence microscopy. In both species fluorescence was observed in four distinct areas including the anterior portion of the head corresponding to the acrosome, a band in the posterior portion of the equatorial segment, and spots at the base and tip of the flagella. Although fluorescence was seen along the length of the flagellum as well, it was usually weak compared to that in the four distinct regions mentioned above. In sperm that had undergone the acrosome reaction, fluorescence was absent from the acrosomal portion of the head but remained in the other three regions. In sperm exhibiting various degrees of acrosomal shedding, the fluorescence associated with the anterior portion of the head corresponded with the partially lifted membrane as viewed by phase contrast. When "free" acrosome caps were found, they were also fluorescent, though their fluorescence was often weaker than that observed in intact sperm.

*Supported by National Science Foundation Grants PCM 80-06166 (B.A.P.) and PCM 79-12316 (M.J.C.).



0077-8923/80/0356-0393 $01.75/0 © 1980, NYAS

In contrast to the distinct fluorescent staining observed, control sperm that were treated with either nonimmune goat IgG, second antibody alone, or antibody preabsorbed with an excess of porcine brain calmodulin exhibited no or greatly reduced fluorescence.

These studies indicate that calmodulin is present in mammalian sperm in high concentrations, that a majority of the calmodulin is localized within the head, and that a considerable portion of the cellular calmodulin is localized around the acrosome. Taken together these data suggest a role for calmodulin in the process of acrosome activation. In addition, immunofluorescence-determined localization of calmodulin near the posterior portion of the equatorial segment and at the base and tip of the flagellum suggest potential roles for calmodulin in calcium-dependent control of sperm-egg fusion and flagellar formation, breakdown, or function.

# FLUORESCENTLY-LABELED CALMODULIN LOCALIZES SPECIFIC SITES ON MITOCHONDRIA

Marcia Kaetzel,* Robert Pardue, B. R. Brinkley, and John Dedman*

*Department of Cell Biology*
*Baylor College of Medicine*
*Houston, Texas 77030*

Calmodulin (CaM) has been implicated as a calcium mediator of several cellular processes including motility, secretion, and chromosome movement. It exerts its effect through a series of activation steps, initiated by the binding of calcium. The induced change in molecular conformation allows CaM to interact with enzymes and other binding proteins. These and other biochemical reactions occur in concert to bring about the proper physiological response. Intracellular localization of CaM, CaM-regulated enzymes and CaM binding proteins reveal areas of CaM activity, as in the mitotic apparatus (Welsh *et al.* 1979. *Proc. Natl. Acad. Sci. USA* **75:** 1867) and the postsynaptic density (Lin *et al.* 1980. *J. Cell Bio.* **85:** 473). Studies using immunochemical techniques identify binding sites occupied by endogenous calmodulin. Such observations are an aid in correlating known biochemical reactions with particular cellular activities. We have directly labeled CaM with rhodamine in order to identify available, unoccupied CaM binding sites. The fluorescent conjugate (CaM-RITC) retained full ability to activate cAMP phosphodiesterase. In mildly fixed cultured 3T3 cells (3% formalin) the biologically active CaM-RITC bound predominantly to mitochondria (FIGURE 1A). Binding was markedly reduced in the presence of 1 mM EGTA (FIGURE 1B). Stelazine, a phenothiozine which binds to calmodulin, prevented the interaction of CaM-RITC with mitochondrial sites. In addition, a ten-fold excess of unlabeled CaM competitively inhibited binding. Fluorescently labeled troponin C and parvalbumin did not bind to mitochondria or any other cellular organelles. Rhodamine (TMRITC) alone did not bind to 3T3 mitochondria. Thus, the interaction observed in FIGURE 1A requires an active $Ca^{2+}$-CaM complex and is specific for CaM. The physiological role of CaM sites associated with mitochondria is not presently understood. Our studies suggest that CaM may regulate outer mitochondrial membrane enzymes. Further evidence suggests the possibility of differential occupation of these sites during varying metabolic states of the cell. It is anticipated that microinjection of CaM-RITC into hormonally responsive living cells will further elucidate the cellular role of the CaM-mitochondria interaction.

*Present address: Department of Internal Medicine, Division of Endocrinology, University of Texas Health Sciences Center at Houston, 6431 Fannin, Houston, Texas 77025.



0077-8923/80/0356-0395 $01.75/0 © 1980, NYAS

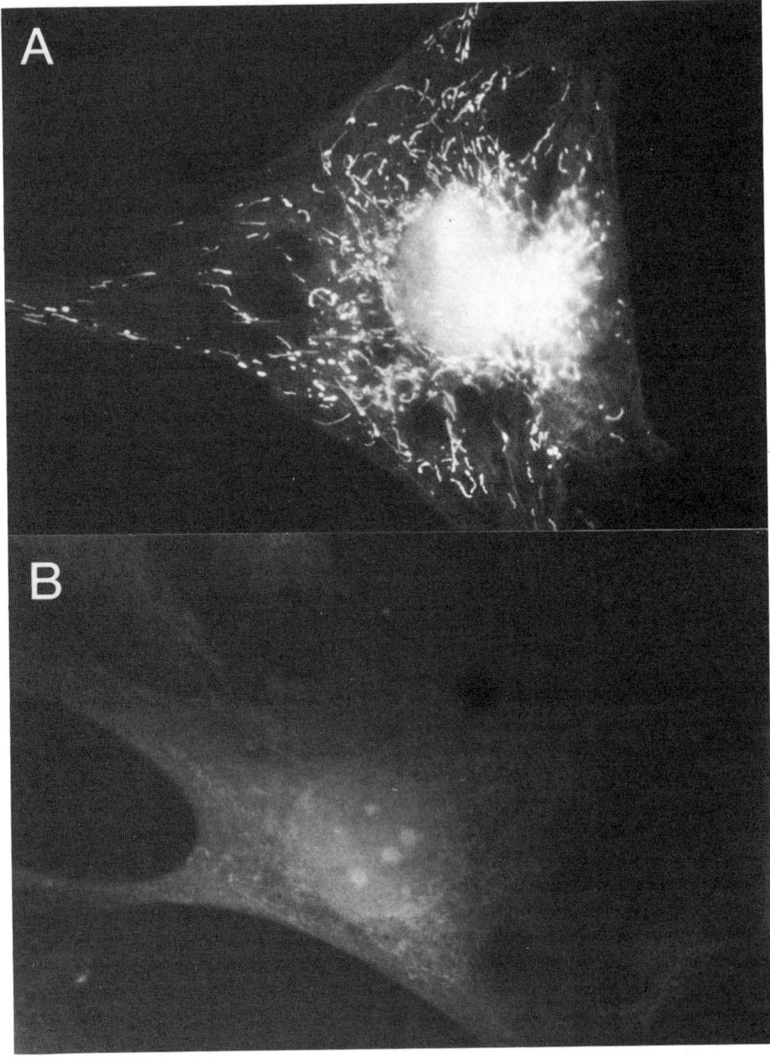


FIGURE 1. In mildly fixed cultured 3T3 cells (3% formalin) the biologically active CaM-RITC bound predominantly to mitochondria (A). Binding was markedly reduced in the presence of 1 mM EGTA (B).

# $^{1}H$- AND $^{113}Cd$-NMR STUDIES OF CALMODULIN

Joachim Krebs,* Ernesto Carafoli,* Eva Thulin,[†]
and Sture Forsén[†]

*Laboratory of Biochemistry
Swiss Federal Institute of Technology (ETH)
CH-8092 Zurich, Switzerland
[†]Department of Physical Chemistry
University of Lund
S-22007 Lund, Sweden

Nuclear magnetic resonance (NMR) is a powerful tool in obtaining structural and dynamical information on biological macromolecules in solution. We have investigated calmodulin isolated from bovine brain and beef testis by using different NMR-techniques, employed under a variety of experimental conditions.

360 MHz $^{1}H$-NMR spectra of bovine brain calmodulin dissolved in $D_2O$, in the presence of 0.2 M KCl show distinct differences with respect to the spectra of $Ca^{2+}$-saturated calmodulin, and with respect to a computed spectrum composed of the amino acid residues of the protein in the random coil form. The spectrum, in the absence of ions, indicates a protein in a constrained conformation, which undergoes a profound rearrangement upon binding of $Me^{n+}$-ions. This is especially pronounced in sequential $Ca^{2+}$ binding experiments. The spectra of bovine brain (FIGURE 1a) and beef testis (FIGURE 1b) calmodulin in the $Ca^{2+}$-saturated form at pH=9.5 provide evidence for the difference in the environmental conditions of a number of animo acid residues, and also for the fact that the calmodulins isolated from the two different sources are identical.

Using a variety of techniques, assignments could be made for tyr 99, tyr 138,

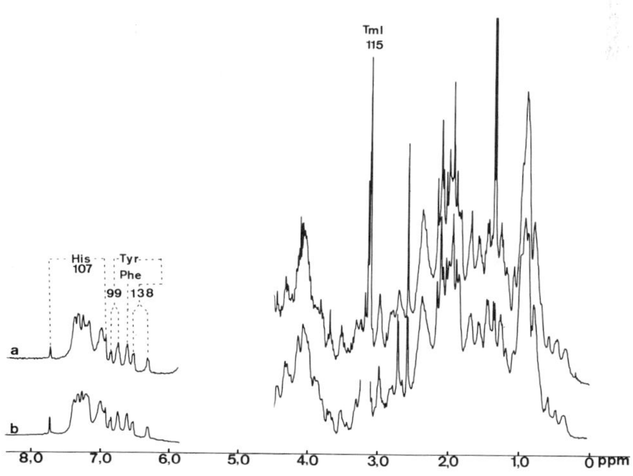


FIGURE 1. $^{1}H$-NMR spectra of calmodulin at 360 MHz. a) $Ca^{2+}$-saturated calmodulin ($1 \times 10^{-3}$M) from bovine brain in 0.2 M KCl ($D_2O$ at pH 9.5, T = 303° K). b) $Ca^{2+}$-saturated calmodulin ($1 \times 10^{-3}$M) from beef testis in 0.2 M KCl ($D_2O$ at pH 9.5, T = 303° K).



0077-8923/80/0356-0397 $01.75/0 © 1980, NYAS

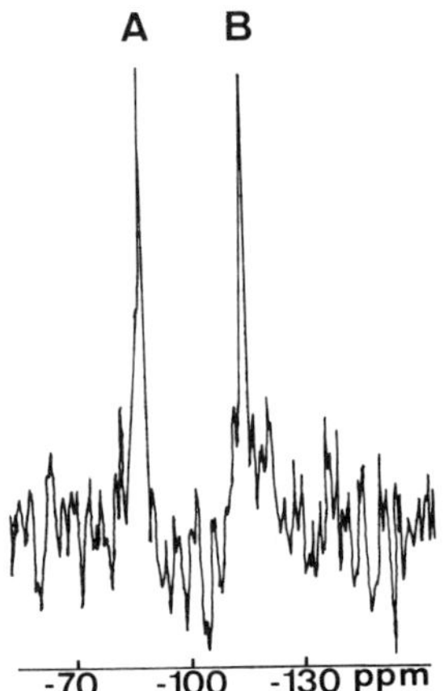


FIGURE 2. $^{113}Cd$-NMR spectrum of calmodulin at 56.55 MHz $Cd^{2+}$-saturated calmodulin ($Ca^{2+}$ free) from bovine brain ($1 \times 10^{-3}$ M) at pH 7.0. Chemical shifts are given relative to 0.1 M Cd $(ClO_4)_2$, shifts to low field taken as positive.

his 107, and trimethyllysine (tml) 115 in the $Ca^{2+}$ free—as well as in the $Ca^{2+}$-saturated state. Full details of these results will be given elsewhere.[1]

The mobility of several amino acids has been investigated. The two tyrosines differ strongly with respect to their environment, especially in the $Ca^{2+}$-saturated form: tyr 99 titrates with a p$K$ of 10.2, tyr 138 with a p$K$ of $> 11.5$, in good agreement with the results obtained by Richman and Klee[2] with other techniques. Nevertheless, for both residues the ortho and meta protons are effectively equivalent. This results from a rapid flipping motion of the tyrosine around the $C_\beta$–$C_\gamma$-bond, producing a rotational isomerism of the respective protons with a flipping rate of $\geq 10^4$ $s^{-1}$. Since tyr 138 is located in a hydrophobic environment, these observations indicate segmental motion in the interior core of calmodulin.

His 107 differs in its free mobility, depending on whether the $Ca^{2+}$ titration has been performed at pH 7.5 or at $\geq 9.5$. In the latter case, his 107 becomes immobilized, whereas in the former it remains mobile. The 2 states are not interconvertible by changing the pH (at T = 303° K), indicating a high energy barrier.

Since $Cd^{2+}$ has often proved to be a fairly good substitute for $Ca^{2+}$ in $Ca^{2+}$ binding proteins, $^{113}Cd$-NMR measurements have been performed. As shown in FIGURE 2, 2 signals at $-88$ ppm and $-115$ ppm can be observed for calmodulin in the $Cd^{2+}$-saturated form, (probably) representing two high affinity binding sites. Similar results have been reported for parvalbumin and troponin C.[3,4] A remarkable difference in the $^{113}Cd$ spin-lattice ($T_1$)-relaxation time between troponin C (1.7 sec) and calmodulin (0.28 sec) could be observed.[5]

## REFERENCES

1. KREBS, J. & E. CARAFOLI. 1980. Submitted for publication.
2. RICHMAN, P. G. & C. B. KLEE. 1979. J. Biol. Chem. **254:** 5372–5376.
3. DRAKENBERG, T., B. LINDMAN, A. CAVE & J. PARELLO. 1978. FEBS Lett. **92:** 346–350.
4. FORSEN, S., E. THULIN & H. LILJA. 1979. FEBS Lett. **104:** 123–126.
5. FORSÉN, S., E. THULIN, T. DRAKENBERG, J. KREBS & K. SEAMON. 1980. FEBS Lett. **117:** 189–194.

# THE ROLE OF CALMODULIN IN THE ACTIVATION OF TRYPTOPHAN HYDROXYLASE BY PHOSPHORYLATING CONDITIONS

Donald M. Kuhn, James P. O'Callaghan, and Walter Lovenberg

*National Institutes of Health*
*National Heart, Lung, and Blood Institute*
*Bethesda, Maryland 20205*

We reported previously[1] that tryptophan hydroxylase, the initial and rate-limiting enzyme in the biosynthesis of the neurotransmitter serotonin, could be activated by ATP and $Mg^{2+}$. This apparent phosphorylation of tryptophan hydroxylase is novel in that it is calcium-dependent and cAMP-independent. We now present evidence indicating that the activation of tryptophan hydroxylase by phosphorylating conditions is totally dependent on calmodulin. The role of calmodulin in the activation process was assessed in two different manners. First, the addition to the phosphorylation reaction mixture (ATP = 0.5 mM; $Mg^{2+}$ = 5.0 mM) of a variety of neuroleptic agents that bind calmodulin[2] completely blocked the activation of tryptophan hydroxylase by ATP-$Mg^{2+}$. The effective concentrations ranged from 10–100 $\mu$M. The drugs used and their order of inhibitory potencies ($IC_{50}$) are pimozide > fluphenazine > trifluoperazine > cis-flupenthixol > haloperidol > penfluridol > chlorpromazine. A variety of other drugs including theophylline, morphine, naloxone, LSD, desmethylimiprimine, and propranolol did not inhibit the ATP-$Mg^{2+}$ effect on tryptophan hydroxylase at concentrations of 100 $\mu$M. The addition of pure calmodulin (0.05–1.0 $\mu$g) to the reaction mixture containing either fluphenazine, trifluoperazine or cis-flupenthixol (50 $\mu$M each) prevented the drug-induced inhibition of the phosphorylation affect. Second, chromatography of brain extracts on an affinity column of fluphenazine-Sepharose 4B[3] to remove endogenous calmodulin, rendered the hydroxylase unresponsive to phosphorylating conditions. The subsequent elution of the fluphenazine-Sepharose 4B column with EGTA (2.5 mM) revealed that a single protein, with electrophoretic and biochemical (membrane phosphorylation) properties identical to those of authentic calmodulin, had bound to the affinity matrix. The readdition of this purified calmodulin after dialysis against 0.1 mM $Ca^{2+}$ to calmodulin-free enzyme completely restored the phosphorylation effect. Restoration of the ATP-$Mg^{2+}$ induced activation of tryptophan hydroxylase by calmodulin was calcium dependent. Taken together these data indicate that the activation of tryptophan hydroxylase by phosphorylating conditions (ATP and $Mg^{2+}$) is dependent on calmodulin. Since we also observed that neither the basal activity of tryptophan hydroxylase nor its stimulation by a $Ca^{2+}$-dependent protease was altered by affinity chromatography on fluphenazine-Sepharose 4B, it appears that a cAMP-independent protein kinase confers calcium-calmodulin dependence on the activation process.

## References

1. KUHN *et al.* 1978. Biochem. Biophys. Res. Comm. **82:** 759.
2. LEVIN & WEISS. 1979. T.P.E.T. **208:** 454.
3. CHARBONNEAU & CORMIER. 1979. Biochem. Biophys. Res. Comm. **90:** 1039.



0077–8923/80/0356–0399 $01.75/0 © 1980, NYAS

# CALMODULIN REGULATION OF RED BLOOD CELL $Ca^{2+}$ TRANSPORT*

F. L. Larsen, S. Katz, and B. D. Roufogalis

*University of British Columbia*
*Vancouver, British Columbia V6T 1W5*
*Canada*

Calmodulin (CaM) is present in the red blood cell and has been shown to stimulate the $Ca^{2+}$ pump[4] and its biochemical expression, the $(Ca^{2+}+Mg^{2+})$-ATPase activity.[1,3] CaM reportedly stimulates the $(Ca^{2+}+Mg^{2+})$-ATPase activity by increasing both the affinity of the enzyme for $Ca^{2+}$ and the $V_{max}$.[2,5] The present study investigated the effect of CaM on the kinetics of active $Ca^{2+}$ transport into inside-out (IO) vesicles. Our aim was to determine whether CaM affects the transport of $Ca^{2+}$ in a manner similar to its effects on $(Ca^{2+}+Mg^{2+})$-ATPase activity.

IO vesicles were made according to the method of Larsen and Vincenzi[4] from freshly drawn blood. Experiments were conducted at 25°C and 37°C in media containing either 0.1 mM EGTA or 2 mM EGTA and (in mM): 40 NaCl, 7.5 KCl, 9 histidine, 9 imidazole, 3 $MgCl_2$, 0.1 ouabain and various amounts of $CaCl_2$ and $^{45}CaCl_2$ at pH 6.9. The reactions were started by adding 3 mM ATP. The vesicles were trapped on filters and counted for $^{45}Ca^{2+}$. Vesicles were depleted of endogenous CaM by incubation at 30°C for 30 min in 40 mM Tris-glycylglycine (pH 7.4) and 1 mM EDTA. ATPase activities were determined by the Fisk-SubbaRow method.

CaM stimulated active uptake of $Ca^{2+}$ and $(Ca^{2+}+Mg^{2+})$-ATPase activity at all $Ca^{2+}$ concentrations except the lowest (approximately 0.2 $\mu$M). The magnitude of stimulation varied from less than 2-fold to more than 15-fold, depending on the free $Ca^{2+}$ concentration. Transport experiments on nonEDTA extracted vesicles showed essentially one affinity for $Ca^{2+}$ (apparent $K_d$ = 2 $\mu$M) in the presence and absence of added CaM. EDTA-extracted vesicles incubated without added CaM showed two distinct $Ca^{2+}$ affinities (apparent $K_d$s = approximately 2 and 30 $\mu$M) for both transport and $(Ca^{2+}+Mg^{2+})$-ATPase activity. The kinetics of EDTA-extracted vesicles could be restored to high $Ca^{2+}$ affinity and high activity by the addition of CaM. Trifluoperazine ($3\times10^{-5}$ M) and CaM-binding protein (20 $\mu$g/ml) inhibited primarily the high affinity component of $Ca^{2+}$ transport and $(Ca^{2+}+Mg^{2+})$-ATPase activity of vesicles incubated in the absence of CaM. In the absence of added CaM, the stoichiometry was 2 $Ca^{2+}$ pumped per $P_i$ liberated at low $Ca^{2+}$ concentrations. At higher $Ca^{2+}$ concentrations (above 40 $\mu$M) or at all $Ca^{2+}$ concentrations with added CaM (1 $\mu$g/ml), the stoichiometry was 1 $Ca^{2+}$ per $P_i$.

Thus, CaM affects $Ca^{2+}$ transport and $(Ca^{2+}+Mg^{2+})$-ATPase activity similarly. CaM activates $Ca^{2+}$ transport by increasing both the affinity of the enzyme for $Ca^{2+}$ and the $V_{max}$. Concomitant analysis of $Ca^{2+}$ transport and $(Ca^{2+}+Mg^{2+})$-ATPase activity suggests a variable stoichiometry that depends on the concentration of $Ca^{2+}$ and CaM.

*Supported by the MRC of Canada and the B.C. Heart Foundation (B. D. R.) and the Canadian Cystic Fibrosis Foundation (S. K.).



0077-8923/80/0356-0400 $01.75/0 © 1980, NYAS

## REFERENCES

1. GOPINATH, R. M. & F. F. VINCENZI. 1977. Biochem. Biophys. Res. Comm. **77:** 1203.
2. HANAHAN, D. J., R. D. TAVERNA, D. D. FLYNN & J. E. EKHOLM. 1978. Biochem. Biophys. Res. Comm. **84:** 1009.
3. JARRETT, H. W. & J. T. PENNISTON. 1977 Biochem. Biophys. Res. Comm. **77:** 1210.
4. LARSEN, F. L. & F. F. VINCENZI. 1979 Science **204:** 306.
5. SCHARFF, O. & B. FODER. 1978. Biochem. Biophys. Acta **509:** 67.

# EVIDENCE FOR CALMODULIN SENSITIVE ADENYLATE CYCLASE IN BOVINE ADRENAL MEDULLA*

Norman C. LeDonne, Jr. and Carole J. Coffee

*Department of Biochemistry*
*School of Medicine*
*University of Pittsburgh*
*Pittsburgh, Pennsylvania 15261*

The adenylate cyclase activity from bovine adrenal medulla is stimulated by calmodulin when assayed in the presence of calcium, with maximal activity occurring at 100 nM free calcium. The activity of a crude particulate fraction, which has been washed once with buffer free of EGTA or EDTA to remove loosely associated calmodulin, is not reduced when assayed in the presence of EGTA at concentrations up to 1 mM. Washing the particulate fraction with buffer containing EDTA will remove some, but not all, of the bound calmodulin. Approximately 30% of the original particulate calmodulin remains associated with the membrane fraction after three washes with an EDTA-containing buffer, with no significant decrease in calmodulin specific activity occurring with consecutive washes. The concentration of ATP in the assay is important in demonstrating stimulation of adenylate cyclase by calcium plus calmodulin. At high (1.2 mM) concentrations of ATP, calmodulin appears to reverse the calcium inhibition of adenylate cyclase activity, and the activity observed in the presence of calcium plus calmodulin is the same as that seen in the presence of EGTA. At low (0.12 mM) concentrations of ATP, calmodulin plus calcium stimulates the adenylate cyclase activity to levels greater than those obtained in the presence of EGTA. The calmodulin stimulation of the adenylate cyclase activity is saturable with respect to added calmodulin; maximal stimulation of the adenylate cyclase is observed with 2 μg of calmodulin, and no further increase in activity is observed with up to 5 μg of calmodulin. Calcium is required for calmodulin stimulation; in the absence of calcium, up to 5 μg of calmodulin has no effect on activity. The stimulation of the adenylate cyclase is specific for calmodulin. None of the low molecular weight proteins that were tested, including parvalbumin and troponin C which are structurally related to calmodulin, will substitute for calmodulin in the stimulation of the adenylate cyclase. GTP will stimulate the adenylate cyclase activity 60 to 70%, and Gpp(NH)p will stimulate the adenylate cyclase activity 70 to 100%, regardless of whether the assays are performed in the presence of EGTA, 1 μM free calcium or 1 μM free calcium plus calmodulin. Calmodulin will stimulate the adenylate cyclase activity 30% in the presence or absence of GTP or Gpp(NH)p; therefore, neither GTP nor Gpp(NH)p will enhance or attenuate the calmodulin stimulation of the adenylate cyclase activity. Both penfluridol and trifluoperazine, drugs which inhibit the calmodulin-induced stimulation of enzymes, will inhibit the adenylate cyclase activity of EDTA-washed particulate fractions in the absence of added calmodulin. The dose response curves and $IC_{50}$ values obtained are the same regardless of whether the assays are performed in the presence of EGTA, 1μM free calcium, or

*Supported by National Institutes of Health (Grant AM 16728).



0077-8923/80/0356-0402 $01.75/0 © 1980, NYAS

1 $\mu$M free calcium plus calmodulin. Promethazine, a drug which is relatively ineffective in displacing calmodulin-bound trifluoperazine, is also less effective as an inhibitor of adenylate cyclase activity. $IC_{50}$ values of 10 $\mu$m for penfluridol, and 1.1 mM for trifluoperazine were obtained. The similar dose response curves and $IC_{50}$ values suggest that some of the calmodulin that remains after washing the particulate fraction with EDTA-containing buffers is tightly bound to the adenylate cyclase in a state that is insensitive to EGTA.

# CALMODULIN REGULATION OF CALCIUM TRANSPORT IN CARDIAC SARCOPLASMIC RETICULUM*

Gary Lopaschuk and Sidney Katz

*Division of Pharmacology and Toxicology*
*Faculty of Pharmaceutical Sciences*
*University of British Columbia*
*Vancouver, British Columbia V6T 1W5*
*Canada*

Previous work in our laboratory has shown that calmodulin stimulates $Ca^{2+}$ transport in cardiac microsomal preparations enriched in sarcoplasmic reticulum in a concentration-dependent manner.[1] $K^+$ and $Na^+$ have also been found to enhance $Ca^{2+}$ transport in these preparations.[2] We have observed that in the presence of $K^+$ (110 mM) and $Na^+$ (110 mM) stimulation of $Ca^{2+}$ transport activity by calmodulin is greatly reduced. This result was obtained at all free calcium concentrations studied. That this was not an ionic strength effect was indicated by the lack of antagonism of calmodulin stimulation by $Li^+$ (110 mM).

Calmodulin stimulation of $Ca^{2+}$ transport into sarcoplasmic reticulum was further investigated in this present study. As a first step, we investigated whether calmodulin was indigenous to the preparations used. Attempts were therefore made to isolate calmodulin from dog cardiac microsomal preparations enriched in sarcoplasmic reticulum by methodology used to isolate calmodulin from other sources.[3,4] These extracts were then tested to determine their ability to stimulate $Ca^{2+}$ transport into sarcoplasmic reticulum using standard incubation conditions.[2] It was observed that neither boiling nor treatment with 0.6 mM EGTA could extract calmodulin from these preparations. This result indicates that the microsomal preparations utilized do not contain indigenous calmodulin.

Since calmodulin does not appear to be a component of the sarcoplasmic reticulum, it was postulated that binding to sites on the membrane must occur in order for calmodulin to augment $Ca^{2+}$ transport. It was also suggested that $K^+$ and $Na^+$ may decrease calmodulin stimulation of calcium transport by altering this binding. Studies were therefore performed using $^{125}$I-labeled calmodulin to determine the degree of binding to microsomal preparations in the presence and absence of $K^+$ (110 mM), $Na^+$ (110 mM), and $Li^+$ (110 mM). Calmodulin was iodinated by the method of Hunter and Greenwood[5] and binding studies performed using incubation procedures similar to those utilized to measure $Ca^{2+}$ uptake in these preparations.[2] Following incubation the samples were centrifuged at 40,000 $\times$ g for 40 min and the resulting pellet washed and counted for radioactivity using a gamma spectrophotometer.

FIGURE 1 indicates that [$^{125}$I]calmodulin binds to microsomal preparations enriched in sarcoplasmic reticulum in a calcium concentration-dependent manner. At calcium concentrations above $10^{-7}$ M, this binding is significantly decreased ($P < 0.05$) in the presence of $K^+$ (110 mM) or $Na^+$ (110 mM). $Li^+$ (110 mM), previously shown not to alter $Ca^{2+}$-transport augmentation by calmodulin,

*Supported by a Grant from the B.C. (Canada) Heart Foundation. G. L. is a Pre-Doctoral Fellow of the Canadian Heart Foundation.



0077-8923/80/0356-0404 $01.75/0 © 1980, NYAS

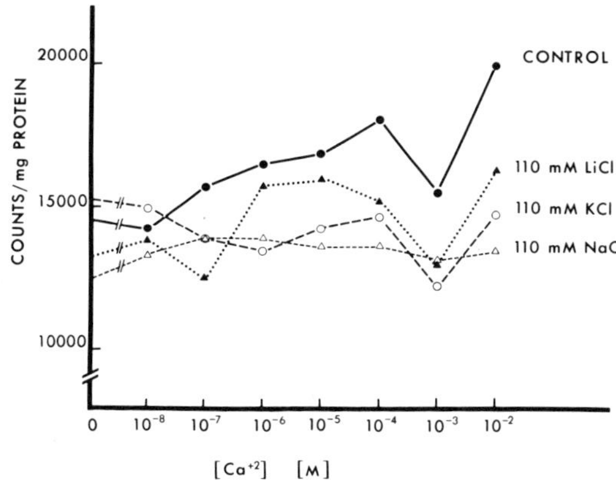


FIGURE 1. Effect of monovalent cations on [$^{125}$I]calmodulin binding to cardiac microsomes enriched in sarcoplasmic reticulum. [$^{125}$I]calmodulin binding, expressed as cpm/mg sarcoplasmic reticulum protein, was determined in the absence of monovalent cations (●——●), and in the presence of 110 mM KCl (O——O), 110 M NaCl (Δ——Δ) or 110 mM LiCl (▲···▲). Result shown is the mean of at least 3 experiments.

does not alter calmodulin binding to a significant extent. $K^+$ and $Na^+$ therefore may inhibit calmodulin stimulation of calcium transport in these preparations by decreasing calmodulin binding to the sarcoplasmic reticulum. The lack of inhibition by $Li^+$ (110 mM) indicates that this result is not due to a nonspecific ionic effect.

Our studies therefore have shown that calmodulin is not an indigenous protein of the sarcoplasmic reticulum preparations used. Calmodulin, though, has been found to bind, in a calcium-dependent manner, to these preparations. This binding is altered by monovalent cations previously shown to inhibit calmodulin stimulation of calcium transport.

## REFERENCES

1. KATZ, S. & M. A. REMTULLA. 1978. Biochem. Biophys. Res. Comm. **83:** 1373.
2. JONES, L. R., H. R. BESCH & A. M. WATANABE. 1978. J. Biol. Chem. **253:** 1643.
3. JUNG, N. S. G. T. Masters Thesis. University of Washington. Seattle, Wash.
4. DEPAOLI-ROACH, A., J. B. GIBBS & P. J. ROACH. 1979. FEBS Lett. **105:** 321.
5. HUNTER, W. M. & F. C. GREENWOOD. 1962. Nature (London) **194:** 495.

# ACTIVATION OF HUMAN RED CELL MEMBRANE ($Ca^{2+}+Mg^{2+}$)-ATPase PHOSPHORYLATION BY CALMODULIN*

Madan G. Luthra,† Richard P. Watts, and Hyun Dju Kim

*Department of Physiology*
*College of Medicine*
*University of Arizona*
*Tucson, Arizona 85724*

Studies independently carried out in our laboratory and in other laboratories established that ($Ca^{2+}+Mg^{2+}$)-ATPase of human erythrocyte membranes is activated by a cytoplasmic, nonhemoglobin protein.[1-3] This protein was purified[4,5] and was found to be similar to calmodulin.[5,6] In this study, the effect of calmodulin on human red cell membrane ($Ca^{2+}+Mg^{2+}$)-ATPase phosphorylation was tested and found that calmodulin increased the rate of $Ca^{2+}$-dependent phosphorylation of 130,000 molecular weight phospho-intermediate of ($Ca^{2+}+Mg^{2+}$)-ATPase.

Membranes prepared by the method of Garrahan *et al.*,[7] were phosphorylated by the method of Rega and Garrahan.[8] In a typical experiment, 1-ml reaction mixture contained 12.5 $\mu$M $MgCl_2$, 50 $\mu$M EGTA, 100 $\mu$M $CaCl_2$, 150 mM Tris-HCl buffer, pH 7.6, 750 $\mu$g membrane protein, and 21 $\mu$g calmodulin whenever present. The reaction was carried out on ice by the addition of carrier free $\gamma[^{32}P]$ATP to give a final concentration of 7.5 $\mu$M. Reaction was stopped at various time intervals with 5 ml of cold solution containing 6% trichloroacetic acid, 50 mM $NaH_2PO_4$ and 1 mM nonradioactive ATP. Membrane precipitate was centrifuged and washed with the same medium and dissolved in 3% sodium dodecylsulfate (SDS). SDS gel electrophoresis was carried out by the method of Fairbanks *et al.*[9] at 4°C and radioactivity was measured in 2 mm gel slices. Only one major peak of radioactivity at 130,000 molecular weight was observed during

TABLE 1

EFFECT OF CALMODULIN ON THE RATE OF PHOSPHORYLATION OF 130,000 MOL. WT. PROTEIN OF RED CELL MEMBRANES

| Time (sec) | $^{32}P$ Incorporation (cpm/500 $\mu$g protein) | |
|---|---|---|
| | Without Calmodulin | With Calmodulin |
| 5 | 300 | 410 |
| 10 | 400 | 660 |
| 20 | 610 | 900 |
| 40 | 850 | 1,300 |
| 60 | 980 | 1,600 |

The specific activity of $\gamma[^{32}P]$ATP was 1.5 Ci/mmole ATP.

*Supported by grants from the American Heart Association (Arizona Affiliate) to M. G. L. and National Institutes of Health (AM17723) to H. D. Kim, and Trainee Grant to the Department of Physiology (HL 07249).

†Present address: Department of Medicine, Baylor College of Medicine, 1200 Moursund, Houston, Texas 77030.



0077-8923/80/0356-0406 $01.75/0 © 1980, NYAS

TABLE 2
EFFECT OF VARYING AMOUNTS OF CALMODULIN ON STIMULATION OF MEMBRANE PHOSPHORYLATION

| Calmodulin concentration ($\mu$g/ml medium) | cpm/mg Protein | % Activation |
|---|---|---|
| 0.00 | 3,675 | 0.0 |
| 1.25 | 3,725 | 2.8 |
| 2.50 | 4,150 | 14.5 |
| 5.00 | 4,675 | 29.0 |
| 10.00 | 5,375 | 48.3 |
| 20.00 | 5,900 | 62.8 |

The specific activity of $\gamma$[$^{32}$P]ATP was 3.0 Ci/mmole ATP.

the course of incubation in the presence and absence of calmodulin. Calmodulin stimulated the rate of incorporation of $^{32}$P into this protein (TABLE 1). At the end of 60 seconds approximately a 1.6-fold increase was observed. The same membrane preparation assayed for ($Ca^{2+}+Mg^{2+}$)-ATPase activity by measuring $P_i$ release showed a 1.6-fold activation, suggesting that calmodulin activated both phosphorylation and dephosphorylation reactions. Membrane phosphorylation was found to depend upon the concentration of calmodulin (TABLE 2), and was inhibited by EGTA.

## REFERENCES


1. LUTHRA, M. G., G. R. HILDENBRANDT & D. J. HANAHAN. 1976. Biochim. Biophys. Acta **419:** 164.
2. BOND, G. H. & D. L. CLOUGH. 1973. Biochim. Biophys. Acta **323:** 592.
3. QUIST, E. E. & B. D. ROUFOGALIS. 1975. Arch. Biochem. Biophys. **168:** 240.
4. LUTHRA, M. G., K. S. AU & D. J. HANAHAN. 1977. Biochem. Biophys. Res. Commun. **77:** 678.
5. JARRET, H. W. & J. T. PENNISTON. 1978. J. Biol. Chem. **253:** 4676.
6. GOPINATH, R. M. & F. F. VINCENZI. 1977. Biochem. Biophys. Res. Commun. **77:** 1203.
7. GARRAHAN, P. J., M. I. POUCHAN & A. F. REGA. 1969. J. Physiol. (Lond) **202:** 305.
8. REGA, A. F. & P. J. GARRAHAN. 1975. J. Memb. Biol. **22:** 313.
9. FAIRBANKS, G., T. L. STECK, & D. F. H. WALLACH. 1971. Biochemistry **10:** 2606.

# CALMODULIN IN THE CILIATES *PARAMECIUM TETRAURELIA* AND *TETRAHYMENA THERMOPHILA**

Nita J. Maihle and Birgit H. Satir

*Department of Anatomy*
*Albert Einstein College of Medicine*
*Bronx, New York 10461*

Calmodulin is a protein of special interest, since it is now known that calcium is involved in the regulation of biological processes such as secretion and ciliary motility.[8] A more detailed account of the physiology of these processes and a possible mechanism for the involvement of calmodulin in secretion are described in Satir *et al.*[8] In this brief report, we present evidence for the presence and localization of calmodulin in cilia, cortex, and calcium containing crystals in *Paramecium tetraurelia* and in cilia and cortex of *Tetrahymena thermophila.* Preliminary reports of these findings for *Paramecium* have been presented at the Society of Protozoologists[2] and at American Society of Cell Biologists.[5]

Indirect immunofluorescent localization was performed using goat anticalmodulin supplied by Drs. J. Dedman and A. Means, and rhodamine-conjugated rabbit anti-goat IgG as the secondary label (Cappel Laboratories, Pa.). All cells were harvested during early stationary phase and were prepared for immunofluorescence following the methodology described in Maihle *et al.*[4] For all experiments, the following controls were performed in order to insure the immunospecificity of the results: 1) incubation in 2° Ab alone, 2) incubation in a nonabsorbed fraction from the affinity column used in the purification of anticalmodulin, and 3) preabsorption of the 1° Ab in a 6 M excess of mammalian calmodulin. Cells were observed with a Zeiss Universal light microscope and all control cells displayed only a diffuse low level of fluorescence.

In experimental aliquots of cells treated with anticalmodulin and 2° Ab, fluorescence is seen in three distinct regions of *Paramecium*[6] and in two distinct regions in *Tetrahymena* (cf. FIGURES 3a & b Satir *et al.*[8]). In *Paramecium*, label is present in calcium containing birefringent crystals which are of uniform diameter. These labeled crystals vary in number within the cell ranging from 0–5/cell. The fluorescent images are typically spherical in shape, measuring approximately 10–15 $\mu$m in diameter and are present in both axenic and monoaxenic cultures of *Paramecium* but absent in *Tetrahymena*. In cilia, both somatic and oral are consistently fluorescent along the entire length of each cilium. Finally, label is present in a linear punctuate pattern of fluorescent spots in both normal and deciliated cells. Based upon the highly organized nature of the cortex of these cells, we currently interpret this pattern of localization to represent the basal bodies of the cell.

Calmodulin is present in both ciliates and has been partially purified from both organisms by heating at 80°C for 6 min and ammonium sulfate precipitation. The protein from both genera exhibits an apparent molecular weight of 17.0 Kd which is lower than that of purified bovine brain calmodulin at 18.5 Kd using SDS 15% PAGE analysis. The partially purified protein exhibits a mobility shift

*Supported U.S. Public Health Service Grants GMS 24724 and 27298.



0077-8923/80/0356-0408 $01.75/0 © 1980, NYAS

in the presence of 10 $\mu$M calcium in SDS PAGE and both preparations stimulate mammalian phosphodiesterase activity.

The ability of *Paramecium* and *Tetrahymena* extracts to stimulate mammalian phosphodiesterase activity, along with the presence of an appropriate ca. 17 Kd heat-stable protein whose apparent molecular weight shifts in EGTA versus $CaCl_2$, and the affinity of both a *Paramecium* and *Tetrahymena* component for anticalmodulin prepared in response to antigen isolated from mammalian tissue,[1] clearly supports the conclusion that calmodulin, chemically and antigenically similar to mammalian calmodulin, is present in both ciliates.

We are currently pursuing the role of calmodulin in *Paramecium* and *Tetrahymena* by determining which proteins in subcellular fractions of these cells interact with calmodulin. We have examined whole cell extracts and ciliary fractions of *Tetrahymena* using immobilized bovine brain calmodulin. There are approximately 12 soluble proteins present in whole cell *Tetrahymena* extracts that bind to the immobilized calmodulin in a calcium-dependent manner.[6] The most prominent of these proteins within the cell appears to originate from the cilia as determined by using both axonemal and ciliary membrane preparations.[3] These data will aid in providing insights into the role of calmodulin in calcium-mediated processes and may yield valuable information regarding the fundamental functional significance of calmodulin in eukaryotic cells.

## References

1. Dedman, J. R., M. J. Welsh & A. R. Means. 1978. $Ca^{2+}$-dependent regulator. Production and characterization of a monospecific antibody. J. Biol. Chem. **253:** 7515–7521.
2. Maihle, N. J. & B. H. Satir. 1979. Indirect immunofluorescent localization of calmodulin in *Paramecium tetraurelia*. J. Protozool. **26**(18): 10a.
3. Maihle, N. J., J. Avolio, J. L. Salisbury, P. Satir & B. H. Satir. 1980. Isolation of calmodulin regulated ciliary proteins from *Tetrahymena thermophila* using affinity chromatography. (In preparation.)
4. Maihle, N. J., J. Dedman, A. R. Means & B. H. Satir. 1980. Biochemical characterization and indirect immunofluorescent localization of calmodulin in *Paramecium tetraurelia*. (Submitted for publication.)
5. Maihle, N. J., R. S. Garofalo & B. H. Satir. 1979. Biochemical characterization and indirect immunofluorescent localization of calmodulin in *Paramecium tetraurelia*. J. Cell Biol. **83**(2): 476, Z2919.
6. Maihle, N. J., J. L. Salisbury & B. H. Satir. 1980. Calmodulin in *Tetrahymena*. Isolation, purification, localization, and affinity chromatography. Second International Congress on Cell Biology. September, 1980.
7. Reed, W. & P. Satir. 1980. Calmodulin in mussel gill epithelial cells: role in cilary arrest. (This volume.)
8. Satir, B. H., R. S. Garofalo, D. M. Gilligan & N. J. Maihle. 1980. Possible functions of calmodulin in protozoa. (This volume.)

# ISOLATION OF A PROTEIN FROM BRAIN USING CALCIUM-DEPENDENT AFFINITY-BASED ADSORPTION CHROMATOGRAPHY ON PHENOTHIAZINE-SEPHAROSE CONJUGATES

Daniel R. Marshak, Linda J. Van Eldik, and D. Martin Watterson

*The Rockefeller University*
*New York, New York 10021*

We have isolated a low molecular weight protein from bovine and chicken brains using a combination of classical techniques of protein chemistry and calcium-dependent affinity-based chromatography. The soluble extract from a whole brain homogenate was brought to 55% saturation with solid ammonium sulfate, and the supernatant from this precipitation was brought to pH 4.0 with 50% sulfuric acid. The pellet from the pH 4 precipitation was resuspended, dialyzed, and chromatographed on DEAE-Sephadex A50 using a 0.2–0.7 M NaCl gradient. A fraction containing calmodulin and the small acidic protein (SAP)

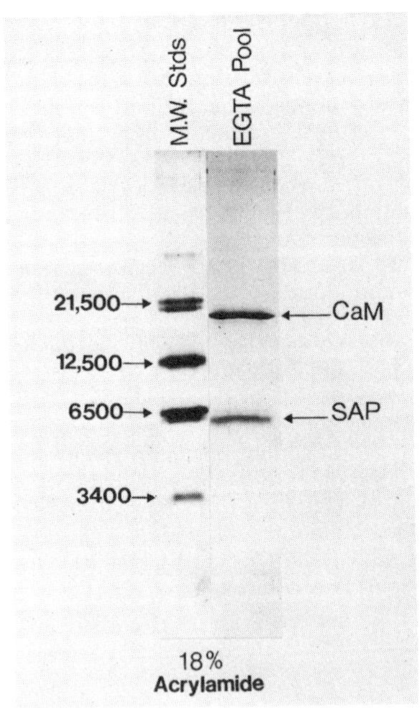


FIGURE 1. Polyacrylamide gel electrophoresis in the presence of 0.1% (w/vol) sodium dodecyl sulfate and 1 mM EGTA. Gels were stained with Coomassie blue. The left lane shows molecular weight standards: soybean trypsin inhibitor (21,500), cytochrome c (12,500), aprotinin (6,500), and insulin B chain (3,400). The right lane is the EGTA-eluted material from phenothiazine-Sepharose chromatography. CaM– calmodulin; SAP– small acidic protein.



0077-8923/80/0356-0410 $01.75/0 © 1980, NYAS

was collected, dialyzed, and lyophilized. This material was applied to a column (1.5 × 18 cm) containing 2-chloro-10-(3-aminopropyl) phenothiazine-Sepharose in buffer containing 2 mM calcium chloride. Calmodulin and the smaller acidic protein bind to the affinity resin in the presence of calcium and elute with buffer containing 2 mM ethylene bis(oxyethylene nitrilo) tetraacetic acid (EGTA). While all the calmodulin eluted with buffer containing EGTA, additional small acidic protein eluted with 4 M urea. SAP from the EGTA-eluted pool separated from calmodulin during a second DEAE-Sephadex A50 chromatography, and SAP was further purified by reverse-phase high-performance liquid chromatography.

During polyacrylamide gel electrophoresis in the presence of sodium dodecyl sulfate, SAP migrated as a single, broad band with an $M_r$ of approximately 6,300. Unlike calmodulin, this band did not show a calcium-dependent shift in electrophoretic mobility.[1] Preparations of other tissues, such as chicken gizzard, did not contain SAP as a readily solubilized protein. The small acidic protein from bovine brain did not activate cyclic nucleotide phosphodiesterase, and it neither augmented nor inhibited calmodulin activation of phosphodiesterase. Bovine brain SAP competed quantitatively with bovine brain calmodulin for binding to an anti-calmodulin antiserum.[2]

In some respects the small acidic protein from brain resembles one of the S-100 proteins[3] which is a small, highly acidic and brain-specific protein, and which has structural homology with calmodulin. However, S-100 does not compete with bovine brain calmodulin for binding to the anti-calmodulin antiserum. In addition, Levin and Weiss[4] have found only a very small amount of calcium-dependent binding of phenothiazine drugs to bovine brain S-100. The data we now present suggest that phenothiazines interact with other brain proteins as well as with intact calmodulin.

## References


1. Burgess, W. H., D. K. Jemiolo & R. H. Kretsinger. 1980. Biochim. Biophys. Acta **623:** 257–270.
2. Van Eldik, L. J. & D. M. Watterson. This volume.
3. Isobe, T. & T. Okuyama. Eur. J. Biochem. 1978. **89:** 379–388.
4. Levin, R. M. & B. Weiss. Biochim. Biophys. Acta. 1978. **540:** 197–204.

# *IN VITRO* METHYLATION OF CALMODULIN

T. J. Murtaugh, A. Sitaramayya, L. S. Wright, and F. L. Siegel*

*Department of Pediatrics and Physiological Chemistry*
*University of Wisconsin*
*Madison, Wisconsin 53076*

Calmodulin (CM) shares considerable amino acid sequence homology with troponin C (TnC), but, unlike TnC, calmodulin contains a single residue of trimethyllysine (TML). Calmodulin has also been reported to serve as a substrate for protein carboxymethyl transferase; the functional significance of these post-translational modifications of CM is not known, nor has their enzymology been investigated. The present report describes our study of calmodulin methylation.

High speed supernatant fractions of several rat tissues were incubated with S-adenosyl-L-[methyl-$^3$H]-methionine ($^3$H-SAM) in 50 mM Tris-HCl and the methylated peptide products were resolved by polyacrylamide gradient (6–20%) SDS gel electrophoresis (PAGE) and identified by fluorography. In all tissues tested (brain, liver, heart, kidney, and testes) calmodulin was a major methylated peptide; the patterns of methylation of other peptides varied markedly from one tissue to another. The inclusion of 2 mM EGTA in the incubation buffer inhibited the methylation of calmodulin, but not of other peptides. The pH of this electrophoretic separation was 8.6; carboxymethyl groups are unstable under these conditions and only N-methylated peptides survive and are seen in the fluorograms. All radioactive bands disappear if pronase is added prior to electrophoresis; the addition of RNase or DNase has no effect, indicating that peptides, rather than nucleic acids are the observed methylated species. The addition of purified exogenous calmodulin to the incubation medium inhibited the N-methylation of calmodulin but not other peptides.

The enzyme responsible for N-methylation of calmodulin has been partially purified by DEAE-cellulose chromatography; when active fractions from this procedure are incubated with boiled substrate (0.3 M NaCl eluate from DEAE-cellulose) and $^3$H-SAM, the only radioactive peptide seen after electrophoresis is calmodulin. Tritiated methyl groups were shown to be incorporated into trimethyllysine, using a new procedure for the rapid chromatographic separation of N-methylated amino acids. The pH optimum of this enzyme, calmodulin lysine N-methyl transferase (CLNMT), was found to be 7.8; the activity is dependent upon the presence of a divalent cation (2 mM $Ca^{2+}$, $Mg^{2+}$, or $Mn^{2+}$). Lineweaver-Burk plots indicate linear kinetics with a $K_m$ for SAM of 11 $\mu$M. Activity of CLNMT increased with calmodulin concentrations in the range of 1 ng to 5 $\mu$g per assay tube, and decreased with greater amounts of calmodulin.

An assay for calmodulin carboxymethyltransferase was developed, utilizing phenothiazine affinity chromatography of $^3$H-methylcalmodulin formed *in vitro*. The pH of this enzyme is 6.0 and it is stimulated by levels of added calmodulin which inhibit calmodulin lysine N-methyltransferase (20 $\mu$g).

*Supported by Public Health Service Grant NS 11652.



0077-8923/80/0356-0412 $01.75/0 © 1980, NYAS

# THE RELATIONSHIP OF THE STRUCTURE TO THE FUNCTION OF CALMODULIN IN THE MYOSIN LIGHT CHAIN KINASE SYSTEM

A. C. Nairn, R. J. A. Grand, C. M. Wall, and S. V. Perry

*Department of Biochemistry*
*University of Birmingham*
*P.O.B. 363 Birmingham B15 2TT*
*United Kingdom*

The myosin light chain kinase (MLCK) system of rabbit fast skeletal muscle consists of a protein of molecular weight 77,000, the catalytic component, and calmodulin. The catalytic subunit has no activity in the absence of calmodulin.[1] A $Ca^{2+}$-dependent interaction was demonstrated between the two subunits using a Sepharose-calmodulin affinity column,[2] but a complex could not be seen on polyacrylamide gel electrophoresis.[1] In contrast, a $Ca^{2+}$-dependent, urea-stable complex, has been shown using gel electrophoresis between calmodulin and troponin I,[3] myelin basic protein,[4] and phosphorylase kinase.[5] The differences in these properties suggested there may be at least two classes of interaction between calmodulin and calmodulin-binding proteins. An attempt was made, therefore, to study the relationship of the structure to the function of the calmodulin molecule with respect to the different types of interaction.

The effect of troponin I and T, the myelin basic protein on calmodulin-activated MLCK was examined. The assays were performed as described by Nairn and Perry[1] with the order of addition such that calmodulin and troponin I, etc. were mixed prior to the addition of the catalytic subunit. A calmodulin to kinase molar ratio of 1.6:1 was chosen, in which case, about 80% of the maximum kinase activity was obtained.[1] Addition of troponin I only inhibited activity when added in excess of 100 moles per mole of calmodulin. Up to 80% of the original activity was inhibited with addition of 1,000 moles of troponin I. With troponin T and myelin basic protein similar results were obtained, although both proteins were even less effective than troponin I.

The effect of proteolytic fragments obtained from calmodulin and troponin C on the activity of MLCK was studied. Significant activation of the kinase was found with a peptide obtained by tryptic or thrombic digestion containing residues 1–106 and a peptide containing residues 1–90 obtained by tryptic digestion. Although complete activation was obtained with the peptide-containing residues 1–106, this was approximately 100 times less active than whole calmodulin on a molar basis. The second active peptide was a further two orders of magnitude less active and was unable to fully activate the kinase at a 3,000-fold molar excess. Several other calmodulin, as well as troponin C, peptides were prepared following cleavage with either trypsin, thrombin, or cyanogen bromide but these were unable to activate the kinase to significant levels.

The lack of the effects of troponin I, etc., suggested the differences in the properties of the two types of interaction may be a function of the region of the calmodulin molecule involved. It is not clear, however, where these specific regions are in the calmodulin molecule. A peptide obtained by cyanogen bromide digestion containing residues 77–124 and which interacted with



0077-8923/80/0356-0413 $01.75/0 © 1980, NYAS

troponin I[6] did not activate MLCK. It is probable, however, that the whole calmodulin molecule is necessary for activation of MLCK, involving a large number of interactions and, therefore, substantial elements of the three-dimensional structure of the protein.

## References


1. Nairn, A. C. & S. V. Perry. 1979. Biochem. J. **179:** 89–97
2. Nairn, A. C. & S. V. Perry. 1979. Biochem. Soc. Trans. **7:** 966–967
3. Amphlett, G. W., T. C. Vanaman & S. V. Perry. 1976. FEBS Lett. **72:** 163–168
4. Grand, R. J. A. & S. V. Perry. 1979. Biochem. J. **183:** 285–295
5. Shenolikar, S., P. T. W. Cohen, P. Cohen, A. C. Nairn & S. V. Perry. 1979. Eur. J. Biochem. **100:** 329–337
6. Vanaman, T. C. & S. V. Perry. 1978. Prog. Eur. Cong. Muscle Mobility, 7th Abstr. p. 39

# NEUROLEPTIC DRUGS ARE NONSTEREOSPECIFIC INHIBITORS OF CALMODULIN-STIMULATED PHOSPHODIESTERASE ACTIVITY

Jon A. Norman

*Friedrich Miescher-Institut*
*CH-4002 Basel*
*Switzerland*

Calmodulin, or the calcium-dependent regulator protein (CDR), is a $Ca^{2+}$ binding protein that satisfies many of the requirements for being an intracellular $Ca^{2+}$ receptor. When calmodulin is complexed with $Ca^{2+}$, it can activate a variety of enzyme systems including the enzymes controlling cyclic nucleotide metabolism and protein kinases.[1] The high affinity $Ca^{2+}$ binding properties of calmodulin in addition to its activating properties for many enzymes form the basis of the hypothesis that calmodulin is the link between intracellular $Ca^{2+}$ availability and the $Ca^{2+}$-dependent activation of these enzymes. If calmodulin is an important regulatory protein, then compounds that can inhibit its stimulation of these enzyme systems may be clinically important or be valuable probes for elucidating the physiological role of calmodulin.

It was reported by Levin and Weiss[2] that micromolar concentrations of a variety of structurally unrelated neuroleptic compounds, including phenothiazines, could inhibit the calmodulin-dependent activation of cyclic AMP phosphodiesterase. Neuroleptic compounds are used clinically as major tranquilizers in the control of schizophrenia and the possibility that these compounds may be clinically effective through a calmodulin-dependent process was proposed by Levin and Weiss.[2] To test this hypothesis, the clinically active and inactive isomers of three neuroleptic compounds butaclamol, flupenthixol, and chlorprothixene, as well as six other neuroleptic compounds for which isomers do not exist, were tested in this study as inhibitors of calmodulin-stimulated phosphodiesterase activity. It was reasoned that if inhibition of calmodulin-stimulated phosphodiesterase activity is related to the clinical efficacy of these drugs, then only the clinically active isomer of these compounds should be inhibitory. The results indicate that both isomers of butaclamol, flupenthixol, and chlorprothixene are equally potent as calmodulin-stimulated phosphodiesterase inhibitors. Therefore, the clinical effects of these compounds cannot be mediated through calmodulin. To further understand the nature of this nonstereospecific inhibition of calmodulin-stimulated phosphodiesterase activity, the octanol: water partition coefficients of these drugs were determined as an estimate of their lipid solubility. It was found that the octanol:water partition coefficients for these compounds correlated very well with the concentration necessary to cause half-maximal inhibition of calmodulin-stimulated phosphodiesterase activity.[3] Therefore, it is possible that these neuroleptic drugs inhibit calmodulin-stimulated phosphodiesterase activity through nonstereospecific hydrophobic interactions with calmodulin. Since the clinically active and inactive isomers of butaclamol, flupenthixol, and chlorprothixene have the same octanol:water partition coefficients, it is unlikely that these isomers could differentially distribute themselves inside the cell and therefore selectively inhibit calmodulin functions as a mechanism for their clinical efficacy.



0077-8923/80/0356-0415 $01.75/0 

Since evidence is lacking for the physiological role of calmodulin and due to the lack of mutants for calmodulin, the neuroleptic agents described above may be potential probes for calmodulin's function *in vivo*. As a precaution, neuroleptic drugs have been shown to influence a vast array of membrane-associated events and enzyme activities[4] and the possibility of *in vivo* artifacts must be considered.

## References


1. Cheung, W. Y. 1980. Calmodulin plays a pivotal role in cellular regulation. Science **207:** 19–27.
2. Levin, R. M. & B. Weiss. 1976. Mechanism by which psychotropic drugs inhibit adenosine cyclic 3′, 5′-monophosphate phosphodiesterase of brain. Mol. Pharmacol. **12:** 581–589.
3. Norman, J. A., A. H. Drummond & P. Moser. 1979. Inhibition of calcium-dependent regulator-stimulated phosphodiesterase activity by neuroleptic drugs is unrelated to their clinical efficacy. Mol. Pharmacol. **16:** 1089–1094.
4. Seeman, P. 1972. The membrane actions of anaesthetics and tranquilizers. Pharmacol. Rev. **24:** 583–655 (1972).

# CALMODULIN-MEDIATED PHOSPHORYLATION OF SYNAPTOSOMAL CYTOSOLIC PROTEINS

James P. O'Callaghan,* Lawrence A. Dunn, and Walter Lovenberg

*Section on Biochemical Pharmacology*
*Hypertension-Endocrine Branch*
*National Heart, Lung and Blood Institute*
*Bethesda, Maryland 20205*

Several subcellular fractions of central nervous system tissue have been found to contain endogenous protein kinase activities that can be regulated by calcium ions. These fractions include: synaptosomes,[1] synaptic membranes,[2,3] synaptic vesicles,[4] and postsynaptic densities.[5] In addition to its effect on protein kinase systems associated with these membrane fractions, calcium may regulate the phosphorylation of specific substrates found in the soluble compartment. That a calcium-regulated phosphorylation system in neuronal cytosol may play a role in synaptic function is indicated by several findings. Enzymes that control the biosynthesis of neurotransmitters, e.g., tyrosine hydroxylase, and tryptophan hydroxylase, are activated under phosphorylating conditions[6,7] and can be regulated in a calcium-dependent manner.[7,8] Calmodulin, the calcium binding protein, is required for the calcium-regulated phosphorylation of several specific neuronal proteins[2-5] and has also been shown to be required for the activation of tryptophan hydroxylase[9,10] when incubations are carried out under phosphorylating conditions and in the presence of calcium. Thus the possibility exists that specific calcium-calmodulin regulated protein kinases and their associated substrates are present in the cytosol and may play an important role in neuronal function. In the present report, we present evidence that calcium regulates the phosphorylation of several specific synaptosomal cytosolic proteins, an effect that was found to be dependent on the presence of calmodulin.

## Materials and Methods

### *Materials*

The following chemicals obtained from commercial sources were used in this study: Sepharose 4B, $CaCl_2$, EGTA, and [$\gamma^{32}$P]ATP (10–40 Ci/mmole), Fluphenazine HCl was generously provided by the Squibb Institute of Medical Research. All other materials were of the highest purity commercially available. Fluphenazine-sepharose was prepared by the means of the bisoxirane technique as described.[11]

*Staff Fellow, Pharmacology Research Associate Program, National Institute of General Medical Sciences, National Institutes of Health



0077-8923/80/0307-0417 $01.75/0 © 1980, NYAS

### Tissue Preparation

The subjects were male Wistar rats (Charles River). Following sacrifice by decapitation, synaptosomes were isolated from a crude $P_2$ fraction of the neostriatum, an area of brain containing a high concentration of calmodulin.[12] A synaptosomal lysate was prepared[13] and centrifuged at 140,000 × g for 90 minutes. The resulting supernatant fraction was used as a source of synaptosomal cytosol. Proteins were assayed according to the method of Bradford.[14]

### Phosphorylation Assay

The standard phosphorylation mixture was as previously described.[15] The incubation time was one minute and, when indicated, EGTA, $CaCl_2$, and fluphenazine were included in the assay at the concentrations listed. In some instances, endogenous calmodulin was removed from synaptosomal cytosol by affinity chromatography on columns previously equilibrated in assay buffer. The column eluates obtained following this procedure were then assayed for phosphorylation. The SDS-polyacrylamide gel electrophoresis (SDS-PAGE) system used in this investigation has been described.[16]

### Autoradiography and Microdensitometry

Following SDS-PAGE, the gels were fixed in acetic acid and stained with Coomassie blue. The gels were then destained by diffusion before drying under heat and vacuum. Autoradiography was performed for 12–24 hours using Kodak

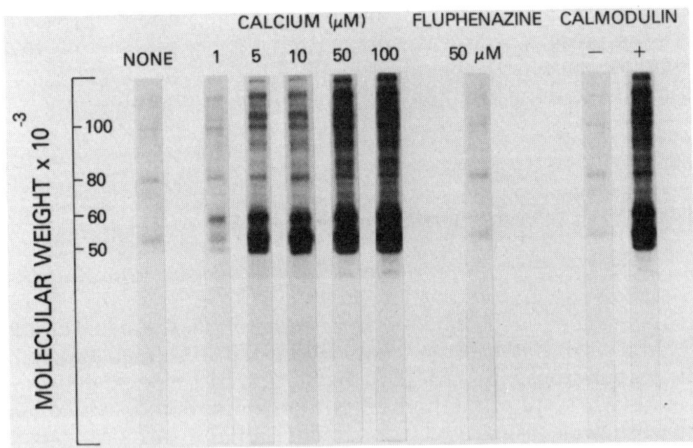


FIGURE 1. Autoradiographs showing the effects of calcium chloride, fluphenazine and fluphenazine-sepharose chromatography on the phosphorylation of synaptosomal cytosolic proteins. Calcium was added to the standard phosphorylation assay at the concentrations listed. Fluphenazine (50 μM) was included in the phosphorylation assay in the presence of 50 μM calcium. The autoradiographs in the last two lanes show the phosphorylation of the cytosol proteins in the presence of calcium chloride (50 μM) following the removal (−) of calmodulin and readdition (+) of exogenous calmodulin (0.5 μg).

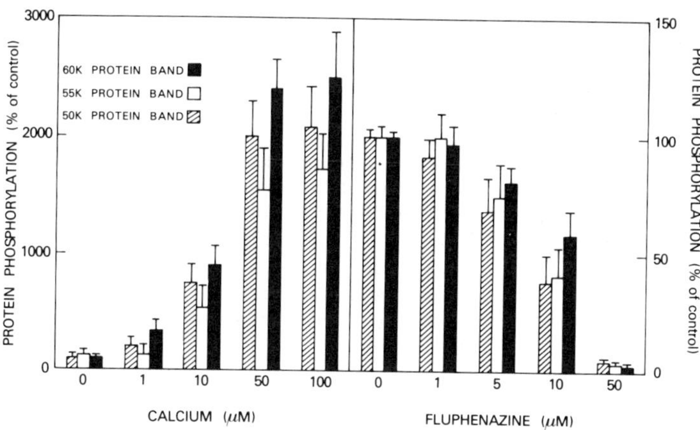


FIGURE 2. Phosphorylation of specific synaptosomal cytosolic proteins as a function of calcium concentration and as a function of fluphenazine concentration in the presence of 50 μM calcium chloride. The net incorporation of phosphate into specific proteins was quantified from the densitometric scans of the autoradiographs and is expressed as a percent of control phosphorylation.

RP X-ray film. Molecular weights of specific protein bands were estimated from molecular weight standards. The incorporation of phosphate into specific protein bands was determined by quantitative microdensitometry as described.[17] The background darkness of the autoradiographs was taken as the baseline value for these determinations.

## Results and Discussion

Autoradiographs were obtained for the synaptosomal cytosolic proteins that were resolved by SDS-PAGE following phosphorylation in the presence and absence of calcium. The data in FIGURE 1 indicates that calcium (1.0–100 μM) stimulated the phosphorylation of several specific synaptosomal cytosolic proteins. These effects of calcium were both concentration and time dependent and were most apparent for proteins with molecular weights ($M_r$) of 50,000, 55,000, and 60,000. Quantitative data for the effects of calcium on the phosphorylation of these three protein bands are presented in FIGURE 2. Calcium caused a linear increase in the phosphorylation of all three protein bands within the 1.0–50 μM range. Maximal stimulation of phosphorylation by calcium was as great as 25-fold above control levels. The calcium concentrations found to affect specific protein phosphorylation were consistent with those that might be encountered intraneuronally following depolarization.[18] In the absence of added calcium, the calcium chelator, EGTA, did not lower the phosphorylation of any protein; however, EGTA was completely effective in blocking calcium-stimulated phosphorylation (data not shown).

Levin and Weiss[19] have demonstrated that, *in vitro*, a number of antipsychotic drugs bind to calmodulin with high affinity in a calcium-dependent fashion. We exploited this property of these drugs in order to examine the role of calmodulin in the calcium stimulated phosphorylation of synaptosomal cytosolic proteins. Two approaches were chosen. In the first, the antipsychotic, fluphenazine, was added to the phosphorylation assay to serve as an antagonist of the effects of calmodulin on protein phosphorylation. In the second, fluphenazine was linked to sepharose in order to serve as an affinity matrix for the removal of calmodulin from synaptosomal cytosol. The addition of fluphenazine (1.0–50 $\mu$M) to the phosphorylation assay caused a concentration-dependent decrease in calcium stimulated phosphorylation of the 50,000, 55,000, and 60,000 protein bands (FIGURE 2). In the presence of 50 $\mu$M fluphenazine, calcium stimulated phosphorylation was reduced to levels equal to those observed in the absence of calcium (FIGURES 1 & 2). Fluphenazine added to the standard assay in the absence of calcium had no effect on phosphorylation (data not shown). When synaptosomal cytosol was subjected to affinity chromatography on a fluphenazine-sepharose matrix it was found that this procedure was effective in removing endogenous calmodulin and resulted in an abolition of calcium stimulated phosphorylation (FIGURE 1). Responsiveness to calcium could be completely restored by the addition of calmodulin (0.5 $\mu$g) to the phosphorylation assay (FIGURE 1) suggesting that fluphenazine does indeed block calcium-regulated phosphorylation by binding to endogenous calmodulin.

The findings of the present investigation satisfy most of the criteria for calmodulin-regulated reactions that have been proposed by Cheung.[20] To understand the function of calmodulin in relation to protein phosphorylation in neuronal cytosol, future efforts must be devoted toward an identification of the calcium-, calmodulin-regulated protein kinases and their corresponding phosphoprotein substrates.

## REFERENCES


1. KRUEGER, B. K., J. FORN & P. GREENGARD. 1977. Depolarization-induced phosphorylation of specific proteins, mediated by calcium ion influx, in rat brain synaptosomes. J. Biol. Chem. **252:** 2764–2773.
2. DELORENZO, R. J., G. P. EMPLE & G. H. GLASER. 1977. Regulation of the level of endogenous phosphorylation of specific brain proteins by diphenylhydantoin. J. Neurochem. **28:** 21–30.
3. SCHULMAN, H. & P. GREENGARD. 1978. $Ca^{2+}$-dependent protein phosphorylation system in membranes from various tissues, and its activation by "calcium-dependent regulator." Proc. Natl. Acad. Sci. USA **75:** 5432–5436.
4. DELORENZO, R. J., S. D. FREEDMAN, W. B. YOHE & S. C. MAURER. 1979. Stimulation of $Ca^{2+}$-dependent neurotransmitter release and presynaptic nerve terminal protein phosphorylation by calmodulin and a calmodulin-like protein isolated from synaptic vesicles. Proc. Natl. Acad. Sci. USA **76:** 1838–1842.
5. GRAB, D. J., K. BERZINS, R. S. COHEN & P. SIEKEVITZ. 1979. Presence of calmodulin in postsynaptic densities isolated from canine-cerebral cortex. Proc. Natl. Acad. Sci. USA **254:** 8690–8696.
6. AMES, M. M., P. LERNER & W. LOVENBERG. 1978. Tyrosine hydroxylase: activation by protein phosphorylation and end-product inhibition. J. Biol. Chem. **253:** 27–31.
7. KUHN, D. M., R. L. VOGEL & W. LOVENBERG. 1978. Calcium-dependent activation of tryptophan hydroxylase by ATP and magnesium. Biochem. Biophysical Res. Comm. **82:** 759–766.
8. SIMON, J. R. & R. H. ROTH. 1979. Striatal tyrosine hydroxylase: Comparison of the activation produced by depolarization and dibutyryl-cAMP. Mol. Pharmacol. **16:** 224–233.
9. KUHN, D. M., J. P. O'CALLAGHAN, J. C. JUSKEVICH & W. LOVENBERG. 1980. Activation of

brain tryptophan hydroxylase by ATP-$Mg^{++}$: Dependence on calmodulin. Proc. Natl. Acad. Sci. USA **77:** 4688–4691.

10. Yamauchi, T. & H. Fujisawa. 1979. Activation of tryptophan 5-mono-oxygenase by calcium-dependent regulator protein. Biochem. Biophys. Res. Comm. **90:** 28–35.
11. Charbonneau, H. & M. J. Cormier. 1979. Purification of plant calmodulin by fluphenazine-sepharose affinity chromatography. Biochem. Biophys. Res. Comm. **90:** 1039–1047.
12. Egrie, J. C., J. A. Campbell, A. L. Flangas & F. L. Siegel. 1977. Regional cellular and subcellular distribution of calcium-activated cyclic nucleotid phosphodiesterase and calcium-dependent regulator in porcine brain. J. Neurochem. **28:** 1207-1213.
13. Jones, D. H. & A. I. Matus. 1974. Isolation of synaptic plasma membrane from brain by combined flotation-sedimentation density gradient centrifugation. Biochim. Biophys. Acta **356:** 276–287.
14. Bradford, M. M. 1976. A rapid and sensitive method for the quantitation of microgram quantities of protein utilizing the principle of protein-dye binding. Anal. Biochem. **72:** 248–254.
15. O'Callaghan, J. P., J. Juskevich & W. Lovenberg. 1980. Stimulation of synaptic membrane phosphorylation by a calcium and calmodulin independent heat stable cytosol factor. Biochem. Biophys. Res. Comm. **95:** 82–89.
16. O'Farrell, P. 1975. High resolution two-dimensional electrophoresis of proteins. J. Biol. Chem. **250:** 4007–4021.
17. Ueda, T., H. Maeno & P. Greengard. 1973. Regulation of endogenous phosphorylation of specific proteins in synaptic membrane fractions from rat brain by adenosine 3':5'-monophosphate. J. Biol. Chem. **248:** 8295–8305.
18. Blaustein, M. P., R. W. Ratzlaff & N. K. Kendrick. 1978. The regulation of intracellular calcium in presynaptic nerve terminals. Ann. N.Y. Acad. Sci. **307:** 195–212.
19. Levin, R. M. & B. Weiss. 1979. Selective binding of antipsychotics and other psychoactive agents to the calcium-dependent activator of cyclic nucleotide phosphodiesterase. J. Pharmacol. Exp. Ther. **208:** 454–459.
20. Cheung, W. Y. 1980. Calmodulin plays a pivotal role in cellular regulation. Science **207:** 19–27.

# TISSUE LEVELS OF CALMODULIN IN EXPERIMENTAL DIABETES*

B. Perez de Gracia, G. Ahluwalia, A. R. Rhoads, and W. L. West

*Howard University*
*College of Medicine*
*Washington, D.C. 20059*

Since calcium has been implicated as a mediator of insulin action and is known to regulate cyclic nucleotide turnover through calmodulin, cellular levels of calmodulin activity were examined in diabetic animals and compared with controls. Sprague-Dawley rats were injected with 60 mg/kg of alloxan via the tail vein and assessed for renal glucosuria after 48 hours. Blood glucose exceeded 200 mg/dl in the diabetic groups. Tissues were homogenized in 0.33 M sucrose buffered with 40 mM Tris-HCl, pH 7.4. Total homogenates were diluted with deionized water, heated at 100°C for 3 minutes and centrifuged to remove denatured protein. Calmodulin was measured in the supernatant by the degree of activation of either calmodulin-deficient phosphodiesterase of bovine cerebrum or the ($Ca^{2+}$ + $Mg^{2+}$)-ATPase of human erythrocyte ghosts. Calmodulin activity was expressed as units (50% activation of target enzyme) per mg of tissue protein in undiluted homogenate. Significant increases ($P < 0.001$) in calmodulin activity were observed in uterus, liver, cardiac ventricle, and adipose tissue of diabetic rats when compared to controls. Both assay procedures yielded essentially the same results. No alterations in calmodulin activity were observed in kidney, diaphragm, and cerebral cortex of diabetic animals. Observed increases in calmodulin activity were lower using the ATPase assay (47–82%) when compared to the phosphodiesterase-based assay (89–170%). From standard curves constructed with pure calmodulin, these observed increases correspond to increments ranging from 15 to 50 ng of calmodulin per mg of tissue protein. Decreased cyclic AMP phosphodiesterase activity was also observed in the same tissues which showed increased calmodulin activity. Aside from tissue responsiveness, relationships between cyclic AMP and cyclic GMP phosphodiesterase and calmodulin activity were not apparent. The calcium-dependent, nondialyzable, and heat-stable properties associated with the observed increases in activity are consistent with increased calmodulin levels in tissues from alloxan-induced diabetic rats.

*Supported by National Institutes of Health Grant AM 19364-01.



0077-8923/80/0356-0422 $01.75/0 © 1980, NYAS

# CALMODULIN IN MUSSEL GILL EPITHELIAL CELLS: ROLE IN CILIARY ARREST*

William Reed and Peter Satir

*Department of Anatomy*
*Albert Einstein College of Medicine*
*Bronx, New York 10461*

Ciliary arrest is an important physiological event in ciliated epithelial cells from the gill of freshwater mussels. *In vivo*, mussel gill lateral (L) cell cilia spontaneously stop beating (arrest) or arrest may be induced experimentally by nervous or direct mechanical stimulation.[1] Arrest, like behavioral controls of ciliary motion in many organisms, has been shown to be mediated by an increase in cytoplasmic $Ca^{2+}$ concentration from ca. 0.1 $\mu$M to $> 1$ $\mu$M, which results from an influx of extracellular $Ca^{2+}$ into the ciliary axoneme.[2-4] The effect of $Ca^{2+}$ has been shown to be directly on the axonemal proteins themselves, since arrest can be reproduced in Triton-treated, membraneless isolated axonemes that have been reactivated by exogenous ATP.[2] Since calmodulin has been implicated as a sensor of cytoplasmic $Ca^{2+}$ transients in many processes, we have begun to look for calmodulin in mussel gill epithelial cells as a potential axonemal $Ca^{2+}$ binding protein that initiates ciliary arrest.

Cells were isolated in a live state from mussel gill (e.g. *Elliptio*) by gentle shaking at 30°C in solutions containing 50 mM KCl for four to six hours. Under these conditions, L-cells and other epithelial cell types detach from the substratum and float free in solution, leaving nerve and muscle attached to the tissue. The epithelial cells are collected by low speed centrifugation and analyzed for the presence of calmodulin.

As is shown in FIGURE 1, homogenates of gill epithelial cells contain a heat-stable factor that will substitute for authentic calmodulin in the stimulation of bovine brain cAMP phosphodiesterase (PDE). Stimulation is blocked in the presence of 3.3 mM EGTA or 50 $\mu$M trifluoperazine (TFP). The dose response characteristics of TFP inhibition of PDE stimulated by homogenates (data not shown) indicate that 14 $\mu$M TFP is sufficient to inhibit 50% of the stimulation and that 25 $\mu$M TFP returns PDE activity to near basal levels.

On alkaline urea polyacrylamide gels, authentic calmodulin in crude extracts appears as a rapidly migrating component in the absence, but not the presence, of $Ca^{2+}$.[5] Gill epithelial cells homogenized in 6 M urea and subjected to alkaline urea polyacrylamide gel electrophoresis in the absence of $Ca^{2+}$ are shown in FIGURE 2 (lane 2). The gel contains two rapidly migrating polypeptides: the lower component comigrates with brain calmodulin (FIGURE 2, lane 1). When the doublet region is eluted from the gel, the eluate is found to be capable of stimulating PDE activity. If $Ca^{2+}$ is included in the homogenization medium, the doublet is not present (FIGURE 2, lane 3). Therefore, both polypetides are apparently acidic $Ca^{2+}$ binding proteins. By these criteria, the lower component has tentatively been identified as gill epithelial cell calmodulin; the identity of the upper component is as yet unknown.

*Supported by U.S. Public Health Service Grants HL 22560 and GM 27859.



0077-8923/80/0356-0423 $01.75/0 © 1980, NYAS

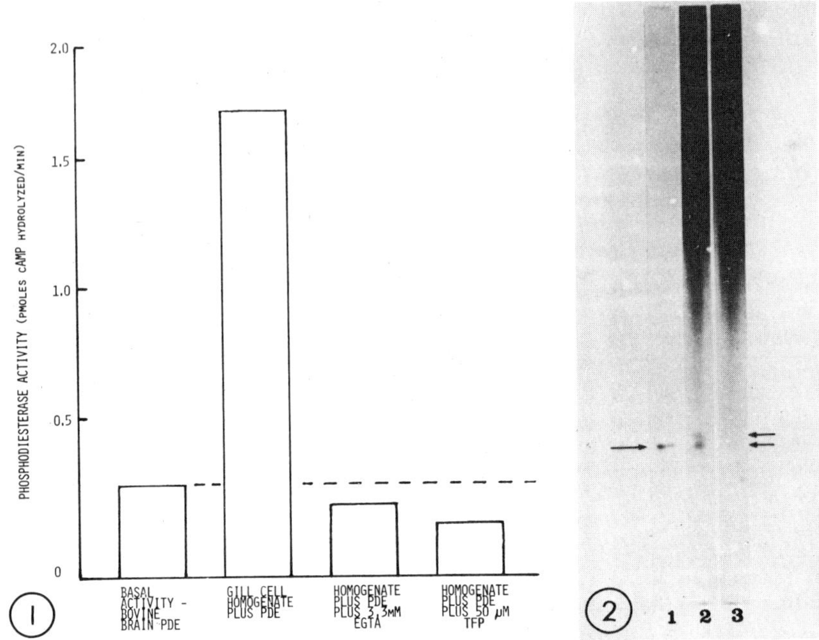


FIGURE 1. Stimulation of cAMP phosphodiesterase by a homogenate of epithelial cells that has been heated for 7.5 minutes at 85°C. Stimulation is blocked by EGTA and trifluoperazine. EGTA and TFP reduce the PDE activity to sub-basal levels.

FIGURE 2. Alkaline urea polyacrylamide gel electrophoresis (pH 8.6, 8 M urea) of bovine brain calmodulin (lane 1, single arrow), urea extract of gill cells in 5 mM EGTA (lane 2), and in 1 mM $CaCl_2$ (lane 3). Double arrows indicate the position of the doublet when it is present.

We have attempted to localize calmodulin to the ciliary axoneme and demonstrate its function by physiological means. If calmodulin mediates the arrest response of L-cell cilia, direct application of appropriate concentrations of TFP to Triton-treated membraneless ciliary axonemes should abolish ciliary arrest in the presence of high $Ca^{2+}$. Such a TFP-shunt for physiological processes has been discussed by B. Satir *et. al.* in this volume. We employed 25 μM TFP for these experiments, since this concentration should give detectable effects on gill calmodulin-mediated processes, while minimizing nonspecific binding to ciliary components.

Cell models were prepared by extracting an aliquot of epithelial cells in 0.02% Triton X-100, 1 mM EGTA, 25 mM HEPES, 75 mM KCl, pH 7.0 for 10–30 seconds and then diluting 10× with 25 mM HEPES, 75 mM KCl, pH 7.0 (wash solution). The permeabilized cells are used within 10 minutes of extraction. A typical L-cell model in wash solution is shown in FIGURE 3a. The cilia are stopped due to the depletion of endogenous ATP as the ciliary membrane is dissolved by the detergent. The disruption of the membrane with Triton ensures that $Ca^{2+}$, $Mg^{2+}$-ATP and TFP can be introduced at known concentrations directly at the axoneme, eliminating the possibility that the observed effects are modified by transmembrane events. When the models are perfused with wash solution containing 2.0 mM $Mg^{2+}$-ATP and 0.1 $\mu$M free $Ca^{2+}$ (as $Ca^{2+}$-EGTA buffers), the cilia generally reactivate (FIGURE 3b). In qualitative terms, normal beat is restored. When, subsequently, the $Ca^{2+}$ is increased by perfusion with stop solution (2.0 mM $Mg^{2+}$-ATP, 50 $\mu$M free $Ca^{2+}$), the cilia stop beating (FIGURE 3c). Ciliary arrest is reversed by perfusion with stop solution plus 25 $\mu$M TFP (compare FIGURES 3c & 3d). In FIGURE 3d, cilia at the top and right side of the model have reactivated (image blurred) although the cluster of cilia to the left has remained inactive. Cilia that do not reactivate in TFP are irreversibly arrested and will not reactivate when TFP is washed out and fresh reactivation solution is perfused in. In general, a significant population of cilia may be re-induced to

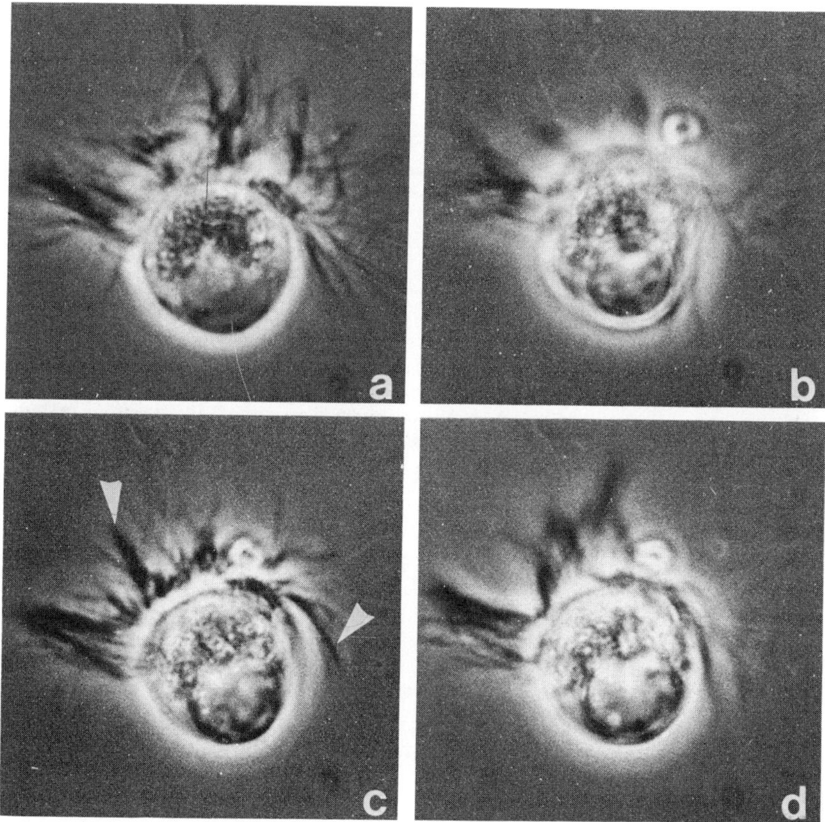


FIGURE 3. Reactivation, arrest, and reversal of arrest of L-cell ciliary models. $Ca^{2+}$ concentrations are adjusted according to the binding constants of ATP and EGTA for $Ca^{2+}$ and $Mg^{2+}$ at pH 7.0 and ionic strength 0.1. A sufficient volume of perfusate is used to ensure that endogenous sources and sinks of ATP and $Ca^{2+}$ do not significantly alter their concentrations in the perfusate. (a) Cell model in wash solution (b) after reactivation in 2.0 mM Mg-ATP, 0.1 $\mu$M free $Ca^{2+}$, (c) after arrest in stop solution containing 2.0 mM Mg-ATP, 50 $\mu$M $Ca^{2+}$, (d) and after reversal of arrest in stop solution containing 25 $\mu$M TFP. Under these photographic conditions, beating cilia appear blurred, stopped cilia are sharp. Arrows indicate two clusters of cilia that reactivate when TFP is added. All 1060 $\times$.

beat in concentrations of $Ca^{2+}$ fifty times greater than the minimum necessary for complete L-cell inhibition in the presence of ATP, if TFP is included in the solution. This suggests that a specific TFP-sensitive axonemal component is the intermediate between $Ca^{2+}$ and the arrest response.

We have shown here that in mussel gill cells, there are calmodulin-like heat-stable, low molecular weight $Ca^{2+}$ binding proteins that stimulate brain PDE. This stimulation is affected by TFP in concentrations appropriate to the TFP-shunt that also reverses ciliary arrest. The finding that calmodulin may be a functional component of gill cilia is compatible with the immunofluorescent localization of calmodulin in *Tetrahymena* cilia (Maihle and B. Satir, this volume).

The $Ca^{2+}$-sensitive switch responsible for arrest of the L-cell ciliary axoneme is thought to be located in the spoke head or central sheath, as part of the system that constrains microtubule sliding in the axoneme. It seems likely that calmodulin will prove localized to these structures, as well as possibly to the dynein arms.

## References

1. Satir, P., I. Fong & S. F. Goldstein. 1972. Laser induced neuroid transmission in a ciliated epithelium. Acta Protozool. **11:** 287–290.
2. Walter, M. & P. Satir. 1978. Calcium control of ciliary arrest in mussel gill cells. J. Cell Biol. **79:** 110–120.
3. Satir, P. 1975. Ionophore mediated calcium entry induces mussel gill ciliary arrest. Science **190:** 586–588.
4. Reed, W. & P. Satir. 1979. Spreading ciliary arrest in gill epithelium: relation to $Ca^{2+}$ influx. (Abstr.) Biophys. J. **25:** 296a.
5. Grand, R. J. A., S. V. Perry & R. A. Weeks. 1979. Troponin C-like proteins (calmodulins) from mammalian smooth muscle and other tissues. Biochem. J. **177:** 521–529.

# IDENTIFICATION OF CALMODULIN IN MITOCHONDRIA FROM RAT LIVER: A POSSIBLE ROLE IN REGULATION OF THE OLIGOMYCIN SENSITIVE ATPase

Larry Ruben, D. B. P. Goodman, and Howard Rasmussen

*Department of Internal Medicine*
*Yale University School of Medicine*
*New Haven, Connecticut 06510*

Although $Ca^{2+}$ is an important regulator of mitochondrial function, it is not clear that mitochondria, like the cytosol, use calmodulin as a coupling agent between $Ca^{2+}$ and the $Ca^{2+}$ response. Part of the problem stems from the lack of information concerning localization of calmodulin within the mitochondrial compartment. Previous cell fractionation studies,[1,2] or antibody localization studies[3] were not sufficient to detect mitochondrial calmodulin. In the present report, we describe the occurrence of calmodulin in mitochondria.

Rat liver mitochondria, isolated in the presence of 0.5 mM EGTA, were boiled and the supernatant passed over 2 ml DEAE-cellulose collumns. The mitochondrial extract that eluted in the presence of 0.4 M NaCl contained a heat-stable, $Ca^{2+}$ and chlorpromazine sensitive stimulator of brain phosphodiesterase which was presumed to be calmodulin ($34 \pm 14$ ng/mg protein, $n = 13$). The calmodulin associated with the mitochondrial fraction was still present after mitochondria were purified on 35%–50% w/w sucrose gradients ($28 \pm 18$ ng/mg protein, $n = 3$). Enzyme marker studies indicated that the plasma membrane contributed 0 ng calmodulin/mg mitochondrial protein while the endoplasmic reticulum contributed at most 3.6 ng calmodulin/mg mitochondrial protein in mitochondria purified on sucrose gradients.

Further evidence that calmodulin resides within mitochondria is derived from the observation that [$^3$H]trifluoperazine, a calmodulin binding agent, comigrated with mitochondria on sucrose gradients. Other calmodulin binding agents,[4] including: penfluridol (16 $\mu$M), pimozide (20 $\mu$M), and chlorpromazine (100 $\mu$M); specifically inhibited ADP-stimulated respiration without affecting uncoupler-stimulated or $Ca^{2+}$-stimulated respiration. This specific effect together with the observation that the inactive analogue, trifluoperazine sulfoxide (1 mM), does not inhibit ADP-stimulated respiration, suggest that calmodulin is involved in regulation of ADP phosphorylation. The site of regulation could occur at the level of adenine nucleotide translocation or the oligomycin-sensitive ATPase. Both processes are known to be affected by $Ca^{2+}$.[5,6] These two possibilities were distinguished by analyzing the distribution of labeled adenine nucleotides in the presence of calmodulin binding agents. Addition of calmodulin binding agents prior to the addition of [$^3$H]ADP (1.8 mM, 10 $\mu$Ci) did not alter the specific activity of the ADP pool, but did diminish by 70% the ratio of [$^3$H]ATP/[$^3$H]ADP. Therefore, it is concluded that mitochondria contain calmodulin, and this calmodulin plays a role in regulation of the oligomycin sensitive ATPase.



0077-8923/80/0356-0427 $01.75/0 © 1980, NYAS

REFERENCES

1. SMOAKE, J. A., S. SONG & W. Y. CHEUNG. 1974. Biochim. Biophys. Acta **341:** 402–411.
2. KAKIUCHI, S., R. YAMAZAKI, Y. TESHIMA, K. UENISHI, S. YASUDA, A. KASHIBA, K. AOBUE, M. OHSHIMA & T. NAKAJIMA. 1978. Adv. Cyclic Nuc. Res. **9:** 253–263.
3. HARPER, J. F., W. Y. CHEUNG, R. W. WALLACE, H. HUANG, S. N. LEVINE & A. L. STEINER. 1980. Proc. Natl. Acad. Sci. USA **77:** 366–370.
4. LEVIN, R. M. & B. WEISS. 1979. J. Pharmacol. Exp. Ther. **208:** 454–459.
5. GOMEZ-PUYOU, A., M. T. GOMEZ-PUYOU, M. KLAPP & E. CARAFOLI. 1979. Arch. Biochem. Biophys. **194:** 399–404.
6. YAMADA, E. W., F. H. SHIFFMAN & N. J. HUZEL. 1980. J. Biol. Chem. **255:** 267–273.

# EVIDENCE FOR THE INVOLVEMENT OF CALMODULIN IN ENDOCYTOSIS*

J. L. Salisbury, J. S. Condeelis, and P. Satir

*Department of Anatomy*
*Albert Einstein College of Medicine*
*Bronx, New York 10461*

Cells are equipped with the molecular machinery enabling them to internalize portions of their environment along several distinct pathways. Using the cultured human lymphoblastoid cell line WiL2, we have been studying two types of endocytosis:[3] (a) *bulk fluid phase internalization* of the surrounding milieu using uptake of exogenous horseradish peroxidase (HRP) as a marker, and (b) *receptor mediated endocytosis* along with a coated vesicle pathway in which cell surface receptor IgM is internalized after cross-linking with anti-IgM antibodies. In this brief report we present several observations that implicate the calcium-dependent regulator protein, calmodulin (CaM), in these two pathways.

WiL2 cells contain a heat-stable acidic protein of 18.5 kd mol. wt. that comigrates with bovine brain CaM (FIGURE 1), activates bovine brain phosphodiesterase (PDE) in a $Ca^{2+}$-dependent manner, and undergoes a $Ca^{2+}$-induced mobility shift on appropriate gels (FIGURE 1). On the basis of these criteria we believe this protein to be WiL2 CaM. This protein is a component of isolated coated vesicles in these cells[3] and elsewhere.[1]

Internalization of exogenous HRP by WiL2 cells is inhibited in a reversible manner by the drug trifluoperazine dihydrochloride (TFP) over a narrow drug concentration range (10–25 $\mu$M) (FIGURE 2). At drug concentrations of 50 $\mu$M or greater the effect on endocytosis is no longer reversible. These results suggest that a TFP-sensitive step is involved in bulk fluid phase endocytosis.

WiL2 cells have receptor IgM as an integral plasma membrane protein. These receptors are present in a uniform diffuse distribution on the cell surface.[3] Upon challenge with anti-IgM antibody the receptors become cross-linked and cluster in the plane of the membrane (FIGURE 3 a,b) and subsequently cap over one pole of the cell. The clustered receptors become associated with clathrin coats which assemble on the cytoplasmic side of the membrane resulting in the formation of coated pits (FIGURE 3c). These coated pits invaginate and vesiculate to form coated vesicles (FIGURE 3d) which subsequently deliver their ligand contents to the lysosomal compartment (FIGURE 3e). Treatment of the cells with the drug TFP (25 $\mu$M) has no effect on the formation of the ligand-receptor complex, or the cross-linking and clustering of receptors or on the redistribution of the clustered receptors into a cap. However, the drug does inhibit the formation of coated membrane opposite the clustered receptors so that less than half the control numbers of coated pits are scored after 15 min of TFP treatment (TABLE 1). Also, TFP nearly completely inhibits the transfer of ligand receptor complexes from the cell surface to the lysosomal compartment. These results implicate a TFP-

*Supported by the American Cancer Society (Grant BC-302) and the U.S. Public Health Service (Grant GM 25813 to J. S. C., Grants HL 22560 and GM 27859 to P.S.). J. S. C. is a Rita Allen Foundation Scholar. J. L. S. is a Leukemia Society of America Fellow.



0077-8923/80/0356-0429 $01.75/0 © 1980, NYAS

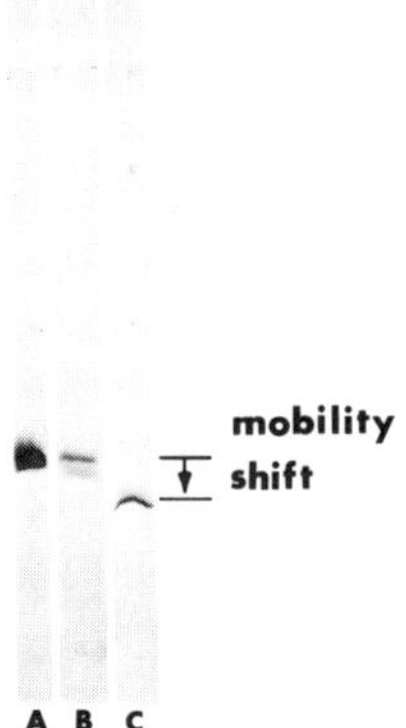


FIGURE 1. Gel electrophoregram (15% polyacrylamide slab gel with 0.1% SDS) of CaM from WiL2 cells illustrating the $Ca^{2+}$ induced mobility shift. Track A: Bovine Brain CaM (EGTA). Track B: WiL2 CaM (EGTA). Track C: WiL2 CaM ($Ca^{2+}$).

sensitive component in the recruitment of clathrin coats to clustered cell surface receptors.

It seems likely that the TFP-sensitive component is WiL2 CaM, especially since the concentration range for TFP inhibition is identical to that in which CaM stimulation of PDE is also inhibited (Weiss, this volume).[6] It is not entirely clear how CaM is involved in either the coated vesicle pathway or in bulk fluid phase endocytosis. Because of the TFP sensitivity, we would anticipate that a CaM-$Ca^{2+}$ complex is an intermediate between stimulus and endocytic response. Two hypotheses that we are currently testing are (1) an indirect link between CaM and endocytosis via the cell's motility machinery, perhaps involving myosin light

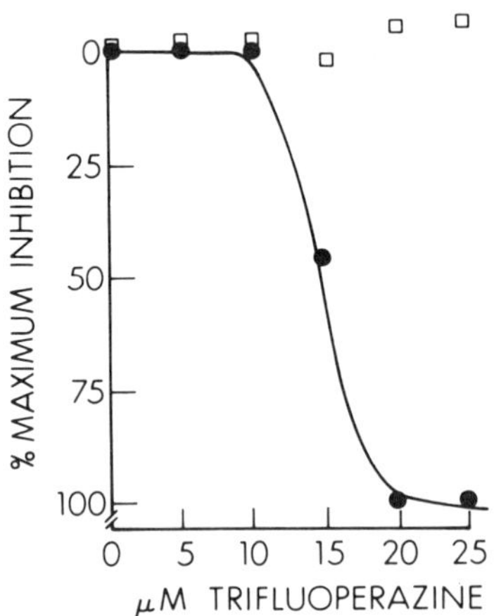


FIGURE 2. Bulk fluid phase uptake of HRP (according to the methods of Steinman[4]). Incubated in TFP at the concentrations indicated for 15 min at 24°C and then either washed three times (□) prior to incubation or incubated directly (●) in HRP and assayed spectroscopically.

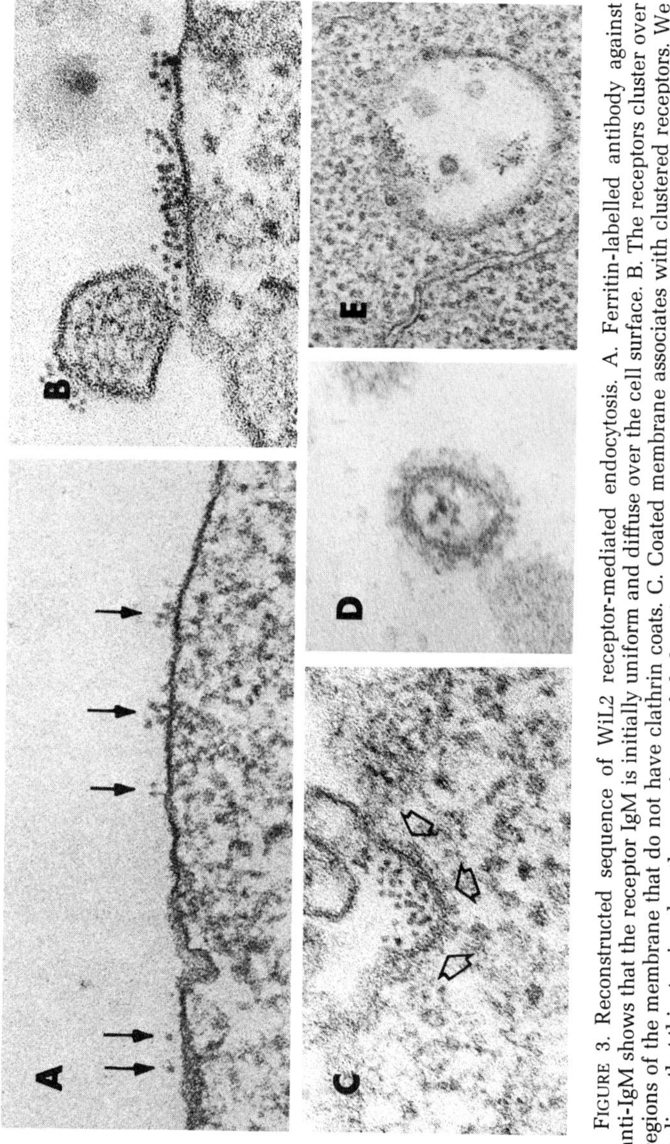


FIGURE 3. Reconstructed sequence of WiL2 receptor-mediated endocytosis. A. Ferritin-labelled antibody against anti-IgM shows that the receptor IgM is initially uniform and diffuse over the cell surface. B. The receptors cluster over regions of the membrane that do not have clathrin coats. C. Coated membrane associates with clustered receptors. We believe that this step involves the recruitment of clathrin from cytoplasmic pools since the total fraction of the cell surface that is coated increases 3 fold upon ligand challenge. This step is sensitive to TFP. D. Coated regions of the membrane invaginate and vesiculate to form coated vesicles in the cytoplasm. E. The ligand is finally delivered to the lysosomal compartment. A–D, 100,000×; E, 50,000×.

TABLE 1

EFFECT OF TRIFLUOPERAZINE ON LIGAND-RECEPTOR DISTRIBUTION RELATIVE TO MEMBRANE SPECIALIZATIONS

| | % Membrane Surface Occupied By Ferritin Clusters | | | |
|---|---|---|---|---|
| | At Cell Surface | | In Cytoplasmic Organelles | |
| | Unspecialized Membrane | Coated Membrane | Coated Vesicle Profiles | Large Vesicles (Lysosomes) |
| Initial Distribution at 37°C | 94 | 5 | 0 | 1 |
| Control After 15 min at 37°C | 34 | <u>36</u> | <u>7</u> | 22 |
| TFP Treated After 15 min at 37°C | 73 | <u>16</u> | <u>4</u> | 4 |

Based on morphometric analysis according to the methods of Weibel.[5] Data excerpted from Salisbury *et al.*[3] In TFP treated cells coated membrane compartment (underlined values) is less than half of the control.

chain kinase and (2) a more direct involvement in the signal for the regulated assembly or recruitment of clathrin coats under clustered receptor complexes by an interaction among CaM, $Ca^{2+}$, and a structural component of the coated vesicle.

## REFERENCES


1. LINDEN, C. D., T. F. ROTH & J. R. DEDMAN. 1979. The association of calmodulin with coated vesicles. J. Cell Biol. **83:** 289a.
2. SALISBURY, J. L., J. S. CONDEELIS & P. SATIR. 1979. Microfilaments and clathrin play a direct role in endocytosis of crosslinked cell surface immunoglobulin. J. Cell Biol. **83:** 72a.
3. SALISBURY, J. L., J. S. CONDEELIS & P. SATIR. 1980. The role of coated vesicles, microfilaments and calmodulin in receptor mediated endocytosis by cultured B lymphoblastoid cells. J. Cell Biol. **87:** 132–141.
4. STEINMAN, R. M. 1976. Horseradish peroxidase as a marker for studies of pinocytosis. *In* Vitro Methods of Cell-Mediated and Tumor Immunity. B. R. Bloom & J. R. David, Eds.: 379–386. Academic Press, Inc. New York, N.Y.
5. WEIBEL, E. R. 1969. Stereological principles for morphometry in electron microscopic cytology. Int. Rev. Cytol. **26:** 225–307.
6. WEISS, B. & R. M. LEVIN. 1978. Mechanism for selectively inhibiting the activation of cyclic nucleotide phosphodiesterase and adenylate cyclase by antipsychotic agents. *In* Advances in Cyclic Nucleotide Research, Vol. 9 W. J. George & L. J. Ignarro, Eds.: 285–304. Raven Press. New York, N.Y.

# NMR STUDIES ON TYROSINE-138 OF CALMODULIN

Kenneth Seamon

*Laboratory of Bioorganic Chemistry*
*National Institute of Arthritis and Metabolic Disease*
*Bethesda, Maryland 20205*

$^1$H-NMR has been utilized to study the metal ion-dependent conformations of calmodulin. The binding of calcium to calmodulin as revealed by spectral shifts of assigned resonances was rationalized in terms of a sequential binding of pairs of calcium ions at two sets of sites.[1] Previous studies by Klee and coworkers have established that tyrosine-138 of calmodulin is in a unique environment that is altered upon calcium binding to calmodulin.[2] These conclusions have been further explored using $^1$H-NMR.

As the first two moles of calcium are added to calmodulin; 1) the tyrosine-138 meta proton resonance appears upfield increasing in its intensity in parallel with a decrease in its intensity at its position corresponding to the apo-protein and, 2) the tyrosine-138 ortho proton resonance shifts upfield ~20 Hz. The perturbations of the tyrosine-138 resonances appear to be the result of more than one change in the local environment around the tyrosine ring during the course of the first conformational transition. The meta resonance exhibits slow exchange behavior which allows one to set an upper limit on the rate constant for the conformational transition from the relationship $k \leq 2\pi\Delta\nu$ where $k$ is the rate constant in $s^{-1}$ and $\Delta\nu$ corresponds to the separation in Hz for the two resonances in slow exchange. It is also possible to place a lower limit on the rate constant derived from the fast exchange behavior of the ortho resonance from the relationship $k \geq 2\pi\Delta\nu$ where $\Delta\nu$ corresponds to the separation in Hz between the initial and final resonance positions. Estimated rate constants which can be associated with the first conformational transition were calculated based on the above considerations and are given in TABLE 1. If it is assumed that binding of calcium to the high affinity sites is diffusion controlled ($k_{on} \sim 10^8$ $M^{-1}s^{-1}$) and that the kinetic constants in TABLE 1 are related to the first order rate constant for the decomposition of the calcium-calmodulin complex, then the thermodynamic association constant for the formation of the complex can be estimated (TABLE 1). These results are in excellent agreement with the association constants of $3 \times 10^5$ $M^{-1}$ and $8.6 \times 10^5$ $M^{-1}$ determined by Crouch and Klee[3] for the two higher affinity binding sites of calmodulin. The kinetic constant associated with the trimethyllysine resonance leads to an overestimation of the association constant for the high affinity sites. This may indicate that the off rate for calcium from the site(s) ultimately affecting the resonance is faster than the actual conformational transition. Since the behavior of resonances in slow exchange places only an upper limit on the rate constant for the environmental change, it is conceivable that the transition affecting the tyrosine-138 meta protons is governed by a much smaller rate constant than that associated with the ortho protons. The results of titrating apo-calmodulin with protons or magnesium ions seem to justify this conclusion. As magnesium is added to apo-calmodulin the ortho proton resonance exhibits an upfield shift and the meta proton resonance shifts slightly downfield. The shift downfield of the meta proton resonances in no way resembles the behavior of the resonance due to calcium binding. Similar behavior is exhibited by the reson-



0077-8923/80/0356-0433 $01.75/0 © 1980, NYAS

TABLE 1

| Resonance | Estimated $k(s^{-1})$ | Estimated $K_{assoc}(M^{-1})$ |
|---|---|---|
| Tyrosine-138 (meta protons) | $\leq 814$ | $\geq 1.2 \times 10^5$ |
| Tyrosine-138 (ortho protons) | $\geq 113$ | $\leq 8.8 \times 10^5$ |
| $\epsilon$-Trimethyllysine-115 | $\leq 40$ | $\geq 2.5 \times 10^6$ |

ances during the course of a proton titration. The meta proton resonance shifts downfield accompanied by a very small upfield shift of the ortho proton resonance.

## CONCLUSIONS

The anomalous chemical shift of the meta proton resonance is due at least in part to an environment determined by free carboxyl groups. Proton binding or magnesium binding to these groups affects the environment around the meta protons resulting in a downfield shift of the resonance closer to its normal unperturbed resonance position. The slow exchange behavior and large change in chemical shift of the meta protons due to calcium binding are therefore associated with a large conformational transition that is not induced by proton or magnesium binding. The upfield shift of the ortho protons is more closely associated with the exchange of metals or protons at a binding site and the kinetics of the environmental change affecting these protons are related to the dynamics of the metal ion exchange at the binding sites.

## REFERENCES

1. SEAMON, K. B. 1980. Biochemistry **19:** 207.
2. RICHMAN, P. G. & C. B. KLEE. 1979. J. Biol. Chem. **254:** 5372.
3. CROUCH, T. H. & C. B. KLEE. 1980. Biochemistry. **19:** 3692.

# CALMODULIN AND NEUROSECRETION

R. M. Sheaves and D. B. Hope

*University Department of Pharmacology*
*University of Oxford*
*Oxford, England OX1 3QT*

Vasopressin is secreted from the posterior pituitary gland by a calcium-dependent exocytotic mechanism. The divalent cation binding kinetics of calmodulin[1] suggest that it could be a possible link between calcium influx and vasopressin secretion.

Calmodulin was purified from bovine posterior pituitary glands and the yield (74 mg/100g wet weight tissue) showed that these glands are a rich source of the protein. A phosphodiesterase assay showed that calmodulin accounts for approximately 0.5% of the total tissue protein. A similar assay on a purified nerve ending fraction showed that calmodulin accounts for approximately 1.0% of the total tissue protein. Participation of the protein in secretory function is thus indicated.

The role of calmodulin in vasopressin secretion from isolated rat posterior pituitaries was studied using the finding of Levin and Weiss[2] that antipsychotic drugs bind with high affinity to calmodulin. These drugs (1 $\mu$M) were found to inhibit the potassium-stimulated release of vasopressin in the following order: Pimozide (22 ± 6.5% of control, $n$ = 5) Haloperidol (54 ± 5.3%, $n$ = 5) Trifluoperazine (63 ± 14.1%, $n$ = 5) Chlorpromazine (71 ± 7.3%, $n$ = 5) Pimozide, which binds strongly to calmodulin, is also a potent inhibitor of release. This effect may be due to the drugs ability to bind with high affinity to calmodulin. Alternatively, the effect could be produced by other actions of these drugs such as via dopamine receptor blockade. However, dopamine had no effect on potassium-stimulated release and did not alter pimozide inhibition of release.

We are investigating further the role of calmodulin in neurosecretion by examining its localization within nerve terminals and by considering the function of calmodulin-sensitive enzymes within particular subcellular compartments.

## REFERENCES

1. WOLFF, D. J., P. G. POIRER, C. O. BROSTROM & M. A. BROSTROM. 1977. J. Biol. Chem. **252:** 4108.
2. LEVIN, R. M. & B. WEISS. 1979. J. Pharmacol. Exp. Ther. **208:** 454.



0077-8923/80/0356-0435 $01.75/1 © 1980, NYAS

# ACTIVATION OF CALMODULIN BY TERBIUM ($Tb^{3+}$) AND ITS USE AS A FLUORESCENCE PROBE*

E. Ann Tallant,† Robert W. Wallace,‡ Michael E. Dockter,†‡ and Wai Yiu Cheung†‡

*‡Departments of Biochemistry*
*St. Jude Children's Research Hospital and*
*†University of Tennessee Center for the Health Sciences*
*Memphis, Tennessee 38101*

Terbium, a trivalent lanthanide, is an excellent probe for studying the interaction of calcium with its receptor proteins; it effectively replaces calcium in several biological systems, including the activation of $\alpha$-amylase and the conversion of prothrombin to thrombin and of trypsinogen to trypsin. The two cations have similar ionic radii and coordination numbers, and both bind preferentially to charged or uncharged oxygen atoms. These similarities compensate for the different valences of the two ions. Terbium has unique physical properties that are amenable to analysis by fluorescence, nuclear magnetic resonsance, and other physical techniques. The terbium fluorescence is of particular interest; it is greatly enhanced upon binding to proteins, a property highly useful to probe protein structure.

Terbium binds to calmodulin and the binding may be followed directly by exciting the terbium at 222 nm, or indirectly by exciting the tyrosine in calmodulin at 280 nm with subsequent fluorescence resonance energy transfer to terbium. Bovine brain calmodulin contains two tyrosine residues—tyr 99 and 138—which are postulated to be within the third and fourth calcium binding domains. Energy transfer from tyrosine to terbium occurs primarily upon binding of the second and third mole of the ion. Preliminary calculations suggest a distance of greater than 5 Å between the first terbium binding site and the tyrosine residue, and a distance of less than 5 Å between the second and third binding sites and the tyrosine. The maximum fluorescence intensity occurs at a ratio of 4 moles of terbium per mole of calmodulin, suggesting that terbium occupies all of the 4 $Ca^{2+}$ binding sites. The apparent dissociation constant of the $Tb^{3+}$·calmodulin complex is 7 $\mu$M.

In the presence of $Tb^{3+}$, calmodulin stimulates phosphodiesterase and forms a complex with troponin I or with CaM-$BP_{80}$. In addition, terbium causes conformational changes in calmodulin, as suggested by its altered mobility in polyacrylamide gel electrophoresis under denaturing and nondenaturing conditions. The change in intrinsic tyrosine fluorescence caused by $Tb^{3+}$ is similar to that by $Ca^{2+}$.

Thus, terbium appears to provide a highly useful tool for studying the interaction between calcium and calmodulin and for probing the structure and environment of the metal ion binding sites. Furthermore, the $Tb^{3+}$·calmodulin complex offers a sensitive fluorescence probe to monitor the interaction between calmodulin and its receptor proteins.

*Supported by an Institutional Cancer Center Support (CORE) Grant CA 21765, Project Grants NS 08059 (W.Y.C.) and AM 21974 (M.E.D.) and by ALSAC. R.W.W. is a recipient of a U.S. Public Health Service Fellowship, AM 05689.



0077-8923/80/0356-0436 $01.75/0 © 1980, NYAS

# REPRODUCIBLE PRODUCTION AND CHARACTERIZATION OF ANTI-CALMODULIN ANTISERA

Linda J. Van Eldik and D. Martin Watterson

*Department of Cell Biology*
*The Rockefeller University*
*New York, New York 10021*

We describe here a reproducible and rapid procedure for producing anticalmodulin (CaM) antibodies in rabbits. Using this protocol, a response was elicited in 11 out of 11 rabbits. In addition, titers were high enough that a quantitative radioimmunoassay could be developed using dilutions of whole sera as well as immunoglobulin fractions and antigen-adsorbed immunoglobulin.

Three different injection schedules were used. Injections were subcutaneous in 4–5 sites along the back of New Zealand White, Pasteurella-free rabbits. The antigen was emulsified in either complete Freund's adjuvant (initial injection) or

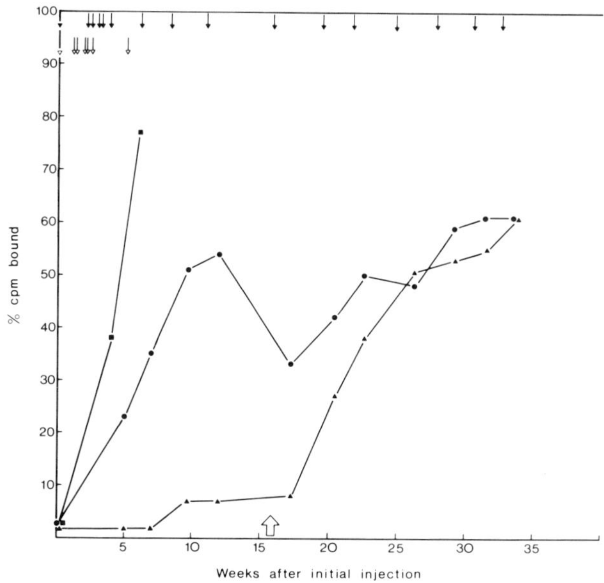


FIGURE 1. Time course of antibody production. Schedule A (■): Rabbits were injected with PAO-CaM at the times indicated by the open arrows. Schedule B (●): Rabbits were injected with PAO-CaM at the times indicated by the solid arrows. Schedule C (▲): Rabbits were injected as in schedule B, except that unoxidized CaM was used as antigen until the time indicated by the large open arrow (day 111). All subsequent injections were with PAO-CaM. Antibody reactivity was assayed using a fixed amount of $^{125}$I-labeled CaM (1 ng; 50,000 cpm) and antiserum (1 = 30 dilution). The percent of counts bound in the absence of antiserum was always less than 4%.



0077-8923/80/0356-0437 $01.75/0 © 1980, NYAS

incomplete Freund's adjuvant (other injections). A total of 0.5–1.0 mg of vertebrate CaM was administered to each rabbit per injection. CaM was performic-acid oxidized (PAO)[1] and iodinated[2] as described elsewhere. Radioimmunoassays were done as described.[3]

FIGURE 1 shows the three injection schedules and the time course of appearance of antiCaM reactivity. In schedule A, rabbits were injected with PAO-CaM on days 0, 7, 9, 13, 15, 17, 34 and bled on days 27 and 41. In schedule B, rabbits were injected with PAO-CaM on days 0, 14, 16, 20, 22, 26 and bled on day 35. Rabbits were boosted approximately every 2 weeks and bled 7–10 days after each boost. In schedule C, rabbits were injected as in schedule B, except that unoxidized CaM was used as antigen until day 111 (open arrow). All subsequent injections were with PAO-CaM.

Those rabbits injected with PAO-CaM (schedules A and B) showed a response at the first bleed, 4–5 weeks after the initial injection. There was no reactivity in any pre-immune sera. Rabbits injected with unoxidized CaM (schedule C) showed little if any reactivity; however, after only two boosts of PAO-CaM, these rabbits produced a response. Even though the antibodies were elicited by the injection of PAO-CaM, the antiCaM serum reacted with unoxidized CaM. The iodinated antigen used in the radioimmunoassays was unoxidized CaM. In addition, unoxidized CaM from vertebrates and plants showed competition curves similar to that of PAO-CaM.

The specificity of the anti-CaM serum was further tested by determining the ability of several structurally and functionally related proteins to quantitatively compete with $^{125}$I-labeled CaM in radioimmunoassay. Parvalbumin and brain S100b showed no reactivity, while skeletal muscle troponin C showed reactivity only at relatively high protein concentrations. Studies on the immunoreactive site(s) in CaM are in progress. The ability to consistently produce large quantities of antiCaM sera will allow further studies on the role of CaM in cell function.

## REFERENCES

1. VAN ELDIK, L. J. & D. M. WATTERSON. 1979. J. Biol. Chem. **254:** 10250–10255.
2. DORVAL, G., K. I. WELSH & H. WIGZELL. 1975. J. Imm. Methods **7:** 237–250.
3. VAN ELDIK, L. J., G. PIPERNO & D. M. WATTERSON. 1980. This volume.

# INVESTIGATION OF CALMODULIN-REGULATED NAD KINASE IN PROLIFERATING FIBROBLASTS*

M. L. Veigl and T. C. Vanaman

*Duke University Medical Center*
*Durham, North Carolina 27710*

The divalent cation, $Ca^{2+}$, serves as a regulatory signal through which eukaryotic cells perceive and respond to both internal and external stimuli. A single protein, calmodulin (CaM), is the primary intracellular receptor for $Ca^{2+}$ in most tissues where it serves as a $Ca^{2+}$-dependent activator of numerous enzyme systems. Various lines of evidence suggest that the deregulation of animal cell growth upon cellular transformation involves the alteration of major cascade regulation systems, possibly those controlled by CaM. The availability of well characterized temperature-sensitive transformation-defective mutants of avian RNA tumor viruses offers an attractive system for identifying and investigating any specific role(s) which CaM plays in normal cell division and/or cellular transformation.

Activation of biosynthetic pathways is required for cell proliferation. The enzyme NAD kinase is a key regulatory activity for many biosynthetic pathways since its product, NADP, is an obligatory cofactor for anabolic enzymes catalyzing oxidation/reduction reactions. NAD kinase from the pea, *Pisum sativum L.* has been shown by other workers to be $Ca^{2+}$-CaM regulated. [Plant Physiol. **59:** 55–60 (1977); Biochem. Biophys. Res. Commun. **84:** 595–602 (1978)]. As CaM is a highly conserved regulatory protein in both structure and function, it seemed likely that NAD kinase from animal systems is also CaM regulated. This study provides a preliminary investigation of the regulation of vertebrate NAD kinase and its potential regulation by CaM in both normal and Rous sarcoma virus-transformed chick embryo fibroblasts (CEF).

An NAD kinase assay was developed, based on assay conditions described by Muto and Miyachi [Plant Physiol. **59:** 55–60 (1977)]. High pressure liquid chromatography (HPLC) methodology was employed to separate and quantitate the NADP produced during incubation.

NAD kinase activity was found in the soluble fraction of both normal (0.5–1.4 nmoles NADP produced/hr/$10^6$ cells) and transformed fibroblasts (2–3 nmoles NADP produced/hr/$10^6$ cells). Transformed cells appeared to contain elevated levels of the enzyme. However a more detailed study is required to confirm this observation, since the enzyme levels in crude extracts vary among cell preparations—probably reflecting the growth rate and the distribution of cells at various stages in the cell cycle.

In both normal and transformed cells the NAD kinase activity was enhanced 2–3 fold upon the addition of EGTA. Further investigation revealed that this apparent EGTA activation simply reflected a slower NADP utilization in the presence of $Ca^{2+}$. A time-course study, in which exogenous NADP was added to the *in vitro* assay mix, revealed a rapid utilization of 1.6–2.1 nmoles of the nucleotide; additional incubation resulted in no further metabolism of NADP.

*Supported by 1F32-NS05863 and 1R01-NS10123.



0077-8923/80/0346-0439 $01.75/1 © 1980, NYAS

These observations suggest rapid depletion of an endogenous substrate in the cellular homogenate which is required for NADP utilization. Attempts are currently underway to identify this substrate and the enzyme(s) that metabolize NADP in a $Ca^{2+}$-dependent manner.

Neither NAD kinase nor the enzyme(s) metabolizing NADP interacted with CaM-Sepharose in a $Ca^{2+}$-dependent manner—the activity appeared in the pass-through and was still EGTA activated. This result suggests that either the enzymes involved in NADP synthesis and metabolism are not CaM regulated or CaM is an integral subunit of these enzymes and this close association prohibits any interaction of the enzyme with either CaM or CAPP (a phenothiazine derivative).

Since NADP is more rapidly metabolized in the presence of $Ca^{2+}$, it is impossible to investigate $Ca^{2+}$-CaM regulation of NADP production in crude supernatants—any $Ca^{2+}$-dependent activation of NAD kinase would be masked by increased destruction of NADP. Therefore, this enzyme is currently being purified and will then be re-examined to establish any potential $Ca^{2+}$-CaM regulation.

# ANTAGONISM OF CALMODULIN BY LOCAL ANESTHETICS, MEPACRINE, AND PROPRANOLOL*

M. Volpi, R. I. Sha'afi, P. M. Epstein, D. Andrenyak, and M. B. Feinstein

*Departments of Physiology and Pharmacology*
*University of Connecticut Health Center*
*Farmington, Connecticut 06032*

Calmodulin (CM) is known to stimulate the $Ca^{2+}$ ATPase activity of the human red cell membrane. We have found that local anesthetics (e.g., dibucaine, QX572, tetracaine, phenacaine), and the antimalarial mepacrine, inhibit the CM-dependent increase in $Ca^{2+}$ ATPase activity. The ATPase activity in the absence of CM (but in the presence of $Ca^{2+}$ and $Mg^{2+}$) is not affected by concentrations of the anesthetics that inhibit the specific CM stimulation by 75–80%. The effect of the local anesthetics is reversible by washing the membranes free of the drugs, but is not overcome by increased $Ca^{2+}$. Increasing CM tends to reverse the effect of the anesthetics, but at high concentrations of anesthetics large excesses of CM (50- to 60-fold over concentrations required for maximal activation of ATPase) do not fully overcome the inhibition. Inhibition of the CM-activation of ATPase by local anesthetics may result from inhibition of the binding of CM to the ATPase since $Ca^{2+}$-dependent binding of $^{125}I$-labeled CM to erythrocyte membranes was inhibited by 80% by the addition of 1 mM dibucaine (DB).

In addition to the membrane bound erythrocyte $Ca^{2+}$ ATPase, we have also found that local anesthetics, mepacrine, and propranolol inhibit the CM stimulation of a soluble cyclic nucleotide phosphodiesterase (PDE) from rat brain. In the presence of 0.5 mM dibucaine, the apparent $K_m$ of PDE for CM increased from 3.5 nM to 170 nM when determined at 0.25 $\mu$M cAMP and from 8 nM to 200 nM when determined at 1 $\mu$M cGMP. Double reciprocal plots indicated that DB acts as a competitive inhibitor with respect to CM, with an apparent $K_i \simeq 10$–$20\ \mu$M. The order of potency for inhibition of CM stimulation of PDE by local anesthetic agents was mepacrine>dibucaine≃propranolol>QX572>tetracaine>phenacaine. As with $Ca^{2+}$ ATPase, inhibition was not overcome by increased $Ca^{2+}$. In contrast to the $Ca^{2+}$ ATPase, unstimulated PDE activity was inhibited to a greater extent by the anesthetics, but the $I_{50}$'s for inhibition of basal PDE were at least several fold greater than that required for inhibition of the CM activated state of the enzyme. Moreover, inhibition of basal PDE by the anesthetics appeared to be competitive with respect to substrate, whereas inhibition of CM stimulated PDE appeared noncompetitive with respect to substrate.

Thus, two effects of the local anesthetic can be distinguished; one at low concentrations, which appears to be directed against calmodulin, is independent of substrate and $Ca^{2+}$ concentration, and the second at high concentrations, which is due to an effect on the affinity of the enzyme for substrate.

These results demonstrate that local anesthetics can antagonize the specific

*Supported by U.S. Public Health Service Grant GM-17356 and grants from The Pharmaceutical Manufacturers Association and The University of Connecticut Research Foundation.



0077-8923/80/0356-0441 $01.75/1 © 1980, NYAS

CM activation of $Ca^{2+}$ ATPase and PDE. Local anesthetics, mepacrine, and many other drugs with local anesthetic-like activity are known to inhibit several $Ca^{2+}$-dependent cellular processes, such as excitation-secretion coupling, smooth muscle contraction, cellular motility, and stimulus-induced release of arachidonic acid from phospholipids. A number of these processes are believed to involve the participation of CM-dependent enzyme (e.g., myosin light chain kinase, phospholipase $A_2$). We suggest that the effects of local anesthetics and drugs with similar actions on $Ca^{2+}$-dependent processes may be due to their effects on specific CM-mediated reactions.

# REGULATION OF ERYTHROCYTE CALCIUM TRANSPORT: ROLE OF CALMODULIN AND PHOSPHATE

David Morton Waisman, Jeffrey M. Gimble,
David B. P. Goodman, and Howard Rasmussen

*Department of Medicine and Cell Biology*
*Yale University School of Medicine*
*New Haven, Connecticut 06510*

Inside-out vesicles (IOV) from human erythrocytes were found to accumulate calcium linearily for at least thirty minutes. This uptake was ATP-dependent and 95% of accumulated calcium could be released from IOV by addition of calcium ionophore A23187.

In untreated IOV, 3.76 ± 1.44 nmol calcium/minute/unit acetylocholinesterase (uACE) were transported compared with 10.57 ± 2.05 (±S.D., $n = 11$) in those treated with calmodulin. The amount of calmodulin necessary for 50% activation of $Ca^{2+}$ accumulation was 60 ± 22 ng/ml (S.D., $n = 4$). The app. $K_m$ ($Ca^{2+}$) for calmodulin stimulated accumulation was 0.8 ± 0.05 $\mu$M (S.D., $n = 5$) using $Ca^{2+}$/EGTA buffers or 25 $\mu$M with direct addition of unbuffered calcium. In the absence of calmodulin these values were 0.4 $\mu$M and 60 $\mu$M respectively.

Millimolar phosphate was found to stimulate the initial velocity of calcium transport three to four fold and increase the $K_m$ (calcium) of calcium transport from 0.3 $\mu$M to 0.8 $\mu$M. A $K_m$ ($P_i$) 1.2 mM for phosphate stimulated calcium transport was measured.

The stimulation of calcium transport by phosphate was due to an ATP and calcium dependent, calmodulin stimulated uptake of phosphate into IOV. The $K_m$ (calcium) of phosphate transport was 1.0 $\mu$M in the presence or absence of calmodulin; 50 ng/ml calmodulin was required to half maximally activate phosphate transport. The $K_m$ ($P_i$) for phosphate transport was 0.6 mM. The rates of phosphate transported (+CAL, 9.5 nmol/min/uACE; −CAL, 2.98 nmol/min/uACE) suggest a one-to-one stoichiometry of phosphate transported with calcium.

Erythrocyte anion exchange protein (band III) inhibitor NAP-Taurine blocked both calcium and phosphate transport. The $K_i$ (NAP-Taurine) was 45 $\mu$M and 80 $\mu$M for phosphate transport and phosphate stimulated calcium transport. Furthermore, preincubation of erythrocytes with 100 $\mu$M SITS (10% hematocrit) for thirty minutes resulted in an 80% inhibition of calmodulin stimulated calcium transport and 77% inhibition of calmodulin stimulated phosphate transport into IOV prepared from the erythrocytes.

These results suggest the possible involvement of band III in the calcium translocation process. It is also suggested that the active transport of calcium by the erythrocyte calcium pump may involve net transport of phosphate.



0077-8923/80/0356-0443 $01.75/1 © 1980, NYAS

# Index of Contributors

(Italicized page numbers refer to comments in Discussions)